MASTERING MATHEMATICS

GEOMETRY AND MEASURES

Series Editor: Roger Porkess

HODDER EDUCATION
AN HACHETTE UK COMPANY

Series contributors:

Bola Adiboye, Caroline Clissold, Ruth Crookes, Heather Davis, Paul Dickinson, Alan Easterbrook, Sarah-Anne Fernandes, Dave Gale, Sophie Goldie, Steve Gough, Kevin Higham, Sue Hough, Andrew Jeffrey, Michael Kent, Donna Kerrigan, Nigel Langdon, Linda Liggett, Robin Liggett, Andrew Manning, Nikki Martin, Chris Messenger, Richard Perring, Grahame Smart, Alison Terry, Sam Webber, Colin White

The publisher would like to thank the following for permission to reproduce copyright material:

Photo credits:

p. 2 © Lisa F. Young – Fotolia.com; **p. 11** t © Andrew Callaghan; b © chartcameraman – Fotolia.com; **p. 20** © Friday – Fotolia.com; **p. 27** © Imagestate Media (John Foxx)/Office V3065; **p. 35** © marchibas – Fotolia.com; **p. 48** © C Squared Studios/Photodisc/Getty Images/European Objects OS44; **p. 57** 1 © florinoprea – Fotolia.com; 2 © Sharon McTeir; 3 © Dorling Kindersley – Getty Images; 4 © Tomislav – Fotolia.com; 5 © burnel11 – Fotolia.com; 6 © gena96 – Fotolia.com; 7 © wong yu liang – Fotolia.com; 8 © Dave White – Thinkstock.com; 9 © Steve Mann – Fotolia.com; 10 © Volodymyr Krasyuk – Fotolia.com; 11 © akulamatiau – Fotolia.com; 12 © Martin Shields/Alamy; 13 © Siede Preis/Photodisc/Getty Images/Tools of the Trade OS48; 14 © Coloures-Pic – Fotolia.com; 15 © Sharon McTeir; 16 © Elaine Lambert; **p. 64** © qingwa – Fotolia.com; **p. 74** © absolut – Fotolia.com (t); © GoldPix – Fotolia.com (b); **p. 76** © indigolotos – Fotolia.com (l); © Rawpixel – Fotolia.com (r); **p. 82** © morchella – Fotolia.com; **p. 83** © Photodisc/Getty Images//Business & Industry 1; **p. 87** © sakura – Fotolia.com; **p. 90** © slava296 – Fotolia.com; **p. 97** a © Africa Studio – Fotolia.com; b © Doves Farm; c © yvdavid – Fotolia.com; **p. 101** a © donfiore – Fotolia.com; b © Olga Kovalenko – Fotolia.com; c © Scanrail – Fotolia.com; d © Anton Maltsev – Fotolia.com; e © yasar simit – Fotolia.com; **p. 102** © Lilya – Fotolia.com; **p. 104** © Shawn Hempel – Fotolia.com; **p. 107** © JOHN KELLERMAN/Alamy; **p. 114** © darkfall – Fotolia.com; **p. 121** © rakijung – Fotolia.com; **p. 127** © Popova Olga – Fotolia.com; **p. 144** © Eric Farrelly/Alamy; **p. 152** © Stockbyte/Photolibrary Group Ltd/Big Business SD101; **p. 160** © Heather Davies; **p. 171** © Heather Davies; **p. 185** © Heather Davies; **p. 192** © Heather Davies; p. 200 © Heather Davies; **p. 208** © Maxim_Kazim – Fotolia.com; **p. 221** a © by-studio – Fotolia.com; b © 2008 photolibrary.com/© Moodboard RF/Photolibrary Group; c © apops – Fotolia.com; d © jitadelta – Fotolia.com; **p. 224** © Hemeroskopion – Fotolia.com; **p. 230** t © Minerva Studio – Fotolia.com; b © Imagestate Media (John Foxx)/Education SS121; **p. 240** © oriontrail – Fotolia.com; **p. 246** © Karramba Production – Fotolia.com; **p. 261** © Hodder Education; **p. 268** © peresanz – Fotolia.com; **p. 270** © Ingram Publishing Limited/Ingram Image Library 500-Sport; **p. 281** © Igor – Fotolia.com; **p. 290** © Gordon Warlow – Thinkstock.com; **p. 293** © traffico – Fotolia.com; **p. 311** © Sharon McTeir; **p. 319** © monstersparrow – Fotolia.com; **p. 320** © yasar simit – Fotolia.com; **p. 328** © Sergey Lavrentev – Fotolia.com; **p. 330** a © darkhriss – Fotolia.com; b © Hemera Technologies – Thinkstock.com; c © scullery – Fotolia.com; d © tnehala – Fotolia.com; e © eldadcarin – Fotolia.com; f © whiteboxmedia limited/Alamy; g © bsanchez – Fotolia.com; h © akiyoko – Fotolia.com; i © Maria Vazquez – Fotolia.com; **p. 337** © Heather Davies; **p. 346** © Anne Wanjie; **p. 359** © actionplus sports images; **p. 357** © Doug Houghton; **p. 363** © Anne Wanjie; **p. 370** © Photographee.eu – Fotolia.com

Orders: please contact Bookpoint Ltd, 130 Milton Park, Abingdon, Oxon OX14 4SB. Telephone: (44) 01235 827720. Fax: (44) 01235 400454. Lines are open 9.00–17.00, Monday to Saturday, with a 24-hour message answering service. Visit our website at www.hoddereducation.co.uk

First published in 2014 by

Hodder Education
An Hachette UK Company,
338 Euston Road
London NW1 3BH

Impression number	5	4	3	2	1
Year	2018	2017	2016	2015	2014

Cover photo © beholdereye – Fotolia

Typeset in 10/11.5pt ITC Avant Garde Gothic by Integra Software Services Pvt. Ltd., Pondicherry, India

Printed in Italy

A catalogue record for this title is available from the British Library

ISBN 978 1471 805875

How to get the most from this book

This book covers the geometry and measures that you need for your key stage 3 Maths course.

The material is split into **six strands**:

- Units and scales
- Properties of shapes
- Measuring shapes
- Geometric construction
- Transformations
- Three-dimensional shapes

Each strand is presented as a series of units that get more difficult as you progress (from Band b through to Band h). In total there are 40 units in this book.

Getting started

At the beginning of each strand, you will find a '**Progression strand flowchart**'. It shows what skills you will develop in each unit in the strand. You can see:

- what you need to know before starting each unit
- what you will need to learn next to progress

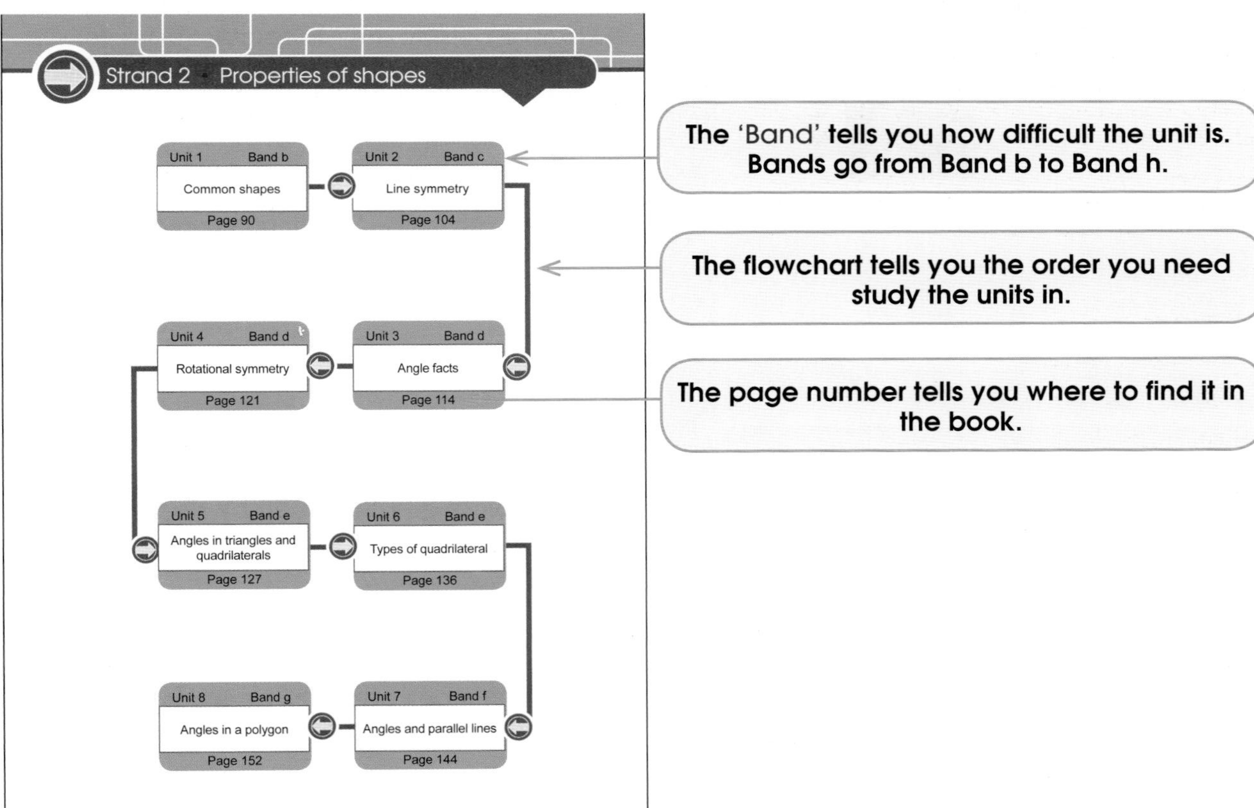

When you start to use this book, you will need to identify where to join each strand. Then you will not spend time revisiting skills you have already mastered.

If you can answer all the questions in the **'Reviewing skills'** section of a unit then you will not have to study that unit.

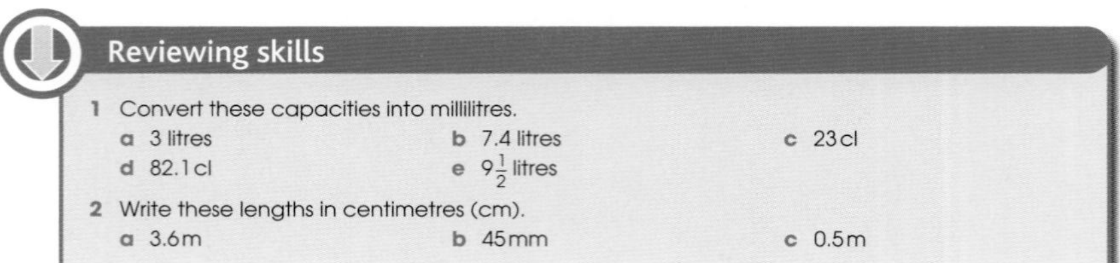

Reviewing skills

1 Convert these capacities into millilitres.

a 3 litres b 7.4 litres c 23 cl

d 82.1 cl e $9\frac{1}{2}$ litres

2 Write these lengths in centimetres (cm).

a 3.6 m b 45 mm c 0.5 m

When you know which unit to start with in each strand you will be ready to start work on your first unit.

Starting a unit

Every unit begins with a **'Building skills'** section:

Unit 1 • Length • Band c

Building skills

Example outside the Maths classroom

Fitting windows

Toolbox

Length is a measure of size. It tells you how long an item is.
To measure or draw a line you use a ruler.
Most rulers are marked in centimetres (cm).
Small divisions are millimetres (mm).
1000 millimetres = 1 metre
1000 metres = 1 kilometre
Before measuring an object, you need to have a rough idea of how big it is so you can choose an appropriate measuring instrument. This could be a ruler, tape measure, trundle wheel, micrometer or odometer (in a car).

This is 3.4 cm or 34 mm

This is 9 cm or 90 mm.

mm 1 2 3 4 5 6 7 8 9 10

Example – Converting between centimetres and millimetres

A calculator is 8 cm long and 60 mm wide.
a How long is the calculator in millimetres?
b How wide is the calculator in centimetres?

Solution

a The calculator is 80 mm long. (There are 10 mm in 1 cm. 8 × 10 = 80)

b The calculator's width is 6 cm. (60 ÷ 10 = 6)

2 Strand 1 • Units and scales

Remember:

- To measure or draw a line, start from the beginning of the scale (the 0 mark, not the end of the ruler!).
- There are 10 millimetres in 1 centimetre.

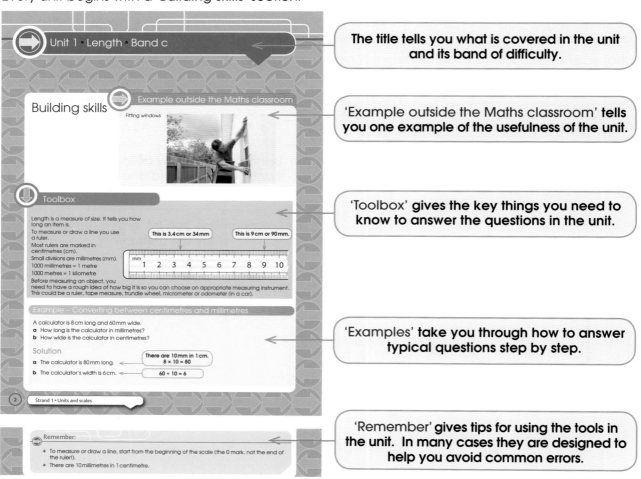

The title tells you what is covered in the unit and its band of difficulty.

'Example outside the Maths classroom' **tells you one example of the usefulness of the unit.**

'Toolbox' **gives the key things you need to know to answer the questions in the unit.**

'Examples' **take you through how to answer typical questions step by step.**

'Remember' **gives tips for using the tools in the unit. In many cases they are designed to help you avoid common errors.**

Now you have all the information you need, you can use the questions to develop your understanding.

Skills practice A

1 Measure each of these lines in centimetres.

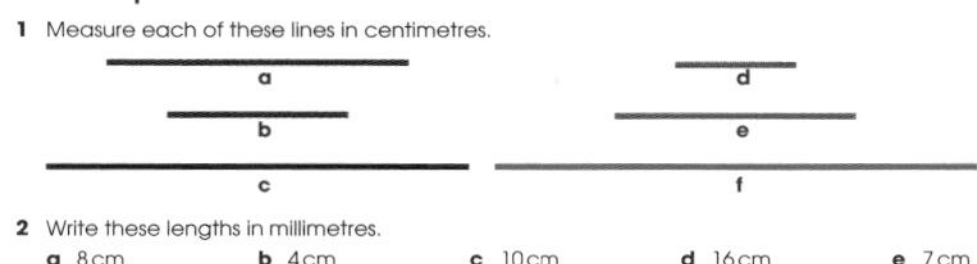

2 Write these lengths in millimetres.
a 8cm **b** 4cm **c** 10cm **d** 16cm **e** 7cm

'Skills practice A' **questions are all about mastering essential techniques that you need to succeed.**

Skills practice B

1 Here are eight baby eels (elvers).

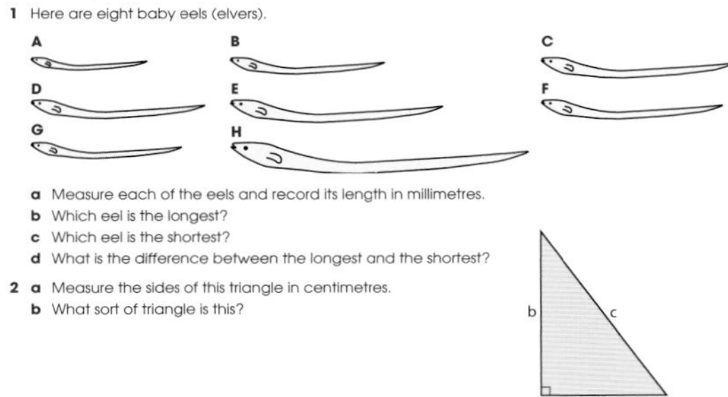

a Measure each of the eels and record its length in millimetres.
b Which eel is the longest?
c Which eel is the shortest?
d What is the difference between the longest and the shortest?

2 **a** Measure the sides of this triangle in centimetres.
b What sort of triangle is this?

'Skills practice B' **questions give you practice in using your skills for a purpose. Many of them are set in context. The later questions are usually more demanding.**

Wider skills practice

1 The total distance around a shape is called its perimeter.
a Measure the perimeter of each of these four coloured squares in millimetres.
b The next square, square **v**, has not been drawn. What is its perimeter?

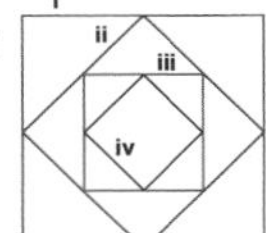

'Wider skills practice' **questions require you to use maths from outside the current unit. In some cases they use knowledge from other subjects or the world outside.**

You can use this section to keep practising other skills as well as the skills in this unit.

Applying skills

Problem solving

1 Match each of the objects or distances in List A with an appropriate measure of length from List B.

List A	List B
The distance from London to Edinburgh	10 centimetres
The length of a swimming pool	20 centimetres
The width of a man's hand	1.6 metres
The height of a fully grown woman	1 millimetre
The width of a sewing needle	700 kilometres
The width of your maths book	50 metres

'Applying skills' **questions give examples of how you will use the Maths in the unit to solve problems:**

- **in the real world**
- **in other subjects**
- **in personal finance**
- **within Maths itself.**

These are more demanding questions, so only one or two are provided in each unit. Together they form a bank of questions.

When you feel confident, use the **'Reviewing skills'** section to check that you have mastered the techniques covered in the unit.

You will see many questions labelled with Reasoning or Problem solving

These are the general mathematical skills that you need to develop. You will use these skills in all areas of Maths.

They will help you think through problems and to apply your skills in unfamiliar situations. Use these questions to make sure that you develop these important skills.

About 'Bands'

Every unit has been allocated to a Band. These bands show you the level of difficulty of the Maths that you are working on.

Each Band contains Maths that's of about the same level of difficulty.

This provides a way of checking your progress and assessing your weaker areas, where you need to practise more.

Moving on to another unit

Once you have completed a unit, you should move on to the next unit in one of the strands. You can choose which strand to work on next but make sure you complete all the units in a particular Band before moving on to the next Band.

A note for teachers

Bands have been assigned to units roughly in line with the previous National Curriculum levels. Here they are, just to help in giving you a reference point.

Band	Approximate Equivalent in terms of Old National Curriculum Levels
b	Level 2
c	Level 3
d	Level 4
e	Level 5
f	Level 6
g	Level 7
h	Level 8

Answers

Answers to all the questions in this book will be available via **Mastering Mathematics Teaching and Learning Resources** or by visiting **www.hodderplus.co.uk/masteringmaths**

Contents

Strand 1 Units and scales 01

Unit 1 Length 02
Unit 2 Mass 11
Unit 3 Time 20
Unit 4 Volume 27
Unit 5 Interpreting scales 35
Unit 6 The metric system 48
Unit 7 Metric–imperial conversions 55
Unit 8 Bearings 64
Unit 9 Scale drawing 74
Unit 10 Compound units 83

Strand 2 Properties of shapes 89

Unit 1 Common shapes 90
Unit 2 Line symmetry 104
Unit 3 Angle facts 114
Unit 4 Rotational symmetry 121
Unit 5 Angles in triangles and quadrilaterals 127
Unit 6 Types of quadrilateral 136
Unit 7 Angles and parallel lines 144
Unit 8 Angles in a polygon 152

Strand 3 Measuring shapes 159

Unit 1 Understanding area 160
Unit 2 Finding area and perimeter 171
Unit 3 Circumference 185
Unit 4 Area of circles 192
Unit 5 Pythagoras' theorem 200

Strand 4 Geometric construction 207

Unit 1 Angles in degrees 208

Unit 2 Constructions with a ruler and protractor 224

Unit 3 Constructions with a pair of compasses 230

Unit 4 Loci 240

Strand 5 Transformations 245

Unit 1 Position and Cartesian co-ordinates 246

Unit 2 Cartesian co-ordinates in four quadrants 261

Unit 3 Translation 270

Unit 4 Reflection 281

Unit 5 Rotation 293

Unit 6 Enlargement 304

Unit 7 Similarity 314

Unit 8 Trigonometry 320

Strand 6 Three-dimensional shapes 327

Unit 1 Properties of 3-D shapes 328

Unit 2 Understanding nets 337

Unit 3 Volume and surface area of cuboids 346

Unit 4 2-D representations of 3-D shapes 357

Unit 5 Prisms 363

Strand 1 • Units and scales

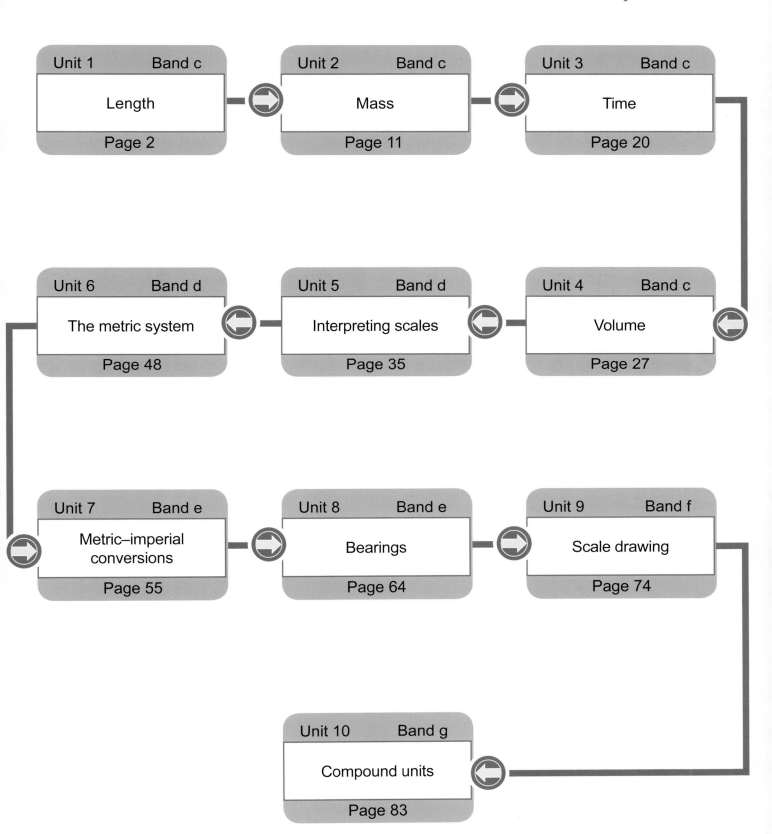

Unit 1	Band c	Length	Page 2
Unit 2	Band c	Mass	Page 11
Unit 3	Band c	Time	Page 20
Unit 4	Band c	Volume	Page 27
Unit 5	Band d	Interpreting scales	Page 35
Unit 6	Band d	The metric system	Page 48
Unit 7	Band e	Metric–imperial conversions	Page 55
Unit 8	Band e	Bearings	Page 64
Unit 9	Band f	Scale drawing	Page 74
Unit 10	Band g	Compound units	Page 83

Building skills

Example outside the Maths classroom

Fitting windows

Toolbox

Length is a measure of size. It tells you how long an item is.

To measure or draw a line you use a ruler.

Most rulers are marked in centimetres (cm).

Small divisions are millimetres (mm).

1000 millimetres = 1 metre

1000 metres = 1 kilometre

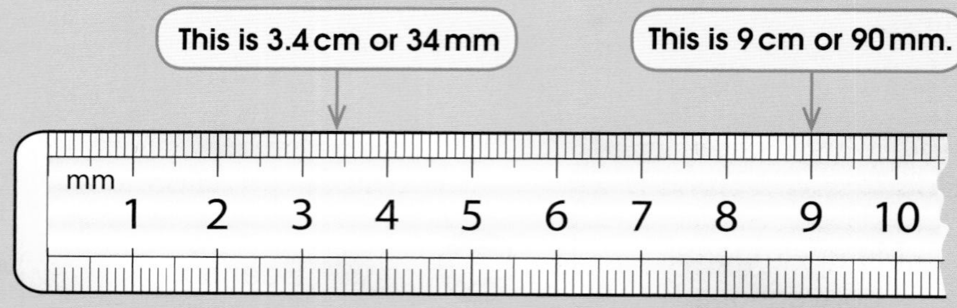

Before measuring an object, you need to have a rough idea of how big it is so you can choose on appropriate measuring instrument. This could be a ruler, tape measure, trundle wheel, micrometer or odometer (in a car).

Example – Converting between centimetres and millimetres

A calculator is 8 cm long and 60 mm wide.

a How long is the calculator in millimetres?

b How wide is the calculator in centimetres?

Solution

a The calculator is 80 mm long. ← **There are 10 mm in 1 cm. 8 × 10 = 80**

b The calculator's width is 6 cm. ← **60 ÷ 10 = 6**

Example – Measuring a line

Bill is measuring his thumb.

a How long is Bill's thumb in centimetres?

b How long is Bill's thumb in millimetres?

?

Solution

a 5 cm

b 50 mm

Example – Choosing an instrument to measure length

What is the most appropriate instrument to measure

a the length of a car

b the length of a needle

c the thickness of a needle

d a football pitch?

Solution

a A tape measure

b A ruler

c A micrometer

d A trundle wheel

Remember:

- To measure or draw a line, start from the beginning of the scale (the 0 mark, not the end of the ruler!).
- There are 10 millimetres in 1 centimetre.

Skills practice A

1 Measure each of these lines in centimetres.

2 Write these lengths in millimetres.

a 8 cm **b** 4 cm **c** 10 cm **d** 16 cm **e** 7 cm

3 Write these lengths in centimetres.

a 20 mm **b** 50 mm **c** 90 mm **d** 140 mm **e** 30 mm

4 Measure each of these lines in centimetres. Which line is the longest?

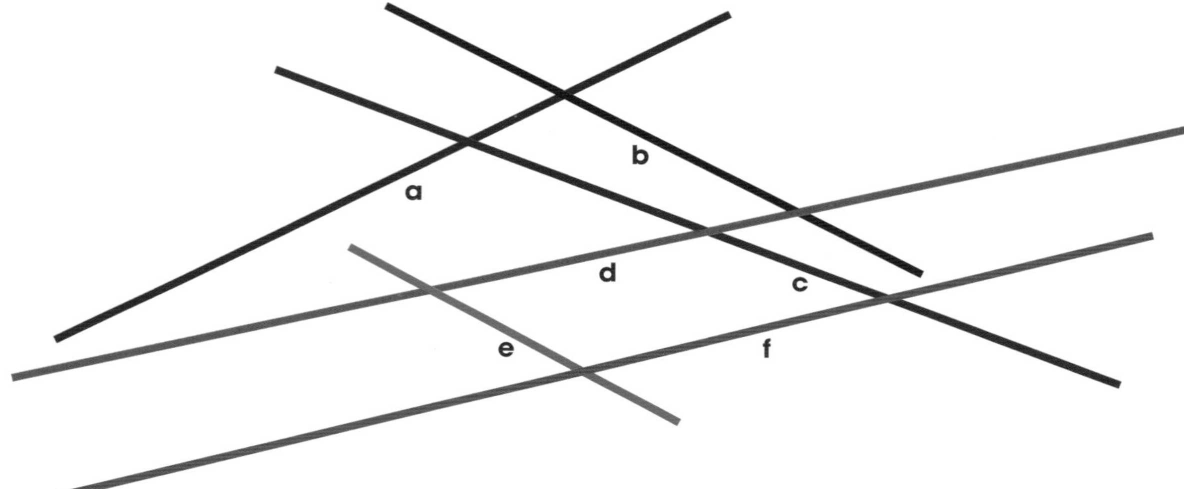

5 Measure the length of the scissors in centimetres.

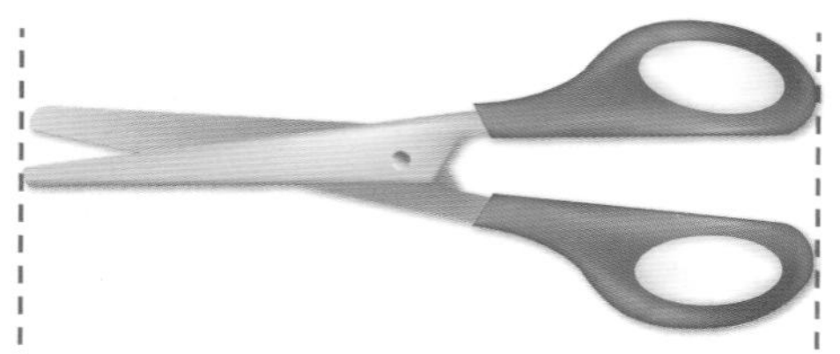

6 Measure each of these fish.

a

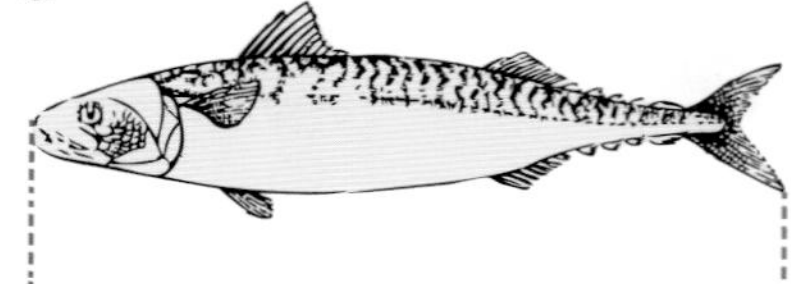

b

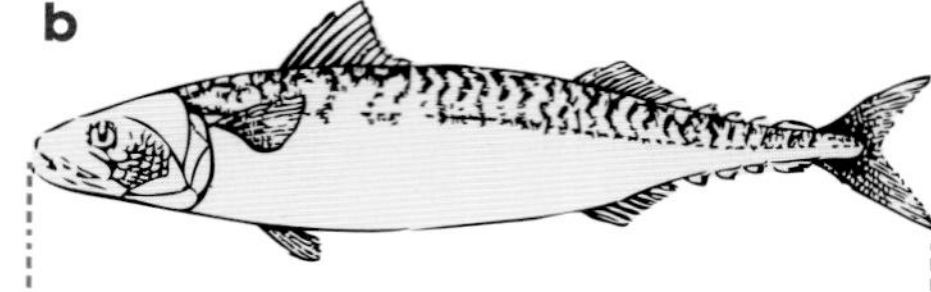

c

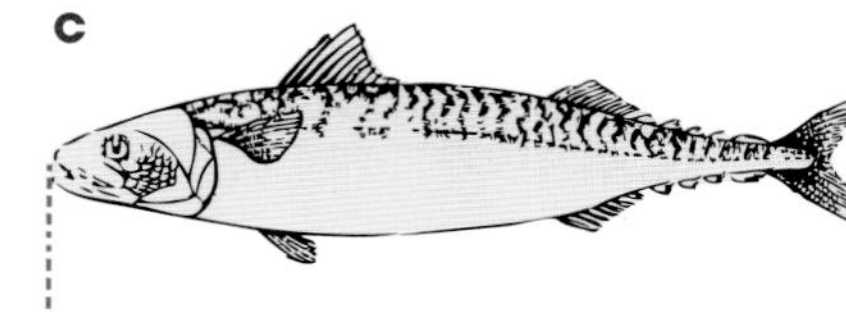

Reasoning

7 Tembo runs the 5000 metres race for his club. How many kilometres is this?

8 Jane has a hamster which is 8 cm long and a mouse which is 70 mm long.
Which one is longer?
Explain your answer.

Skills practice B

1 Here are eight baby eels (elvers).

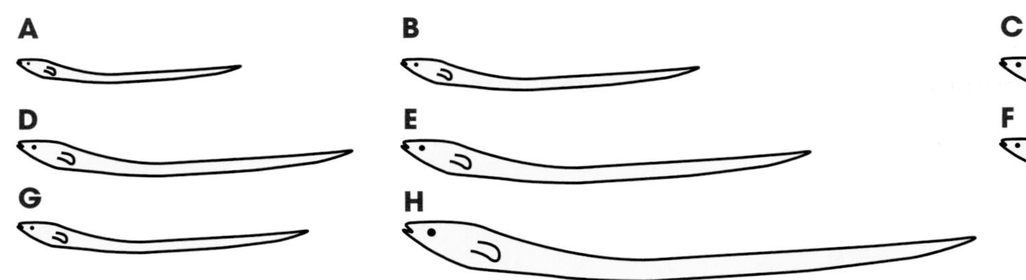

a Measure each of the eels and record its length in millimetres.

b Which eel is the longest?

c Which eel is the shortest?

d What is the difference between the longest and the shortest?

2 a Measure the sides of this triangle in centimetres.

b What sort of triangle is this?

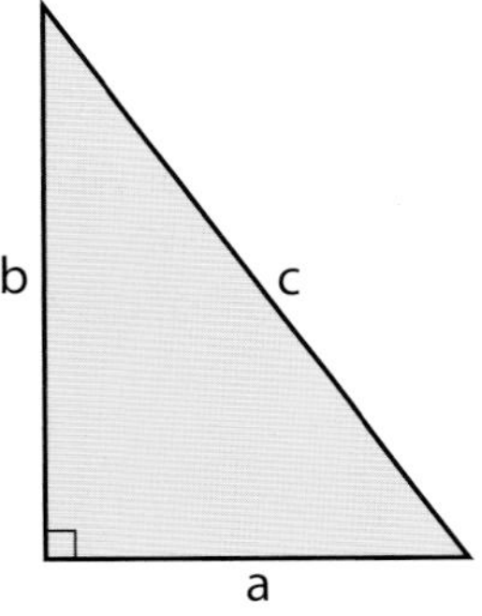

Reasoning

3 Here is a map of an area of London.

a Which is nearer to Victoria Station: Buckingham Palace or the Houses of Parliament?

b You arrive at Waterloo Station.
You want to see Trafalgar Square, Piccadilly Circus, Bow Street Police Station and the Houses of Parliament.
Which route is shortest?

c Which railway station is nearer to Trafalgar Square: Victoria or Waterloo?

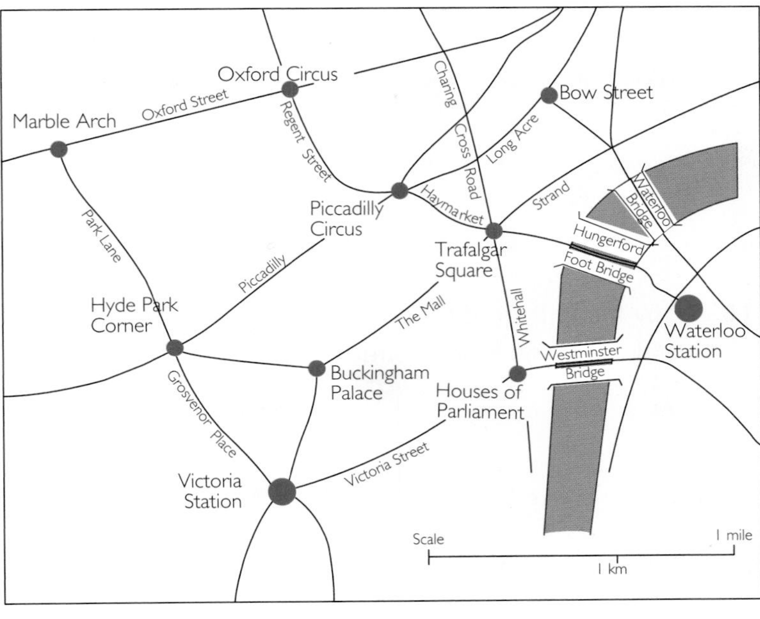

4 This ruler is marked in centimetres.

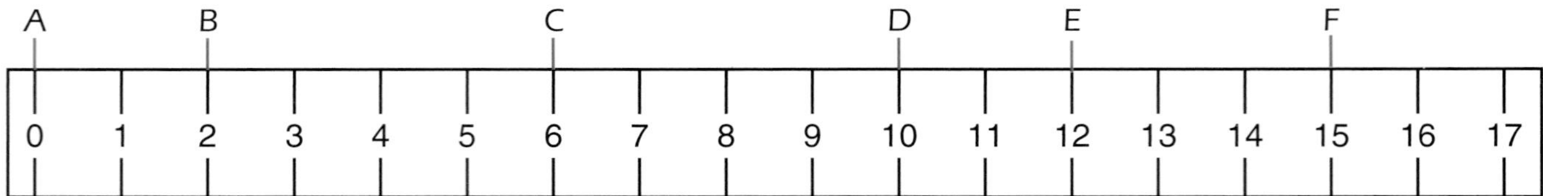

How long is it between these points?

a A to B **b** A to F

c B to C **d** B to E

e C to F **f** D to F

5 Laxmi plays for a football team. As part of their training they do this routine:

Jog 100 metres, run 50 metres and then sprint 50 metres over and over again. One day they covered 4 kilometres. How many times did they do the routine that day?

6 Write these lengths in millimetres.

a 12.5 cm **b** 9.6 cm **c** 27 cm **d** 2.1 cm **e** 7.8 cm

7 Write these lengths in centimetres.

a 75 mm **b** 32 mm **c** 83 mm **d** 111 mm **e** 29 m

8 **a** Measure the lines AB, BC, CD and DA in centimetres.

What do you notice?

b Measure the lines AM, BM, CM and DM in millimetres.

What do you notice?

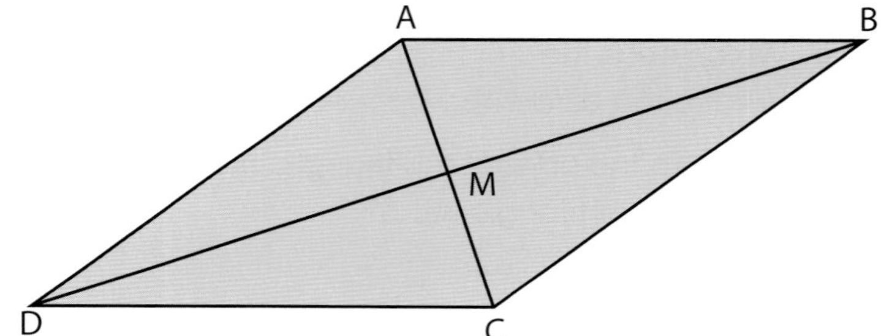

9 **a** Measure the lines PQ, QR, RS and SP in millimetres.

What do you notice?

b Measure the lines OP, OQ, OR and OS in millimetres.

What do you notice?

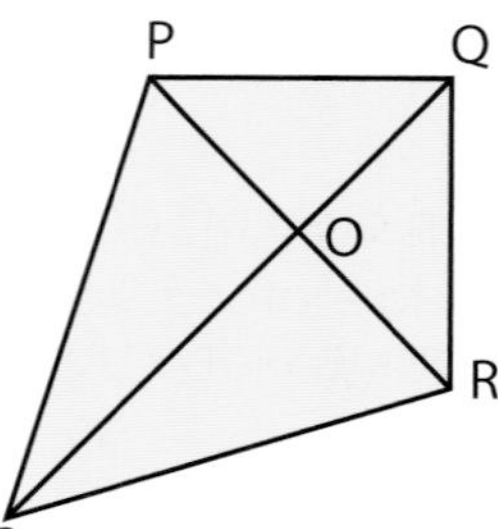

Reasoning

10 Jenny thinks that 35 mm is longer than 4.5 cm because 35 is bigger than 4.5.
Explain why she is not correct.

11 Which instrument is most appropriate to measure each of the things below?
Choose from this list.

micrometer	30 cm ruler	1 metre ruler	odometer	trundle wheel

a The length of a cat's tail
b The width of a window
c The thickness of a card

12 Here is a list of some objects.

A finger nail	A railway carriage	A tall man	The width of this book

Match them to these (approximate) lengths.

1 cm	20 cm	2 metres (200 cm)	20 metres

13 Give examples of two things that would be measured with a tape measure.

Reasoning

14 State some difficulties you would find if you measured the distance from your home to your school with a ruler.

15 Estimate the value of each of these quantities and state a suitable measuring device for them.
a The height of a kitchen
b The width of a biscuit.

Wider skills practice

1 The total distance around a shape is called its perimeter.
a Measure the perimeter of each of these four coloured squares in millimetres.
b The next square, square **v**, has not been drawn.
What is its perimeter?

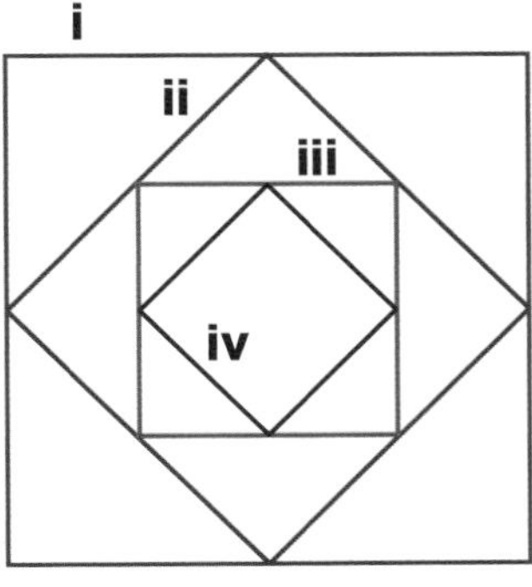

2 **a** Measure OA, OB and AB in centimetres.

b Measure OP, OQ and PQ in centimetres.

c What do you notice?

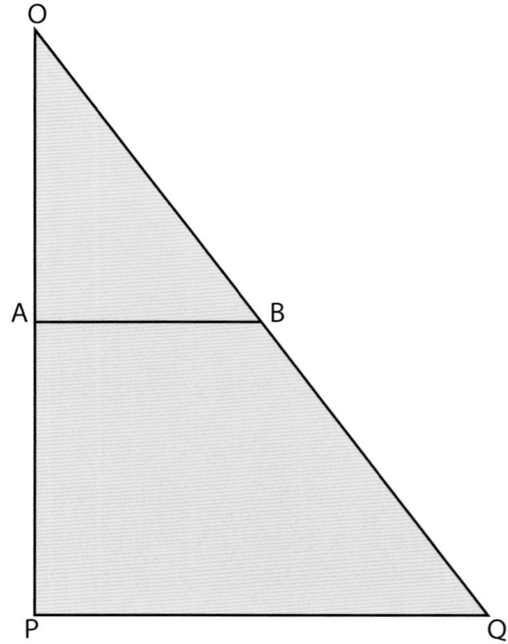

Reasoning

3 Distances from 1 cm to 10 cm are represented by seven marks on a strip.

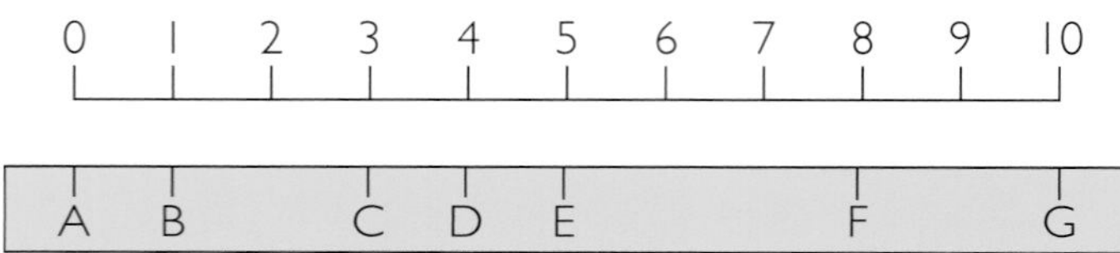

For 1 cm you use AB.
For 2 cm you use BC.
For 3 cm …

a Copy and complete this sequence.

b Find a way of marking the strip so that all the measurements can be made with fewer marks. What is the minimum number you need?

Applying skills

Problem solving

1 Match each of the objects or distances in List A with an appropriate measure of length from List B.

List A	**List B**
The distance from London to Edinburgh	10 centimetres
The length of a swimming pool	20 centimetres
The width of a man's hand	1.6 metres
The height of a fully grown woman	1 millimetre
The width of a sewing needle	700 kilometres
The width of your maths book	50 metres

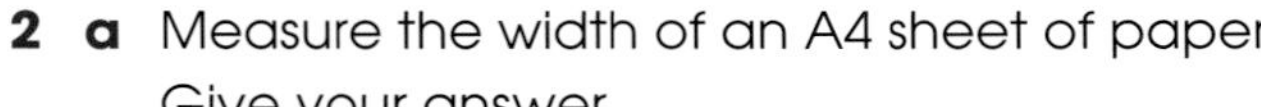

Problem solving

2 a Measure the width of an A4 sheet of paper.
Give your answer
i in centimetres
ii in millimetres.

b Without doing any more measurements, write down the length of an A5 sheet of paper.

Problem solving

3 Pick six people in your class.
a Measure and record how far each one can jump from standing.
b Put the lengths of the jumps in order of size.
c What will you use to measure their jumps?
Who can jump the furthest?

Reviewing skills

1 a Measure the lengths AB, BC and CA in the triangle ABC.
Give your answers in centimetres.
b Measure the lengths XY, YZ and ZX in the triangle XYZ.
Give your answers in millimetres.
c Use your measurements to compare triangles ABC and XYZ.

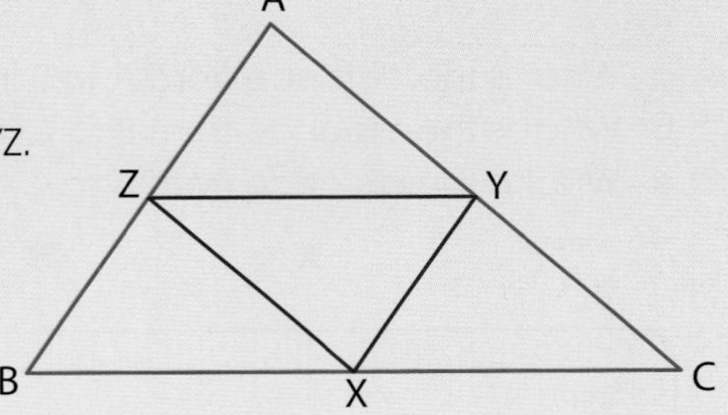

2 a The leaves shown below are drawn to scale. Measure the lengths and widths of each of these scale drawings in centimetres.
b Which type is the longest?
c Which type is the widest?

Lime

Birch

Oak

3 A mail aeroplane visits all these towns in the Hebrides.

a Which part of the journey is the longest?

b Which part of the journey is the shortest?

c List the parts of the journey in order of size.

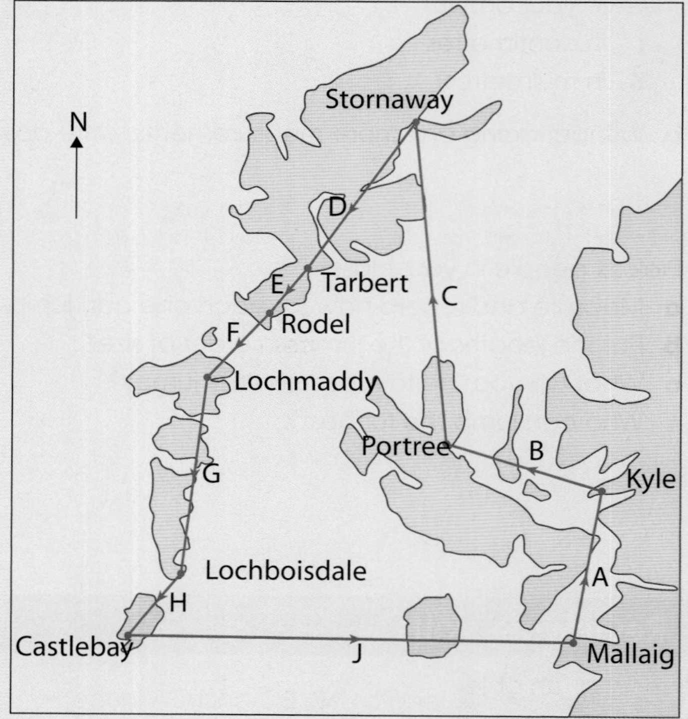

4 **a** What is the distance from A to B in millimetres?

b What is the distance from B to C in centimetres?

c What is the distance from A to C in centimetres?

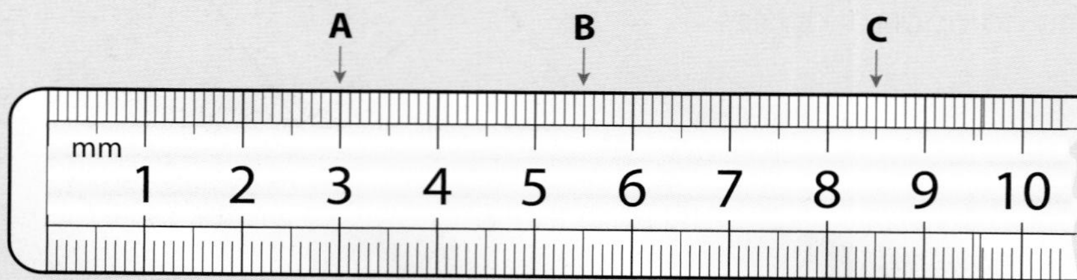

5 **a** Write these lengths in centimetres.

i 40 mm **ii** 72 mm

b Write these lengths in millimetres.

i 4.6 cm **ii** 10.3 cm

6 **a** Estimate

i the width of a pound coin

ii the distance from your school gate to the nearest bus stop

b What instrument is most appropriate to measure each of **i** and **ii** in part **a**?

Unit 2 • Mass • Band c

Example outside the Maths classroom

Building skills

Cooking

9 Almond biscuits

Ingredient	Makes 30
Butter	200g
Icing sugar	80g
Soft flour	200g
Salt	Pinch
Whole egg	25g
Ground almonds	100g
Granulated sugar	100g

1 Rub together the butter, icing sugar, flour and salt until the mixture resembles fine breadcrumbs.
2 Add the egg and work to a smooth paste.
3 Roll into a cylinder 2cm in diameter.
4 Brush with egg white and roll in a mixture of equal quantities of granulated sugar and ground almonds.
5 Refrigerate to firm up the paste.
6 Cut into discs and bake at 180°C until lightly golden.

Variation
Almond biscuits can also be flavoured with lemon zest or chopped glacé cherries.

Toolbox

The mass of an object is how heavy it is.

Weight is used to measure the mass of an object.

Scales are used to weigh objects.

milligrams (mg), **grams** (g) and **kilograms** (kg) are all units of mass.

1000 milligrams (mg) = 1 gram (g)

1000 grams (g) = 1 kilogram (kg)

You need to have a rough idea of the mass of an object so you can choose an appropriate **measuring instrument**. This could be a scientific balance, kitchen scales or bathroom scales.

Example – Measuring masses

Here are two scale readings.

Andy is measuring some bricks on one scale and a bean on the other scale.

a What is the mass of the bricks?

b What is the mass of the bean?

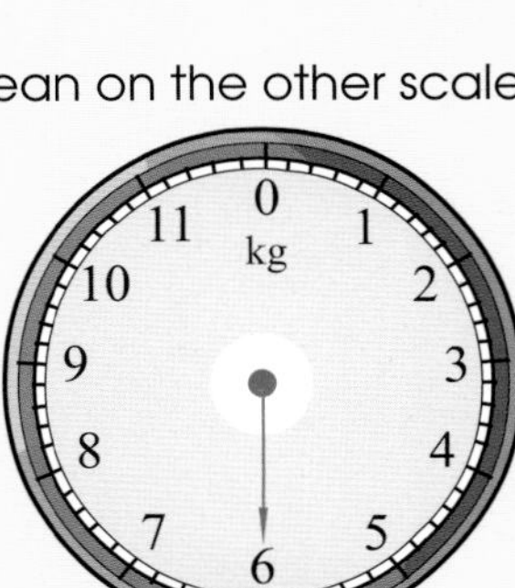

Solution

a Look at the units given on the scales.

You would weigh bricks in kilograms so the scale on the left is weighing the bricks.

The pointer tells you the mass is 6 kg.

b You would weigh a bean in grams so the scale on the right is weighing the bean.

The pointer is half way between the 3 and the 4 so the mass is $3\frac{1}{2}$ g.

Example – Converting between grams and kilograms

The mangos in a supermarket weigh about 100 grams each.
Usha wants 1 kg of mangos.
How many should she buy?

Solution

There are 1000 grams in 1 kilogram.

So the number of mangos she needs is $\frac{1000}{100} = 10$.

Example – Choosing an instrument to measure mass

What is the most appropriate instrument to measure the mass of

a yourself
b a bar of chocolate
c a spider's egg?

Solution

a Bathroom scales
b Kitchen scales
c Scientific scales

Remember:

- There are 1000 milligrams in 1 gram.
- There are 1000 grams in 1 kilogram.

Skills practice A

1 What is the mass shown on each of these scales?

a

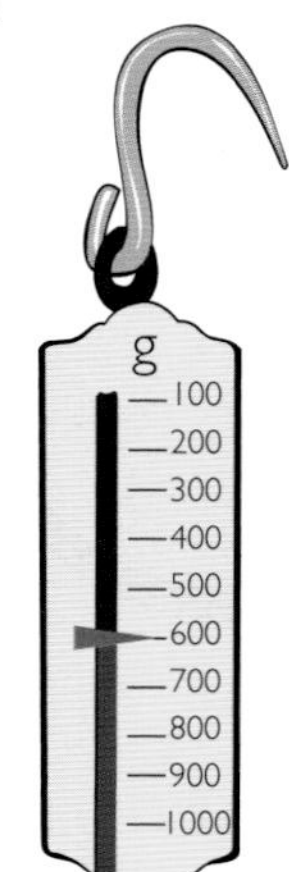

b

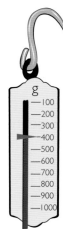

2 What is the mass shown on each of these scales?

a

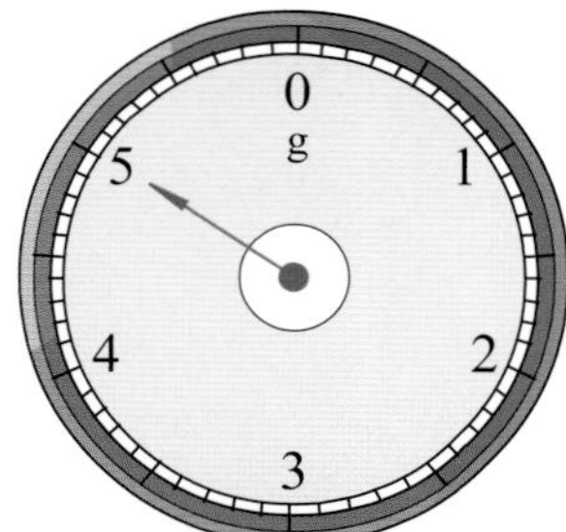

b

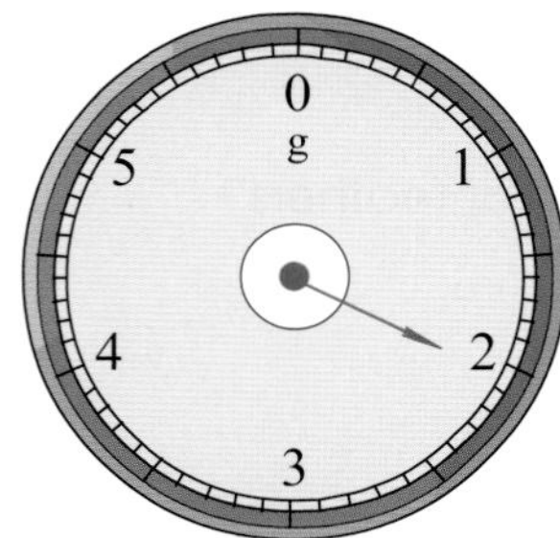

c

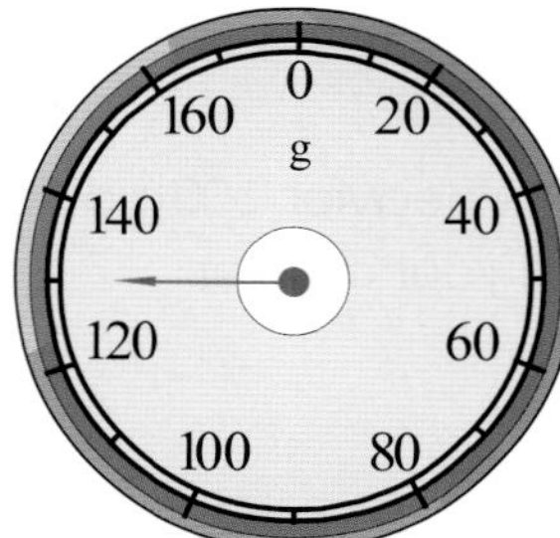

d

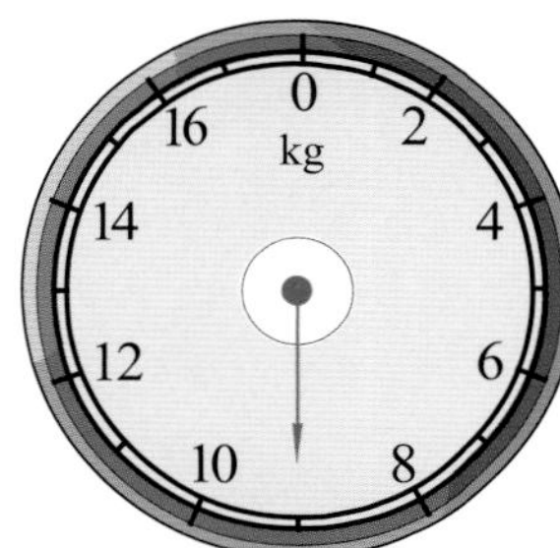

e

3 Eric has two cats.
Macavity weighs 3000 grams.
Growltiger weighs 5 kg.
Which cat is heavier and by how much?

Reasoning

4 Jem has these weights.

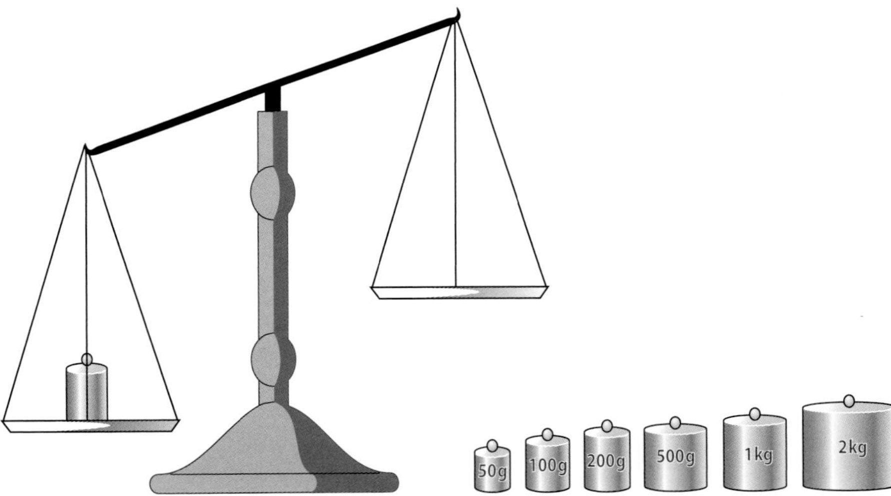

How can he weigh something that weighs

a 800 g **b** 3.5 kg **c** 2200 g **d** 900 g?

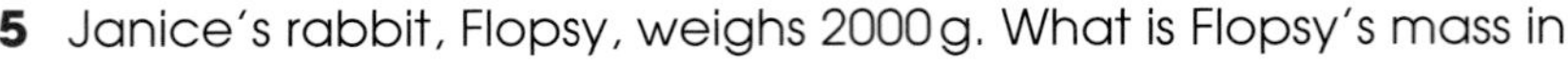

5 Janice's rabbit, Flopsy, weighs 2000 g. What is Flopsy's mass in
  i kilograms
  ii milligrams?

Reasoning

6 Which is heavier, 300 g of lead or 300 g of feathers?
Explain your answer.

Skills practice B

1 a How many grams are there in a kilogram?
  b How many milligrams are there in a gram?

2 a Write these masses in grams.

i	5 kg	ii	20 kg
iii	25 kg	iv	$\frac{1}{2}$ kg

  b Write these masses in milligrams.

i	3 g	ii	8 g
iii	$\frac{1}{2}$ g	iv	5 kg

3 An average cat needs 50 g of a brand of cat food each day.
Kate has two cats. How many days will a 2 kg bag of cat food last?

4 A bag of 20 apples weighs 2 kg.
  a How many grams does each apple weigh?
  b How many apples are there in a bag weighing 3 kg?

5 Six friends are weighing fish.

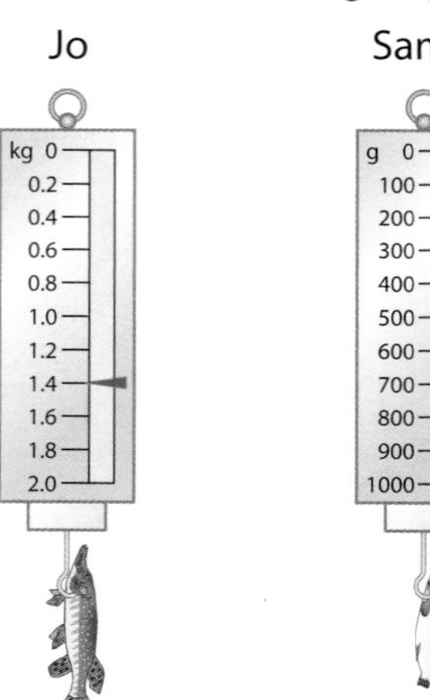

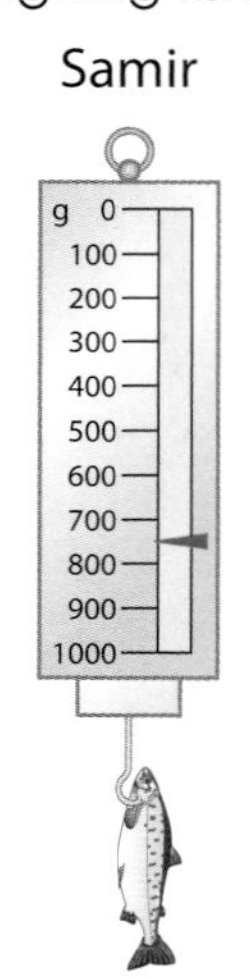

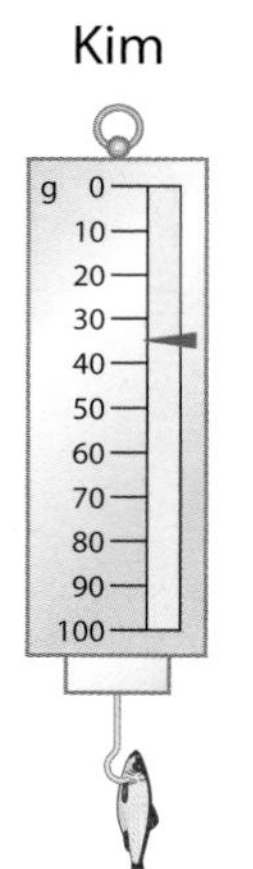

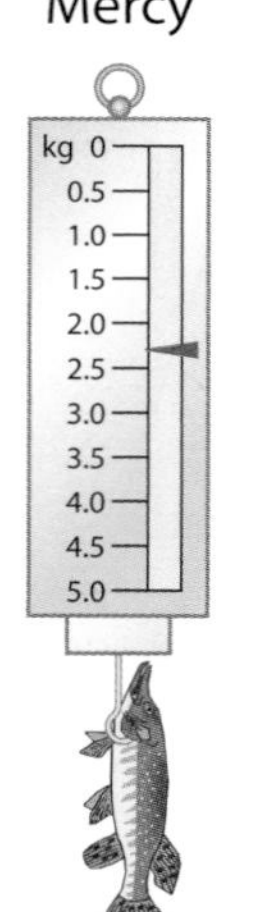

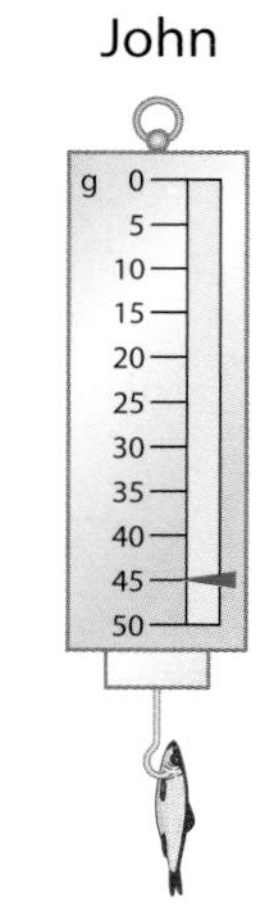

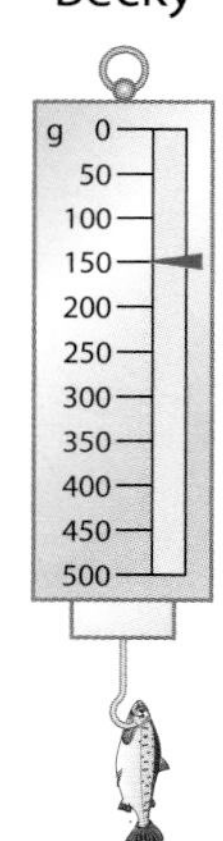

  a Write the masses of the fish in order, smallest first.
  b What is the total mass of all the fish on Jo, John and Becky's scales?

6 Which instrument is most appropriate to measure each of the things below?
Choose from this list.

scientific balance	kitchen scales	bathroom scales	industrial scales

a A lorry
b A child
c A grain of sand
d An apple

7 **a** Estimate the value of each of these quantities and state a suitable measuring device.
i The mass of a cabbage
ii The mass of a guineapig
b What difficulties would you find in **ii** of part **a**?

Reasoning

8 A bag of sugar weighs 1000 g (1 kg).
Copy the table below.

Less than 1000 g (1 kg)	More than 1000 g (1 kg)

Put the objects in the correct column.

Tin of beans

Mobile phone

Vacuum cleaner

Apple

Dog

Doll

Textbook

Computer

Pencil

Desk

Reasoning

9 This table gives information about British coins.

Value	Mass	Thickness	Diameter
1p	3.6g	1.6mm	20.0mm
2p	7.1g	2.0mm	25.9mm
10p	6.5g	1.9mm	24.5mm
20p	5.0g	1.7mm	21.4mm
50p	8.0g	1.8mm	27.3mm
£1	9.5g	3.2mm	22.5mm
£2	12.0g	2.5mm	28.5mm

a Look at this number line.
It shows the thickness of some of the coins in millimetres.

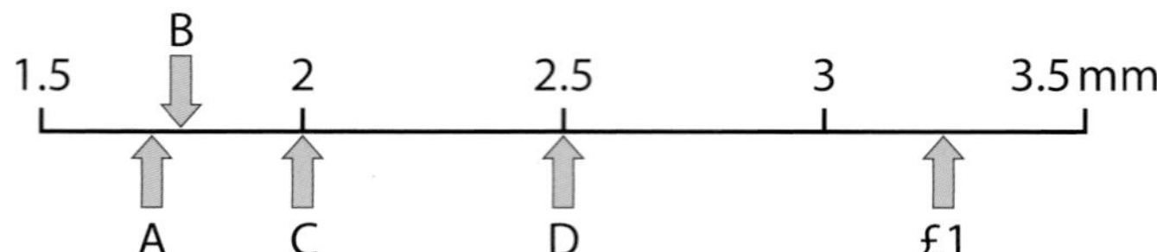

Which coins are at A, B, C and D?

b List the coins in order of diameter.
Mark them on a copy of this number line.

20 mm ... 30 mm

c List the coins in order of their mass, starting with the lightest.

d Ali and Mark are discussing the information in the table.

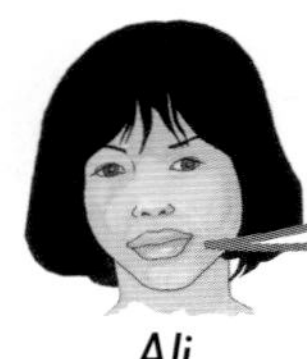

In the table the 50p is heavier than 20p.

Ali

I think the more valuable a coin is, the heavier it is!

Mark

e Which coins make Mark's statement false?

f Each of these bags of coins holds £1.

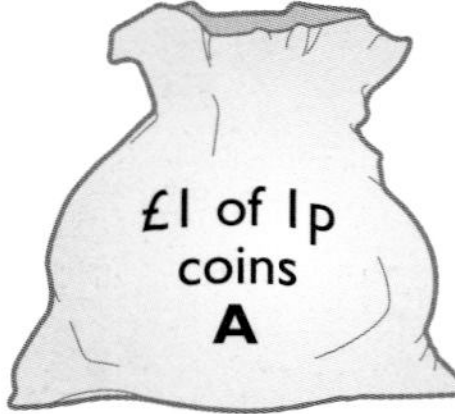

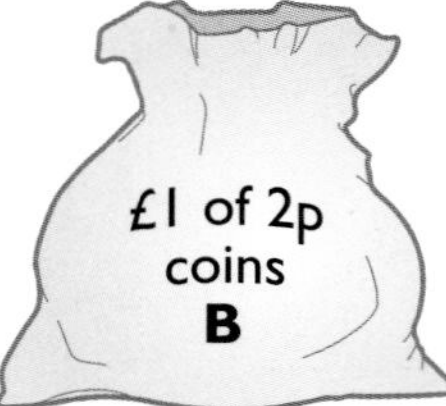

i How heavy is each bag?

ii List the bags in order of their mass.

Wider skills practice

Reasoning

1 The table shows the cost of posting parcels in a certain country.

Weight	Cost
up to 0.25 kg	$1.00
0.5 kg	$1.50
0.75 kg	$2.00
1.0 kg	$3.00
1.5 kg	$4.00
2 kg	$5.00
3 kg	$6.00
Over 3 kg, $1.00 for every extra 1 kg or part of 1 kg.	

a How much does it cost to send this parcel?

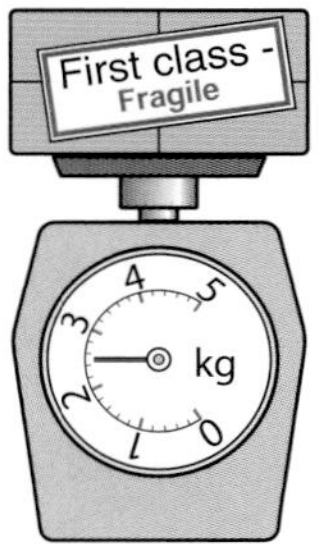

b How much does it cost to send each of these parcels?

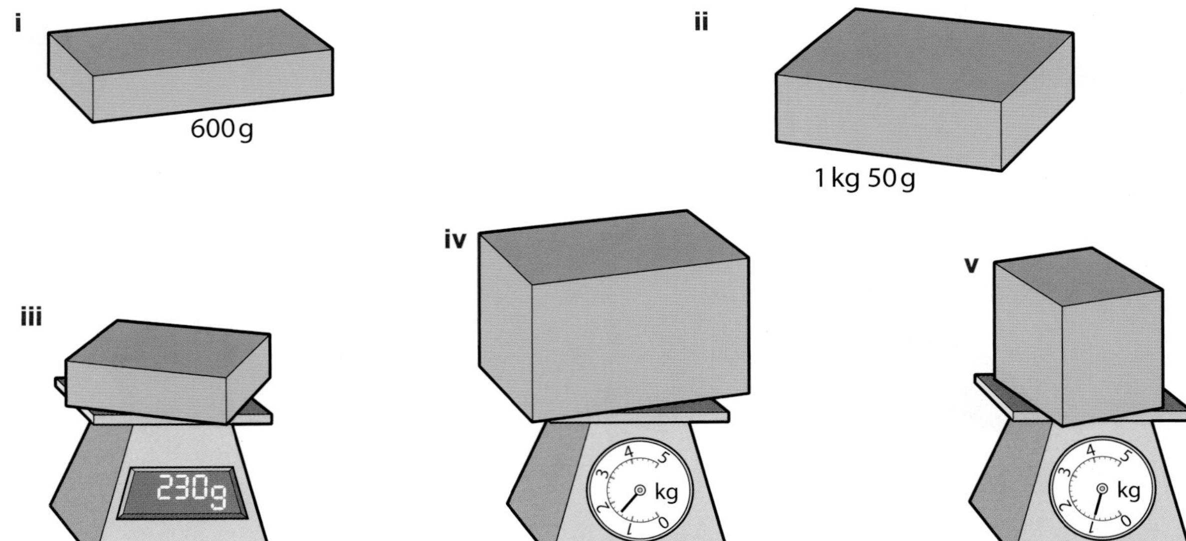

c All the parcels in part **b** are then wrapped together in one very large parcel.

i How much does it cost to send them in this way?

ii How much does this save?

d Put the parcels in part **b** together into two big parcels.
How much does each parcel cost to post?

e Find the cheapest way of posting the parcels in part **b** as two big parcels.

Applying skills

Problem solving

1 a Amanda's school is a health-promoting school.
All the children from Nursery through to Year 2 are given a piece of fruit each day.
There are 27 children in Amanda's class.
They are each given an apple weighing 100 g.

i What is the total weight of apples in grams?

ii Convert your answer to kilograms.

b Here are some portions of fruit and vegetables.

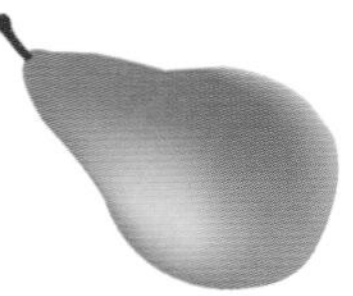
1 pear 120 g

1 orange 110 g

1 banana 160 g

1 handful grapes 100 g

4 strawberries 130 g

1 apple 100 g

2 plums 100 g

3 dried apricots 60 g

1 carrot 120 g

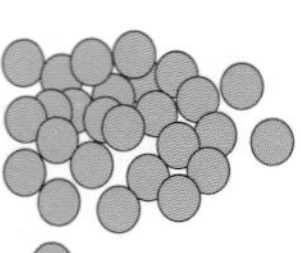
1 handful peas 150 g

1 tomato 150 g

1 handful broccoli 130 g

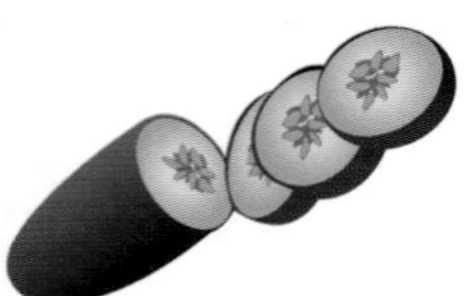
1 courgette 140 g

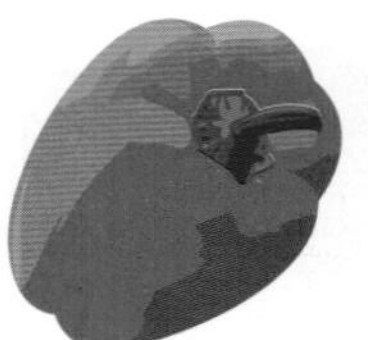
1 red pepper 120 g

Health guidelines say you should eat five portions of fruit or vegetables each day.

i Choose five portions and work out how much they weigh.

ii Sam likes to eat a variety of fruit and vegetables.
He wants to eat at least 600 g of fruit and vegetables.
Design some possible menus for Sam.

Reviewing skills

1 What is the mass shown on each of these scales?

a

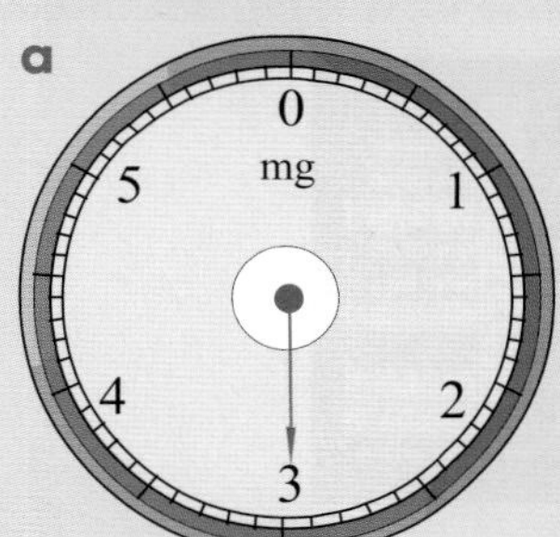

b

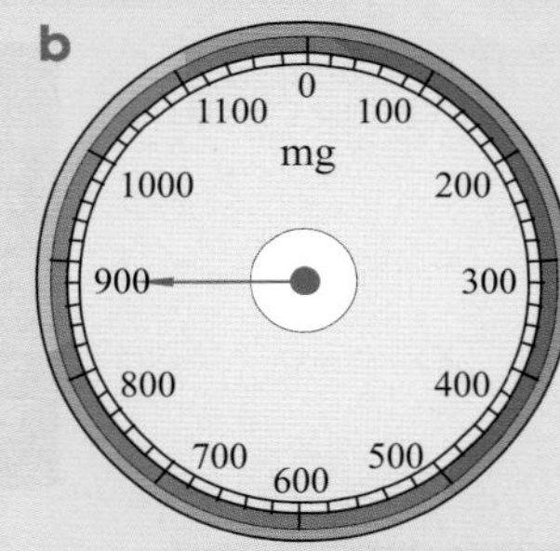

2 Match each of these objects to one of the weights below.

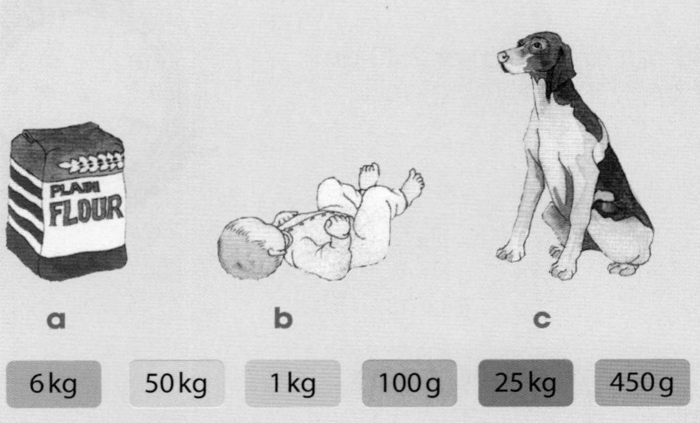

a **b** **c** **d** **e** **f**

6 kg | 50 kg | 1 kg | 100 g | 25 kg | 450 g

3 Look at the ingredients for a flapjacks recipe.
It makes flapjacks for 12 people.

a How much will the flapjacks weigh?

b How much golden syrup would you need for six people?

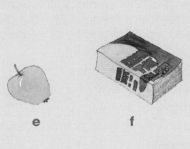

Flapjacks

75 grams of sugar
75 grams of butter
50 grams of golden syrup
150 grams of rolled oats
35 grams of dried fruit

4 **a** Write these masses in grams.

i 9 kg **ii** 4 kg **iii** $2\frac{1}{2}$ kg

b Write these masses in milligrams.

i 6 g **ii** 2 g **iii** $5\frac{1}{2}$ g

Unit 3 • Time • Band c

Building skills

Example outside the Maths classroom

Making plans

0110	Frankfurt	QF	5	C24	
2320	London-Heathw	BA	16	C18	Gate Closed
2325	Tokyo-Narita	NH	902	D35	Gate Closed
2325	London-Heathw	QF	9	C13	Gate Closing
2330	Paris-CDG	DL	8377	C22	Gate Closed
2330	Tokyo-Narita	AA	5832	D44	Gate Closed
2355	Osaka/Kansai	JL	722	D40	Gate Closing
0130	London-Heathw	QF	31	C26	
0015	Beijing	CA	970	D30	Boarding
0050	Moscow-Domod	UN	516	C23	Gate Open

Gates close 10 mins before departure   No boarding calls will be made   01 Mar 2008, 23:44

Toolbox

This clock shows 20 minutes to three. This could be in the morning, or the afternoon.

Using a 12-hour clock, this would be written 2:40 am (morning) or 2:40 pm (afternoon). Using a 24-hour clock, 0240 would be in the morning and 1440 would be in the afternoon.

60 seconds = 1 minute

60 minutes = 1 hour

24 hours = 1 day

7 days = 1 week

Each month has a different number of days.

Month	Number of days
January, March, May, July, August, October, December	31
April, June, September, November	30
February	28 or 29

Every fourth year is a leap year. In a leap year, February has 29 days.

12 months = 1 year

Example – Reading a timetable

Tim catches the train from London to Norwich.

a How long does the journey take that starts at 1150?

b How long does the journey last that ends at 1420?

Station			
London..........	1125	1150	1210
Norwich..........	1330	1340	1420

Solution

a 1150 ⟶ 1250 ⟶ 1300 ⟶ 1340

1 hour, 10 mins, 40 mins

1 hour + 10 mins + 40 mins = 1 hour 50 minutes

So the journey takes 1 hour 50 minutes in total.

b 1210 ⟶ 1410 ⟶ 1420

2 hours, 10 mins

2 hours + 10 mins = 2 hours 10 minutes

So the journey takes 2 hours 10 minutes in total.

Example – Converting between minutes and seconds

A dentist says that you should brush your teeth for $2\frac{1}{2}$ minutes.
How many seconds is this?

Solution

There are 60 seconds in 1 minute.

2 minutes = 120 seconds ← **$2 \times 60 = 120$**

$\frac{1}{2}$ minute = 30 seconds ← **$\frac{1}{2} \times 60 = 30$**

So $2\frac{1}{2}$ minutes is 150 seconds.

Remember:

- Use a.m. and p.m. for the 12-hour clock and four digits for the 24-hour clock.
- You cannot add and subtract times as if they were ordinary numbers. You must deal with minutes and seconds separately.

Skills practice A

1 Draw these times on an ordinary 12-hour clock face.

a 7:15 **b** 11:05 **c** 1:35 **d** 2055 **e** 0440 **f** 1350

2 Write these times in words and then in numbers.

a
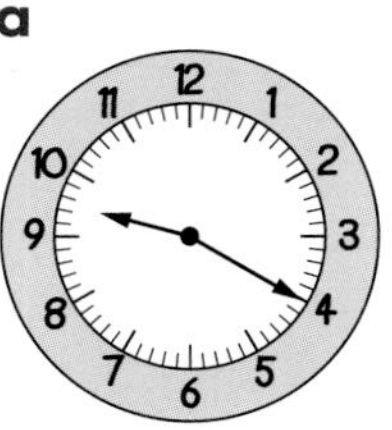

b
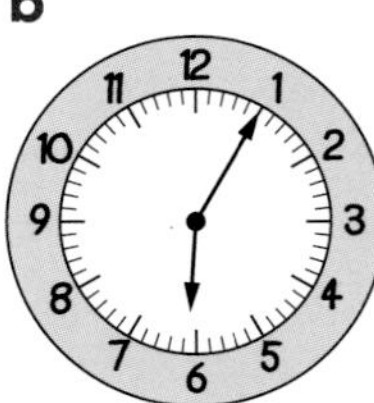

c
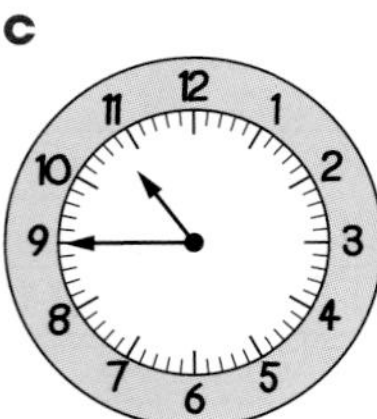

d
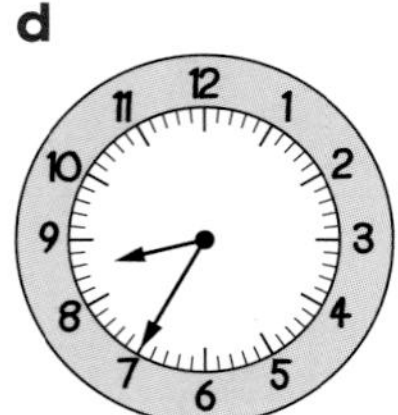

e
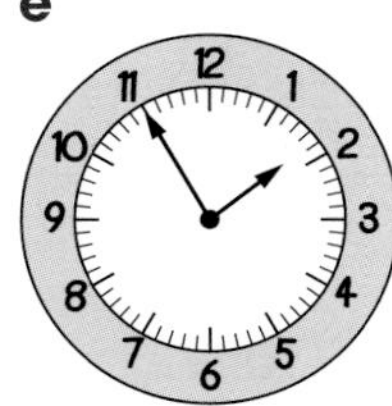

3 Write these times using the 24-hour clock.

a 9.25 a.m.
b 12.15 p.m.
c 12.15 a.m.
d Ten to four in the afternoon

4 Write these times using a.m. or p.m.

a 0728 **b** 1305 **c** 2311

5 Write these times in numbers.
Use the 24-hour clock.
- **a** Half past 7 at night
- **b** 20 minutes past 11 in the morning
- **c** 5 minutes to 9 at night
- **d** 5 minutes past 2 in the afternoon
- **e** 25 minutes to 6 in the morning

6 Look at this TV guide.

TV Guide

9.00 pm	**News and Weather**
9.25 pm	**The X-Files**
10.10 pm	**Sports Highlights**
11.25 pm	**Late Night Film**

- **a** How long do the *Sports Highlights* last?

Rosa wishes to record *The X-Files* and the late night film.
- **b** What is the start time for these programmes using the 24-hour clock?
- **c** Rosa has 180 minutes recording time left on her hard drive. Is there enough time left for her to record them both?

7 The time required to cook a turkey is 20 minutes, plus 40 minutes for each kilogram in weight.

- **a** How long will a 5 kg turkey take to cook?

Peter wants to have the turkey ready for 1.00 p.m.
- **b** When should he start cooking?

8 Ali takes the fast train from Newcastle to York.
It leaves Newcastle at 2000.
- **a** Write this time using the 12-hour clock.

It arrives at York at 2115.
- **b** Write this time using the 12-hour clock.
- **c** How long does her journey take?

9 Look at the opening times for a fish and chip shop.

FRIARS FISH AND CHIPS
OPENING TIMES

	Lunch	Evenings	
Monday	–	5.00 – 10.30	
Tuesday	11.30 – 1.30	6.00 – 11.00	
Wednesday	11.30 – 1.30	6.00 – 11.00	
Thursday	–	4.30 – 6.30	8.00 – 11.00
Friday	11.30 – 1.30	4.30 – 11.00	
Saturday	11.30 – 1.30	6.00 – 11.00	

- **a** On which days is the fish and chip shop closed at lunch time?
- **b** Can you buy chips at 7.00 p.m. on a Thursday?
- **c** For how many hours a week is the shop open?
- **d** On which evening is the shop open for the longest time?

Reasoning

10 Jason says that the film at the cinema starts at 0350.
Norman says that the film starts at 1550.
Who is more likely to be correct?
Explain your answer.

Skills practice B

1 Look at this calendar.

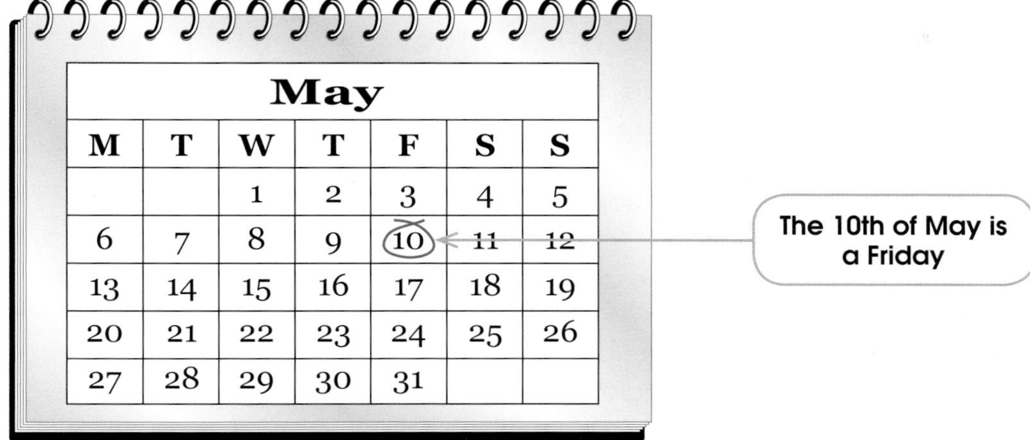

a What day of the week is May 17th?
b How many Wednesdays are there in May in this year?
c How many weekends are there in May?
d What day of the week is May 1st?
e What day of the week is the last day of April?
f The 1st Monday in May is a holiday.
What date is it?

2 John has four weeks to rehearse with his band.
What date is it now?

3 Add two weeks to each of these dates.

a 20 March
b 24 September
c 24 December
d 22 February (in a leap year)
e 22 February (not in a leap year)

4 Copy and complete this table.

Flight from	Time due	Minutes late/early	Time now due
Barcelona	2055	100 minutes late	
New York		135 minutes late	2340
Sydney	2205		0035
Tokyo	2240	50 minutes early	

5 Which is longer?

a 2 years or 500 days
b 200 minutes or 3 hours
c 35 months or 3 years
d 600 hours or the month of June

6 The Ross family are going on holiday to Portugal.
Here is their travel itinerary.

Flight:	MON7180	
Check-in:	Sunday 18 Jul 13 at 1300	Manchester
Take-off:	Sunday 18 Jul 13 at 1530	
Arrive:	Sunday 18 Jul 13 at 1845	Faro

a How long is it between check-in and take-off?
b How long is the flight?
c There is a problem with the aircraft and the flight is delayed by 50 minutes.
What time does it take off?
d They arrive in Faro at 1910.
How much shorter is the flight than scheduled?

Reasoning

7 This is part of the timetable for the number 35 bus in Cannock.

Cannock - Hednesford - Rose Hill														**35**
Monday to Saturday														
Cannock Bus Station	—	—	—	—	0925	1025	1125	1225	1325	1425	1525	1535	1625	1725
Stagborough Way	—	—	—	—	0933	1033	1133	1233	1333	1433	1533	–	1633	1733
Hednesford Bus Station	—	—	—	—	0937	1037	1137	1237	1337	1437	1537	1555	1637	1737
Bracken Close	—	—	—	—	0943	1043	1143	1243	1343	1443	1543	1601	1643	1743
Rose Hill	0739	0811	0816	0851	0951	1051	1151	1251	1351	1451	1551	1609	1651	1751
Bracken Close	0747	0819	0824	0859	0959	1059	1159	1259	1359	1459	1559	1617	1659	1759
Hednesford Bus Station	0753	0825	0830	0905	1005	1105	1205	1305	1405	1505	1605	1623	1705	1805
Stagborough Way	0757	–	0834	0909	1009	1109	1209	1309	1409	1509	1609	1627	1709	1809
Cannock Bus Station	0805	0845	0842	0917	1017	1117	1217	1317	1417	1517	1617	1633	1717	1817

a How often do the buses run from Cannock to Rose Hill?
b How many minutes does it take for the bus journey from Cannock Bus Station to Rose Hill?
c What happens when the bus reaches Rose Hill?

8 Which device is most appropriate to measure each of the things below?
Choose from this list.

stop watch	mobile phone	wall clock	calendar

a The number of days until Christmas
b Your time in a 100-metre race
c The time left in this lesson
d The length of a phone call

9 What unit would you use to measure each of these?
a Your grandmother's age
b How long you spend taking a shower
c How long you are in school each day
d How long it takes you to work out 25×4
e The length of a holiday in Spain.

Reasoning

10 Tim is working out how long his CD single lasts. Here are his calculations.

4   45
6   25
+8   +35
18 minutes   105 seconds

Total time = 19 minutes 45 seconds.

a Sophie says '105 seconds is 1 minute and 5 seconds'.
She is wrong. What is the right answer?
Explain Sophie's mistake.

b Explain Tim's working out.

Reasoning

11 Pete says the time is 25 minutes to seven.
Which of these would represent 25 minutes to seven in the evening?

a 1835   **b** 6:35   **c** 0635
d 6.35 am   **e** six thirty five   **f** 6.35 pm

Wider skills practice

1 It takes 4 years for the planet Yus to complete an orbit of its sun.

a How long would it take to complete two orbits?
b How long would it take to complete five orbits?
c How many orbits does Yus complete in 28 years?
d Explain why it cannot complete a whole number of orbits in 18 years.

It takes 3 years for the planet Xil to complete an orbit around the same sun.

e How long would it take to complete three orbits?

In 2012, both the planets were lined up with their sun.

f When will be the next year that the planets are both in the same positions?
g Will they both be in the same positions in 2072? Explain.
h Will they both be in the same positions in 2076? Explain.

Applying skills

1 The time for the Earth to orbit the Sun is not exactly 365 days.
Every four years we have a leap year with 366 days.

a How many days are there in 4 years?
b What is the true length of a year?

Reasoning

2 A flight from London to New York takes 7 hours.
It leaves London at 0900 and arrives in New York at 1100.
Explain this.

Reviewing skills

1 What unit would you use to measure each of these?

a How long you take to run 3 miles
b The age of a tree
c How long until your birthday
d The time taken to fly to America
e How long you can hold your breath

2 Write these times in words and then in numbers.

a

b

c

d

3 Write these times in numbers. Use the 24-hour clock.

a Half past 7 at night
b 20 minutes past 11 in the morning
c 5 minutes to 9 at night
d 5 minutes past 2 in the afternoon
e 25 minutes to 6 in the morning

4 Here is part of a bus timetable.

Tipton...............	1005
Ottery..............	1020
Honiton...........	1055
Feniton............	1115

I get on the bus at Ottery and get off at Honiton.
Harry

My journey takes 20 minutes.
Michelle

I get on at Ottery and get off at Feniton.
John

a How long is Harry's journey?
b Where does Michelle get on and off the bus?
c How long is John's journey?

Building skills

Example outside the Maths classroom

Taking health advice

Toolbox

The **volume** of an object is the amount of space it takes up.

The **capacity** of a container is the amount inside it.

The volume of a liquid can be measured using a scale on a container or measuring cylinder.

Many containers, such as bottles, cans and measuring spoons, are designed to hold an exact amount.

1000 millilitres (ml) = 1 litre (l)

100 centilitres (cl) = 1 litre (l)

You need to have a rough idea of the volume of an object so you can choose an appropriate **measuring instrument**. This could be a measuring spoon, cylinder and jug, burette or tank.

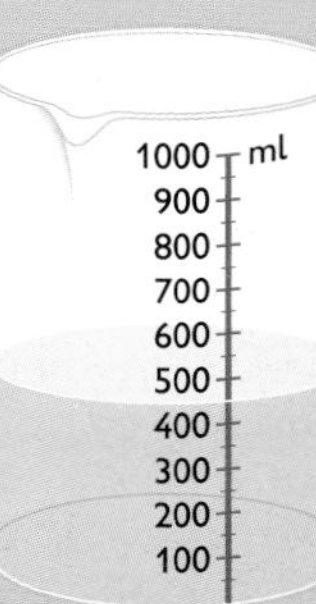

Example – Measuring capacity

Jathika is measuring the volume of liquids for a science experiment.
What is the volume of the liquid in each of these beakers?

Solution

a The volume is 4 litres.

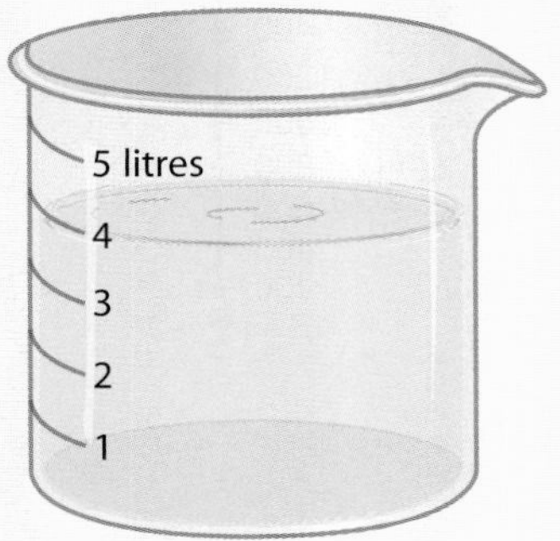

b The volume is 700 ml.

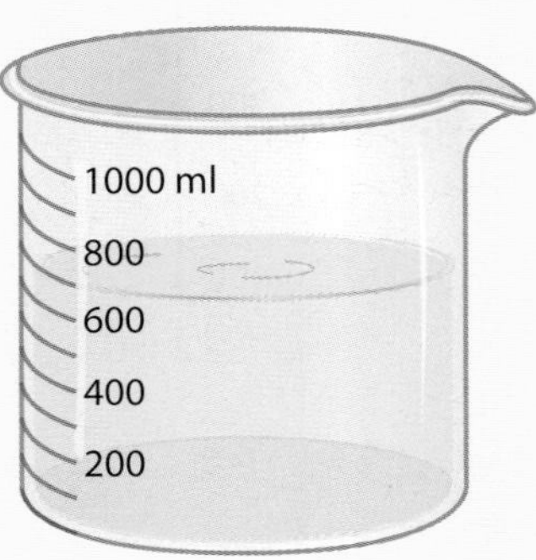

Not drawn to scale

Example – Converting between litres and millilitres

Here is the liquid for another of Jathika's experiments.

a What is the volume of liquid in this beaker?

b How many millilitres is this?

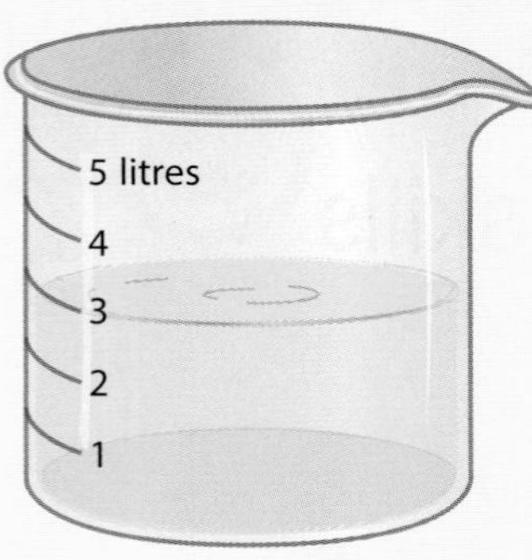

Solution

a The volume is 3 litres.

b 3 litres is 3000 ml.

1 litre = 1000 ml
3 litres = 3 × 1000 = 3000 ml

Remember:

- When you read the volume from a scale, the actual measurement might not be labelled, but you can still estimate it between two labels.
- 1000 ml = 1 litre

Skills practice A

1 How much liquid is in each of these beakers?

a

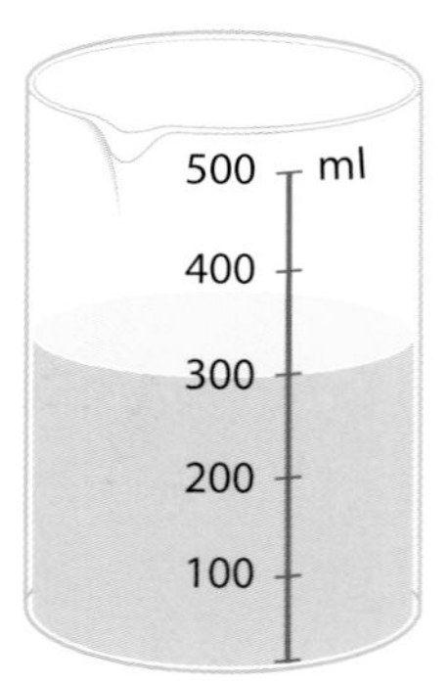

b

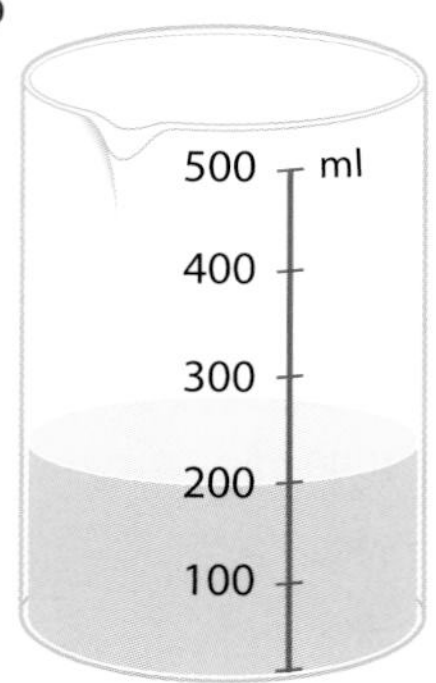

c

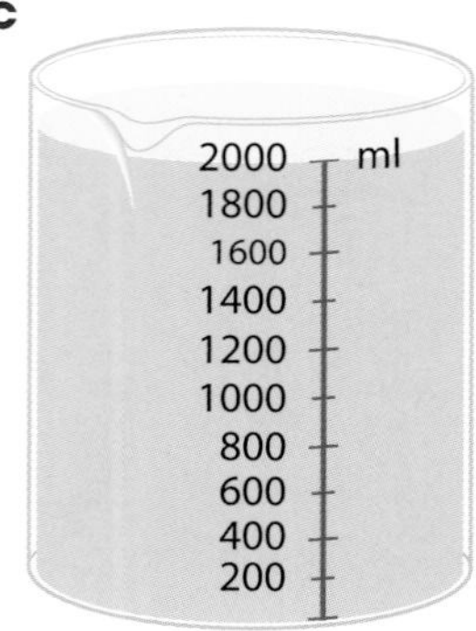

d

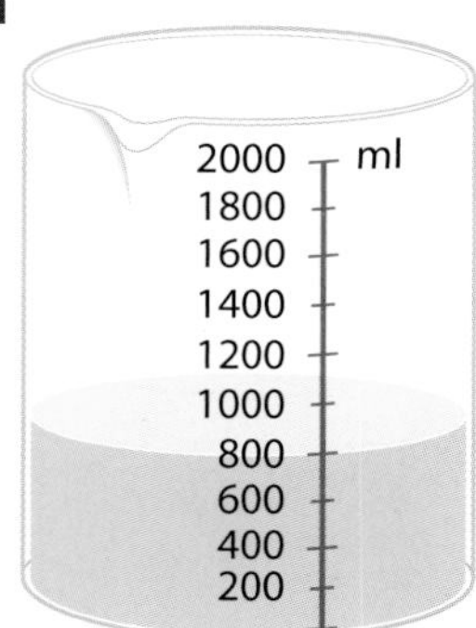

2 This cube is filled with water.
The amount of water is 1 millilitre (1 ml).
What is the volume of water in each of these diagrams?

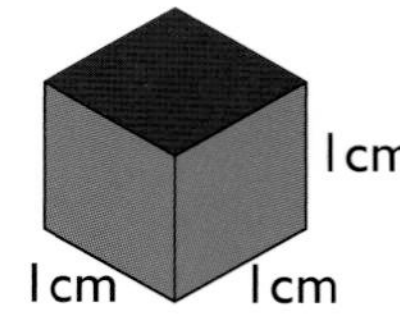

a

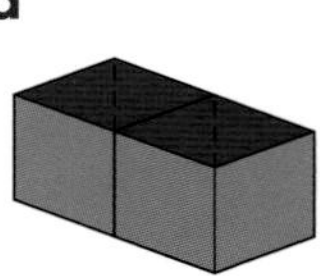

b

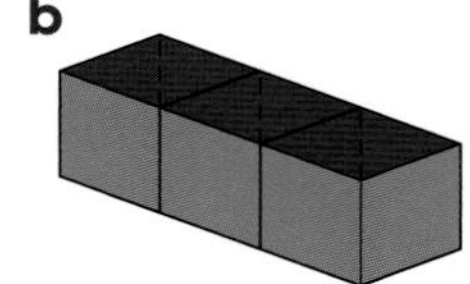

c

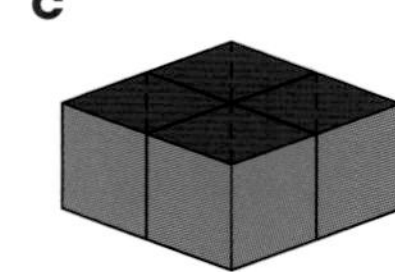

3 Here is a glass.
Its capacity is 250 ml.

a Look at these sets of glasses.
What is the capacity of each set?

i

ii

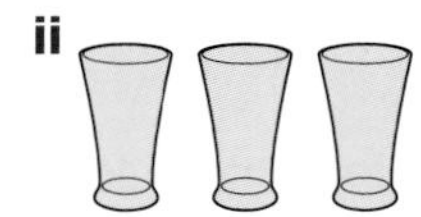

iii

b Julia has one 500 ml bottle of cola.
How many glasses can she fill?

c Chris has two 500 ml bottles of cola.

i How many millilitres of cola are there in the two bottles?

ii How many glasses can he fill?

4 **a** How much water is in each of these jugs?

i

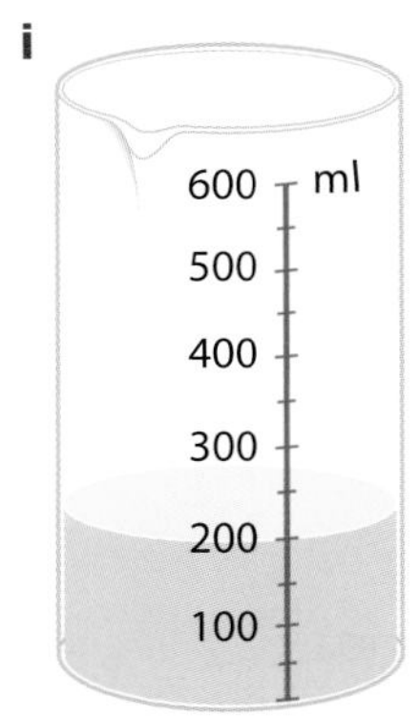

ii

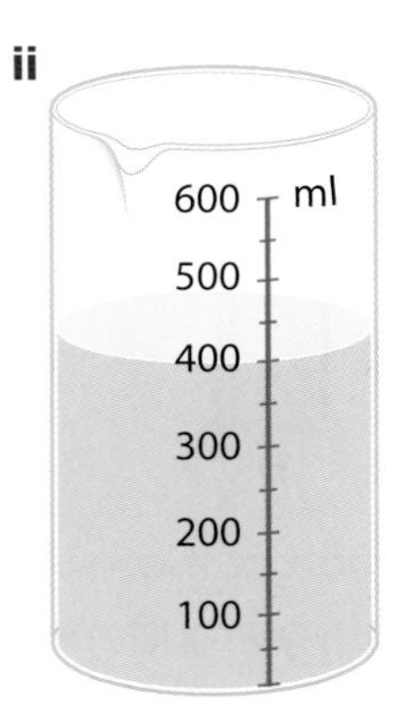

iii

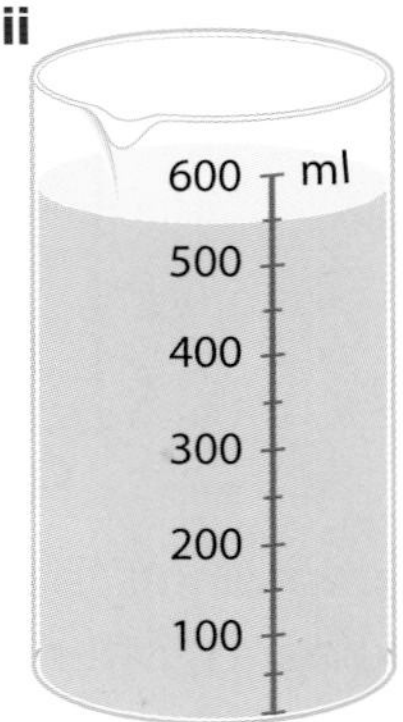

b How much water is there in total?

c Annie needs 2000 ml of water.

i How many litres is this?

ii How much more water does Annie need?

5 Which of these are units of volume?

metre	millilitre	gram	centimetre
kilogram	kilometre	litre	millimetre

6 **a** Convert these quantities into litres.

- **i** 2000 ml **ii** 3000 ml
- **iii** 7000 ml **iv** 2500 ml

b Convert these quantities into millilitres.

- **i** 4 litres **ii** 6 litres
- **iii** 9 litres **iv** 3.5 litres

7 You should drink 2 litres of water every day.

a How much water should you drink in a week?

b What about in a year?

8 An average person uses about 150 litres of water every day.

a How many people live in your household?

b How much water is used in your household every year?

Skills practice B

1 A doctor prescribes 5 ml of medicine to be taken three times per day.
There are 240 ml of medicine in the bottle.

a How much medicine would be taken in a week?

b How long will it take until the medicine in the bottle runs out?

Reasoning

2 Zak has four containers as shown below.

50 ml
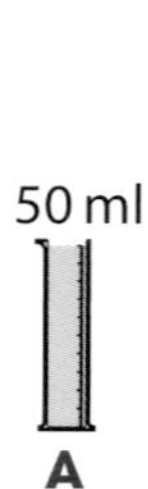
A

100 ml
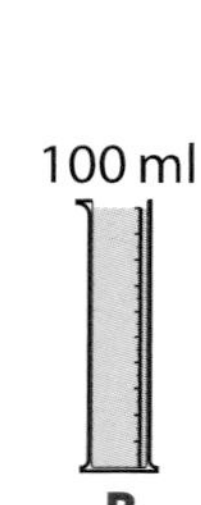
B

500 ml
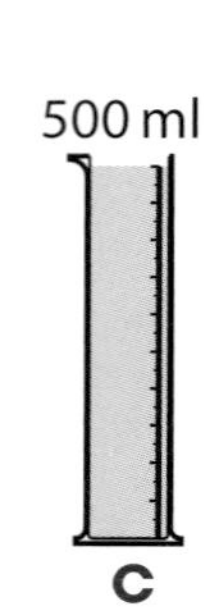
C

1000 ml
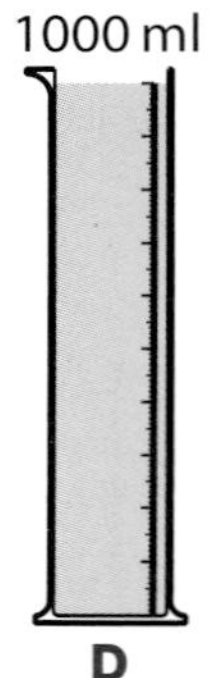
D

How can you use three containers to measure out each of these volumes?
You may use each container as often as you wish but you must measure three times.

a 650 ml **b** 1600 ml **c** 1550 ml **d** 3 litres **e** 250 ml

3 Melissa buys a 1 litre bottle of shampoo. She uses 25 ml each wash.
How many times can she wash her hair?

4 Wine is often measured in centilitres.
Jake's father is making a drink.
He needs exactly 250 cl of a special type of wine.
The wine is only sold in 70 cl bottles and 30 cl bottles.
How many of each type of bottle does he need to buy?

5 A bottle of lemonade holds 2 litres.

a How many millilitres is this?

Katie invites 17 people to a party.
She thinks everyone will drink two glasses of lemonade.
Each glass has a capacity of 200 ml.

b How many bottles of lemonade should Katie buy. (Don't forget Katie herself.)

c Estimate how much lemonade will be left over.

6 Which of these is most appropriate for measuring each of the things below?
Choose from this list.

medicine spoon	measuring jug	the gauge on a pump	a 5-litre can

a The quantity of milk for a recipe

b Petrol for your mower

c A dose of cough medicine

d The volume of petrol you put in your car

7 Estimate the value of each of these quantities and state a suitable measuring device.

a The capacity of a carton of orange juice

b The capacity of a glass of water.

8 Copy this diagram of a measuring cylinder.
Mark on it the scale and the level of the liquid so that it is

a a 1 litre beaker containing 250 ml of liquid

b a 2 litre beaker containing 1.5 litres of liquid.

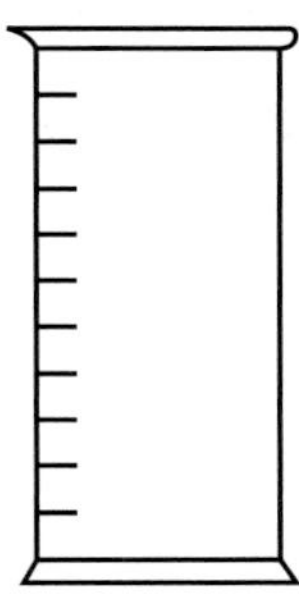

9 Look at the objects.
Match each object to its capacity below.

Bathroom sink 

Teaspoon 

Bath 

Mug 

Kettle 

Bottle of ketchup 

5 ml	400 ml	80 litres	15 litres	250 ml	$1\frac{1}{2}$ litres

Wider skills practice

1 Look at the ingredients for a recipe for Fruity Bread.

a Which ingredients are measured by weight?

b Which ingredients are measured by volume?

2 Harmony works in a café.
She thinks that at least a fifth of the drinks are wasted. This annoys Harmony.
She measures how much is left in the glasses of the next ten customers.

25 ml	0 ml	100 ml	130 ml	0 ml
55 ml	45 ml	80 ml	100 ml	65 ml

a What is the total volume of drink wasted?

Each glass has a capacity of 200 ml.

b Do the customers waste at least one fifth of their drinks?
Explain your answer.

3 The ingredients in this recipe make pancakes for three people.
Graham wants to make pancakes for 30 people.

a Write out the ingredients to make enough for 30 people.
Give the volumes in litres and the masses in kilograms.

b There are six eggs in a box.
How many boxes of eggs are needed?

c Look at the weighing scales.

How much more flour is needed?

d Graham has a 2 litre measuring jug.
It is marked every 250 ml.
How does he use this to get the right amount of milk?

Problem solving

4 **a** Choose suitable units from the list to complete the story below.

litre millilitre kilogram gram kilometre metre centimetre millimetre

Michelle's grandfather is a large man. He is 1.85 ☐ tall and weighs 90 ☐.
He is building a shed in his garden.
The shed is 3 ☐ long and 2 ☐ wide.
To build it he is using 7 ☐ nails and planks of wood that are 3 ☐ long and 10 ☐ wide.
He has a pond in his garden. It holds 300 ☐ of water.
There are a few large fish in the pond that weigh about 4 ☐ each.
For his lunch, Michelle's grandfather usually walks to the pub.
It is 1.7 ☐ away so sometimes he takes the bus.

Applying skills

1 The capacity of a car's fuel tank is 80 litres.
On average, the car uses 125 ml of fuel per mile.

a How much fuel does this car use when it travels 8 miles?

b How far can the car travel on a full tank?

When the amount of fuel in the tank falls to 6 litres, a warning light comes on.

c The warning light comes on.
The driver says 'I can still drive 50 miles'.
Is he right?

Problem solving

2 With a 5 litre beaker and a 3 litre beaker (which have no markings), you can measure 2 litres by filling the 5 litre beaker and pouring as much as possible into the 3 litre beaker.

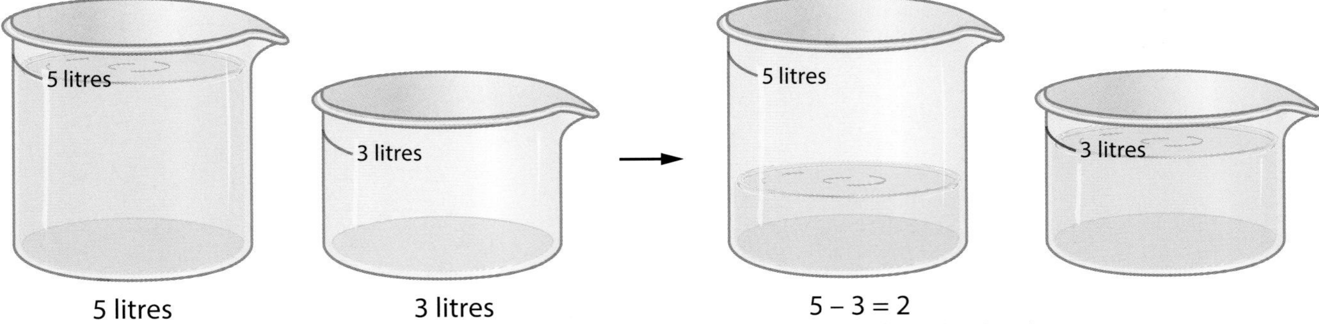

a **i** How can you measure 1 litre using a 4 litre beaker and a 9 litre beaker?

ii How can you measure 4 litres using a 3 litre beaker and a 5 litre beaker?

iii Show how you can measure any whole number of litres from 1 litre to 7 litres using just two beakers measuring 2 litres and 5 litres.

b Design a pair of beakers that could be used to measure anything from 1 to 10 litres.

Reviewing skills

1 a How much liquid is in each of these beakers?

i

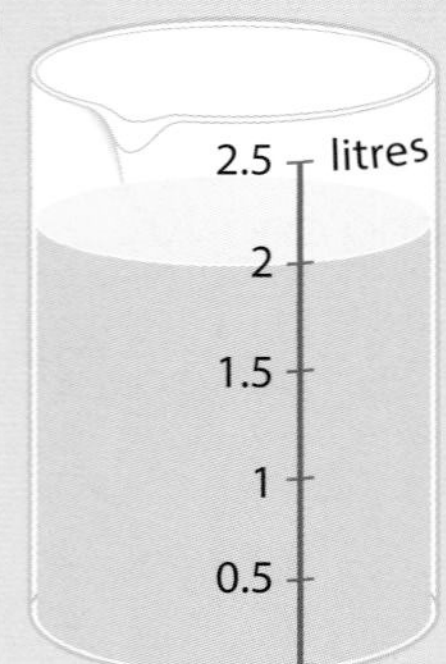

ii

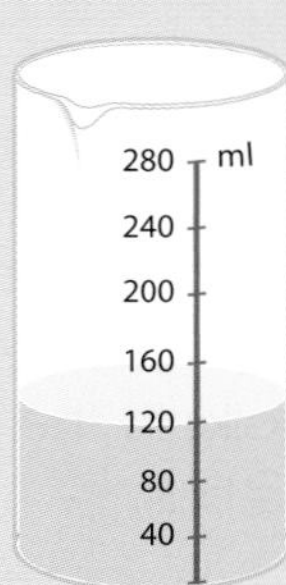

b Asaph pours as much liquid as he can from beaker **ii** into beaker **i** above. How much remains in beaker **ii**?

2 Pete has a jug which holds 750 ml of liquid.
He pours 20 jugs of water into his fish tank.
There were already 25 litres of water in the fish tank.
The fish tank is now full.
What is the capacity of the fish tank?

Building skills

Example outside the Maths classroom

Making comparisons

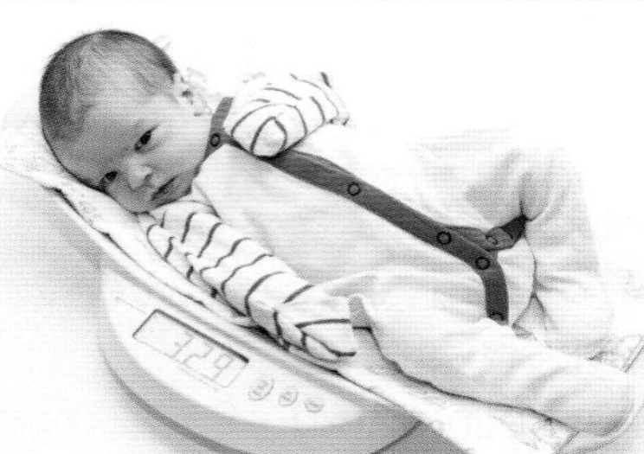

Toolbox

Scales are divided into large divisions and small divisions.
The scale on the right is from a set of scales.
The large divisions are numbers to show 30, 40 and 50 grams.
The small divisions are not labelled but you can work them out.
There are 10 grams between the labelled numbers.
There are five divisions between the labelled numbers so each division represents $10 \div 5 = 2$ grams

Example – Reading scales

The three gauges shown below measure different things.
What is the reading on each of the gauges?

a

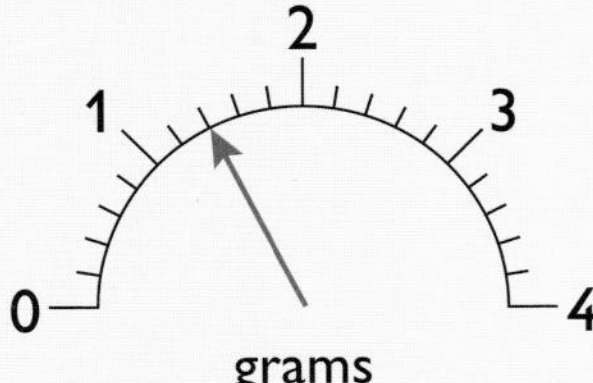

b

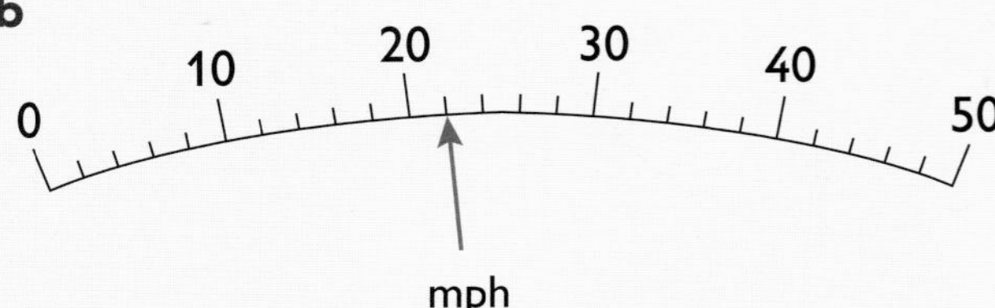

c

Solution

a Each large division is 1.
There are five small divisions between 1 and 2.
Each small division is 1 ÷ 5 = 0.2.
The needle is pointing to 1 + 2 × 0.2 = 1.4 mg.

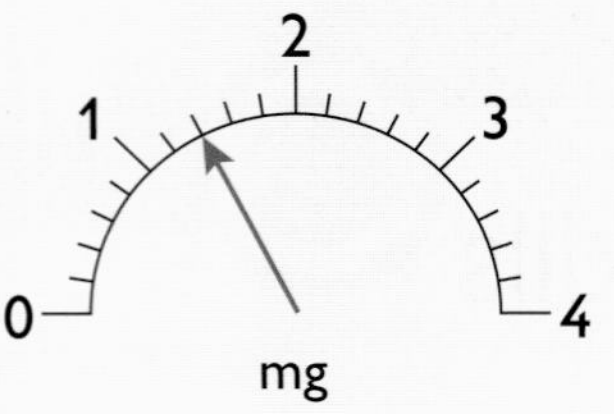

b Each large division is 10.
There are five small divisions between 20 and 30.
Each small division is 10 ÷ 5 = 2.
The needle is pointing to 20 + 2 = 22 mph.

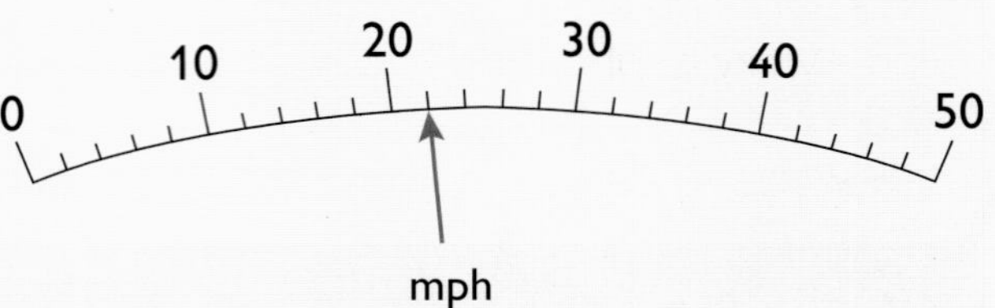

c Each large division is 10.
There are five small divisions between 50 and 60.
Each small division is 10 ÷ 5 = 2.
The needle is pointing to 50 + 4 × 2 = 58 °F.

Temperature, °Fahrenheit

Example – Making comparisons

Oliver and Vicky each weigh themselves in kilograms.
Vicky is lighter than Oliver.

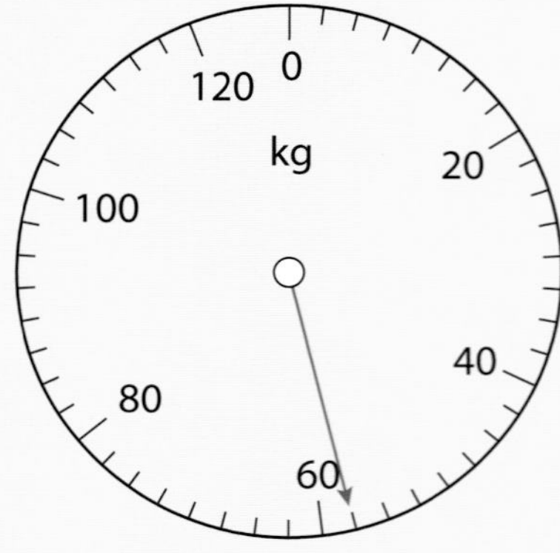

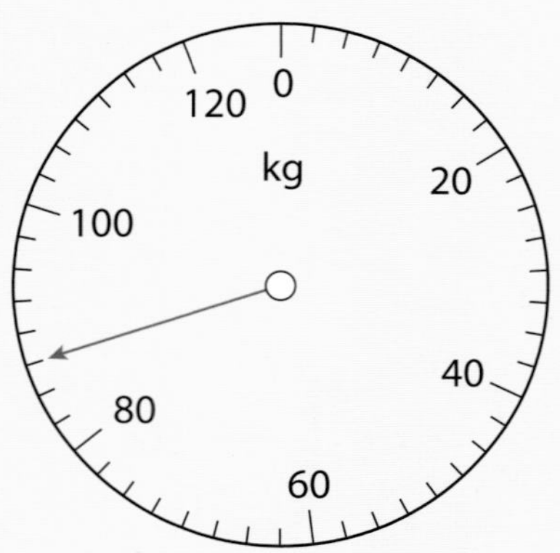

a How much does Vicky weigh?
b What is the difference between their weights?

Solution

The scales are both the same.
Each large division is 20 kg.
There are eight small divisions between the large divisions.
Each small division is $20 \div 8 = 2.5$ kg.

a The left-hand scale reads $60 - 1 \times 2.5 = 57.5$ kg.
The right-hand scale reads $80 + 3 \times 2.5 = 87.5$ kg.
Since Vicky is lighter than Oliver, she must weigh 57.5 kg.

b $87.5 - 57.5 = 30$
Oliver is 30 kg heavier than Vicky.

Remember:

- Count up the number of small divisions.
- Work out the difference between one large division and the next.
- Make sure you read the scale in the correct direction.

Skills practice A

1 The dials show speeds in miles per hour.
What are the speeds shown?

a

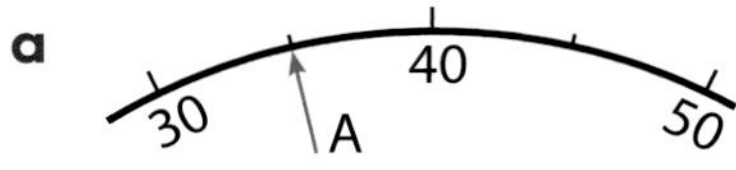

b

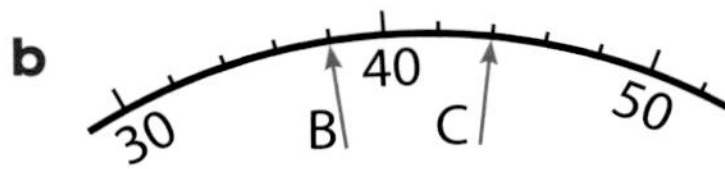

c

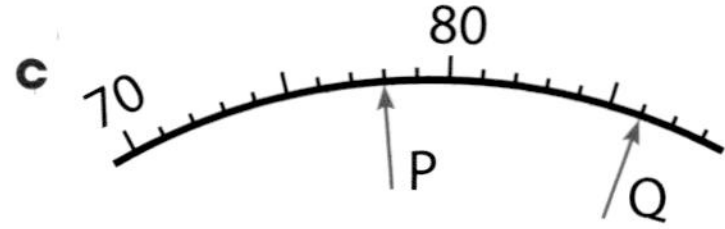

d

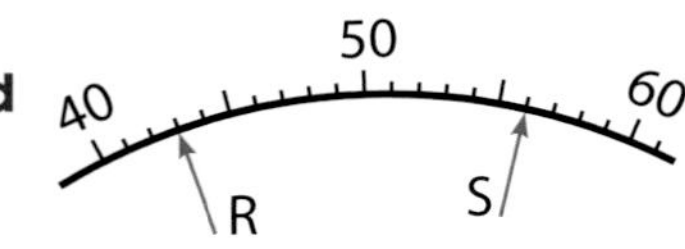

2 Normal body temperature is 37 °C or 98.6 °F

a What does a thermometer measure?
Look at this thermometer.

b What are the units on this thermometer?

c What is one big division?

d What is one small division?

e What is Yuri's temperature?

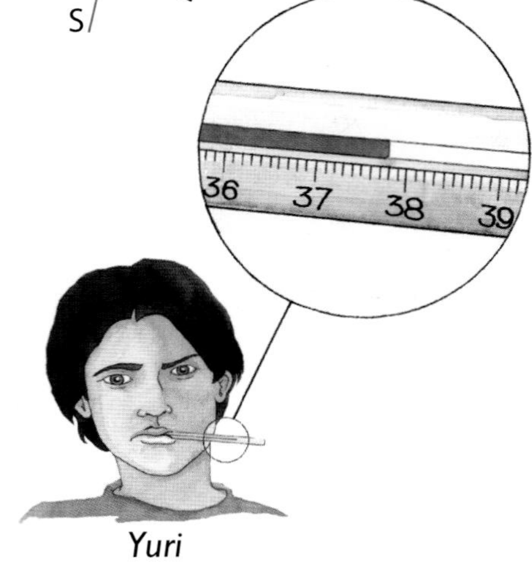

Yuri

Reasoning

4 Harry weighs two pears.

a How much does each pear weigh?

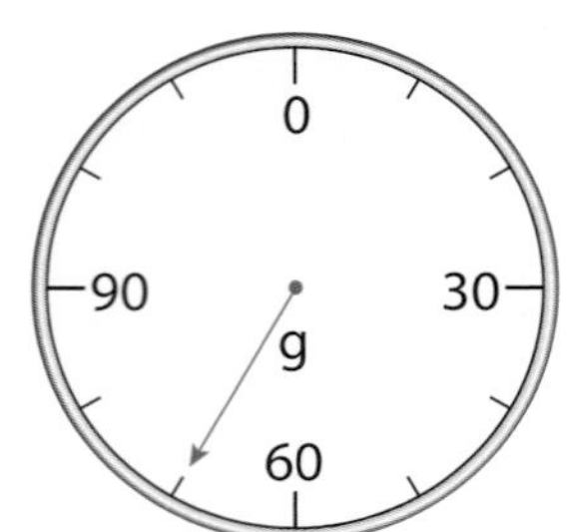

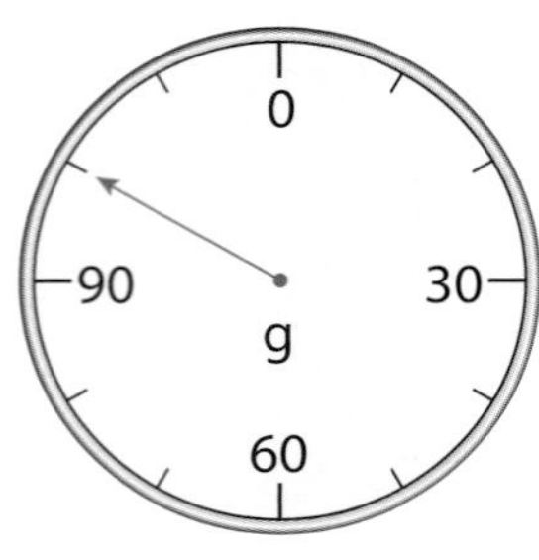

b Harry's mum asks him how much the pears weigh.
Harry is wrong.
Explain his mistake.

They weigh 91 grams and 61 grams.

Harry

5 What are the readings on these scales?

a

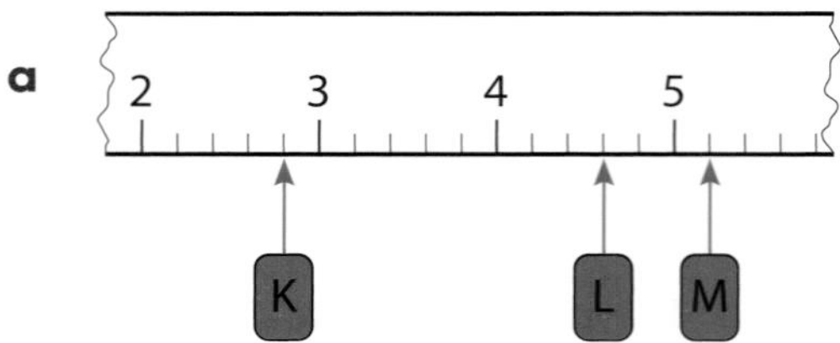

b

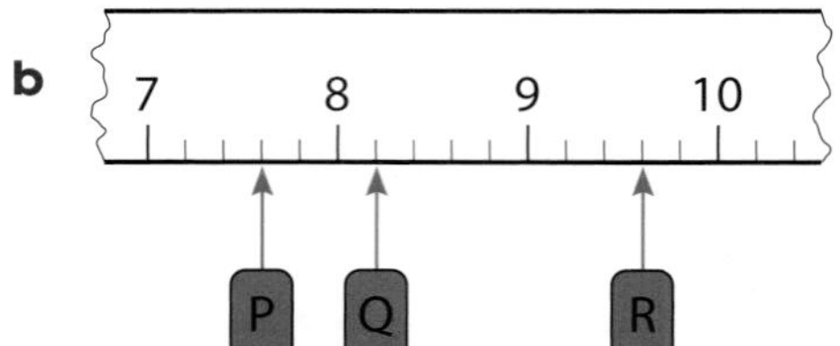

c

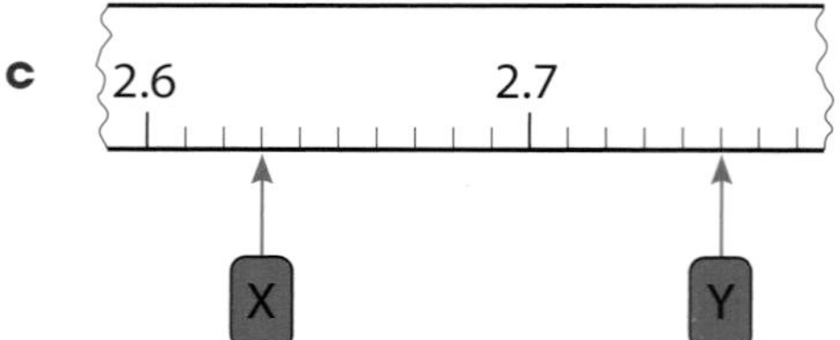

d

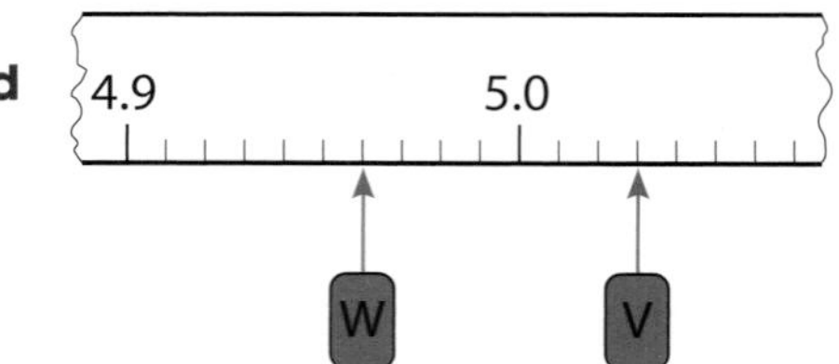

6 How much does each of these bags of apples weigh?

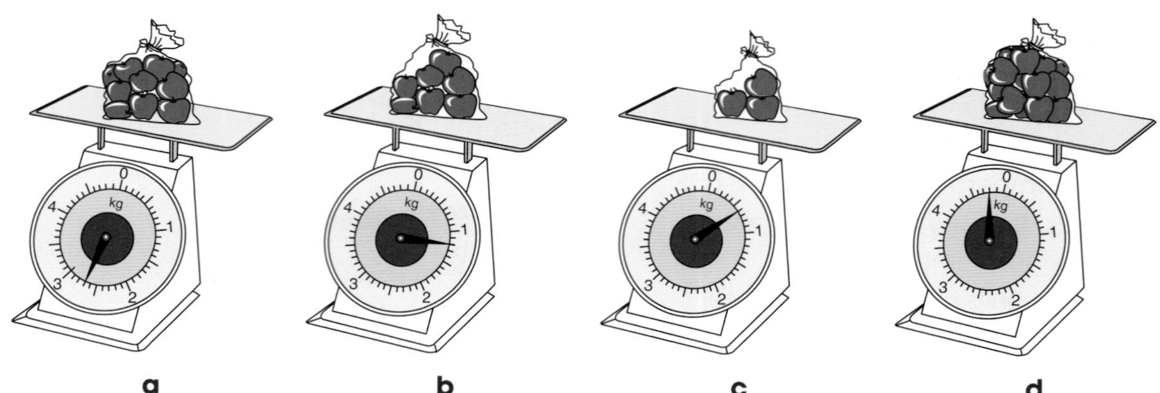

a **b** **c** **d**

7 **a** Some of the scales below show the same readings.
Find four pairs.

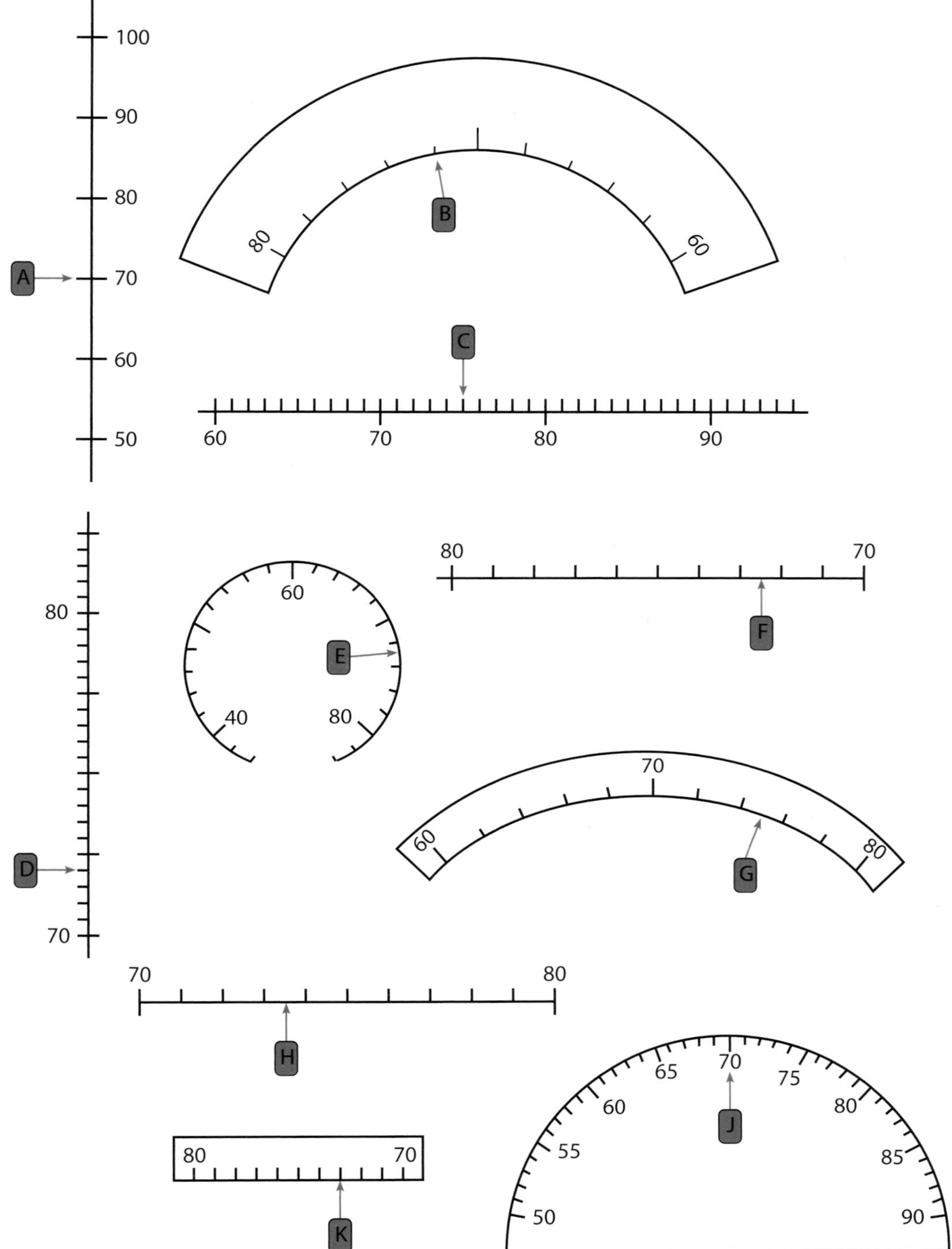

b Draw a matching scale for the scales which are not in a pair.

8 Look at the clock in your classroom.

a What is one small division?

b How many small divisions are there on the clock?

c Why is that?

9 The thermometers show temperatures in degrees Fahrenheit (°F).
What temperatures do they show?

a 97 98 99

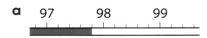

b

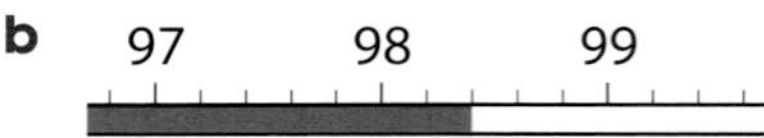

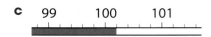

c 99 100 101

d

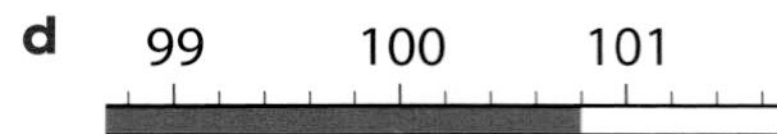

Normal body temperature is 98.4 °F.

e Which of these thermometers shows normal body temperature?

Skills practice B

1 a Copy the number line below and mark on these numbers.

i 0.43 **ii** 0.47 **iii** 0.52 **iv** 0.58

0.4 0.5 0.6

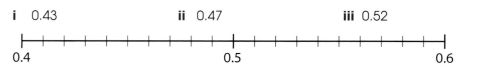

b Copy the number line below and mark on these numbers.

i 0.024 **ii** 0.036 **iii** 0.042 **iv** 0.048

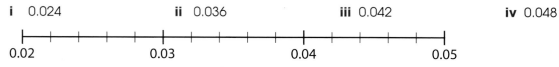

0.02 0.03 0.04 0.05

2 Look at this speedometer from a car.

a How much is each small division?

b What is the speed of the car?

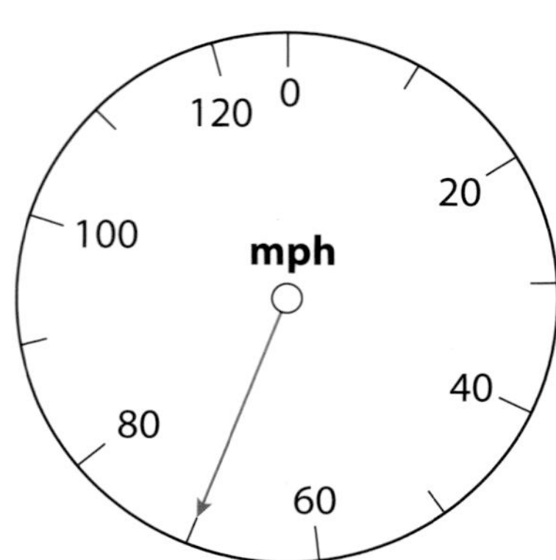

3 Look at this number line.

a How much is each small division?

b What number does each letter show?

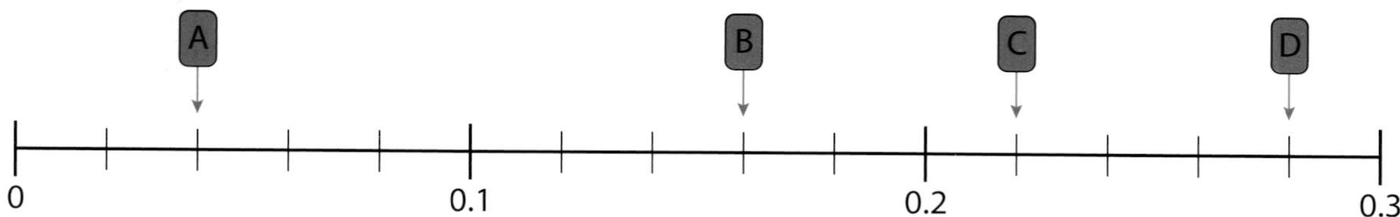

4 Peter got on the scales.
The units are kilograms.
How much does he weigh in kilograms?

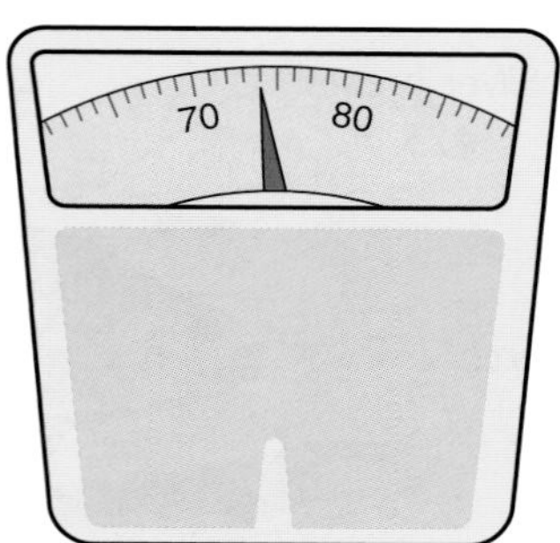

Reasoning

5 Three students were asked to read some values from a scale.
Their answers are shown in the table.

	W	X	Y	Z
Jacqui	50	54	62	73
Ben	50	50.8	60.4	70.6
Samuel	50	58	64	76

Mark their answers.
Give each of them some advice.

6 Look at this thermometer.

a What temperature does the mixture need to reach to make

i caramel

ii jam

iii toffee?

b Between which temperatures is

i yoghurt made

ii milk sterilised?

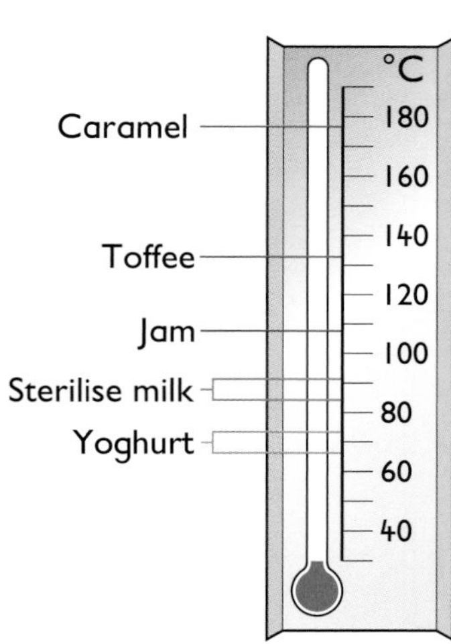

Reasoning

7 Look at these three scales.

a How are they different?

b On copies of the scales, mark these record fish in the correct positions to show their masses.

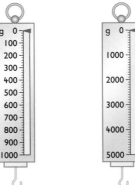

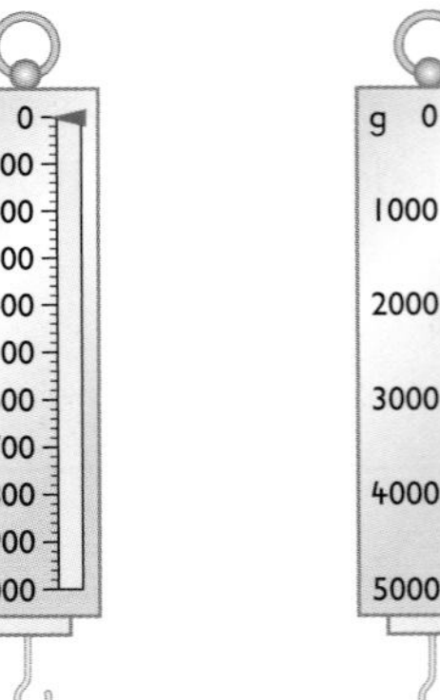

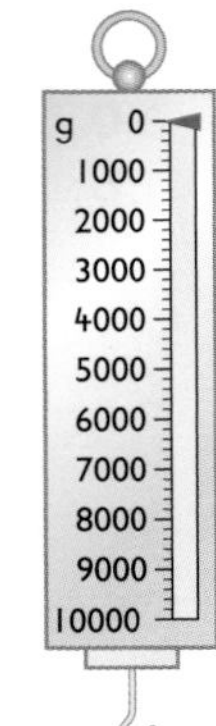

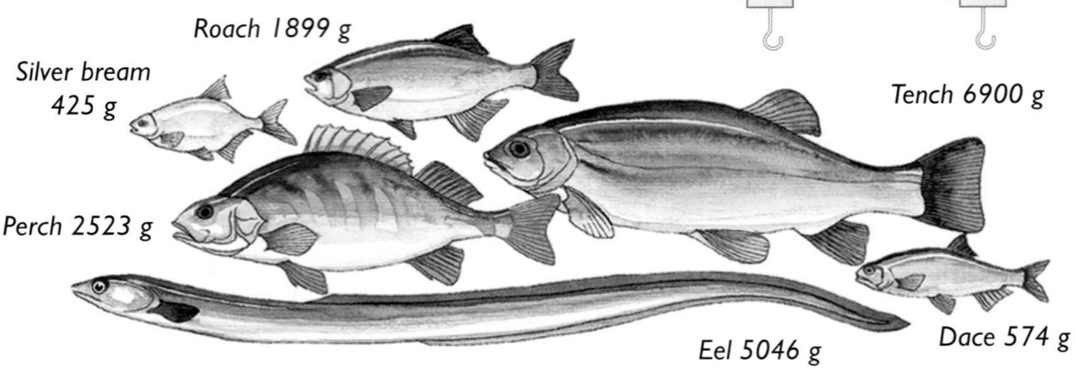

8 These thermometers show both degrees Fahrenheit (°F) and degrees Celsius (°C). Write down the temperatures shown by these thermometers

a in degrees Fahrenheit (°F)

b in degrees Celsius (°C).

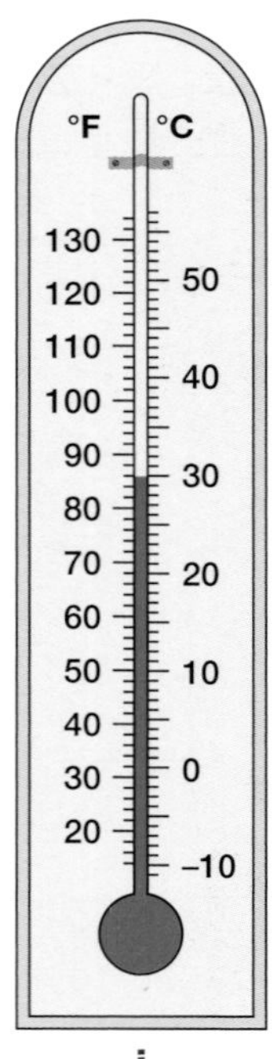

i

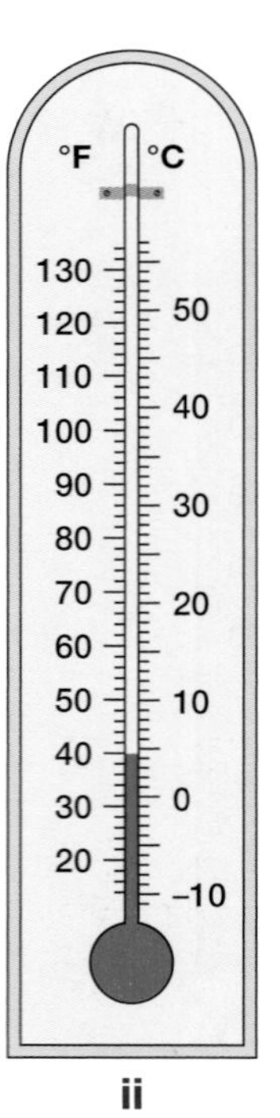

ii

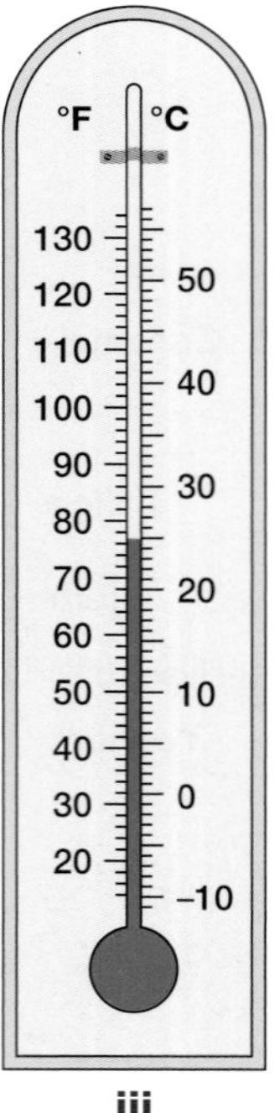

iii

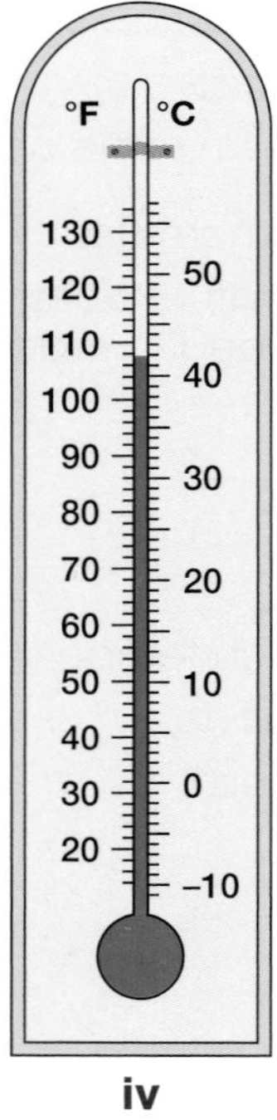

iv

Wider skills practice

1 Most modern utility meters have digital displays, but sometimes you will meet other displays.
This one uses gear wheels.
Some of the dials go clockwise and some go anticlockwise.

a John says this meter reading is 0703.2

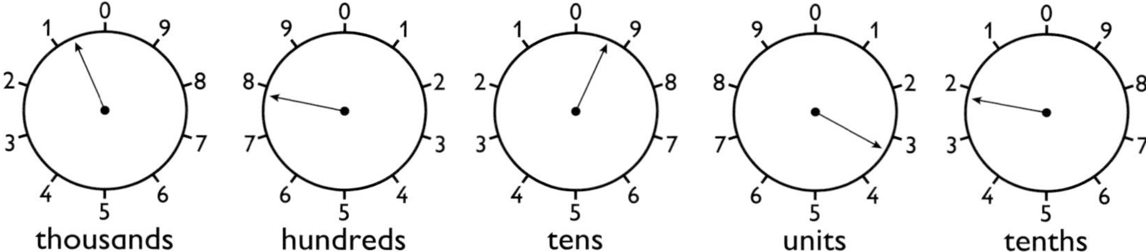

i Why is John wrong?

ii What should the meter reading be?

b Read this meter.

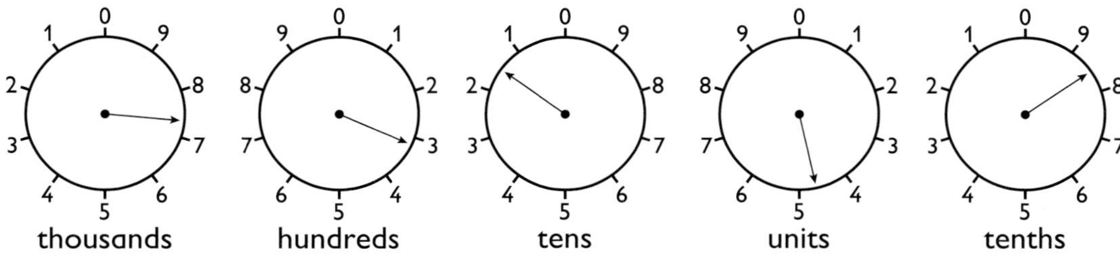

3 Michelle is buying a piece of lamb.

a On a copy of the scales, use a ruler to draw lines to mark 3.2 kg and 4.8 kg.
Lightly shade the area between them.

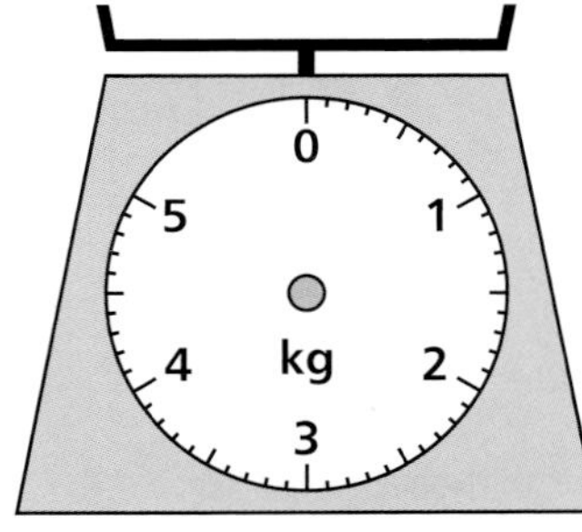

b Draw lines on your diagram to show the weight of each of these pieces of lamb.

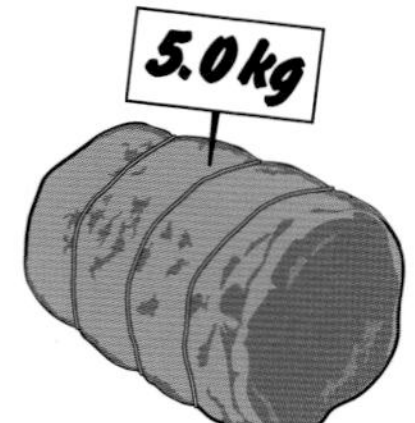

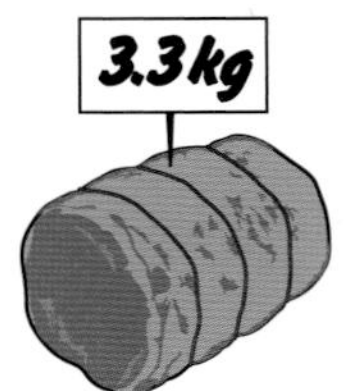

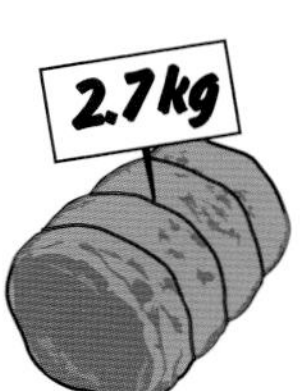

c Which piece of lamb should Michelle choose?

4 This picture shows the dials in Humza's dad's car.

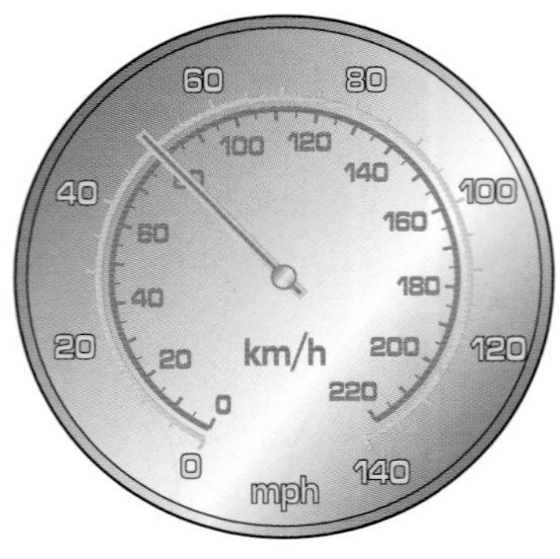

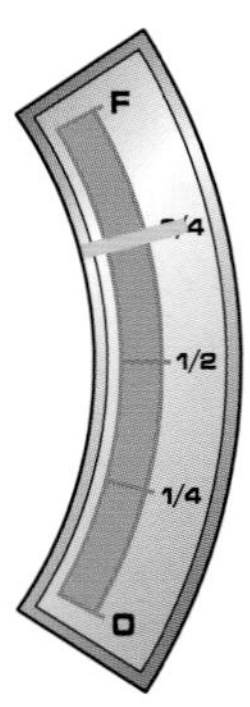

a What speed is the car going

i in miles per hour

ii in kilometres per hour?

b What other information is shown on these dials?

Applying skills

1 a Diane works in a factory.
One of her jobs is to make sure the water containers in the machines have enough water in.
Here are the water containers in three of the machines.
How much water does Diane need to add to fill each container to the maximum level?

i

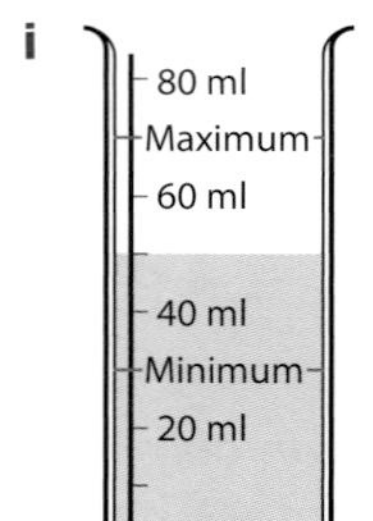

ii

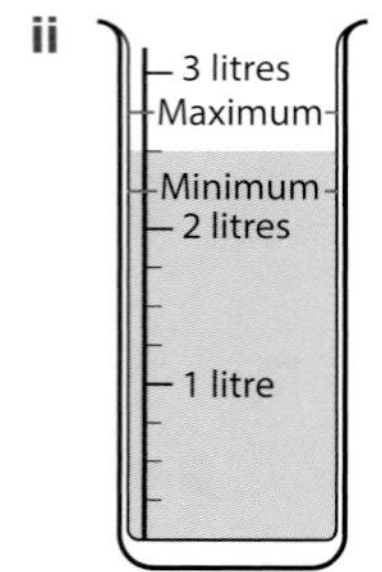

iii

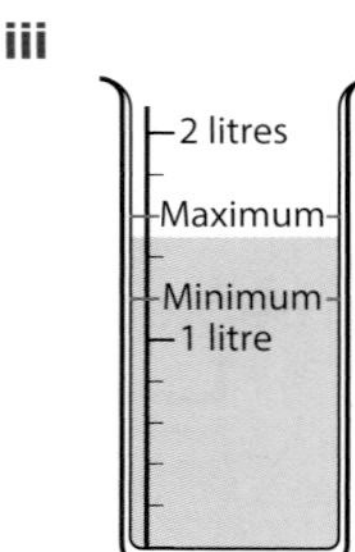

b This is the water container in another machine.
The minimum water level is 1.4 litres.
The maximum water level is 1800 ml.
How much water does Diane add to fill the container to the maximum level?

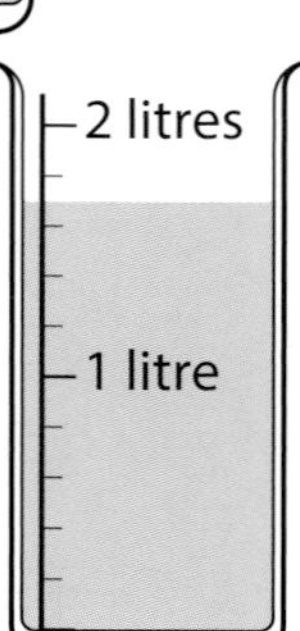

2 The cockpit of an aeroplane is full of dials.
The pilot must read them quickly.

a An **altimeter** shows the height of the aeroplane in 1000 m.
At what heights are aeroplanes A, B and C flying?

A

B

C

b An **air speed indicator** shows the air speed in kilometres per hour.
At what speeds are aeroplanes D, E and F flying?

D

E

F

c Look at the **fuel gauge**. It is full at 3000 kg.

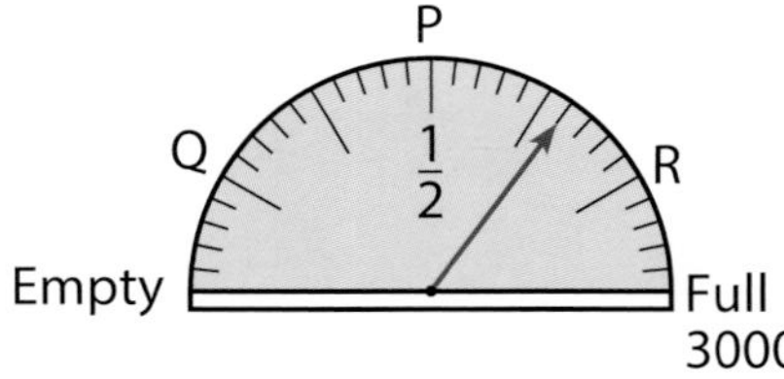

How much fuel is there when the pointer is at

i P

ii Q

iii R

iv The position shown in the picture?

d What other measuring devices does a pilot use?

3 It's a busy day in July at the airport and everyone is keen to get away. Suitcases are crammed full of luggage and some people's suitcases are way too heavy!

Any suitcase that weighs over the ticket allowance is charged as Excess Baggage.

Baggage Allowance

Number of bags :	One
Weight per bag :	Up to 20 kg
Dimensions per bag :	Up to 90 × 75 × 43 cm / 35.5 × 29.5 × 16 ins
	Excess baggage
	£5 per kg when paid online
	£11 per kg at the airport

a The scales show the weight of baggage for five passengers.
How much did each of them pay online for Excess Baggage?

i

ii

iii

iv

v

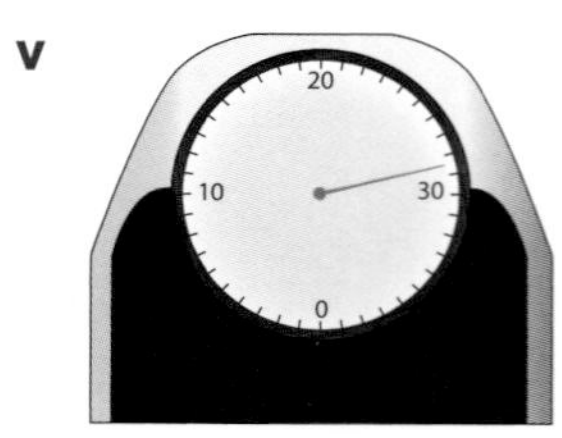

b How much did each passenger save by paying online for their excess baggage before they left for the airport?

c Is this a fair way to charge for luggage? What suggestions would you make?

d These charges are for short haul journeys to places like Europe.
What would be a fair excess baggage charge per kilogram for a long haul flight to Australia?

Reviewing skills

1 Look at this measuring cylinder.

a What is its capacity?

b What does one small division measure?

c How much liquid is in it?

2 How much liquid is in each of these measuring cylinders?

a

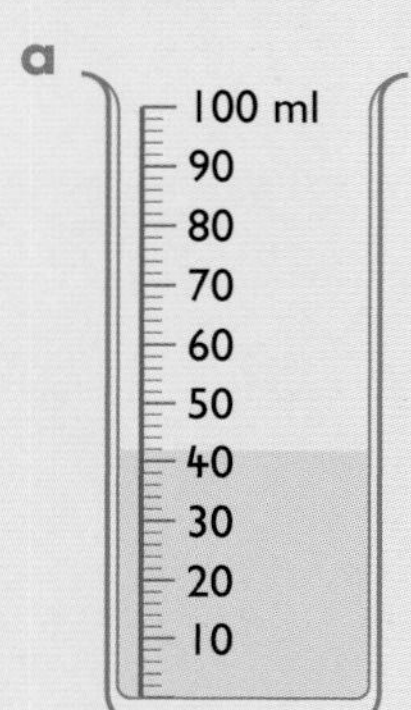

b

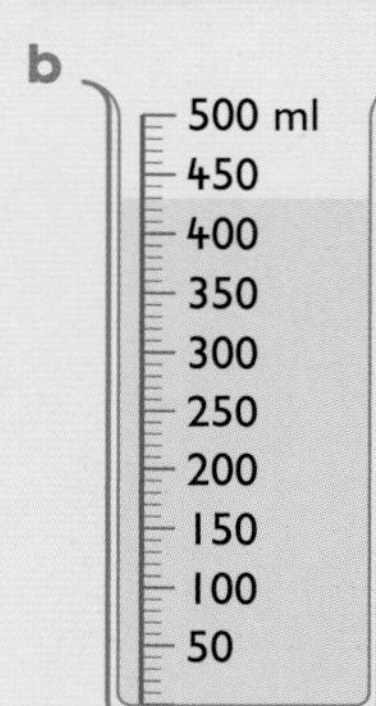

c

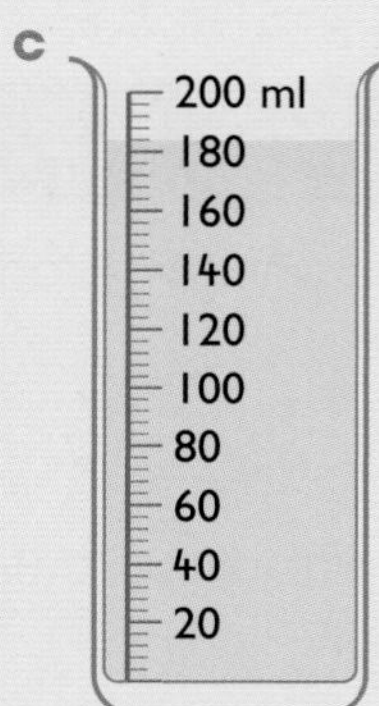

3 This picture shows the dials in Ruth's car.

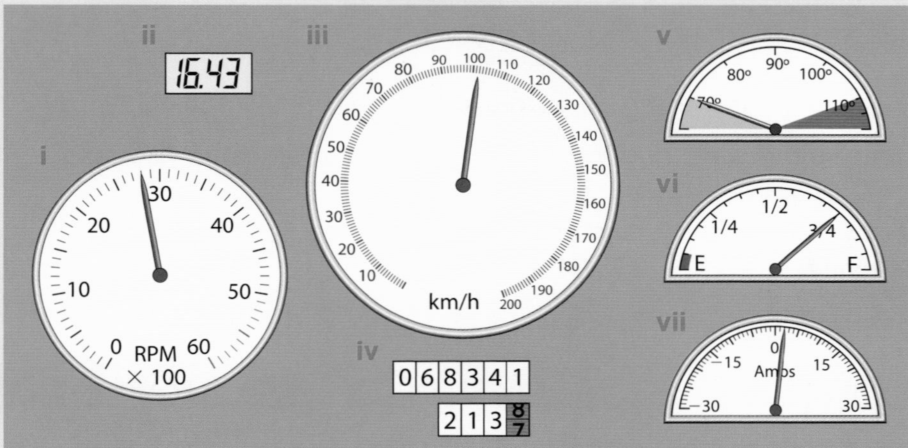

a What are the readings?

b What do all the dials measure?

Unit 6 • The metric system • Band d

Example outside the Maths classroom

Building skills

Taking luggage

Toolbox

In the metric system there is a basic unit for each quantity.

length	1 **metre**
mass	1 **gram**
volume	1 **litre**

A prefix is used to show a fraction or multiple of the basic unit.
Those commonly used are:

milli- $\frac{1}{1000}$ ***kilo-*** 1000 ***centi-*** $\frac{1}{100}$

So 1 kilometre = 1000 metres, 1 kilogram = 1000 grams, etc.

1 millimetre = $\frac{1}{1000}$ metre
or
1000 millimetres = 1 metre

1 milligram = $\frac{1}{1000}$ gram
or
1000 milligrams = 1 gram

The same pattern is used for all metric units and there are many other prefixes, for example *deci-* for $\frac{1}{10}$.
However, *milli-*, *centi-* and *kilo-* are the most commonly used.
Another unit for mass is the **tonne**.
1 tonne = 1000 kilograms

Example – Changing units

The label on a packet of orange juice says 'Content 75 cl'.
What is the content in

a millilitres
b litres?

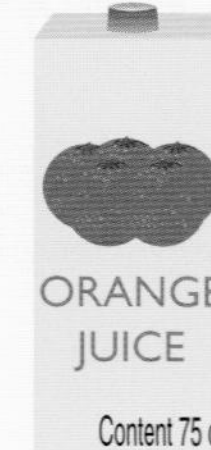

Solution

a The content is 750 ml. ← **1 cl is the same as 10 ml**
75 × 10 = 750

b The content is 0.75 ml. ← **100 cl = 1 litre**
75 ÷ 100 = 0.75
You could also use the fact that 1000 ml = 1 litre and your answer to part a.
750 ÷ 1000 = 0.75

Remember:

- ✦ If you are converting to a smaller unit, you multiply.
- ✦ If you are converting to a bigger unit, you divide.

Skills practice A

1 Copy and complete this conversion diagram.

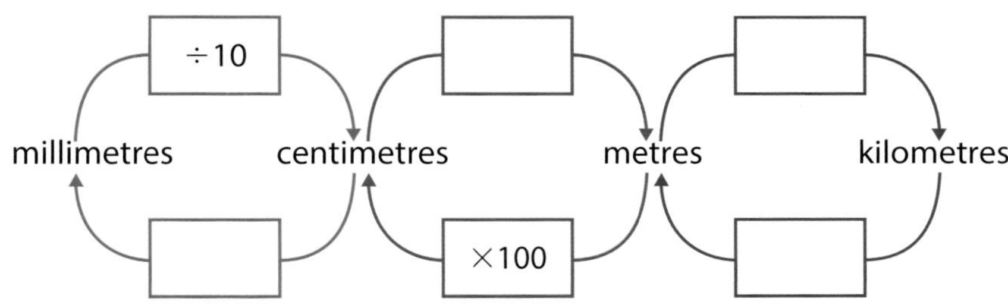

2 Write these lengths in centimetres (cm).

a 2 m　　**b** 3 m　　**c** 1.5 m

3 Write these lengths in metres (m).

a 212 cm　　**b** 300 cm　　**c** 450 cm

4 **a** How many centimetres are there in 1 metre?

b How many millimetres are there in 1 centimetre?

5 Convert these distances from metres to kilometres.

a 3000 m　　**b** 1500 m　　**c** 800 m

6 Convert these weights from grams to kilograms.

a

b

c

7 Convert these volumes from millilitres to litres.

a

2000 ml

b

c

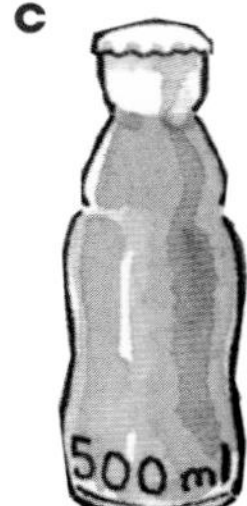

8 Copy and complete these.

- **a** 12 cm = ☐ mm
- **b** 450 mm = ☐ cm
- **c** 700 cm = ☐ m
- **d** 2000 g = ☐ kg
- **e** 3 litres = ☐ ml
- **f** 6000 m = ☐ km

9 **a** Write the measurement 350 cm in

- **i** millimetres
- **ii** metres.

b There are 10 decimetres in 1 metre.
Write 350 cm in decimetres.

10 Write down the larger value in each of these pairs of measurements.

- **a** 14 cm and 130 mm
- **b** 28 mm and 3 cm
- **c** 4 km and 3800 m
- **d** 76 cm and 1 m
- **e** 8100 g and 8 kg
- **f** 3 litres and 2900 ml
- **g** 990 mm and 1 m
- **h** 4100 g and 4 kg

11 Match each measurement in List A to the same measurement in List B.

List A
a 85 millilitres × 10
b 850 grams × 10
c 1400 cm
d 360 metres × 10
e 8500 g ÷ 10
f 360 centimetres

List B
i 140 cm × 10
ii 36 cm × 10
iii 85 g × 10
iv 8500 ml ÷ 10
v 3600 m
vi 85 g × 100

Reasoning

12 State a suitable metric unit for each of these.

- **a** The temperature of a cup of tea
- **b** The capacity of a glass of water
- **c** The mass of a guinea pig
- **d** The height of a kitchen
- **e** The time taken to assemble a piece of furniture
- **f** The width of a biscuit

Skills practice B

1 Look at this domino trail.
The right end of the green domino (A) matches the left end of the pink domino (B).

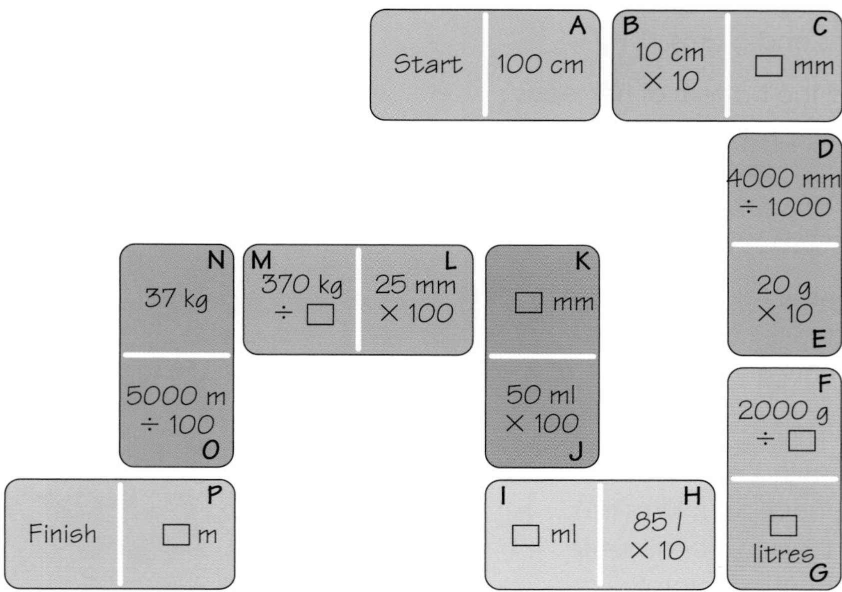

Copy and complete the domino ends C, F, G, I, K, M and P.

2 The length of Sam's pace is 60 cm.
How many paces does she take in 3 kilometres?

Reasoning

3 Convert these measurements to more convenient units.

a 256 000 cm **b** 567 000 m **c** 75 000 ml **d** 0.002 km

4 Look at this recipe for Party Punch.
It makes enough for six people.

Ingredients

2 litres of lemonade
450 ml orange juice
300 ml grapefruit juice
250 ml pineapple juice

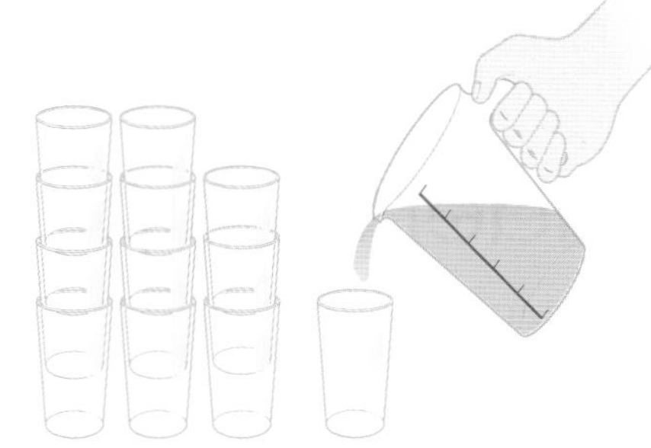

a How many millilitres of Party Punch are there?

Callum uses the recipe to make enough drink for 60 people.

b How much of each ingredient does Callum use?

c Each glass holds 250 ml.
How many glasses does
  i the original recipe make
  ii Callum make?

5 Jack has a row of 30 books.
Each book is 6 cm thick and weighs 520 g.
- **a** How long is the row?
  Give your answer in metres and centimetres.
- **b** How much do Jack's books weigh?
  Give your answer in kilograms and grams.

6 Sophie is sewing sequins around the bottom of her jeans.
- **a** It is 48 cm around the bottom of each leg of her jeans.
  What is the length, in millimetres, around both legs?
- **b** Each sequin has a diameter of 8 mm.
  How many sequins does Sophie need?

7 There are 1000 kilograms in 1 tonne.
Convert these masses to kilograms.
- **a** 3 tonnes
- **b** 7 tonnes
- **c** $4\frac{1}{2}$ tonnes
- **d** 0.5 tonnes

8 The metric prefix *micro*- means $\frac{1}{1000000}$.
So there are 1 million micrometres in 1 metre.
- **a** How many micrometres are there in 1 millimetre?
- **b** How many micrometres are there in 1 centimetre?
- **c** How many micrograms are there in 1 gram?

9 The metric prefix *mega*- means 1 million.
So 1 megametre is 1 million metres.
Write these masses in megagrams.
- **a** 5 000 000 grams
- **b** 24 000 kilograms
- **c** 1 tonne

10 1 tonne is 1000 kg.
A type of brick weighs 1250 g.
Find the weight of 8000 of these bricks in tonnes.

Wider skills practice

1 A formula for cooking a chicken is $T = \frac{1}{2}M + 2$, where T is the time in hours and M is the mass of the chicken in kilograms.
- **a** Alfie has a chicken that weighs 3000 g.
  For how long should he cook his chicken?
- **b** Benji has a chicken that weighs 500 g.
  For how long should he cook his chicken?
- **c** Crystal cooks her chicken for 3 hours.
  What is its mass in
  - **i** kilograms
  - **ii** grams?

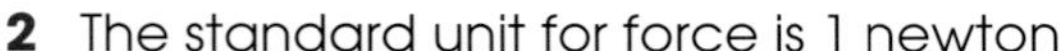

2 The standard unit for force is 1 newton.
  a How many newtons are there in 1 kilonewton?
  b How many millinewtons are there in 1 newton?

Applying skills

Problem solving

1 Look at this list of ingredients.
It makes flapjacks for 12 people.

Flapjacks
75 grams of sugar
75 grams of butter
50 grams of golden syrup
150 grams of rolled oats
35 grams of dried fruit

  a Rajani is catering for a wedding.
  She wants to make flapjacks for 600 people.
  How much of each ingredient does she need?
  b Phil is catering for a school.
  He wants to make 1800 flapjacks.
  He has:
  10 kg of sugar
  10.3 kg of butter
  4 kg of golden syrup
  16.9 kg of rolled oats
  450 g of dried fruit
    i How much of each ingredient does Phil need?
    ii How much more does he need to buy, or will he have some left over of each ingredient?

2 The diagram shows part of a fence.

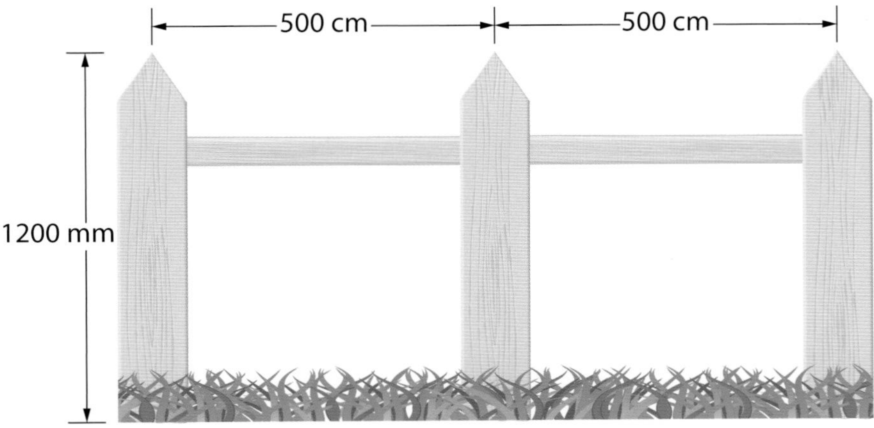

Each post is 1200 mm long and weighs 1800 g.
Posts are connected by a single bar 500 cm long.
There are 301 posts.
Find
  a the total mass of the posts in kilograms
  b the total length of the posts in metres
  c the length of the fence in kilometres.

Reviewing skills

1 Convert these capacities into millilitres.

a 3 litres
b 7.4 litres
c 23 cl
d 82.1 cl
e $9\frac{1}{2}$ litres

2 Write these lengths in centimetres (cm).

a 3.6 m
b 45 mm
c 0.5 m

3 Write down the smaller measurement in each of these pairs.

a 17 mm and 3.2 cm
b $2\frac{1}{2}$ cm and 24 mm
c 3 m and 200 cm
d $\frac{1}{2}$ m and 48 cm
e 4000 m and 6 km
f $1\frac{1}{2}$ km and 1200 m

4 Margaret has bought her five grandchildren identical presents for Christmas.
Each present weighs 1.3 kg.

a What is the total weight in kilograms?
b What is the total weight in grams?

5 A bottle of lemonade holds 2000 ml.

a How many litres is this?

Jo has bought eight bottles of this lemonade for a party.
Her glasses each hold 25 centilitres of lemonade.

b How many glasses of lemonade does she get from the eight bottles?

Unit 7 • Metric–imperial conversions • Band e

Building skills

Example outside the Maths classroom

Following recipes

SCONES

INGREDIENTS
8 oz. flour
½ level tsp. salt
2 level tsps. cream of tartar
1 level tsp. bicarbonate of soda
1-2 oz. butter
¼ pint milk
1 egg

UTENSILS
Teaspoon, palette knife, fork
Sieve, mixing bowl
Wooden spoon
Measuring jug
Flour sifter
2-inch cutter
Baking tins, cake rack
Pastry brush

Toolbox

Metric units are units such as metres, grams and litres.
Imperial units include feet, pounds and pints.
You need to be able to convert between the different imperial units.

Mass	Volume (or capacity)	Distance or length
16 ounces = 1 pound	20 fluid ounces = 1 pint	12 inches = 1 foot
14 pounds = 1 stone	8 pints = 1 gallon	3 feet = 1 yard

You can also convert between metric and imperial units.
Here is a table of all the common conversions for mass, volume and length.

Approximate conversions	Mass		Volume (or capacity)		Distance or length			
Metric	1 kg	28 g	1 litre	5 litres	8 km	1 m	30 cm	2.5 cm
Imperial	2.2 pounds	1 ounce	$1\frac{3}{4}$ pints	1 gallon	5 miles	39 inches	1 foot	1 inch

Example – Converting lengths

The world's largest butterfly is the Queen Alexandra Birdwing.
It has a wingspan of 28 cm.
Write the wingspan in inches.

Solution

2.5 cm is about 1 inch.
So the conversion factor is 2.5
You need to convert cm to inches so you are converting to a larger unit.
This means you need to divide by the conversion factor, 28 ÷ 2.5 = 11.2
The wingspan is about 11.2 inches.

Example – Reading a conversion graph

This is a conversion graph between grams and ounces.

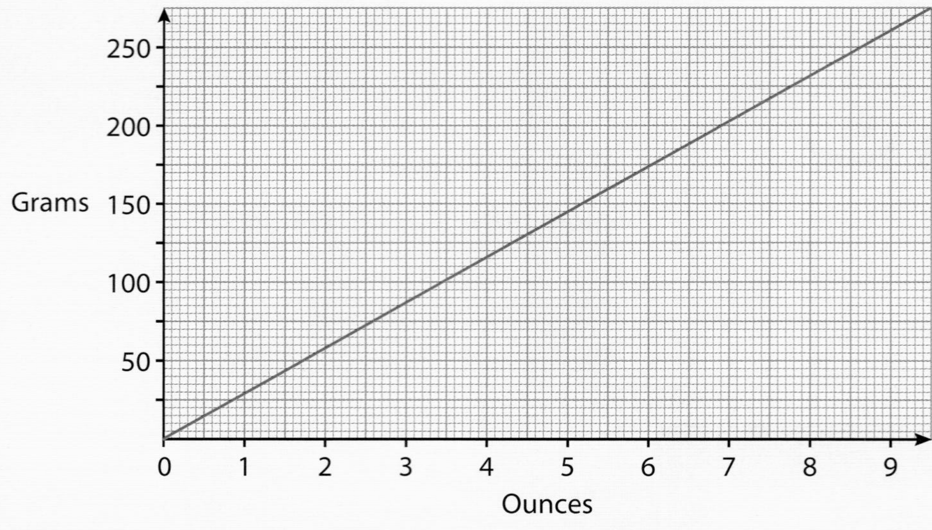

a A packet of butter is labelled 250 g.
What is this in ounces?

b Paul buys 8 ounces of cheese.
What is this in grams?

Solution

a The blue line on the graph shows 250 g is about 9 ounces.

b The green line on the graph shows that 8 ounces is about 225 grams.

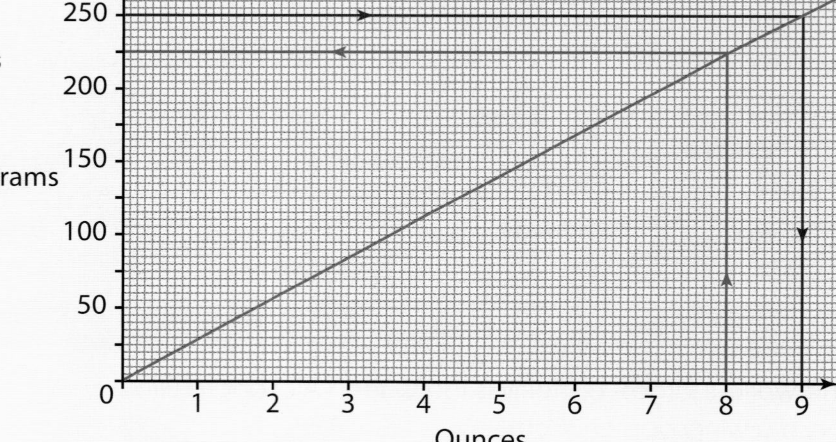

Remember:

- Find the metric–imperial conversion factor.
- If you are converting to a smaller unit, you multiply.
- If you are converting to a bigger unit, you divide.

Skills practice A

1 Here are pictures of a number of measuring devices.
In each case

a identify the instrument, choosing from the list below

b state what it measures: length, mass, capacity, time

c suggest a suitable unit for the measurement.

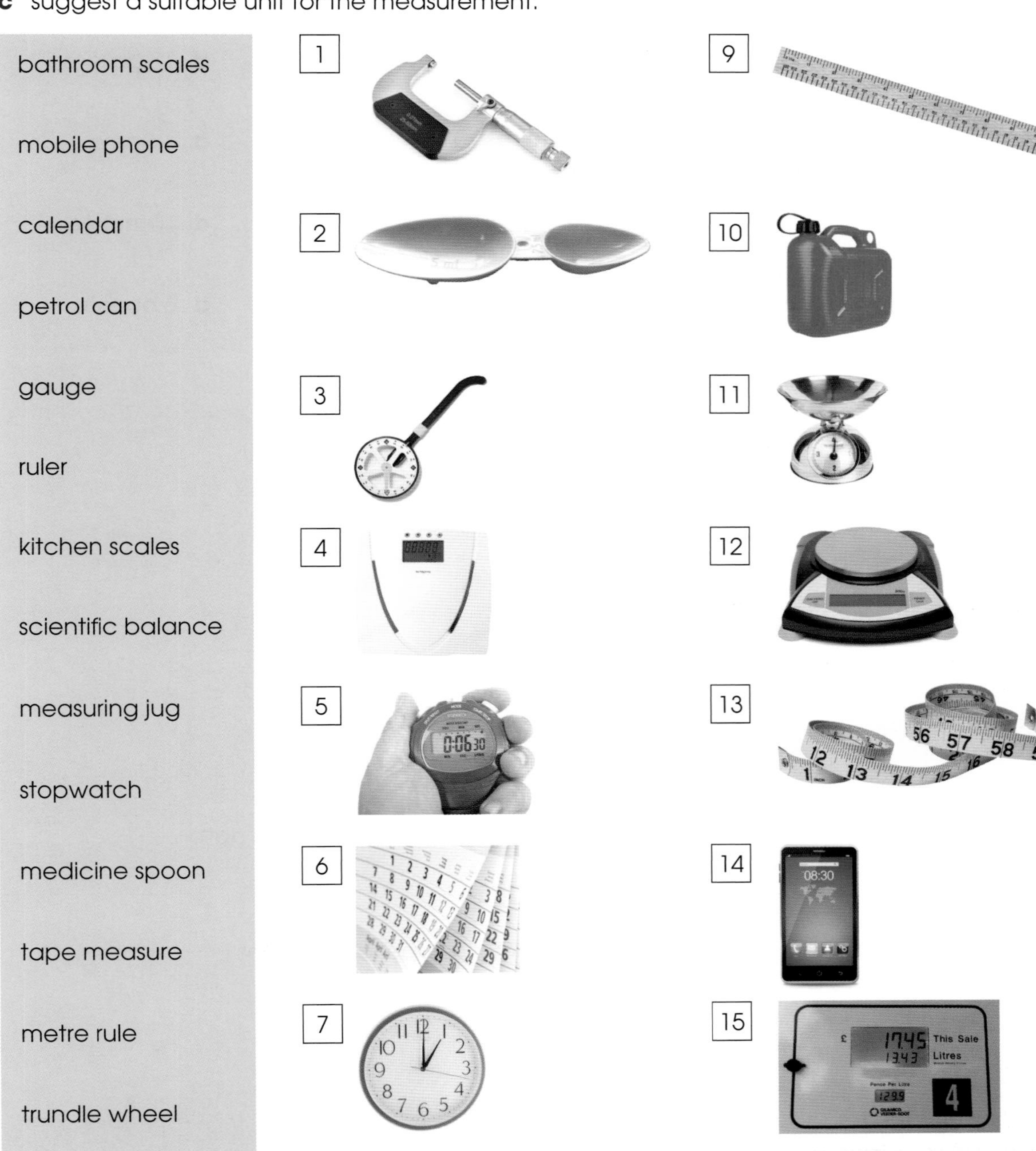

In these exercises, use the conversions given in this table.

Approximate conversions	Mass		Volume (or capacity)		Distance or length			
Metric	1 kg	28 g	1 litre	5 litres	8 km	1 m	30 cm	2.5 cm
Imperial	2.2 pounds	1 ounce	$1\frac{3}{4}$ pints	1 gallon	5 miles	39 inches	1 foot	1 inch

2 Convert these volumes to litres.
a 2 gallons **b** 4 gallons **c** 6 gallons **d** 10 gallons

3 Convert these lengths to inches.
a 5 cm **b** 20 cm **c** 50 cm **d** 7.5 cm

4 Convert these lengths to centimetres.
a 2 inches **b** 10 inches **c** 20 inches **d** 40 inches

5 Convert these weights to grams.
a 2 ounces **b** 5 ounces **c** 9 ounces **d** 6 ounces

6 Jack is on holiday in France.
Convert these distances to miles.

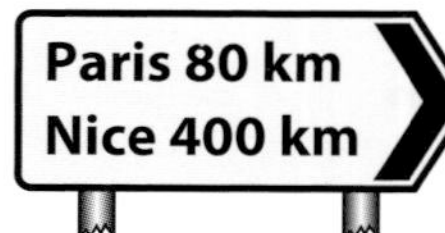

a 80 km
b 800 km
c 400 km

7 Look at the list of ingredients for making pickled red cabbage.
Convert the measurements to metric units.

Pickled red cabbage
5 ounces mixed whole spices
1 large red cabbage
$3\frac{1}{2}$ pints vinegar
2 ounces sugar, optional

8 A small whale contains about 400 gallons of pure oil. There are 8 pints in a gallon.
a How many pints is 400 gallons?
b Change this to litres.

9 Convert the metric measures in the statements below to imperial measures.
a It is 160 kilometres to Dover.
b The village is 240 kilometres from Calais.
c I am 2 metres tall.
d I weigh 60 kilograms.
e My room is 3 metres by 4 metres,
f We need 140 grams of flour, 250 millilitres of milk and 2 eggs.
g The frying pan is 30 cm in diameter.
h I have eaten 100 g of chocolate.

Reasoning

10 This graph converts between pints and litres.

a Explain how to use the graph to convert

i 4 pints into litres

ii 12 pints into litres.

b Convert

i 7 pints into litres

ii 2 litres into pints

iii 2.8 pints into litres

iv 1.5 litres into pints.

Litres

4

3

2

1

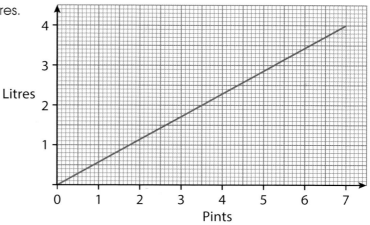

0 1 2 3 4 5 6 7

Pints

Skills practice B

1 Rewrite this story replacing the imperial units with their metric equivalent.

> Sue and Jason were going on holiday to France. They set off early in the morning with their cases packed to bursting – there wasn't an inch to spare!
>
> After travelling 60 miles they stopped at a service station and filled the car with 7 gallons of petrol. Sue went into the shop and bought 3 pints of water to drink and 1 pound of chocolate. She also bought a book that was an inch and a half thick!
>
> They continued on their way at an average speed of 75 mph, arriving at the airport in plenty of time for the flight. The car park was busy and the car had to be squashed into a small space with only 2 feet on each side to spare!
>
> In the airport building they checked in their luggage. It weighed 33 pounds! The final job to do before they could relax was to change their money into Euros. At last they could enjoy their holiday!

2 **a** Which is longer?

i 1 metre or 1 yard

ii 30 miles or 50 km

b Which is greater?

i 2 pints or 1 litre

ii 6 gallons or 28 litres

c How many pints are there in 2 litres?

d How many gallons are there in 48 pints?

e How many litres are there in 5 gallons?

f How many kilometres are there in 4 miles?

g How many metres are there in 390 inches?

h How many pounds are there in 8 kg?

3 Kim's waist is 25 inches.
Do the trousers fit her?

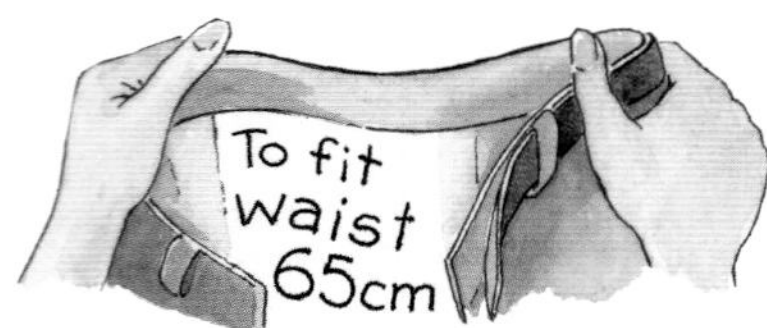

4 Christopher is measuring where to put a basketball net.

> **Rule 10a**
> The top of the basketball net must be 10 feet from the ground.

His tape measure is marked in metres.
How many metres is 10 feet?

5 **a** Is a metre longer than a yard? Explain.
b There are 1760 yards in a mile.
- **i** How many inches are there in 1 mile?
- **ii** How many inches are there in 1500 metres?
- **iii** Which race is longer, the 1500 metres or the mile?

6 Sylvia is driving across France.
She has enough petrol left for 18 miles.
a The next petrol station is 28 km away.
Convert 28 km into miles.
b Sylvia gets to the petrol station. She fills her petrol tank.
The tank takes 9 gallons.
How many litres is 9 gallons?

Reasoning

7 Joanne wants to buy some trousers.
They are 80 cm long.
Joanne's size is 30 inches.
Will the trousers fit Joanne?

8 Georgina is 5 feet 4 inches tall.
Convert 5 feet 4 inches to centimetres.
Does the coat fit Georgina?

9 Look at the list of ingredients in a recipe for Caramel Pear.
Convert the measurements to metric units.

Caramel Pear

4 pears
10 ounces cream
4 ounces brown sugar
4 vanilla pods
$\frac{1}{2}$ pint of water

10 This conversion graph is for changing between pounds and kilograms.

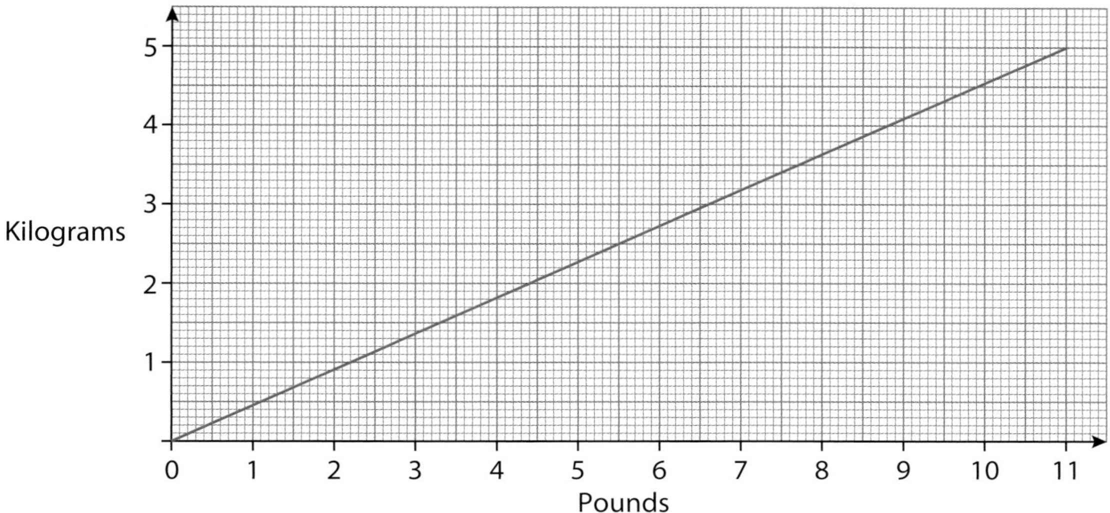

Use the graph to answer the following questions.

a A new-born baby weighs 3 kg.
What is this in pounds?

b A turkey is labelled 5 kg.
What is this in pounds?

c Laura wants to buy 8 pounds of potatoes.
What is this in kg?

d A cat basket is suitable for weights of up to 10 pounds.
What is this in kg?

Reasoning

11 There are 12 inches in 1 foot.
Find a rough conversion for centimetres and feet.

12 **a** Convert these distances into kilometres.

i Barnstaple to Bristol: 160 miles

ii Oxford to Hereford: 80 miles

b Convert these distances into miles.

i Ledbury to Malvern: 16 km

ii Birmingham to London: 192 km

13 The egg of a kiwi weighs approximately 0.5 kg.
A fully grown female bird is about 4 times this weight.

a What is the weight of a fully grown female kiwi?

b Convert this to pounds.

14 £60 is the same as €90.

  a Draw a conversion graph between pounds Sterling and Euros.

  b A hire car costs €70 for one day.
  How much is the price in pounds?

  c Andre buys a meal for £25.
  How much does this cost in Euros?

  d Mercy buys a present for €53.
  Convert the price into pounds Sterling.

Wider skills practice

1 Pete is building a fence around his lawn.
How many metres of fencing does Pete need?

2 The following are more accurate conversion factors than the ones given at the beginning of this unit.

1 inch ≈ 2.54 cm   1 mile ≈ 1.609 km   1 kg ≈ 2.205 lb

Use these values to answer the following questions.
Give your answers to 1 decimal place.

  a I have 8 kg of potatoes.
  How many pounds is this?

  b Convert 1 yard to centimetres.

  c What is 4 miles in kilometres?

  d John fell 40 feet.
  How many metres is this?

Applying skills

Problem solving

1 Richard and Amanda are travelling to Calais.

Once they get to Calais, they will take the train to Folkestone and then drive a further 80 miles to their home.

They stop immediately at a petrol station.

Petrol costs €1.70 per litre.

They buy enough petrol to get them home.

  a How much does it cost?

  b If €1 ≈ 88p, how much does the petrol cost in pounds?

We only have one gallon left.
We can go for 50 miles on one gallon.
CALAIS 160 km

Reviewing skills

1 a An elephant has teeth weighing up to 4.5 kg.
It has four teeth of this size.
What is their total weight?
Convert this into pounds.

b Each of these teeth is just over 25 cm long.
Is a tooth more than a foot long?

2 Joanne makes a wooden car.
She uses an old plan marked in imperial units.

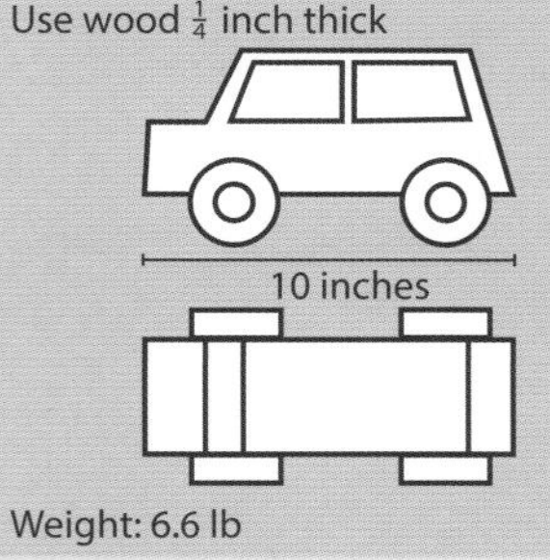

a How long is the car in centimetres?

Joanne has two choices of wood: 60 mm thick or 80 mm thick.

b Which is closer to $\frac{1}{4}$ inch?

c Convert 6.6 pounds to kilograms.

3 Petrol is sold in litres.
It used to be sold in gallons.
This conversion graph can be used to convert between litres and gallons.

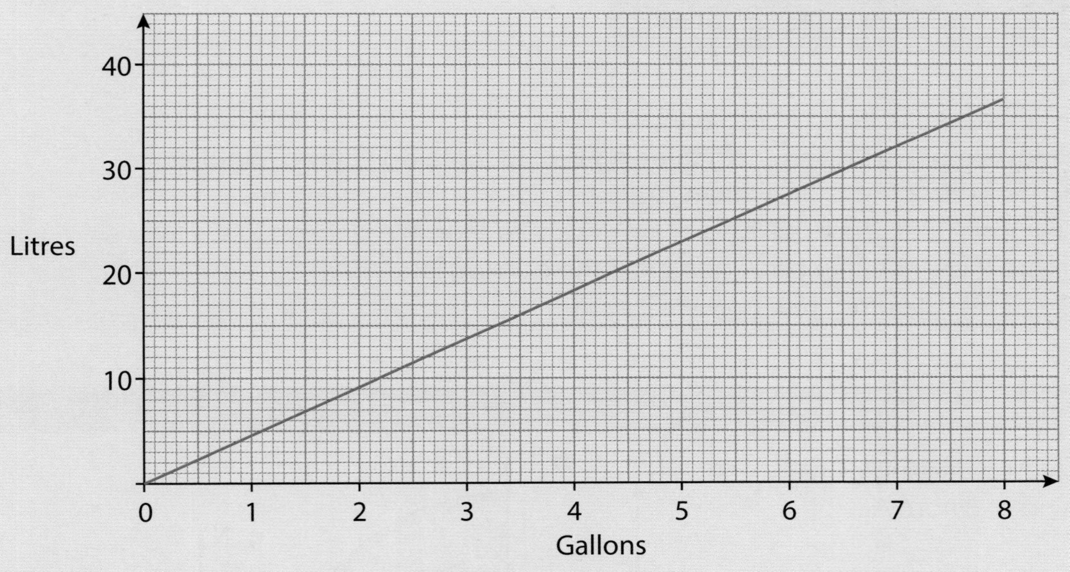

a Michelle buys 35 litres of petrol.
How many gallons is this?

b Terry's car takes 7.5 gallons of petrol.
Use the graph to convert this to litres.

c 1 gallon is equal to 8 pints.

i How many gallons are there in 10 litres?

ii How many pints are there in 10 litres?

iii How many pints are there in 1 litre?

Building skills

Example outside the Maths classroom

Navigating

Toolbox

A **bearing** is given as an **angle** direction.
It is the **angle** measured clockwise from North.
Compass bearings are always given using three figures, so there can be no mistakes.
A **back bearing** is the direction of the return journey.
If a bearing is less than 180°, the back bearing is 180° more than the bearing.
If a bearing is more than 180°, the back bearing is the bearing less 180°.
North is 000°, East is 090°, South is 180° and West is 270°.

Example – Drawing a bearing

B is 3 cm from A on a bearing of 120°.
Draw the bearing of B from A.

Solution

a Draw a North line at A.
b Measure an angle of 120° clockwise from North at A.
c Draw a line at this angle from A.
Use a ruler to mark point B 3 cm along this line.

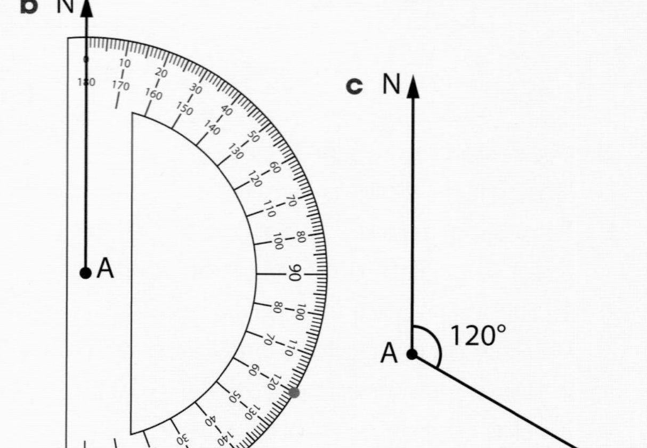

Example – Measuring a bearing

Here is a map of Great Britain.
Find the bearing of

a London from Penzance
b Birmingham from London.

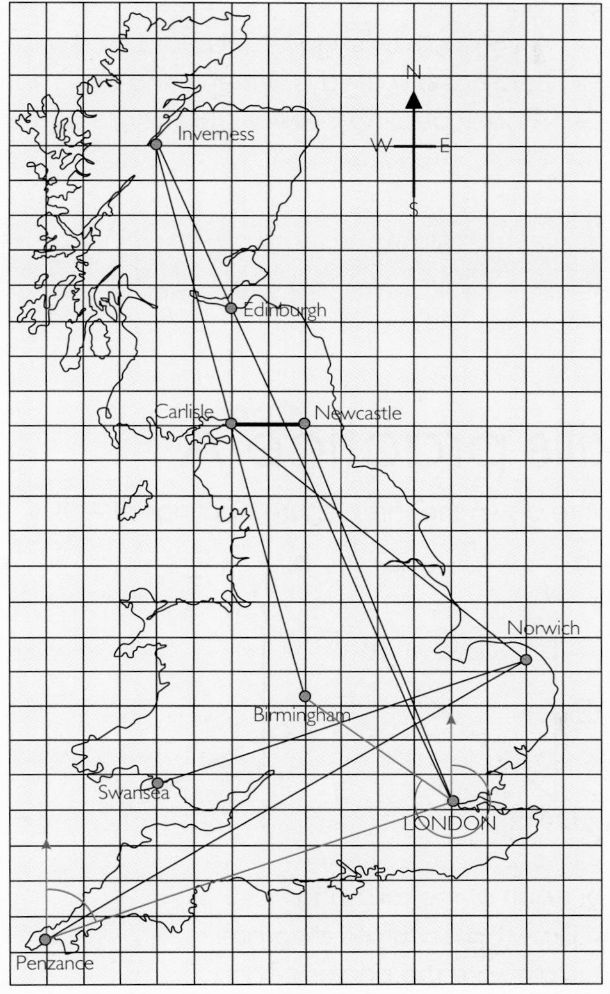

Solution

a The bearing is from Penzance so place your protractor on the map with the centre at Penzance and the zero line vertically up.

Use the grid lines to help you.

Measure the angle clockwise from North.

As 70° only has two digits, write your answer with a zero in front.

070°

b The bearing from London to Birmingham is a reflex angle.

Place your protractor on the map with the centre at London and the zero line vertically up.

If you have a 360° protractor you can measure the angle directly.

If you have a 180° protractor, measure the angle anticlockwise from North.

Subtract this angle from 360°.

360° – 53° = 307°

307°

Example – Calculating a back bearing

The line from a hill to a radio mast is shown in the diagram.

a What is the bearing of the mast from the hill?
b What is the back bearing to return from the mast to the hill?

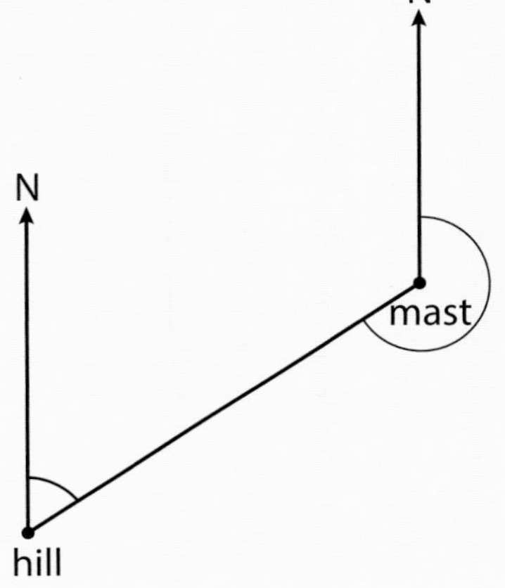

Solution

a Measure the marked angle at the hill. It is 57°.

Remember to write the answer with three figures.

057°

b The back bearing is 180° more than the bearing.

You can check your answer by measuring.

057° + 180° = 237°

Remember:

- A compass bearing is measured clockwise from North.
- Bearings are always written using three figures.
- A back bearing is always 180° more or 180° less than the original bearing.

Skills practice A

1 Write down the three-figure bearings of A, B, C and D from O.

a

N
A
O
310°

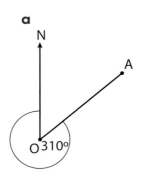

b

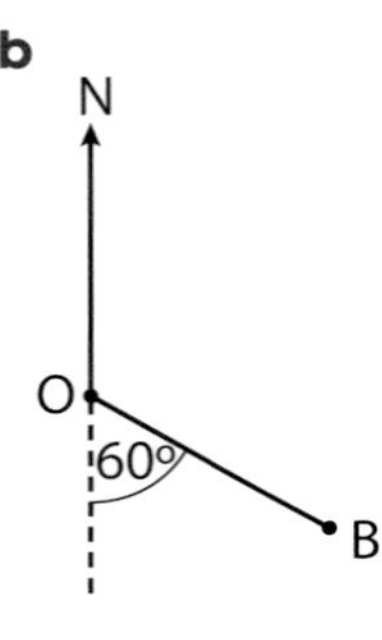

c

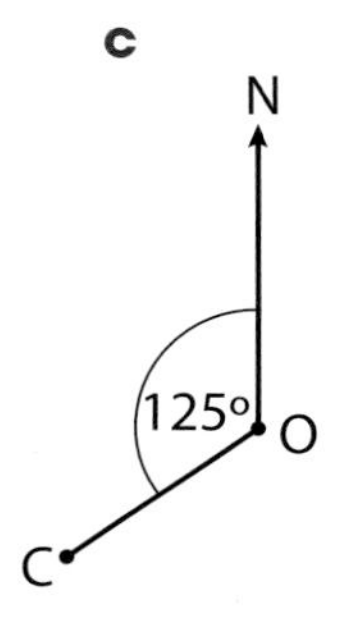

d

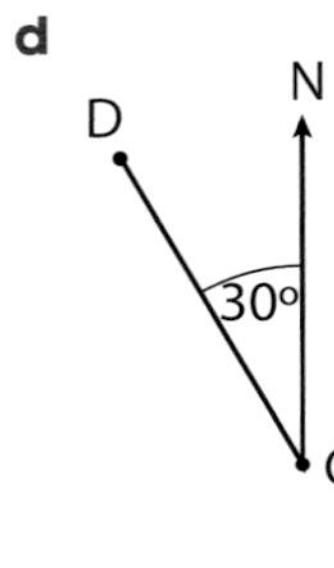

2 For each of the following:

i Draw an accurate diagram.
ii Calculate the angles x and y.
iii Give the bearing from A to B.
iv Give the back bearing from B to A.

a

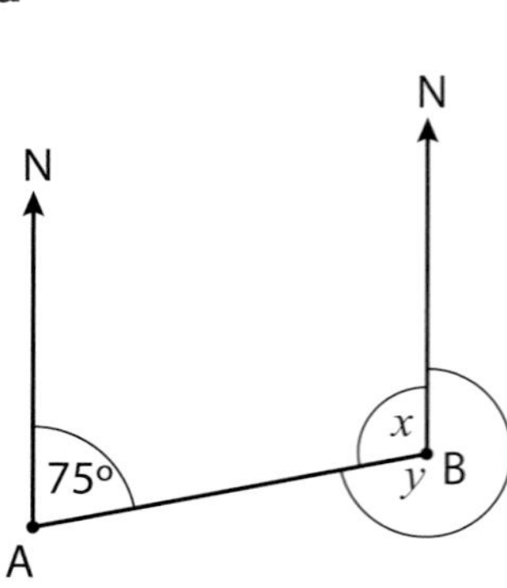

b

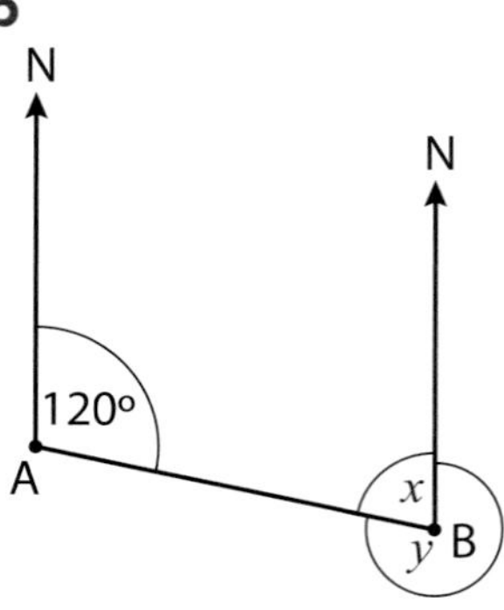

c

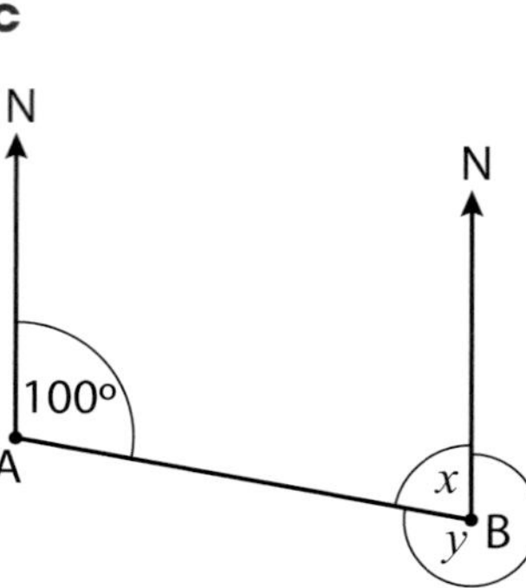

3 **a** Draw point A with a North line starting at point A.

Point B is on a bearing of 030° from point A.
Point B is 5 cm from point A.

b Use a protractor and a ruler to plot point B.

c Draw a North line at point B.

Point C is on a bearing of 150° from point B.
Point C is 5 cm from point B.

d Use a protractor and a ruler to plot point C.

e What is the distance between C and A?

f What is the bearing of A from C?

g What kind of triangle is ABC?

4 Draw accurate diagrams to show these bearings.

a The bearing from A to B is 045°. B is 5 cm from A.

b The bearing from C to D is 135°. D is 6.8 cm from C.

c The bearing from E to F is 260°. F is 2.6 cm from E.

d The bearing from G to H is 320°. H is 72 mm from G.

5 Write down the bearing for each of these compass directions.

a North **b** West **c** South-East **d** North-West

6 Write down the compass points with these bearings.

a 180° **b** 045° **c** 225° **d** 090°

7 The bearing from a ship to a port is shown in the diagram.

a Measure the bearing of the port from the ship.

b What is the back bearing to travel from the port to the ship?

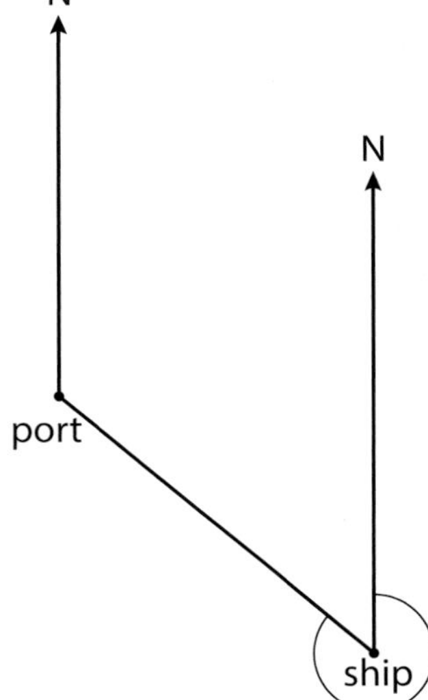

8 Point U is 5 km from V on a bearing of 140°.
Point T is 5 km from U on a bearing of 220°.

a Draw an accurate diagram showing the three points.

b Find the distance between T and V.

c What is the bearing of T from V?

d What is the bearing of V from T?

9 **a** A pilot flies from Edinburgh to Newcastle on a bearing of 146°.
Calculate the back bearing for the return journey.

b The flight path from Carlisle to Inverness is 345°.
Calculate the bearing of Inverness from Carlisle.

Skills practice B

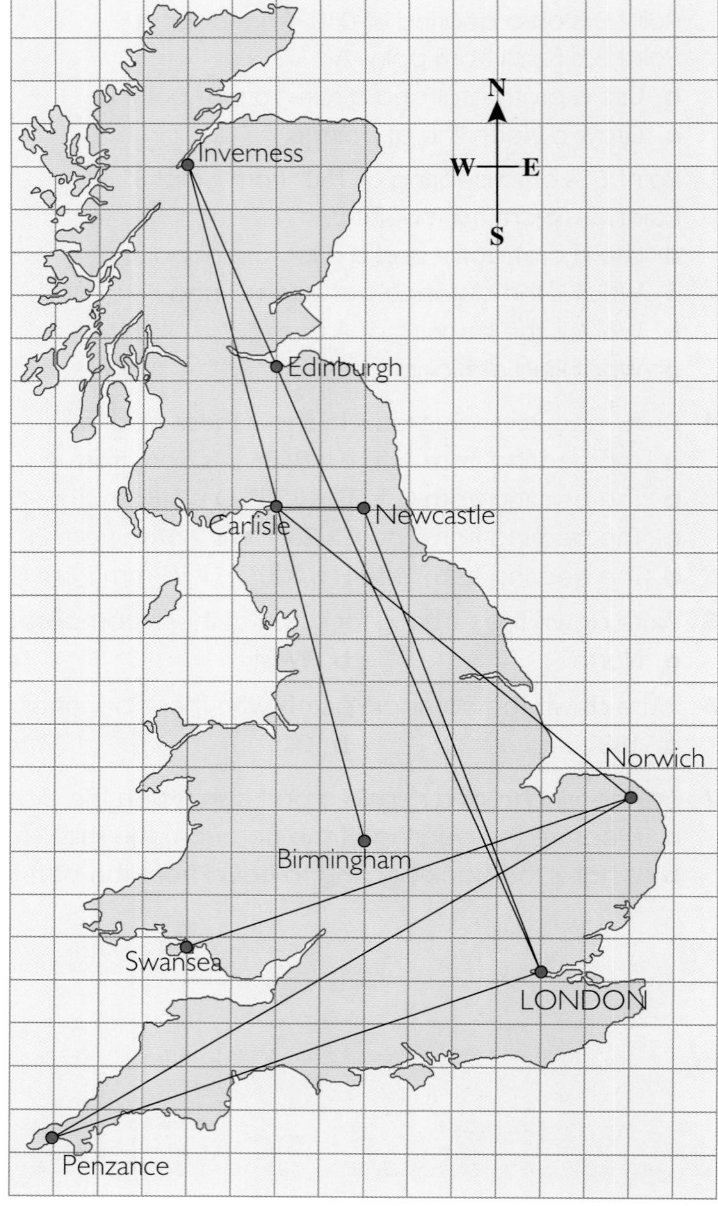

1 Measure these bearings.

- **a** Norwich from Penzance
- **b** Swansea from Norwich
- **c** Carlisle from Newcastle
- **d** London from Newcastle
- **e** Norwich from Carlisle

Reasoning

2 The bearing of Edinburgh from Inverness is 157°.
The bearing of London from Edinburgh is also 157°.

- **a** What can you say about the positions of these three places?
- **b** What is the bearing of London from Inverness?

3 Inverness, Carlisle and Birmingham all lie on the same straight line.

- **a** Measure the bearing of Birmingham from Carlisle.
- **b** From your answer to part **a**, calculate the bearing of Inverness from Carlisle.

4 Dave and Ann are walking on Kinder Scout in the Peak District.

Their route is Seal Stones to Kinder Downfall, then to Noe Stool, then back to Seal Stones.

a Calculate angle A.

b Calculate the bearing of Kinder Downfall from Seal Stones.

c Write down the bearing of Noe Stool from Kinder Downfall.

d What bearing must they follow to return from Noe Stool to Seal Stones?

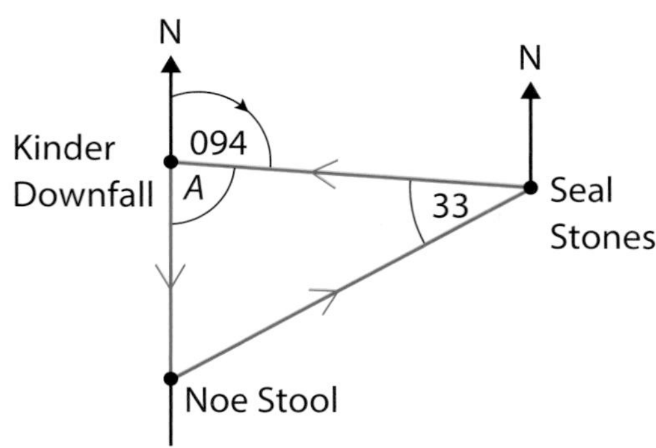

5 Sailing ships cannot sail directly into the wind.

However, they can sail towards the wind but only at an angle to its direction.

Captain Kipper's square rigger cannot sail closer than 67° into the wind.

It can sail in all other directions (shown by the green arc).

In the diagram, the wind is *from* the North.

It is a 'North wind'.

Between which bearings can Captain Kipper *not* sail if the wind is

a a North wind (as in the diagram)

b a South wind

c an East wind

d a South-West wind?

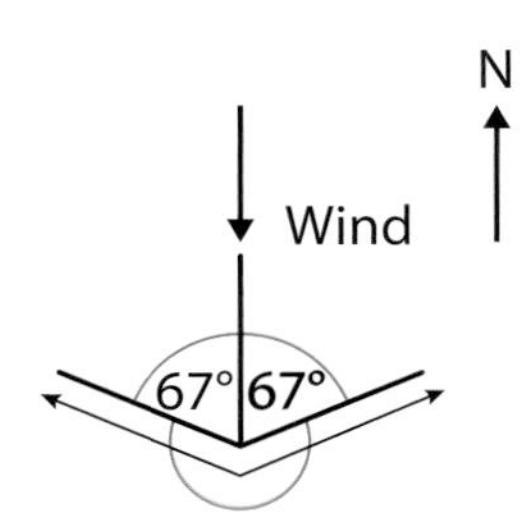

Reasoning

6 Four sailors set off at noon from the same port.

Jack travelled on a bearing of 310°.
Helen travelled on a bearing of 144°.
Ophelia travelled on a bearing of 047°.
Barakat travelled on a bearing of 218°.

At 2 p.m., each one changed course.

Jack travelled on a bearing of 130°.
Helen travelled on a bearing of 314°.
Ophelia travelled on a bearing of 227°.
Barakat travelled on a bearing of 038°.

This took three of them back to the port where they had supper.

Which sailor did not return to the port?

Explain your answer.

Wider skills practice

Reasoning

1 **a** Measure the bearings of these journeys, shown on the map.

- **i** Filwood to Durton **ii** Durton to Borley
- **iii** Filwood to Egwell **iv** Borley to Filwood
- **v** Carham to Filwood **vi** Borley to Carham

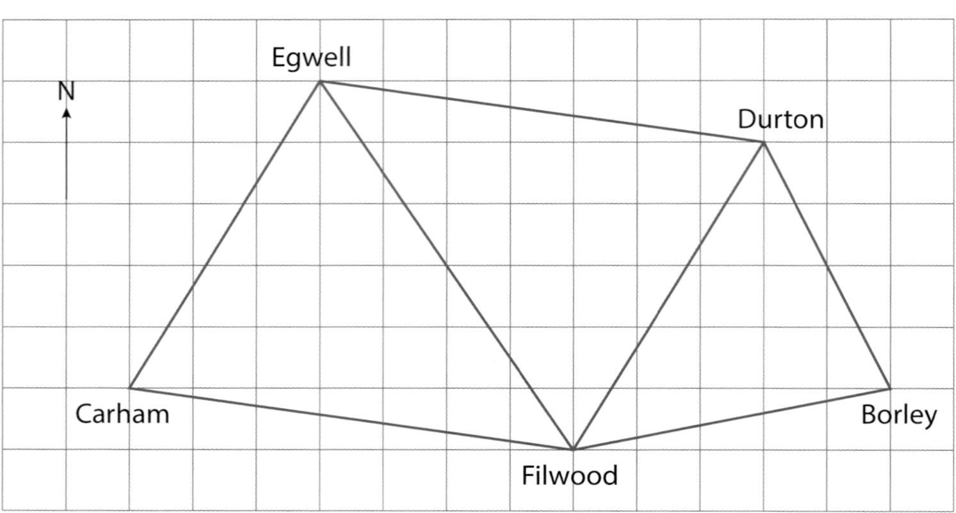

b Calculate the back bearing of each journey in part **a**.

c Which journeys drawn on the map have the same bearings?

d Explain how you can tell from the gridlines on the map that they have the same bearings.

2 Leroy and Marlene leave Homeport in their boat.

They sail for 5 km on a bearing of 070° to Parrot Island.

After some time they sail a further 7 km, on a bearing of 140° to Paradise Bay.

a Draw an accurate diagram.

b Find the bearing for their return journey to Homeport.

c How far is it from Paradise Bay to Homeport?

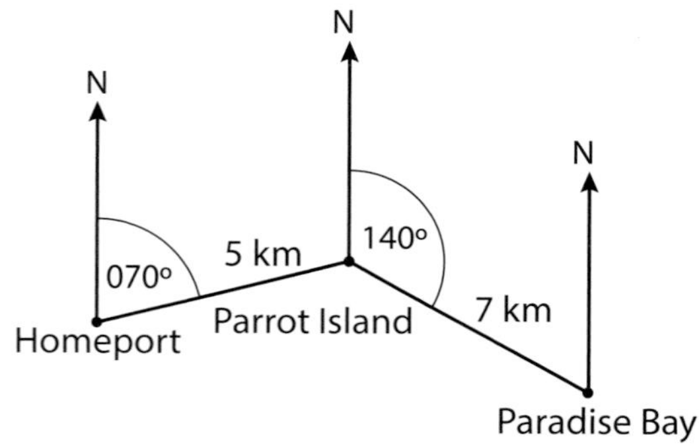

Problem solving

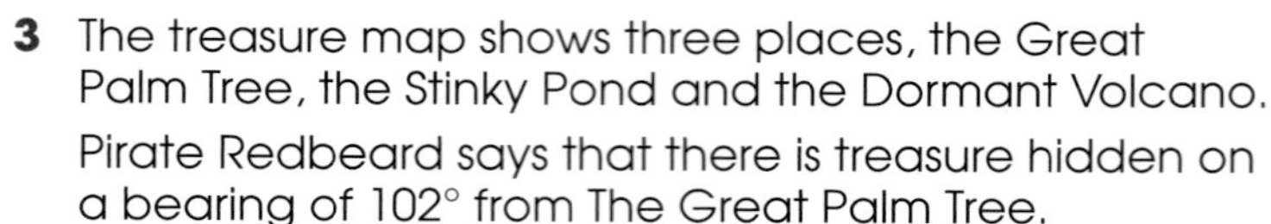

3 The treasure map shows three places, the Great Palm Tree, the Stinky Pond and the Dormant Volcano.

Pirate Redbeard says that there is treasure hidden on a bearing of 102° from The Great Palm Tree.

Pirate Bluebeard says that the same treasure is hidden on a bearing of 300° from the Stinky Pond.

Pirate Blackbeard says that there is treasure hidden on a bearing of 026° from the Dormant Volcano.

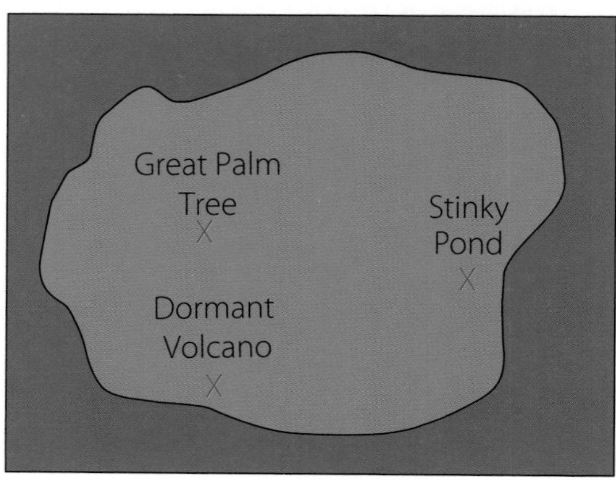

a Can they all be correct?

b What can you say about the location of the treasure if

- **i** two of the pirates always tell the truth and the other one always lies
- **ii** one of the pirates always tells the truth and the other two always lie?

Problem solving

Applying skills

1 Paula has her own aeroplane.

She makes a ten-leg tour of England, Wales and Scotland, starting at Penzance. See the map on the next page.

a Copy and complete the table below.

b What places does she visit?

c How far does she fly in total?

Leg	Start	End	Distance (km)	Course
1	Penzance		200	
2				019°
3		Carlisle		
4	Carlisle			330°
5				359°
6			135	
7			370	
8				125°
9			285	242°
10				237°

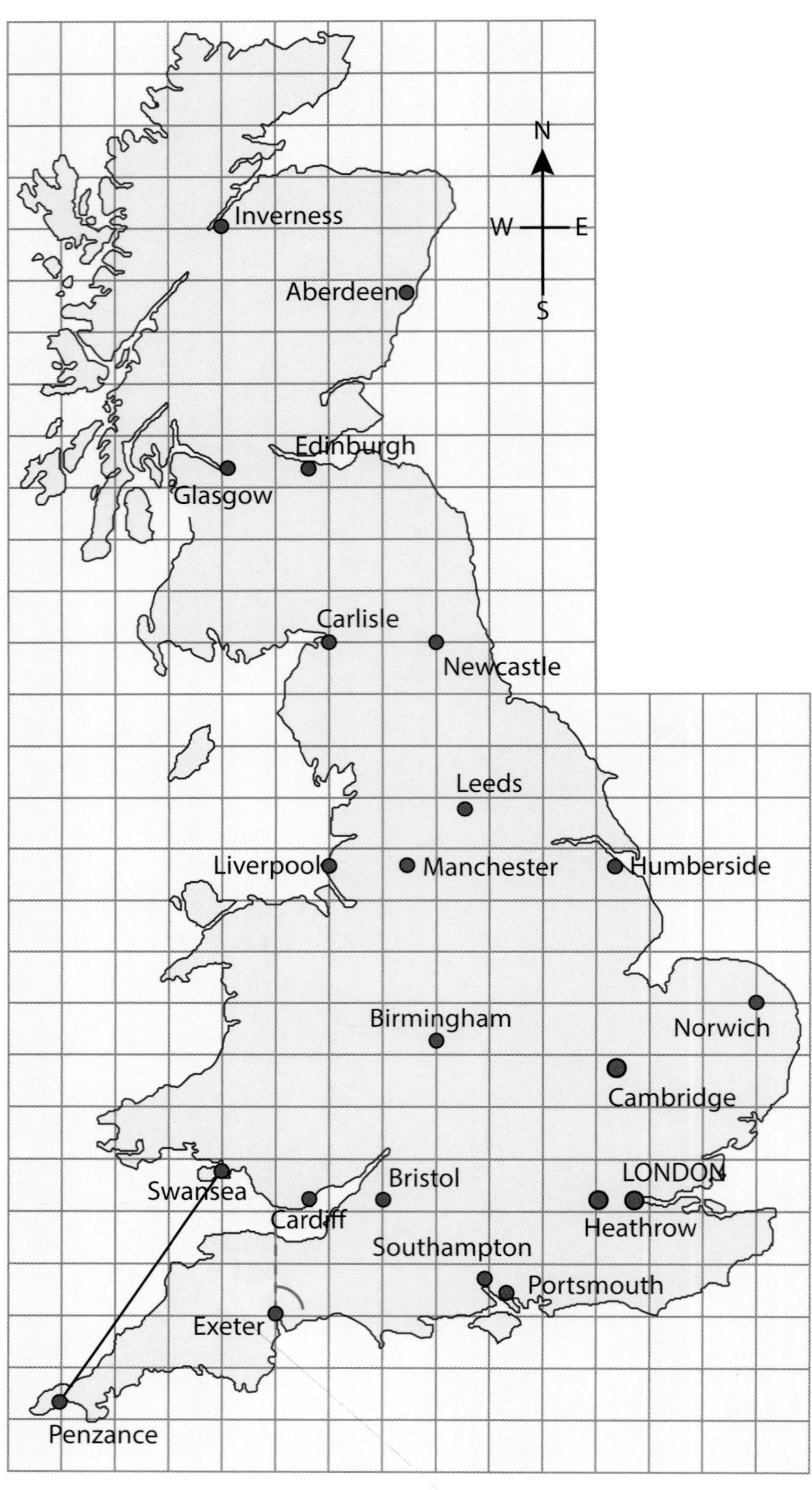

Scale 1 cm represents 50 km

Reviewing skills

1 **a** Write these compass points as bearings between 0° and 360°.

- **i** South-West
- **ii** East

b Write down the compass point with these bearings.

- **i** 000°
- **ii** 315°

2

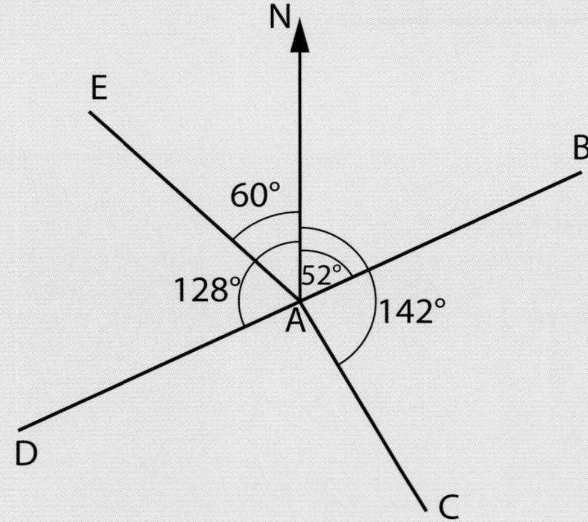

a Measure the bearing of

- **i** B from A
- **ii** A from D.

What can you say about the points D, A and B?

b Measure the bearing of C from B.

Describe the bearing of B from C

- **i** in words
- **ii** in numbers.

c The bearing of E from A is 300°.

Write down

- **i** the angle EAN
- **ii** the bearing of A from E.

Unit 9 • Scale drawing • Band f

Building skills

Example outside the Maths classroom

Designing a house

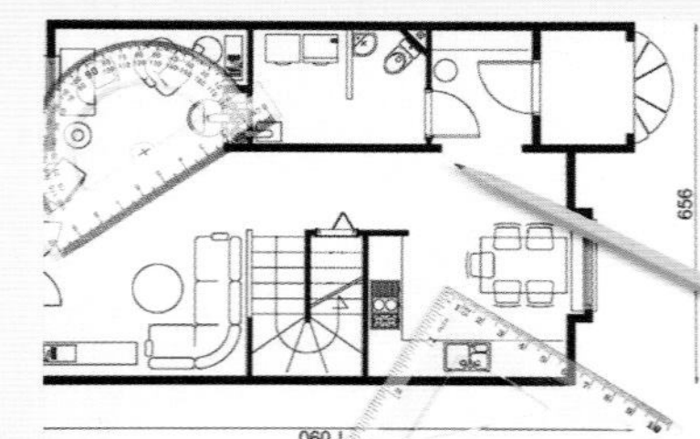

Toolbox

A scale drawing is the **same shape** as the original but a **different size**.

All the lengths are in the same **ratio**.

On a scale drawing where 1 cm on the scale drawing represents 2 m on the actual object, the scale can be written as

$\frac{1}{200}$ or

1 cm = 2 m or

1 : 200

$\frac{1}{200}$ is sometimes refered to as the **scale factor**.

Warning: Be careful with areas and volumes.

For example, 1 cm = 10 mm but 1 cm^2 = 10^2 = 100 mm^2 and 1 cm^3 = 10^3 = 1000 mm^3.

Example – Using a scale

Kitty has a toy car on a scale of $\frac{1}{50}$th.

The toy car is 6 cm long.

The real car is 1.7 m wide.

a How long is the real car?

b How wide is the toy car?

Solution

The toy car is $\frac{1}{50}$ the size of the real car.

So the real car is 50 times the size of the toy car.

a Length of real car = 6 × 50 cm ← **To find the length of the real car, multiply by 50.**

= 300 cm

The real car is 300 cm or 3 m long.

b 1.7 m = 170 cm ← **It is easier to convert to cm first.**

Length of toy car = 170 ÷ 50 cm ← **To find the width of the toy car, divide by 50.**

= 3.4 cm

The toy car is 3.4 cm long.

Example – Reading a map

Adebola is working out distances between places on a map.
The map has a scale of 1 : 50 000.

On the map, the bowling alley is 4.7 cm from the aqueduct.

a How far is the bowling alley from the aqueduct in kilometres?

Adebola's house is 3.8 km from her school.

b How far is her house from her school on the map?

Solution

a 4.7 × 50 000 = 235 000 cm ← **To find a distance on the ground, multiply by the scale factor of 50 000.**

235 000 ÷ 100 000 = 2.35 km ← **There are 100 000 cm in 1 km.**

b 3.8 × 100 000 = 380 000 cm ← **Convert to centimetres first.**

380 000 ÷ 50 000 = 7.6 cm ← **To find a distance on the map, divide by the scale factor of 50 000.**

Remember:

- On a scale drawing, all the sides are in the same ratio compared with the original.
- Scales can be written in different ways, such as 1 cm = 2 m, 1 : 50 000 and $\frac{1}{200}$.
- Units must be the same before using scales.

Skills practice A

Reasoning

1 A shop sells toy train sets for 3–5 year olds.
There are different sizes.
The small set is exactly half the size of the standard set.
The scale, small set : standard set, is 1 : 2.

a The track in the small set is a circle with a 30 cm diameter.

- **i** What shape is the track in the standard set?
- **ii** How far across is it?

b This is an accurate drawing of the engine from the small set.

- **i** How long is the engine?
- **ii** How tall is the engine?
- **iii** Make an accurate drawing of the engine from the standard set.

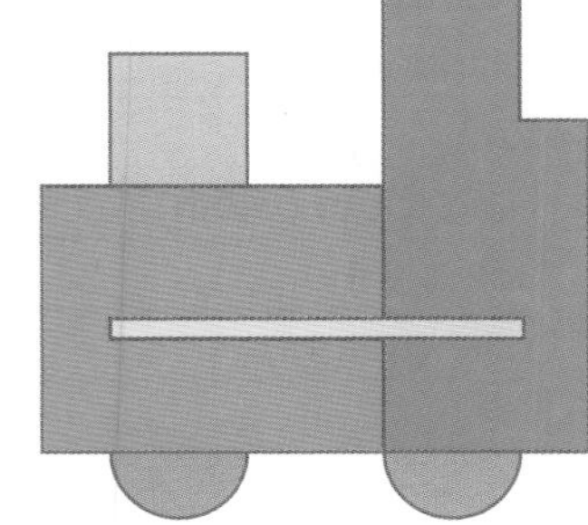

c The shop also sells a giant sized train set.
It is 10 times bigger than the standard set.

- **i** How long is the giant engine?
- **ii** How tall is it?
- **iii** Can children ride on it?

2 Three sorts of creatures live on the planet Grunge: zats, humans and blobbits.
The creatures look like this.

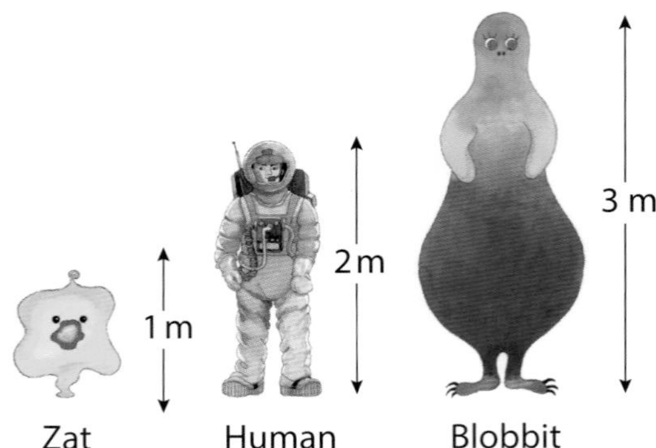

a Find these scales.

i The height of a human : the height of a blobbit

ii The height of a blobbit : the height of a zat

iii The height of a zat : the height of a human

b Find the widths of the three creatures.

3 These two photographs have been taken from catalogues.
Below each photograph is the real measurement.
Find the scale of each photograph.

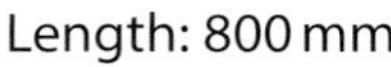

Length: 800 mm

Length: 3400 mm

4 A doll's house is to be built to a scale of $\frac{1}{100}$.
Copy and complete this table showing the lengths of various items in the house.

Item	True length	Length on model
Settee (length)	2 m	
Bookcase (height)		1.8 cm
Hall (length)	13.5 m	
Plate		2 mm
Living room (width)	5.8 cm	
Toothbrush	20 cm	
Bottle (height)		3 mm

5 The diagram shows the plan of a garden.
The scale is 1 cm = 2 m.
Copy and complete the table to show the true measurements of the garden.

Item	Plan measurement	True measurement
Length of patio		
Width of patio		
Width of vegetable plot		
Length of vegetable plot		
Width of pond		
Length of pond		
Length of house		
Width of house		

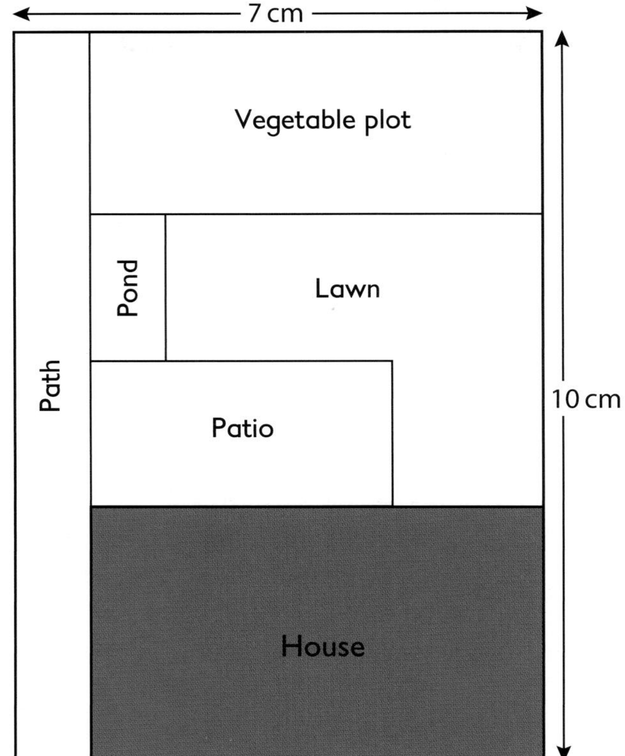

Reasoning

6 Lucy sets out from High Trees.
She walks 5 km South-East then 7 km North-East to reach the Lagoon at point P.

a Make a scale drawing to find

i how far point P is from High Trees

ii the bearing from High Trees of point P.

b What is the bearing of High Trees from point P?

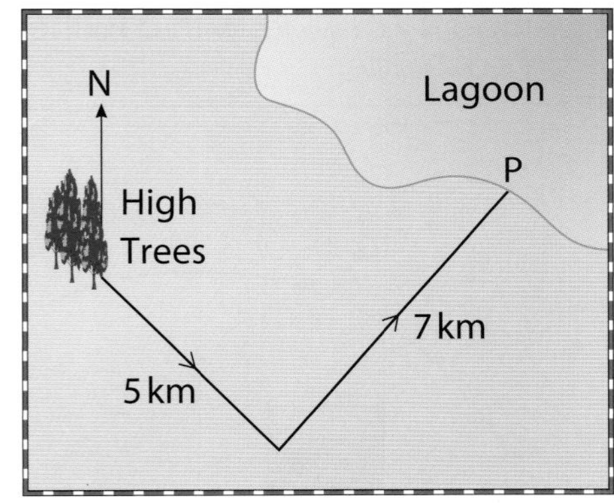

Skills practice B

1 The plan of the house shown is not drawn to scale.
Using the measurements given on the plan, draw an accurate scale drawing.
Use the scale 1 cm = 2 m.

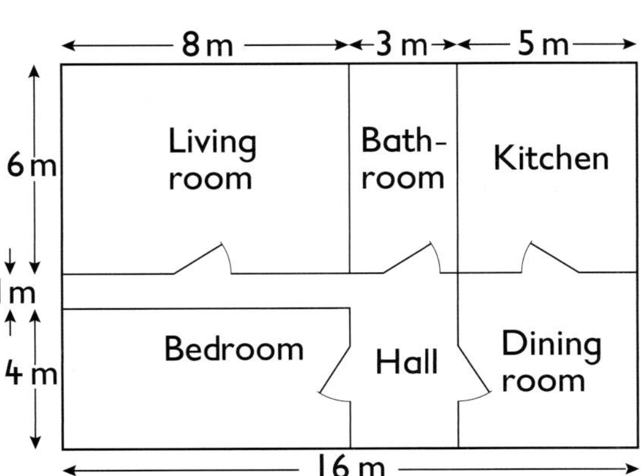

Reasoning

2 Three friends set off on a map-reading activity.
Each person travels for an hour.
Jane walks on a bearing of 030° at 6 kilometres per hour.
Sue cycles due East at 16 kilometres per hour.
Don walks on a bearing of 150° at 10 kilometres per hour.
At the end of the hour, what are the distances between

a Jane and Sue

b Jane and Don

c Don and Sue?

Reasoning

3 Gordon is walking on the moors.
To check his position he takes compass bearings of two masts.
From a map he knows that the TV mast is 13.5 km North of the Youth hostel.
The satellite dish is 4.1 km West of the Youth hostel.

a In what direction must he walk to reach the hostel?

b Gordon reckons he can walk at 4 km per hour.
It is 2 p.m. when he checks his position.
Will he reach the hostel in time for dinner at 6 p.m.?

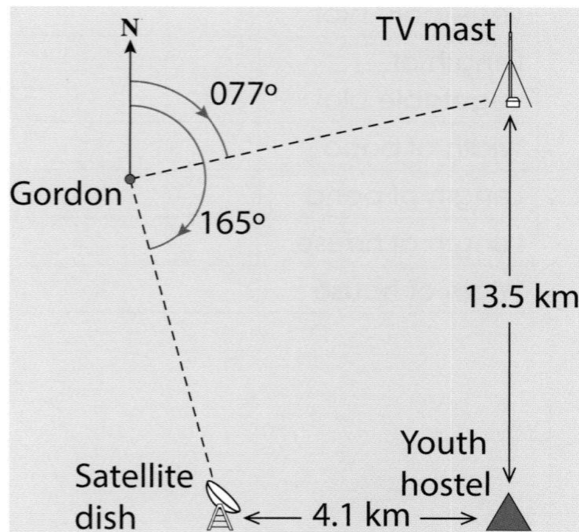

Reasoning

4 A road is 7 m wide, with a further 2 m pavement on each side.

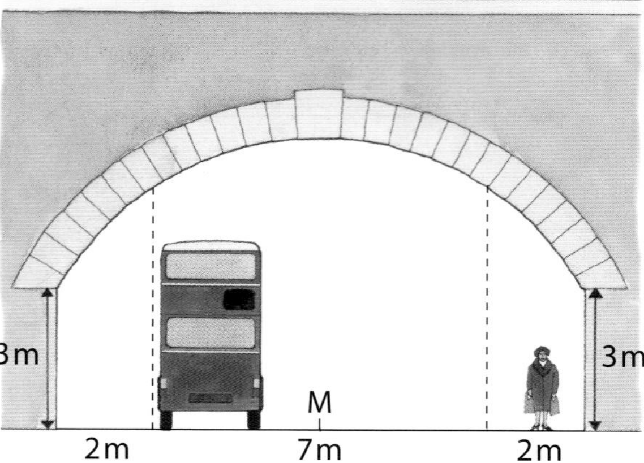

A bridge over the road has vertical sides 3 m high and its arch is part of a circle.
The centre of the circle is at the midpoint M of the road.

a Make a drawing of the bridge on a scale of $\frac{1}{100}$.

b What is the greatest height of the arch?

c What height restriction should be placed on vehicles driving under the arch on the road (shown by the broken lines)?

Wider skills practice

Reasoning

1 Marcia's rugby team have just played a match.

I have no sympathy. You should not have been wearing a ring in a match.
I have lost my ring on the pitch.
We will divide the pitch up into search areas.
We will help you look.

There are 15 players in the team and the touch judge. That makes 16 in all.

They divide the pitch into rectangles, using the lines marked on it. See the diagram on the next page.

Each person will search one rectangle.

a Make an accurate scale drawing of the pitch.

b Find 16 rectangles for the members of the team to search.

Use the lines marked on the pitch.

Everyone should have about the same area to search.

There is no 'right' answer.

The best answer is the fairest – nobody's area is much too big and nobody's is much too small.

c The referee has a different way to search the pitch.

You are being silly. You need to form a line, shoulder to shoulder. Then walk up and down the pitch with your eyes on the ground.

Estimate how long his method would take.

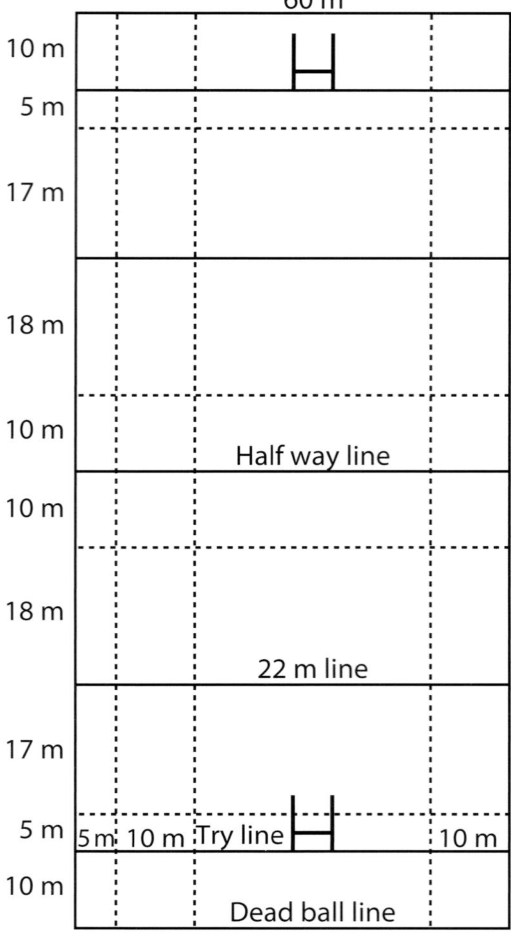

2 Look at this site plan.

a How wide is Beechwood Avenue?

b How many plots are there?

c The edge of the road on the corner of Cedar Close with Beechwood Avenue is an arc of a circle.

  i How many degrees is the arc?

  ii What is the radius of this circle?

d What are the dimensions of plot 3?

e What percentage of plot 2 is covered by the house?

f **i** Work out the area of plot 9.

  ii What is the ratio, area representing plot 9 on the site plan : area of plot 9.

g The local council want the developers to build a playground, covering an area of $1200\,m^2$, at the end of Cedar Close.
What area would represent this on the site plan?

Plot 1 Plot 2 Plot 3 Plot 4 Plot 5 Plot 6
BEECHWOOD AVENUE
CEDAR CLOSE
Plot 12 Plot 11 Plot 10 Plot 9 Plot 8 Plot 7
Scale
1:1000

3 Two sailing boats set sail from a port 12 km due West of a lighthouse.
The Swift sails on a bearing of 067° at a steady speed of 6.5 kilometres per hour.
The Reliant sails on a bearing of 143° at a steady speed of 10 kilometres per hour.

a Make a scale drawing to show their position after 2 hours.

b Find

i the distance between the Swift and the lighthouse

ii the bearing of the lighthouse from the Swift

iii the distance between the Reliant and the lighthouse

iv the bearing of the lighthouse from the Reliant

v the distance between the two boats after 2 hours.

c The two boats are in radio contact and decide to meet.
They sail directly towards each other.
What happens?

Applying skills

Problem solving

1 Kim and her family are moving to a new house.

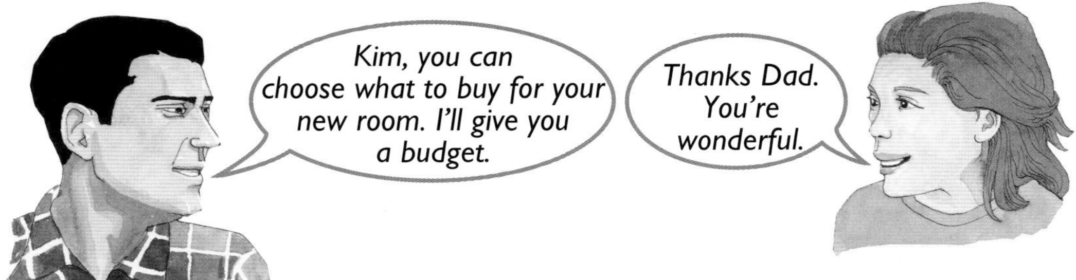

Kim, you can choose what to buy for your new room. I'll give you a budget.
Thanks Dad. You're wonderful.

This is Kim's plan for her ideal room. The room measures 4 metres by 3 metres.

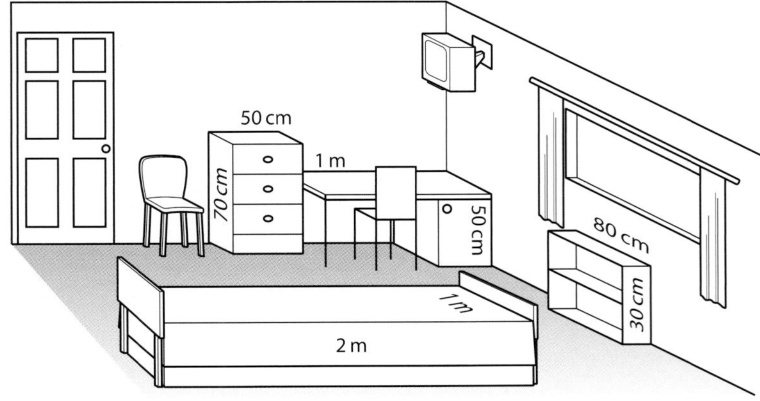

Problem solving

Look at these price lists.

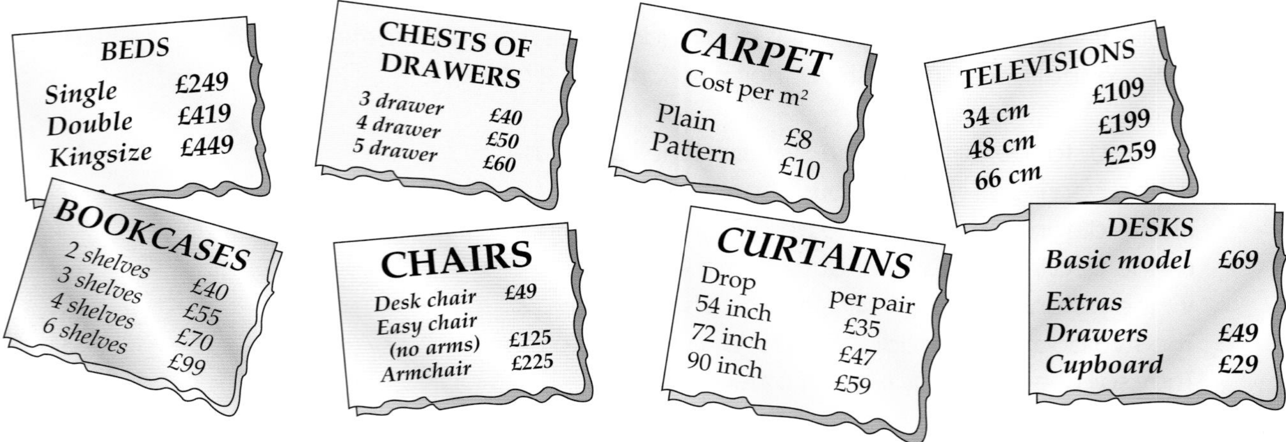

a Make a scale drawing of Kim's room, including the furniture.

b Identify which items are in Kim's room and work out the total cost of her plan.

c Kim's father gives her a budget of £800.
Is she within her budget?

Reviewing skills

1 An Ordnance Survey map has a scale of 1 : 50 000.

a Copy and complete this sentence.
1 : 50 000 on a map means 1 cm on the map represents a true length of ☐ cm on the ground.

b Write this as 1 cm to ☐ km.

c A distance on the map is measured as 2.5 cm.
How far is this?

d What distance on the map would represent 10 km?

2 A 5 metre ladder rests against a wall.
The foot of the ladder is 1.5 metres from the wall.
Make a scale drawing and use it to work out

a how far the ladder reaches up the wall

b the angle between the ladder and the ground.

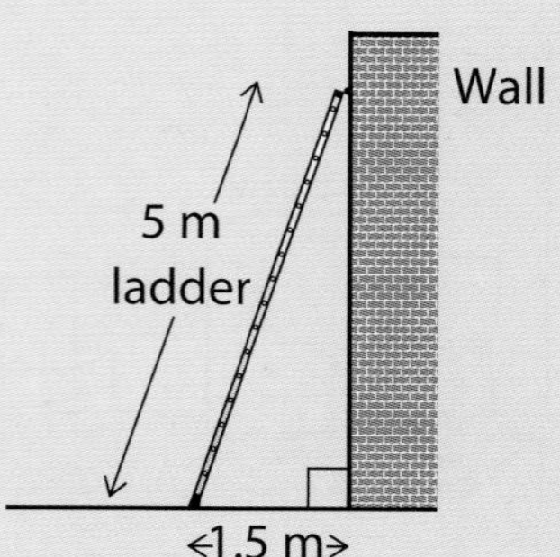

3 A model car is made to a scale of 1 : 25.
The model is 12 cm long.

a How long is the real car?

The wheel on the real car is 75 cm across.

b How far across is the wheel on the model car?

Building skills

Example outside the Maths classroom

Planning a journey

Toolbox

A **compound measure** is a measure involving two quantities.

An example is **speed**, which involves distance and time.

$$\text{speed} = \frac{\text{distance}}{\text{time}}$$

The unit of distance might be km and the unit of time might be hours. In this case the compound unit for speed will be kilometres per hour (km/hr or km hr^{-1}).

Other examples are, grams per cubic centimetre for density and Newtons per square metre for pressure.

Example – Working with density

A metal block has a volume of 1000 cm^3.

It has a mass of 8.5 kg.

Using the formula

$$\text{density} = \frac{\text{mass}}{\text{volume}}$$

find the density of the block in kilograms per cm^3.

Solution

$$\text{density} = \frac{\text{mass}}{\text{volume}}$$

$= 8.5 \div 1000$

$= 0.0085$ kilograms per cm^3

This unit can also be written as kg cm^{-3} or kg/cm^3.

Example – Working with speed

Jason leaves London on a motorbike at 06:00 and travels to his parents' house 432 km away. He arrives at 10:00.

a Find his average speed

i in kilometres per hour

ii in kilometres per minute

iii in metres per second

b Write the unit 'metres per second' in two other ways.

Solution

a His journey takes from 6 o'clock to 10 o'clock so it lasts 4 hours.

$$\text{Average speed} = \frac{\text{distance}}{\text{time}}$$

i $\frac{\text{distance}}{\text{time}} = \frac{432 \text{ kilometres}}{4 \text{ hours}}$

$= 108$ kilometres per hour

ii $\frac{\text{distance}}{\text{time}} = \frac{432 \text{ kilometres}}{240 \text{ minutes}}$ ← **4 hours = 4 × 60 = 240 minutes**

$= 1.8$ kilometres per minute

iii $\frac{\text{distance}}{\text{time}} = \frac{432\,000 \text{ metres}}{14\,400 \text{ seconds}}$ ← **432 kilometres = 432 × 1000 = 432 000 metres** ← **4 hours = 4 × 60 × 60 = 14 400 seconds**

$= 30$ metres per second

b Other ways of writing 'metres per second' include m s^{-1} and m/s.

Remember:

- A compound measure is a measure involving two quantities, such as speed.
- The units used in the simple measures tell you what the compound units will be.

Skills practice A

1 **a** John drives for 3 hours. He travels a distance of 240 km.
Work out his average speed.

b A snail travels for 40 minutes. It travels 36 metres.
Work out its speed in

i metres per minute **ii** cm s^{-1}.

c Andrew swims for 2 hours. He covers 72 lengths, each 25 metres.
What is his speed?

2 Work out the average speed of each of these trains.

- **a** An express train travelling a distance of 300 km in a time of 2 hours.
- **b** A freight train travelling 180 km in three hours.
- **c** A local train travelling 40 km in half an hour.

3 Sebastian sprints 100 m in 10 seconds.
Work out Sebastian's average speed in

- **a** metres per second
- **b** metres per hour
- **c** kilometres per hour.

4 Paul and Nina live 1000 m from school.
This travel graph shows their journey to school one day.

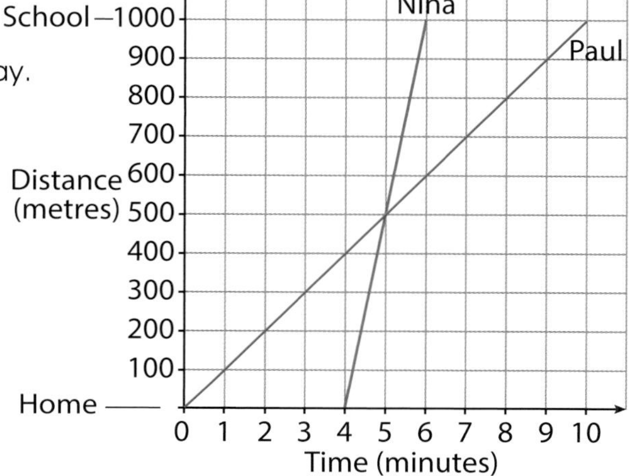

- **a** How far does Paul walk in
  - **i** 2 minutes
  - **ii** 5 minutes?
- **b** What is Paul's speed in
  - **i** metres per minute
  - **ii** kilometres per hour?
- **c** What is Nina's speed in
  - **i** metres per minute
  - **ii** kilometres per hour?
- **d** One of them cycles and the other walks.
  Who cycles?

5 A block of metal A weighs 130 g and has volume 26 cm^3.
A block of metal B weighs 24 000 kg and has volume 2 m^3.
Use the formula

$$\text{density} = \frac{\text{mass}}{\text{volume}}$$

to calculate the density of each block.
In each case, write the units in three different ways.

6 Jane and Sarah have part-time jobs.
In one week, Jane gets £24 for 4 hours' work.
Sarah works 7 hours and gets £38.50.
Find their rates of pay.
Who is the better paid?

7 A sailfish in Florida dragged a fishing line 90 m in 3 seconds.
How fast was this fish swimming, in kilometres per hour, during this time?

Skills practice B

1 Bob's train departs at 13:26 and arrives at its destination 192 miles away at 16:06. What is the train's average speed?

2 Katherine and Elizabeth go to different garages to fill their cars with diesel.
Katherine pays £26.20 for 20 litres.
Elizabeth pays £56.80 for 40 litres.
Work out the cost of diesel in pounds per litre at each garage.
Who gets the cheaper diesel?

3 The speed of music is measured in beats per minute (bpm).

- **a** A rock record's speed is 120 bpm.
It lasts for 3 minutes.
How many beats does it have?
- **b** A rap record has 450 beats.
It lasts for 5 minutes.
What is its speed?
- **c** An indie record has a speed of 170 bpm.
The record has 850 beats.
How long does it last?

4 The mass of the block of oak in the diagram is 390 g.

- **a** Calculate the density of oak in g cm^3.
- **b** Find the mass of a 20 centimetre cube of oak in kilograms.

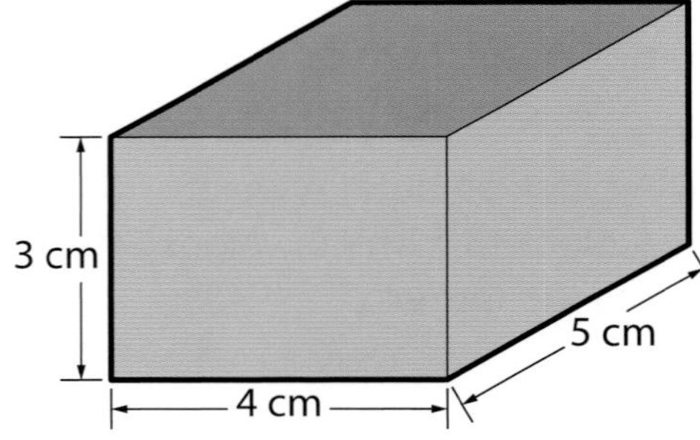

Reasoning

5 Peter is riding a motorbike at 35 miles per hour.
Pierre is driving his car at 12.5 metres per second.
Convert both of these speeds into kilometres per hour, using the fact that 5 miles is approximately 8 kilometres.
Who is travelling faster?

Reasoning

6 The national speed limit for cars on most roads is 60 miles per hour.
A bird can fly at 45 metres per second.
Which is faster, the car or the bird?
Explain your answer.

7 The official world land speed record is 1223.7 kilometres per hour.
It was set on 15 October 1997 by Andy Green in the jet-engine car *ThrustSSC*.
What is this speed in metres per second?

8 A secretary typed a 4000 word document in 1 hour 40 minutes.

- **a** Calculate her typing speed in words per minute.
- **b** Working at this speed, estimate how long it would take this secretary to type a six-page article with about 480 words on each page.

9 A black mamba snake of Eastern Africa can move at up to $5\,m\,s^{-1}$.
What is this speed in miles per hour?
(Use the fact that 8 kilometres is approximately 5 miles.)

Wider skills practice

1 Sound travels 1 kilometre in 3 seconds.
Alana sees a fork of lightning.
Five seconds later she hears the thunder.
How many metres away is the lightning?

2 A light year is the distance that light travels in one year.
The speed of light is 300 million metres per second.
How many kilometres are there in one light year?

3 In weightlifting, competitors can be placed roughly in order by finding the weightlifting fraction for each competitor.

$$\text{Weightlifting fraction} = \frac{\text{weight lifted}}{\text{body weight}}$$

The results for one competition are given in the table.

Name	John	Stephen	Simon	Julian	Kit	Lee
Body weight	61 kg	92 kg	69 kg	84 kg	72 kg	52 kg
Weight lifted	122 kg	220 kg	185 kg	270 kg	175 kg	157 kg
Weightlifting fraction						

a Find the weightlifting fraction for each competitor then copy and complete the table.
b The competitor with the largest weightlifting fraction is the winner.
Put the competitors in order, starting with the winner.
c What is the unit for the weightlifting fraction?

Applying skills

Problem solving

1 A farmer has a field.
He is paying an expert to fertilise the field.
The expert fertilises $10\,m^2$ of the field per minute.
The diagram shows the shape and dimensions of the field.
The farmer pays the expert at a rate of £160 per hour for the first 30 minutes work and £2 per minute for any extra time taken.
How much does the farmer pay?

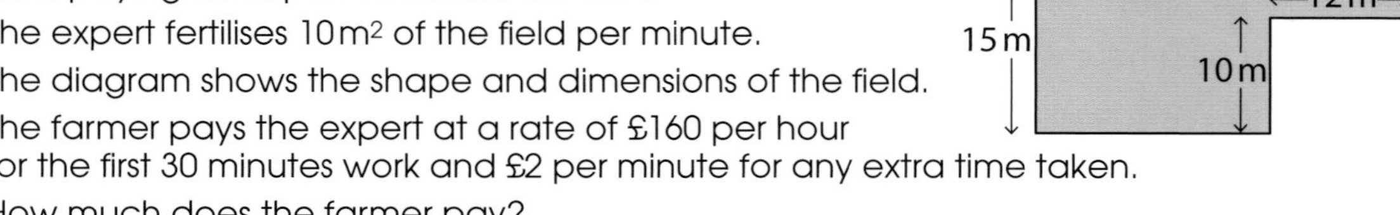

Problem solving

2 A wind speed of 333 km per hour was recorded in Greenland in 1972.
How fast is this in metres per second?

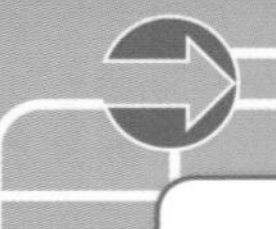

Problem solving

3 Geoff buys this car.

There are many things to pay for when you run a car and one of the most expensive is fuel. Geoff has seen two cars and would like to buy one of them.

a The fuel consumption data have the units mpg.

What do these letters stand for?

b Which car will use the least fuel?

Fuel
Petrol
Fuel consumption
Urban: 29 mpg
Extra-urban: 50 mpg
Combined: 39 mpg

Fuel
Diesel
Fuel consumption
Urban: 35 mpg
Extra-urban: 50 mpg
Combined: 43 mpg

c Geoff uses his car to go to work and back from Monday to Friday. He does about 25 miles on a variety of different roads in the evenings and weekends too.

How many **gallons** of fuel will each car need every week?

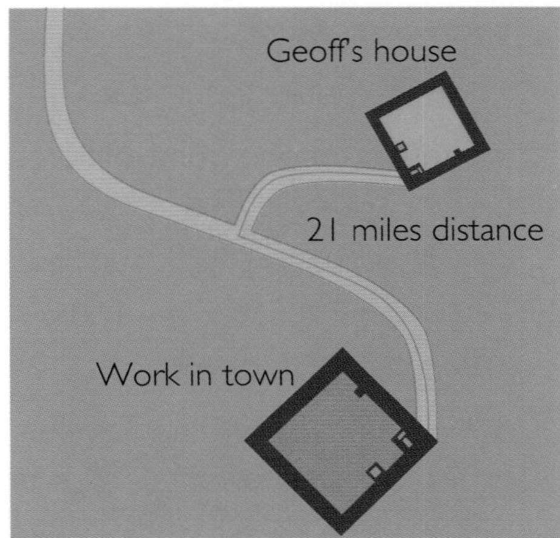

Reviewing skills

1 Gita is paid £22.50 for $2\frac{1}{2}$ hours work.

What is her rate of pay

a in pounds per hour

b in pence per minute

c in pounds for a 40-hour week?

2 An aeroplane flies at 1080 kilometres per hour.

a How far would it travel, at this speed, in

i 6 minutes

ii 1 minute

iii 15 seconds

iv 5 seconds?

b Using the fact that 5 miles is approximately 8 kilometres, convert the aeroplane's speed into miles per hour.

Strand 2 • Properties of shapes

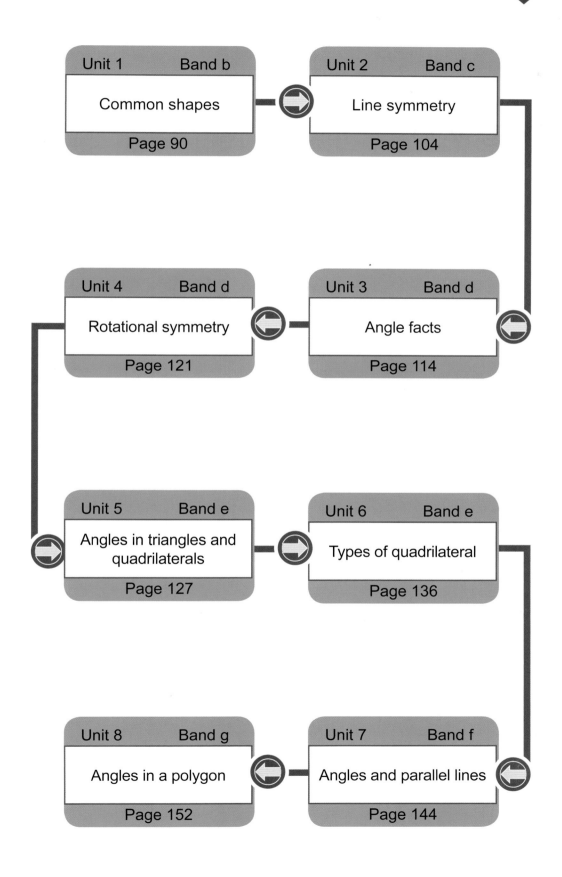

Unit 1	Band b	Common shapes	Page 90
Unit 2	Band c	Line symmetry	Page 104
Unit 3	Band d	Angle facts	Page 114
Unit 4	Band d	Rotational symmetry	Page 121
Unit 5	Band e	Angles in triangles and quadrilaterals	Page 127
Unit 6	Band e	Types of quadrilateral	Page 136
Unit 7	Band f	Angles and parallel lines	Page 144
Unit 8	Band g	Angles in a polygon	Page 152

Building skills

Example outside the Maths classroom

Diagrid buildings

Toolbox

An **angle** is a turn. It is usually measured in degrees.

An **acute angle** is less than 90°

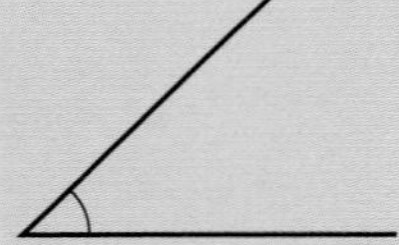

A **right angle** is exactly 90°

An **obtuse angle** is between 90° and 180°

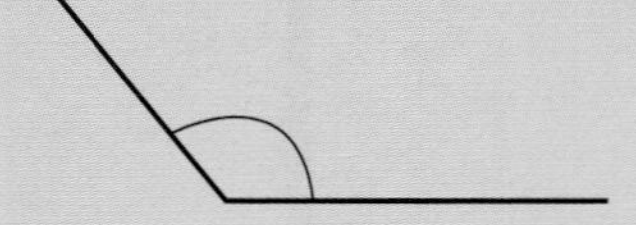

Straight line 180°

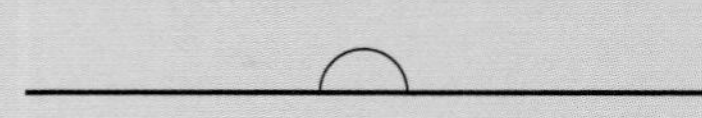

A **reflex angle** is between 180° and 360°

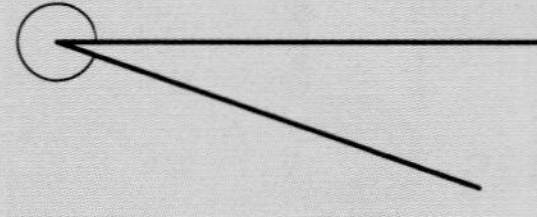

A **full turn** is 360°

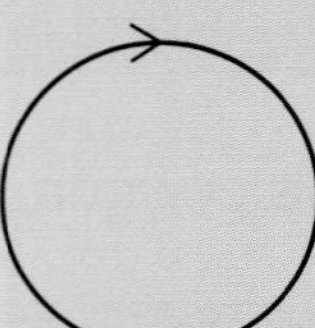

Lines

Parallel lines run in the same direction.

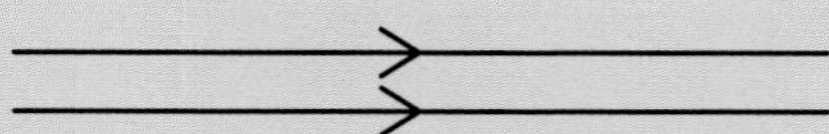

Perpendicular lines are at right angles.

Triangles

Equilateral triangle

All sides and angles are equal.

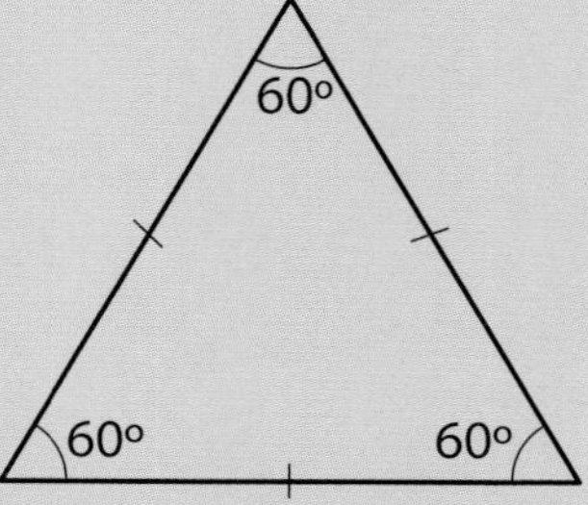

Isosceles triangle

Two sides and two angles equal.

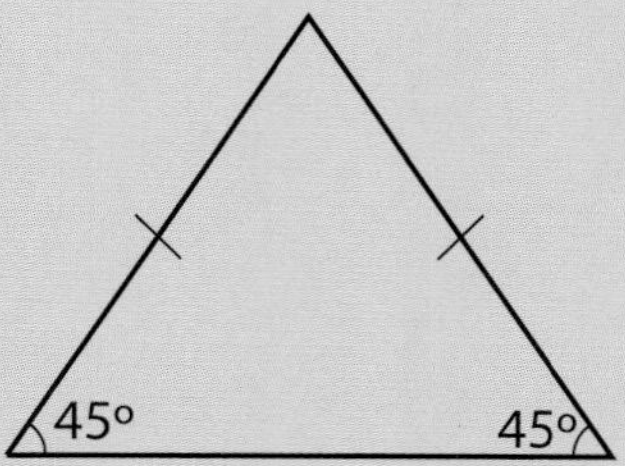

Scalene triangle

All sides and angles different.

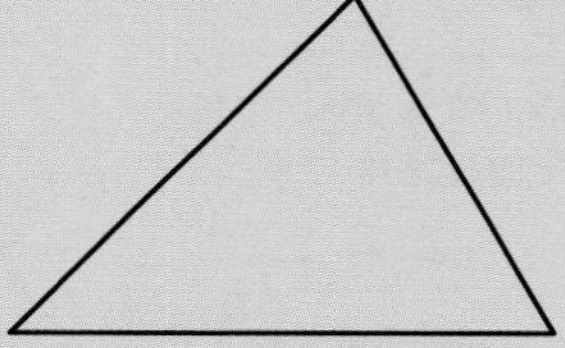

Right-angled triangle

One angle is 90°.

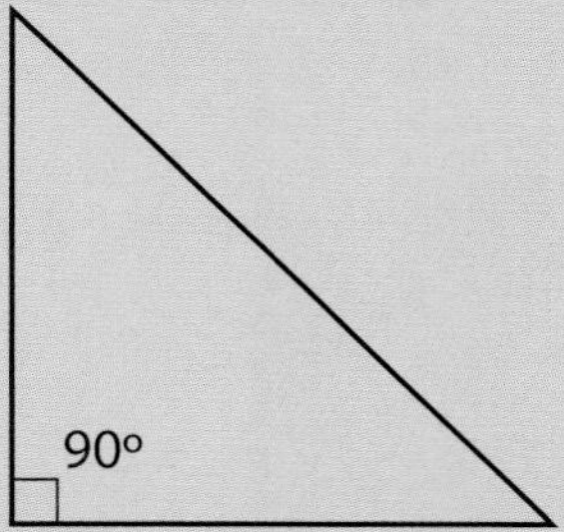

Quadrilaterals

A quadrilateral is any shape with four sides

Square

All four sides equal, all angles 90°.

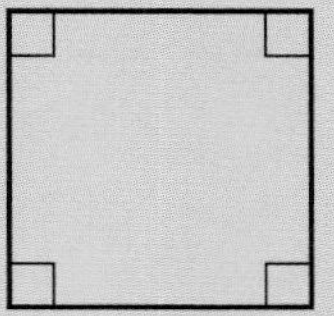

Rectangle

Opposite sides equal, all angles 90°.

Parallelogram

Opposite sides parallel and equal.

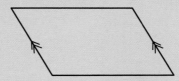

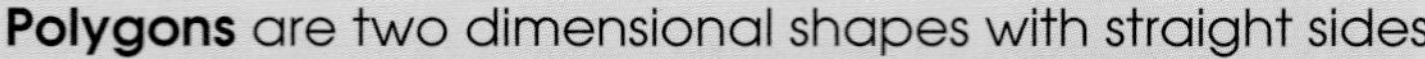

Polygons are two dimensional shapes with straight sides.

Name	Number of sides	Number of vertices
Triangle	3	3
Quadrilateral	4	4
Pentagon	5	5
Hexagon	6	6
Octagon	8	8
Decagon	10	10

A vertex is where two sides (or edges) meet.

In a **regular** polygon, all the sides and angles are equal.

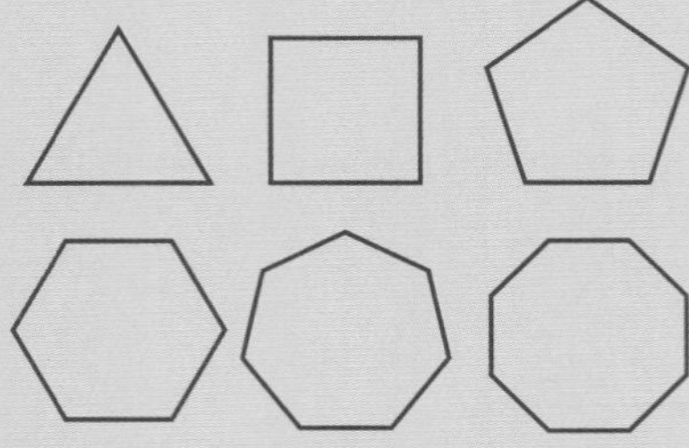

Figures which are the same shape and size as each other are **congruent**.

Three-dimensional shapes

A **cube** has six faces, eight vertices and 12 edges.
The faces are all squares.

A vertex is a corner.
An edge joins two vertices.
A face is a surface.

A **cuboid** (below) also has six faces, eight vertices and 12 edges.
The faces are all rectangles.

edge
face
vertex

edge
vertex
face

Circles

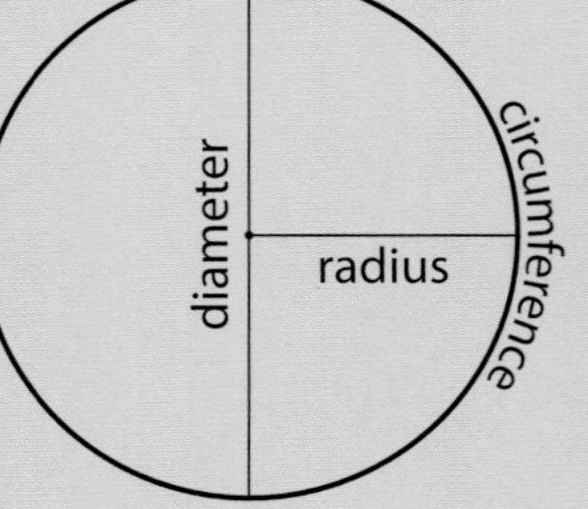

A **sphere** is a 3-D shape like a ball.

Example – Describing shapes

a Name the vertices and the sides of this triangle.
b Describe the shape of the triangle.
c Describe the angles of the triangle.

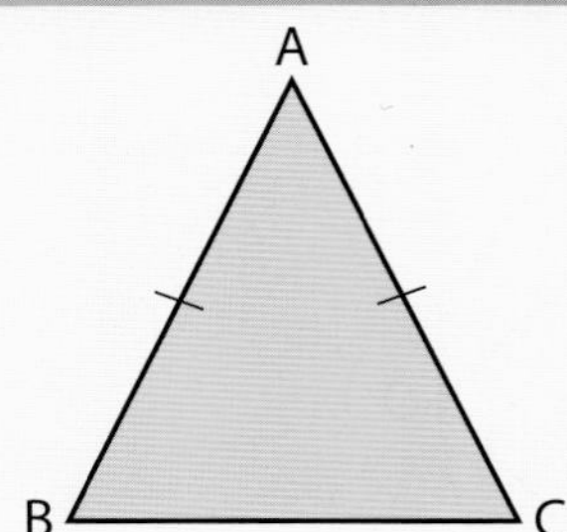

Solution

a The three vertices are A, B and C.
The three sides are AB, BC and AC.
b Sides AB and AC are equal.
The triangle is isosceles.
c All three angles are acute. ← **They are less than 90°.**
The angles B and C are also equal.

Example – Working with a circle and polygons

Look at this circle.

The 12 points marked on the circumference are equally spaced.

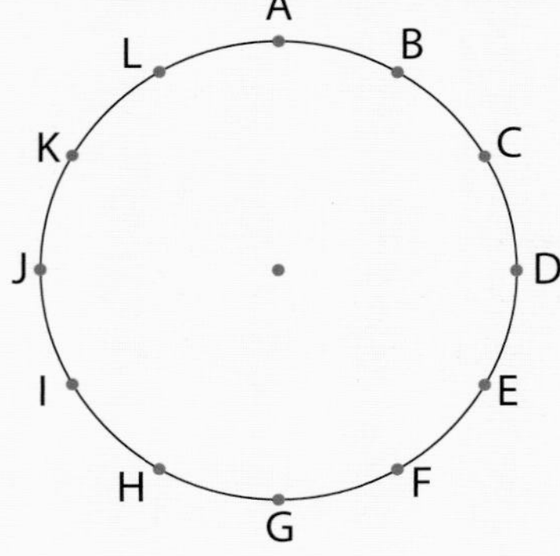

a Measure AG.
Write down the radius of the circle.
b Join AE, EI and IA.
What shape have you drawn?
c Starting at A, join up points to form
  i a square
  ii a rectangle that is not a square
  iii a regular hexagon
  iv a triangle that is isosceles but is not equilateral.

Solution

a AG = 4 cm
AG is a diameter of the circle.
The radius is half the diameter.
Radius = 4 ÷ 2 = 2 cm

b AEI is an equilateral triangle.

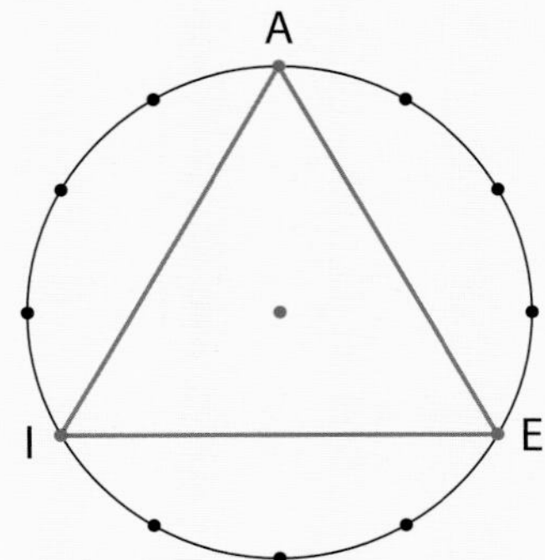

c i A square

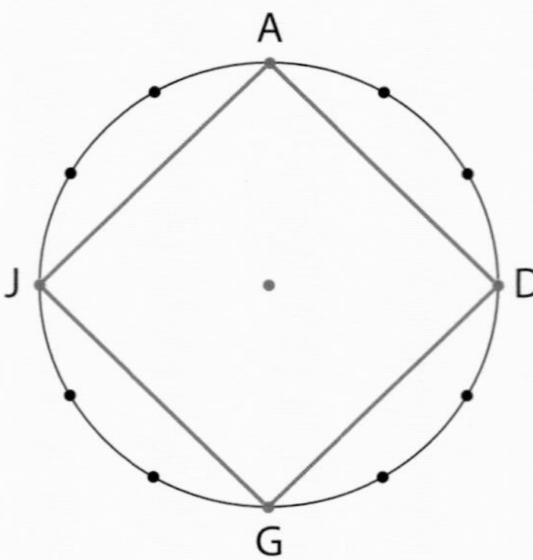

ii A rectangle that is not a square

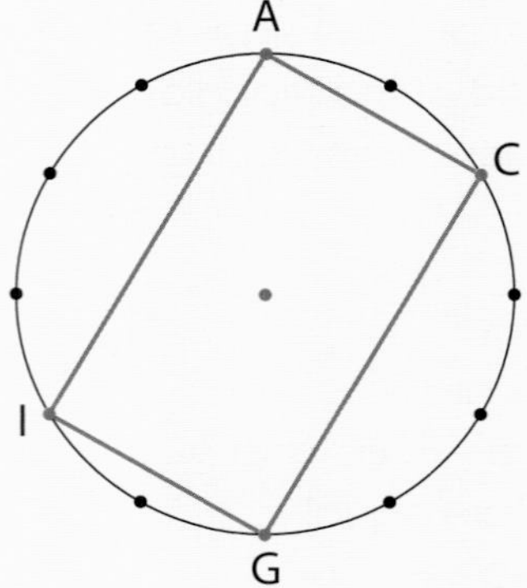

(There are other possible answers.)

iii A regular hexagon

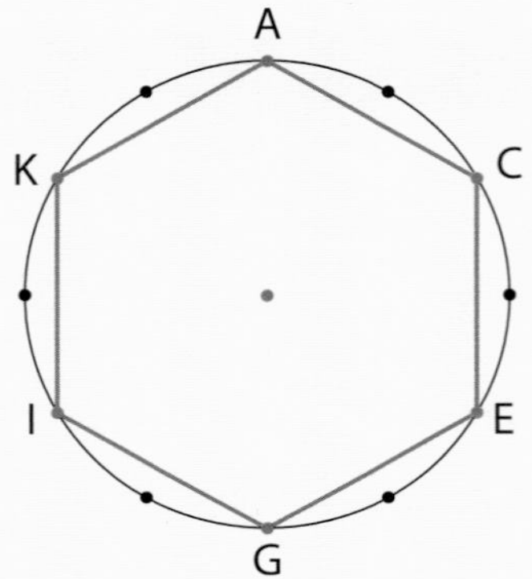

iv An isosceles triangle that is not equilateral

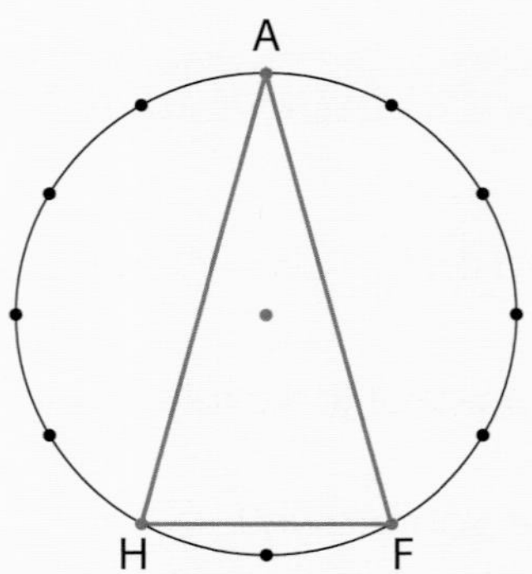

(There are other possible answers.)

Example – Working with 3-D shapes

The diagram shows a cuboid but a cube has been removed from one corner.

How many faces, edges and vertices does this shape have?

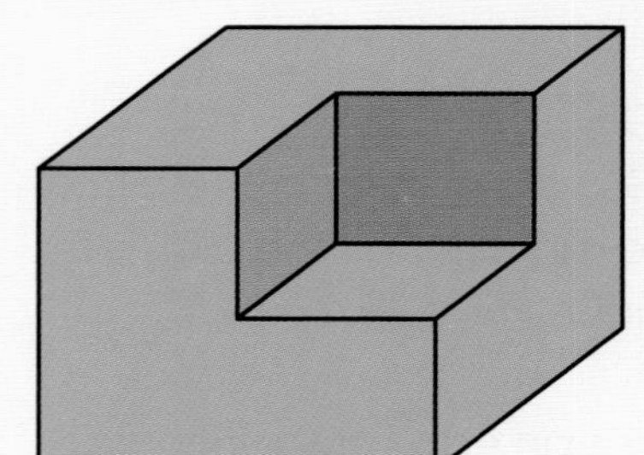

Solution

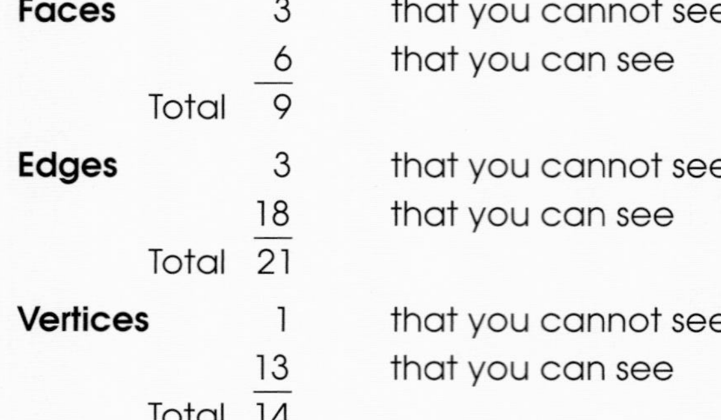

Faces	3	that you cannot see
	6	that you can see
Total	9	
Edges	3	that you cannot see
	18	that you can see
Total	21	
Vertices	1	that you cannot see
	13	that you can see
Total	14	

Remember:

- Using precise language helps to avoid mistakes.
- When naming 2-D shapes think about the number of sides and vertices.

Skills practice A

1 Look at these angles.

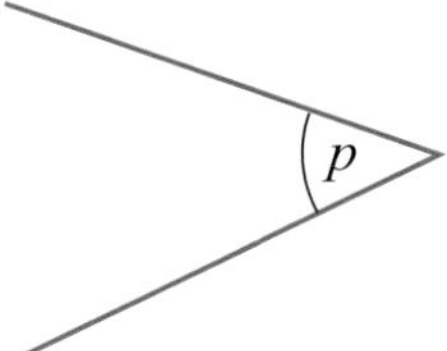

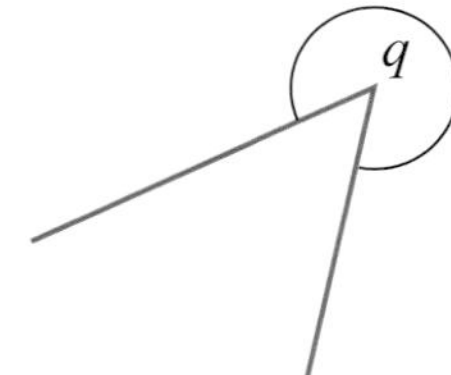

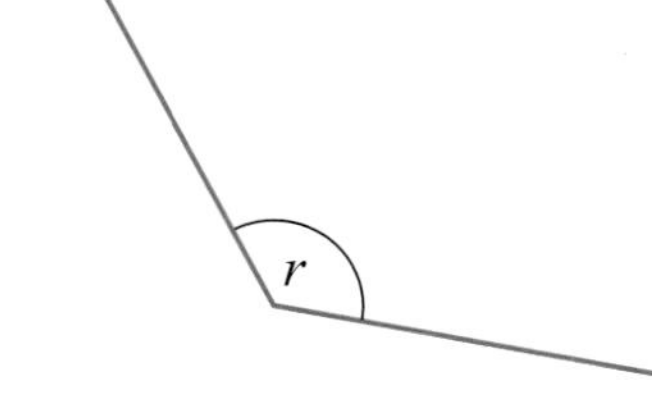

Which angle is

a acute **b** obtuse **c** reflex?

2 Write these angles in order of size, smallest first.

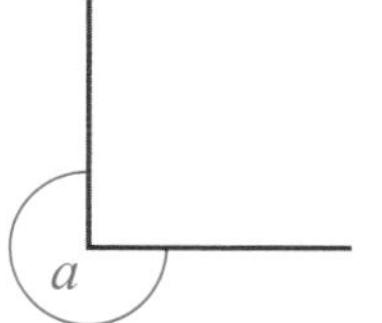

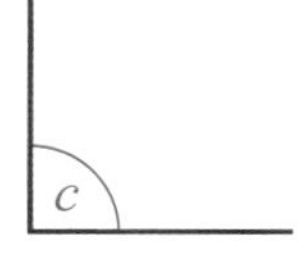

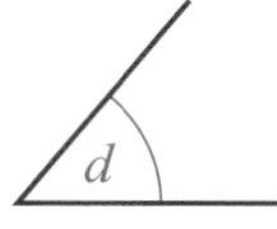

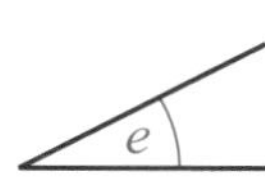

3 Look at these angles.

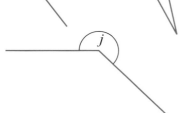

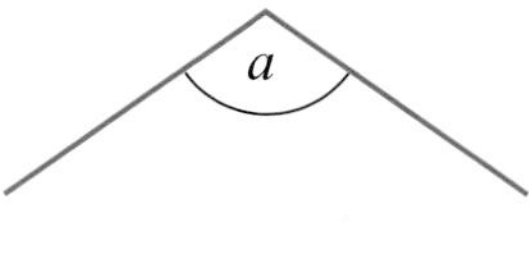

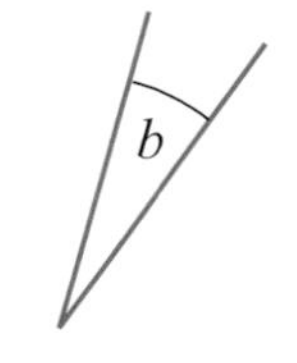

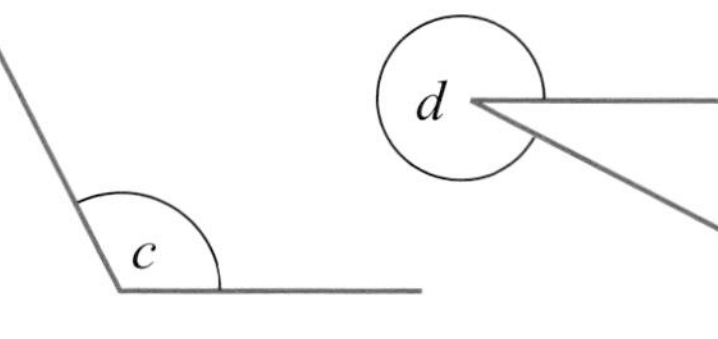

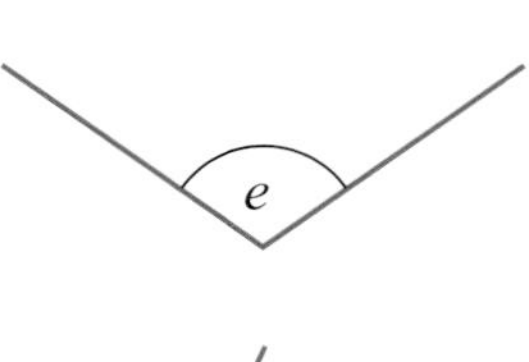

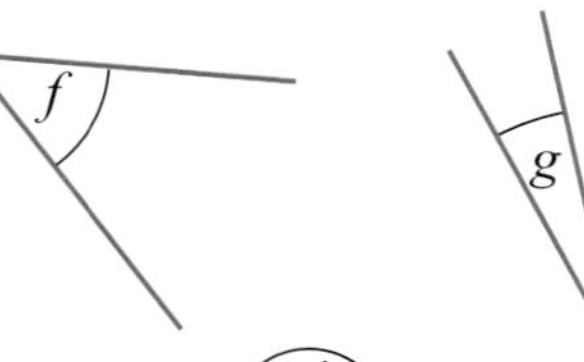

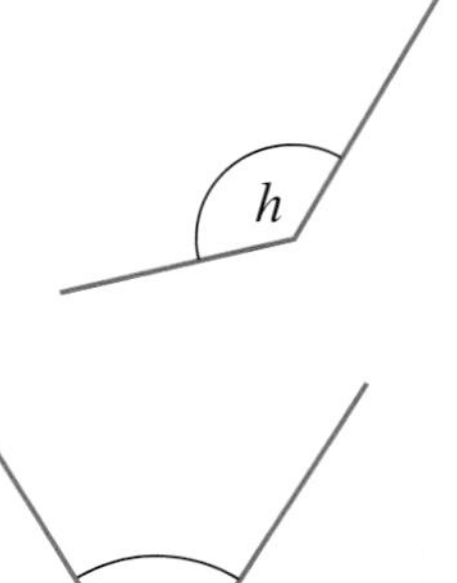

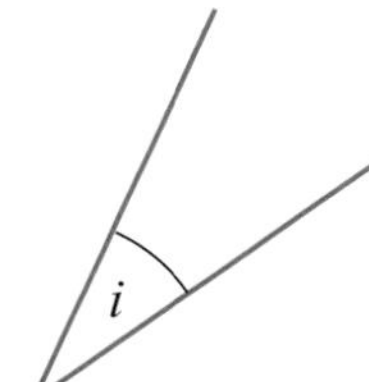

Which angles are **a** acute **b** obtuse **c** reflex?

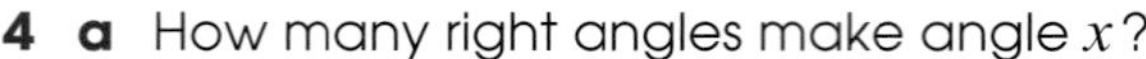

4 a How many right angles make angle x?

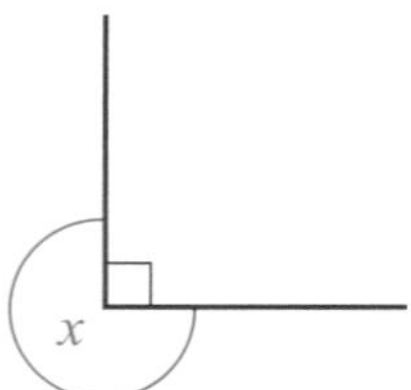

b How many right angles make a whole turn?

5 Look at these triangle labels.

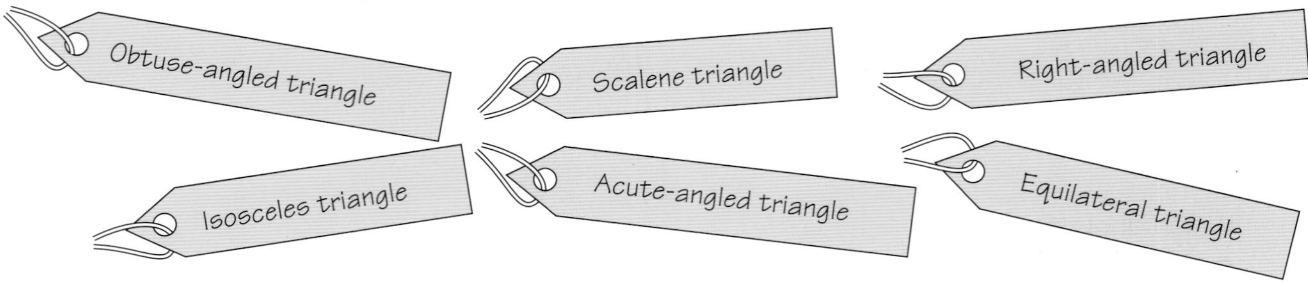

Look at these triangles.
Match two labels to each triangle.

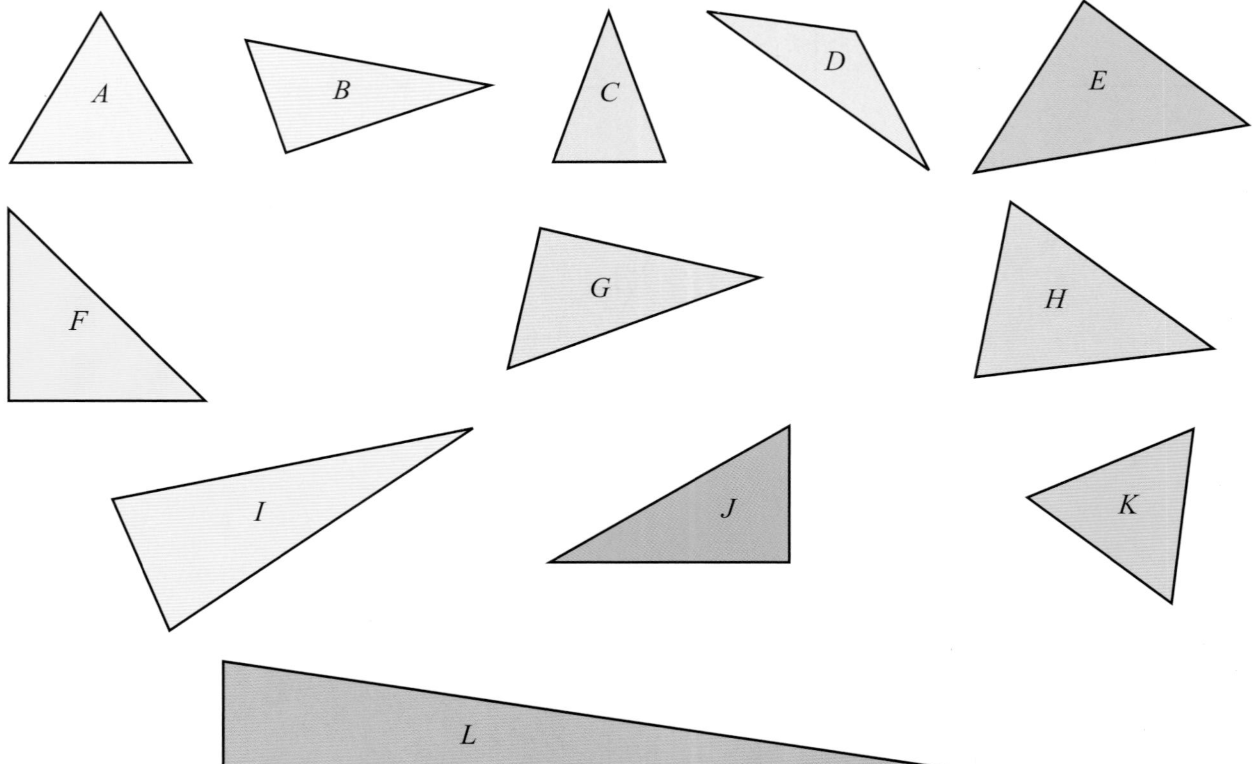

6 **a** Look at this rectangle.
It is divided into four triangles.
i Describe the triangles as fully as you can.
ii Which triangles are congruent?
b A square is divided up in the same way.
Describe the triangles as fully as you can.

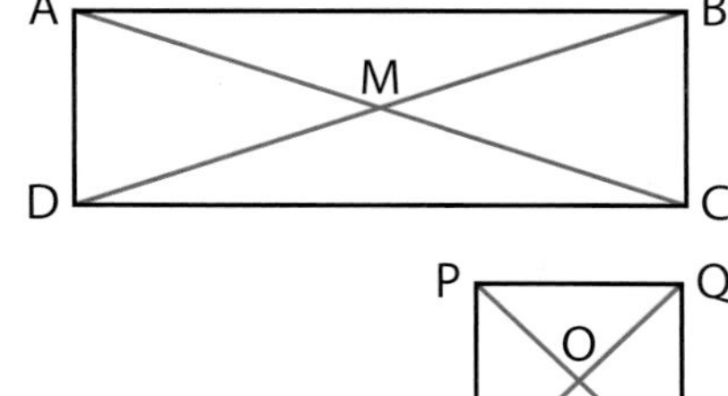

7 **a** Measure the radius of each of these circles.
Use your answer to find the diameter of the circle.

i

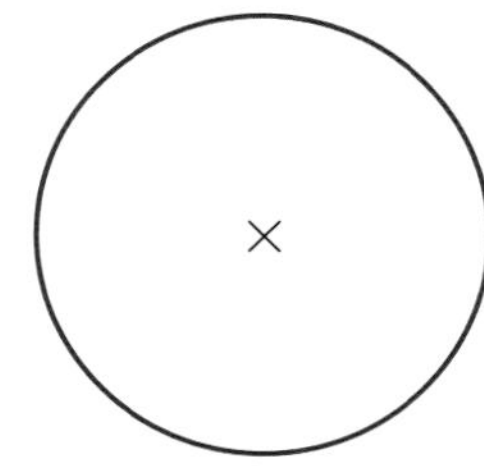

ii

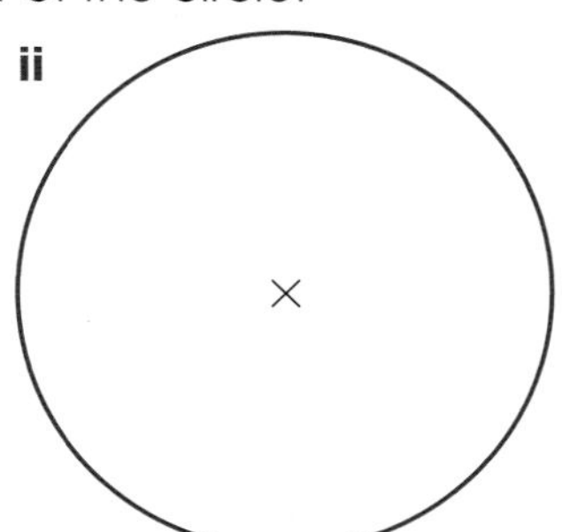

b Measure the diameter of this circle.
Use your answer to find the radius of the circle.

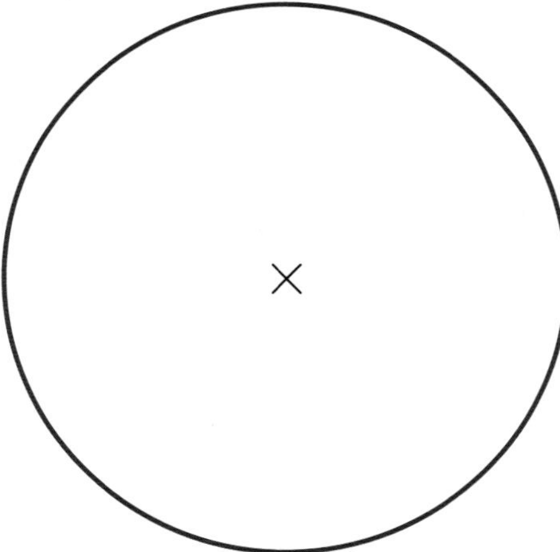

8 Give the shape of each of these objects their mathematical names.

a

b

c

9 Which of these statements are true and which are false?
a A triangle must have an obtuse internal angle.
b A triangle must have an acute internal angle.
c A quadrilateral must have four sides.
d The number of sides of a polygon is always the same as the number of vertices.
e The radius of a circle is twice the diameter.
f Parallel lines never meet.

Skills practice B

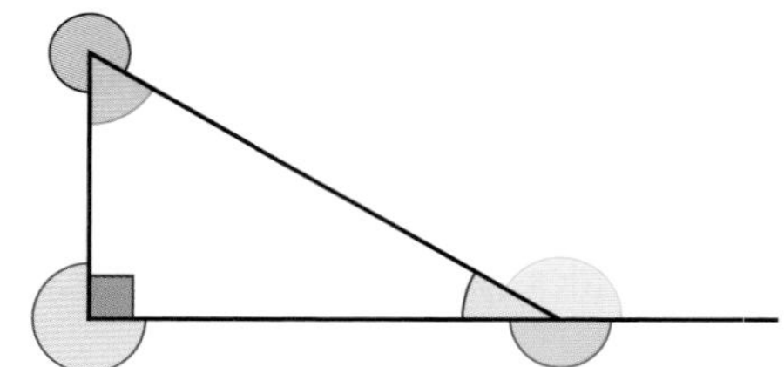

1 Look at this diagram.
There are seven coloured angles.
For each colour, state whether the angle is acute, obtuse, reflex, a right angle or a straight line.

2 a Measure the lines in this shape in mm.
b Name the triangles that are
- **i** equilateral
- **ii** isosceles but not equilateral
- **iii** right-angled
- **iv** scalene.

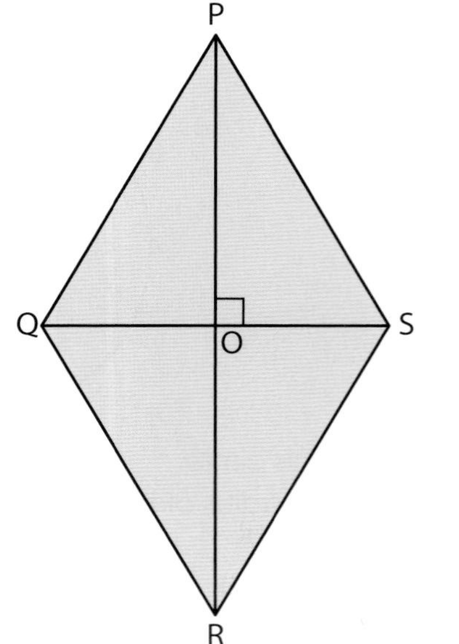

3 Copy and complete this table by sorting these shapes into 2-D and 3-D shapes.

2-D shapes	3-D shapes

pentagon

square

equilateral triangle

cuboid

hexagon

octagon

cube

isosceles triangle

circle

rectangle

sphere

parallelogram

4 Kieran is a tour guide at a castle.
He draws a plan of the castle to help the visitors.
Look at the plan below.

a What shape is the plan?

b The angle a is acute.
What are the special names for the other marked angles?

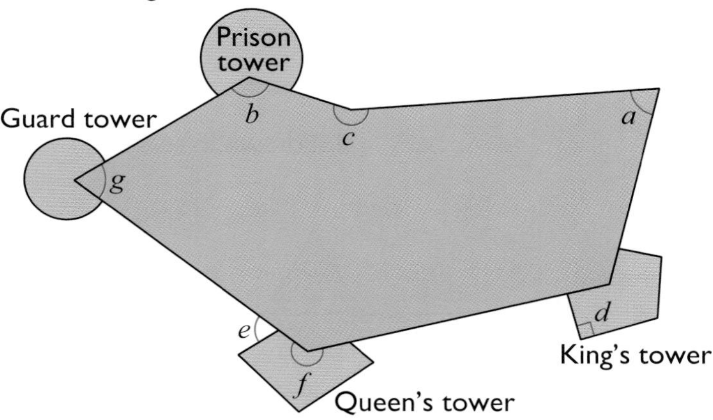

5 Look at this diagram.

a How many squares are there?

b How many rectangles that are not squares are there?

c How many rectangles are there altogether?

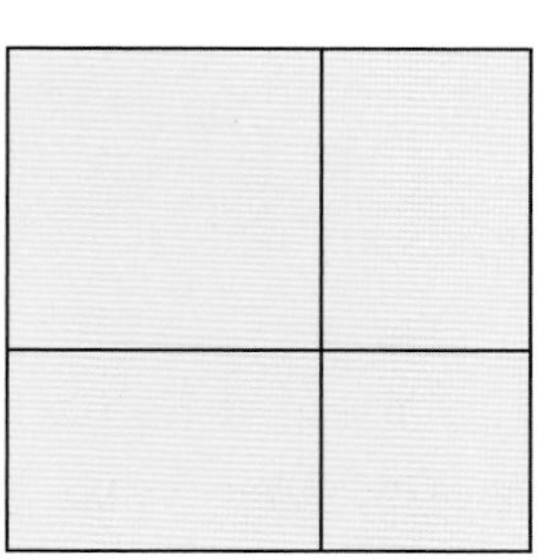

Reasoning

6 **a** Are Lucy and Angus right?

What is the difference between a square and a cube?

A square is flat. It is two-dimensional (2-D).

A cube is a solid. It is three-dimensional (3-D).

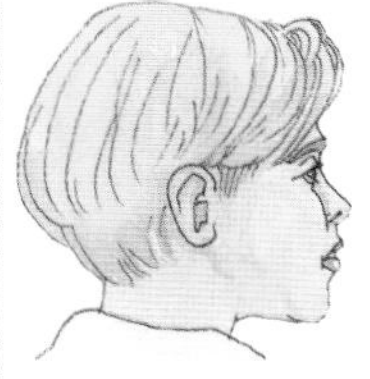

Boy

Lucy

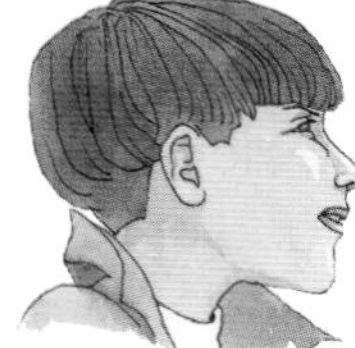

Angus

b Describe the difference between a rectangle and a cuboid.

c What is the 2-D equivalent of a sphere?

7 This card is used on a machine.
ABCD is a square with sides 4 cm long.
The other shapes are semicircles.
Find the height and width of the card.

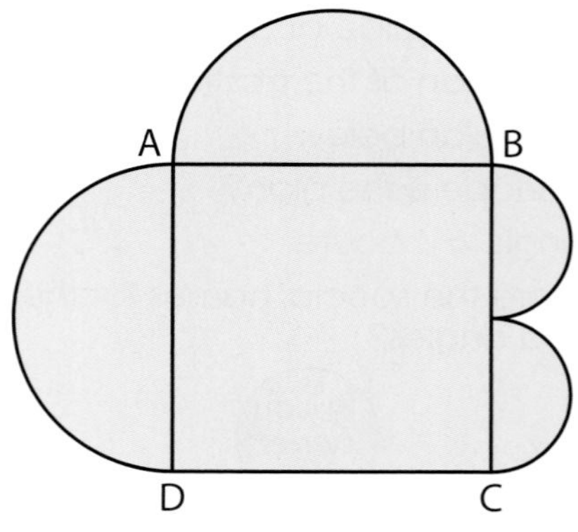

Not drawn to scale.

Reasoning

8 Look at this diagram.

a **i** How many equilateral triangles are there?

ii Which of them are congruent to each other?

b Name another regular polygon in the diagram.

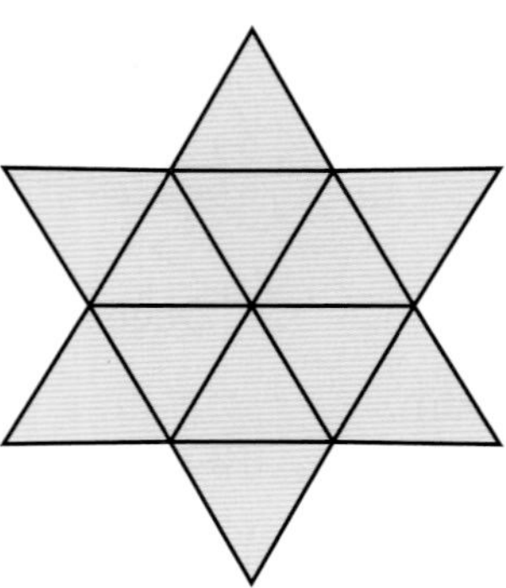

Reasoning

9 Which of these statements are true and which are false?
If false, say why.

a A triangle can have two parallel sides.

b All squares are congruent.

c The letter M is a quadrilateral.

d All cubes are cuboids.

e A sphere is a special sort of circle.

f A cuboid has six faces.

Reasoning

10 Imagine a loaf of bread which is cuboid shaped.

a What shape are the slices?

b What shape is the face of each slice where the cut has been made?

Wider skills practice

Reasoning

1 Look at this regular octagon.

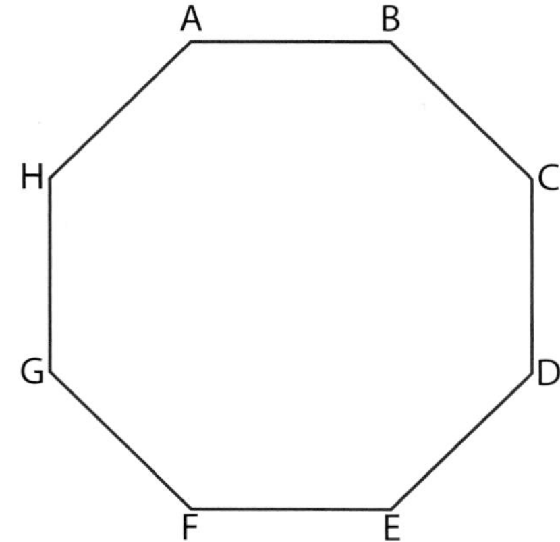

a On one copy of the diagram, join four vertices together to make a square.

What other vertices could you join to make a square?

b On a second copy of the diagram, join four of the vertices to make a rectangle.

What other vertices could you join to make a rectangle?

Reasoning

2 Look at these solids.

i **ii** **iii**

a How many faces (F), vertices (V) and edges (E) does each solid have?

b Work out $F + V - E$ for each solid.

c What do you notice?

d Investigate whether you get the same result with other three-dimensional shapes.

3 Which of these solid shapes are cuboids, which are cubes and which are spheres?
Which are none of these?

a

b

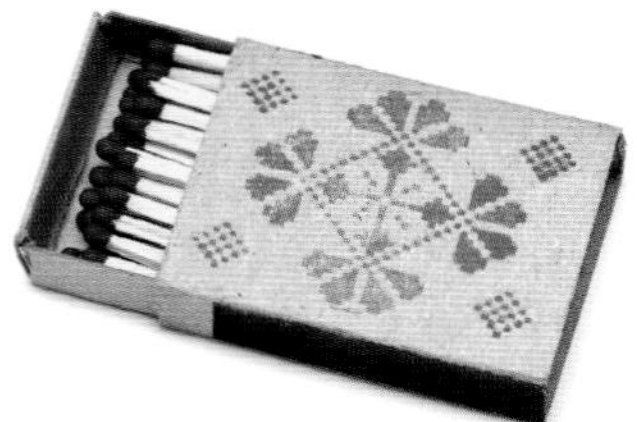

c

d

e

Applying skills

Problem solving

1 **a** Measure the sides of the seven main shapes in this diagram in mm.

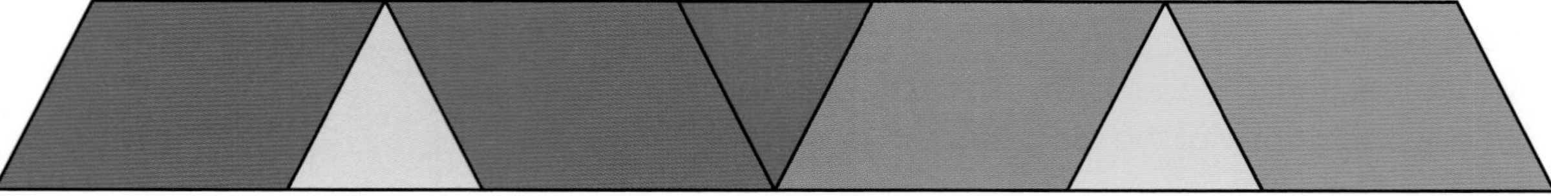

b Name the seven shapes.

c Is it true that they are all congruent?

d If you cut out the shapes, is it possible to make a square out of some or all of them? (No folding allowed!)

Problem solving

2 Look at this honeycomb.
It is made of cells fitted together.
The cells are hexagons.

a How many sides does a hexagon have?

b What can you say about

- **i** the shape of the cells
- **ii** the size of the cells?

c There are no gaps between any of the cells.
How far can the pattern continue?

One shape repeated to cover a surface is called a **tessellation**.
Here is one way to tessellate a triangle.

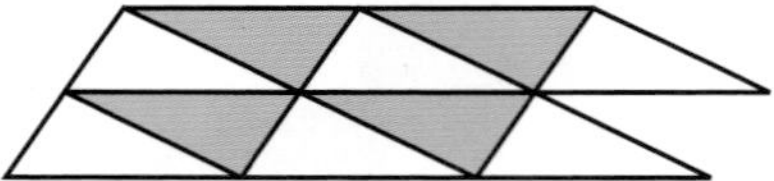

d Find another way to tessellate this triangle.

e Copy this rectangle on to squared paper.
Make a tessellation using this shape.
Colour the rectangles to show the pattern.

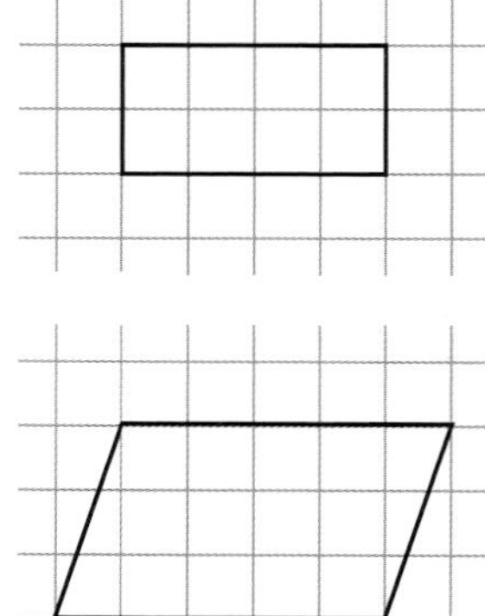

f **i** Copy this parallelogram on to squared paper.
Make a tessellation using this shape.
Colour the parallelograms to show the pattern.

ii Copy the parallelogram again.
Make another **different** tessellation using this shape.

Reviewing skills

1 Match the angles in the red boxes to the descriptions in the yellow boxes.

Obtuse angle	Smaller than a right angle
Reflex angle	Larger than a right angle, smaller than two right angles
Acute angle	Larger than two right angles

2 Describe and name the shapes in this picture of an animal as accurately as you can.

3 **a** Measure the diameter of this circle.
What is its radius?

b On a copy of the diagram, join points 1, 7, 13, 19 and back to 1.
What shape do you get?

c What other regular polygons can you draw using the points on this circle?

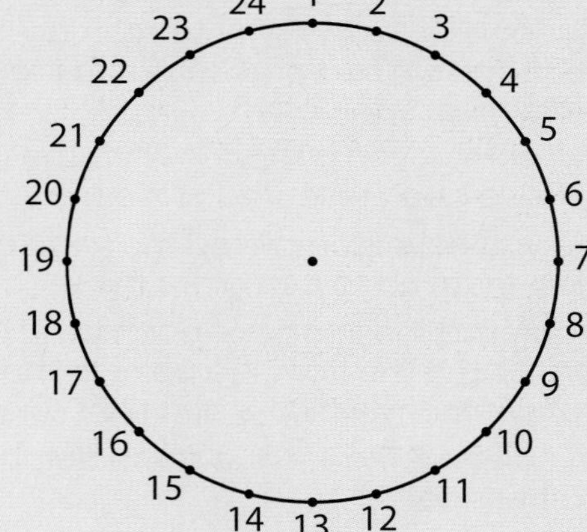

Unit 2 • Line symmetry • Band c

Building skills

Example outside the Maths classroom

Art and nature

Toolbox

When looking for line symmetry, imagine folding the object or shape.
If the shape will fold precisely onto itself then that fold is a **line of symmetry** and the shape has line (or reflective) symmetry.

When completing a shape with a line (or lines) of symmetry find where each vertex will go when they have been reflected in the line.

Once you have completed the vertices, just join them up to complete the shape.

You can find the image of each point by drawing a line from the point to the line of symmetry which is at a right angle.

Then continue the line the same distance the other side.

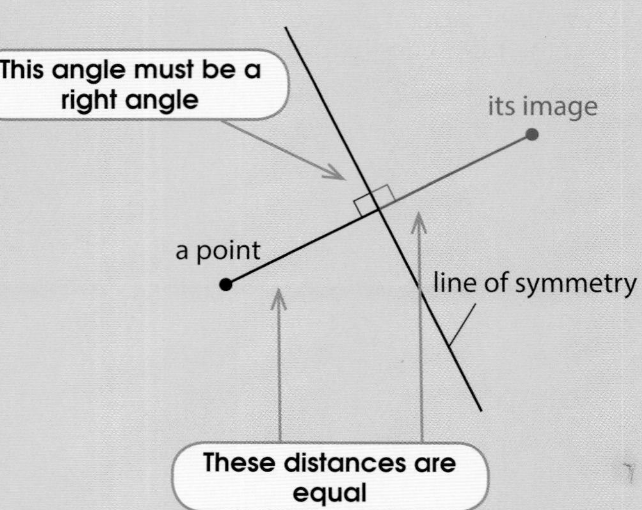

Example – Identifying lines of symmetry 1

How many lines of symmetry do these shapes have?
Draw the lines of symmetry on a copy of each shape.

a

b

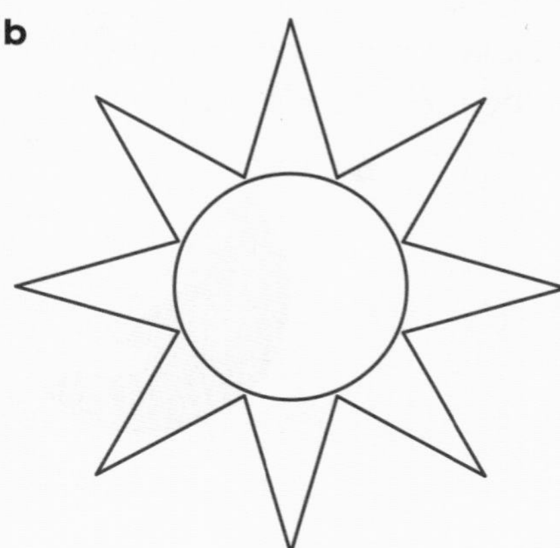

Solution

a A rectangle has two lines of symmetry.

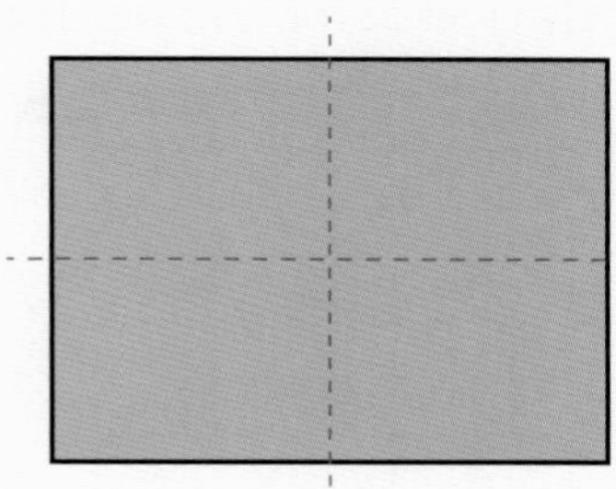

b This shape has four lines of symmetry through the tips of the triangles and four more between each triangle so there are eight lines of symmetry.

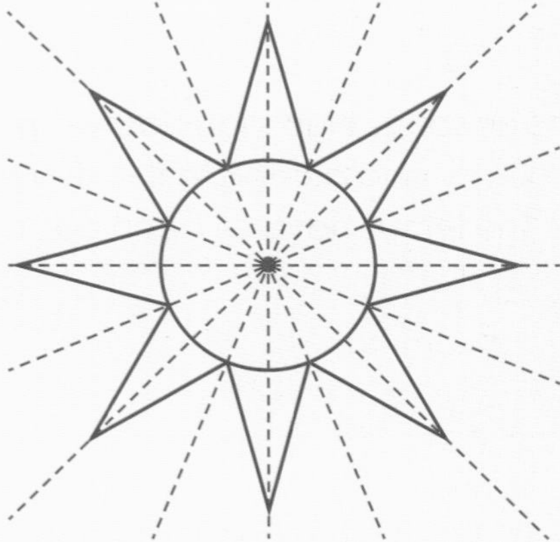

Remember:

- When looking for a line of symmetry try to imagine folding the shape.
- The reflection of each point is the same distance from the mirror line as the original point.

Skills practice A

1 Each of these road signs has a line of symmetry.
The dotted line is the line of symmetry.

i Copy and complete each road sign by making it symmetrical.

ii Find out what the road signs mean.

a

b

c

2 Copy and complete each symmetrical shape.

a

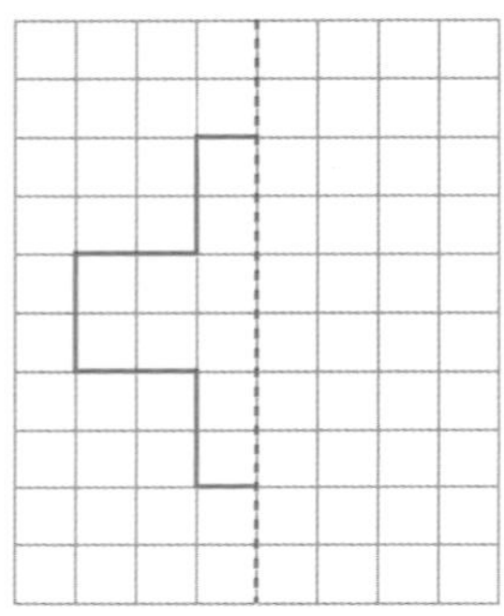

b

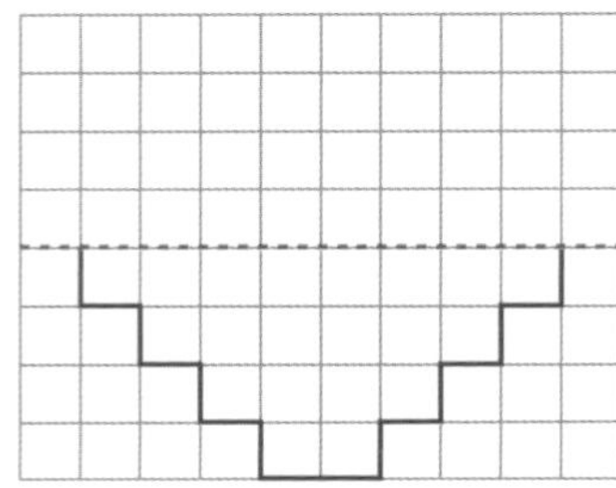

c

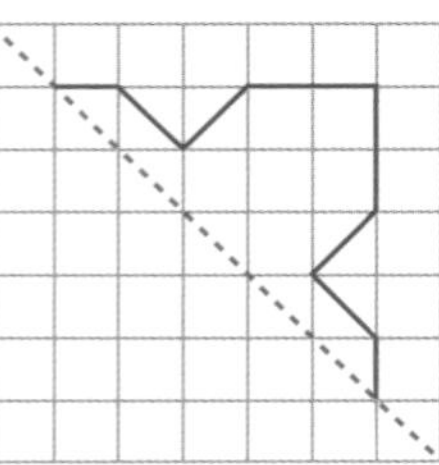

3 This sports club logo has two lines of symmetry.
Copy and complete the design.
The letters have all been put in for you.

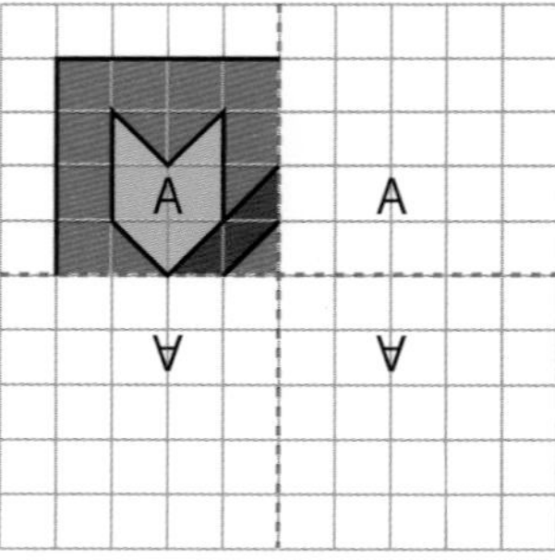

4 **i** What is the name of each of these shapes?

ii Copy the shapes and draw all the lines of symmetry.
How many lines of symmetry does each shape have?

a

b

c

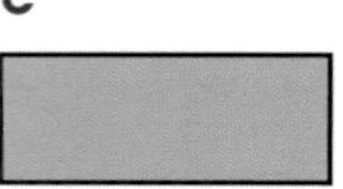

d

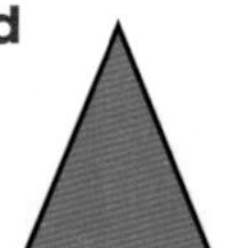

e

5 How many lines of symmetry does each leaf or flower have?

a

Holly leaf

b

Daisy

c

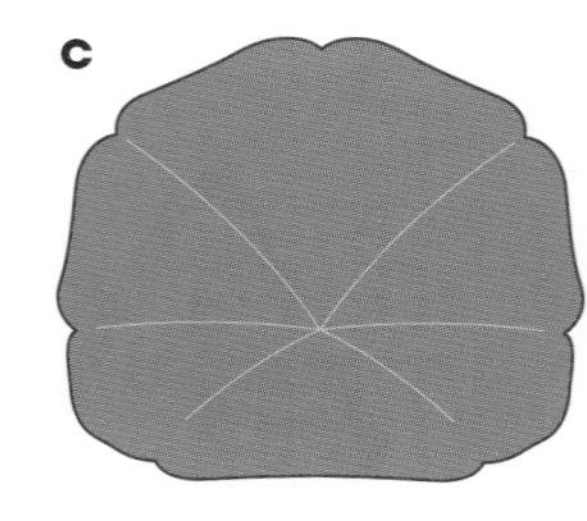

Nasturtium leaf

d

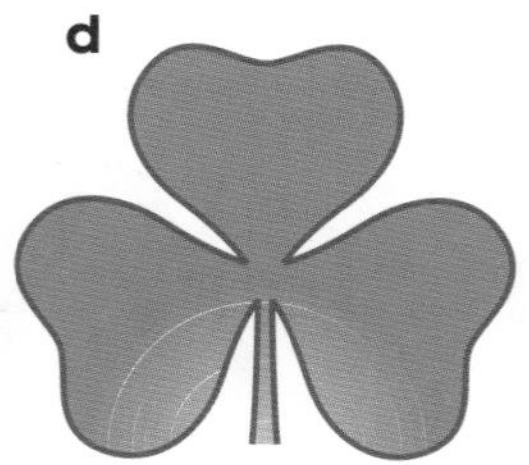

Clover leaf

6 This is a drawing of Mark's house.

a How many lines of symmetry does it have?

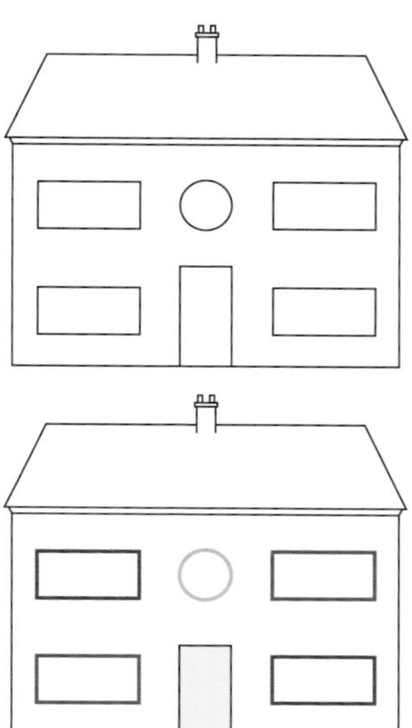

This is Mark's house after it was painted.

b How many lines of symmetry does it have now?

7 Here is a photograph of Vaux-le-Vicomte, a chateaux outside Paris in France.

Find three things about it which are *not* symmetrical.

Skills practice B

1 Which of these road signs have line symmetry?

a 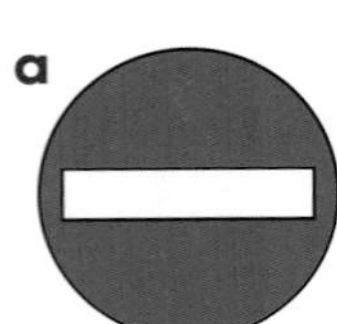

b 

c 

d 

e 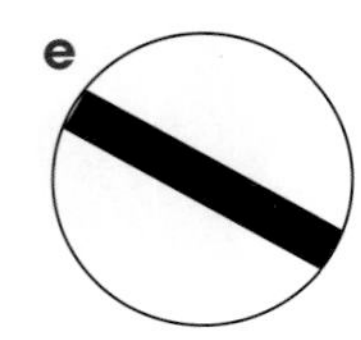

2 a Draw an isosceles triangle like this one.
   b Mark on its line of symmetry.
   c What does the line of symmetry tell you about the angles p and q?

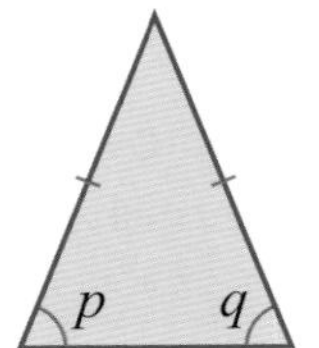

3 Copy or trace each of these shapes and draw their lines of symmetry.
Write down how many lines of symmetry each shape has.

a 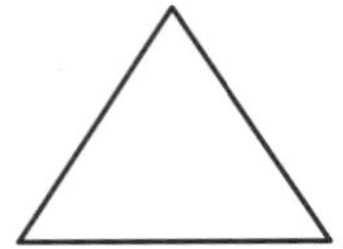

Equilateral triangle

b 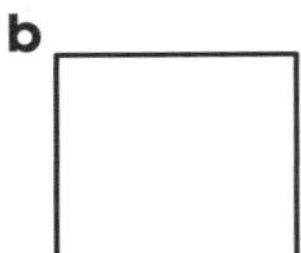

Square

c 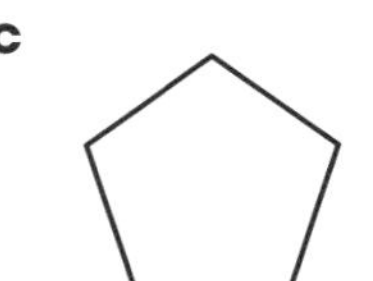

Regular pentagon

d 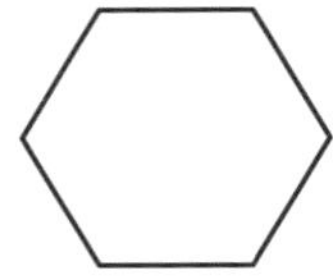

Regular hexagon

e 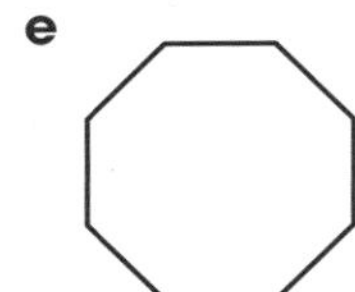

Regular octagon

4 Copy these capital letters carefully and draw their lines of symmetry.

A B C D E F G H

5 Copy each of these shapes on to squared paper.
Draw its reflection in the line or lines of symmetry shown.

a 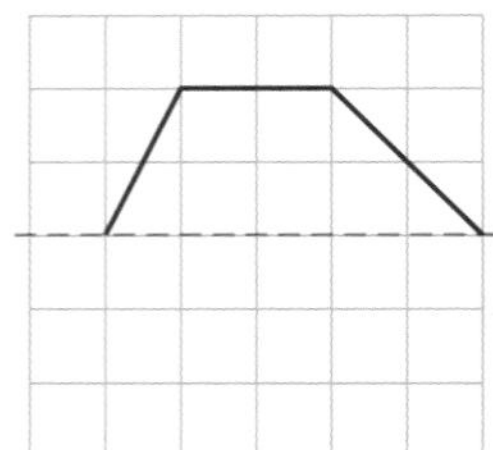

b 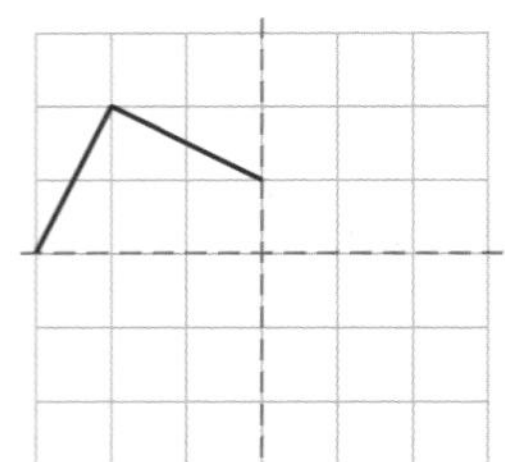

c 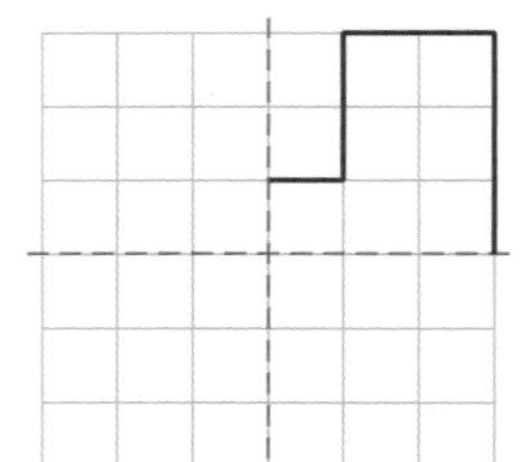

d 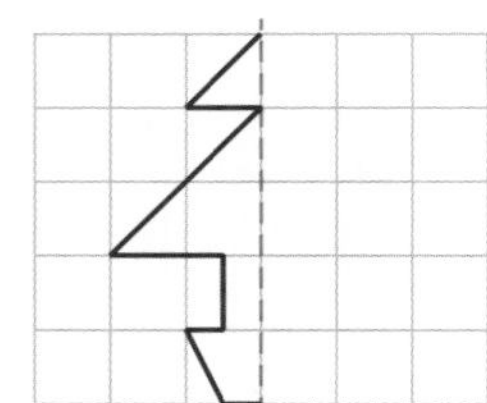

Reasoning

6 a This shape is almost symmetrical.
On a copy of the shape, add one square to make it symmetrical.

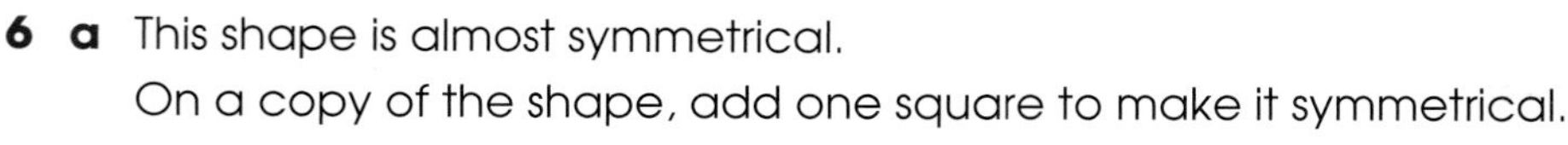

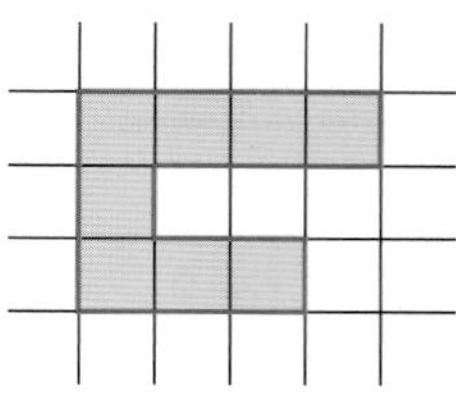

b i Add a square to a copy of each of these shapes to make them symmetrical.

Draw the line of symmetry each time.

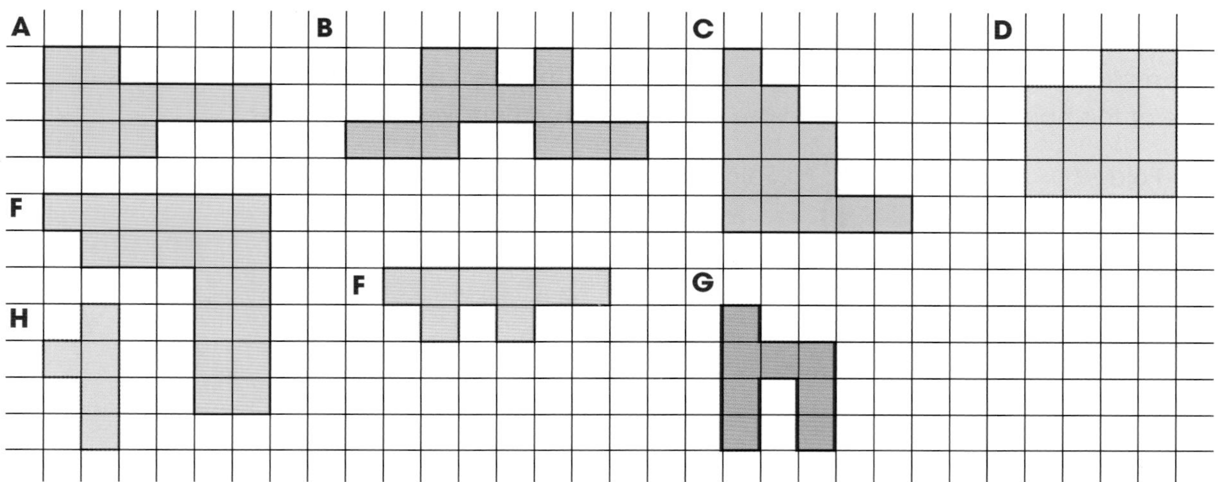

ii There are three different answers for one of the shapes. Which one?

iii Now make each of the shapes symmetrical by removing one square.

Reasoning

7 Peter folds a square piece of paper twice.
He then cuts out some shapes so that it looks like the diagram on the right.

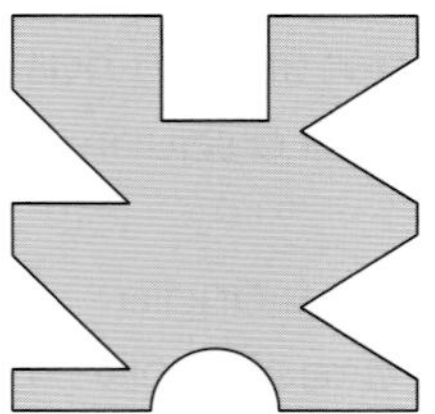

Which of the diagrams below show what the paper will look like when Peter opens it out again?

a **b** **c** **d**

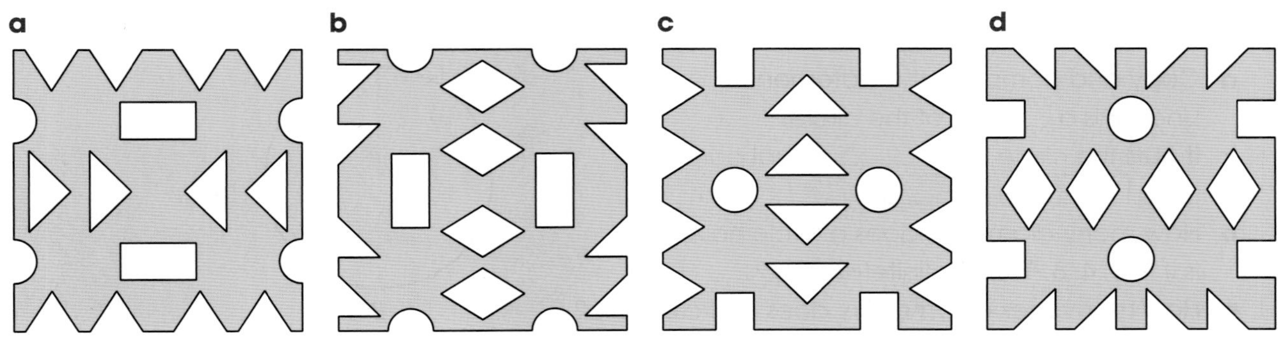

Reasoning

8 Here are six square sheets of paper, each one folded twice.

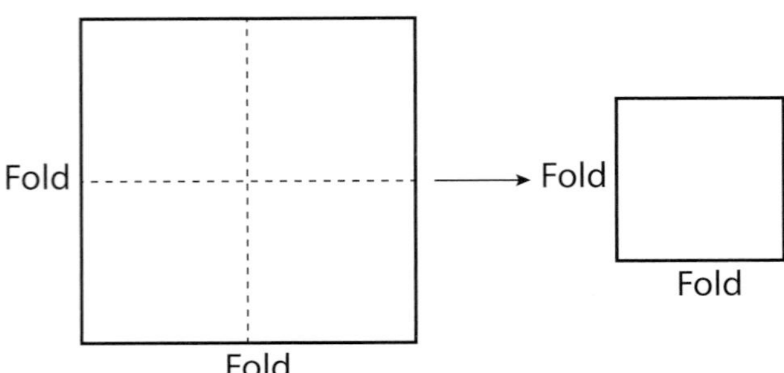

Peter makes cuts in them so that when the paper is unfolded each sheet contains holes. Some of the holes are circles, some are squares and some take other shapes.

A Fold Fold

B Fold Fold

C

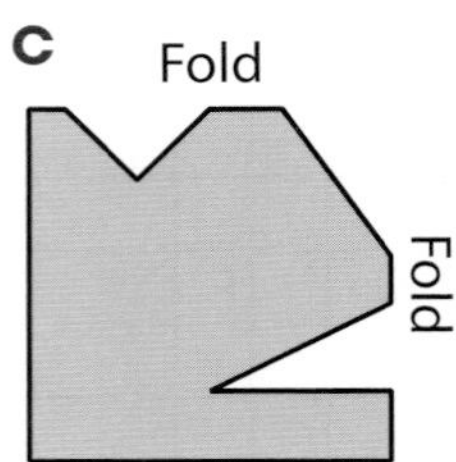

D Fold Fold

E

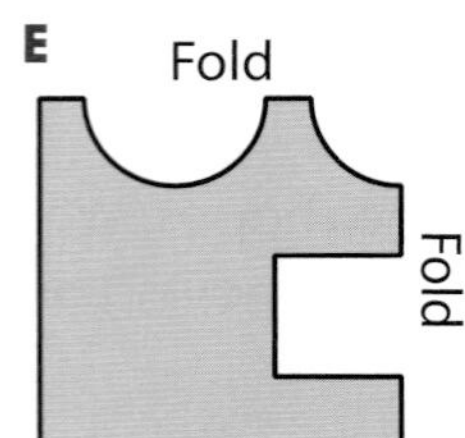

F

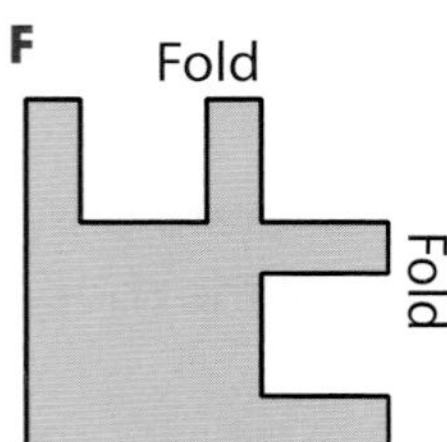

Which sheet of paper (if any) has three of its holes as follows:

a two rectangles and a square

b three circles

c two squares and a triangle

d two circles and a rectangle?

Wider skills practice

1 Look at this circle.

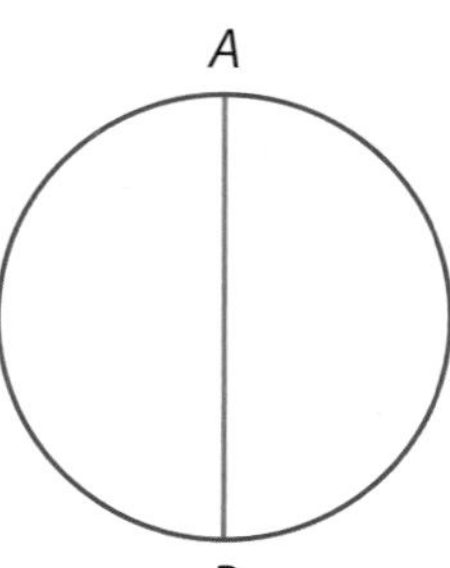

a Copy and complete this statement.
Line AB is a ☐ of the circle.
It cuts the circle into ☐ equal parts.
It is a ☐ of ☐.

b How many diameters does a circle have?
Is it 1, 2, 4, 8, 100 or infinitely many?

c How many lines of symmetry does a circle have?

Reasoning

2 Draw shapes that have

a at least one pair of parallel sides

b at least one pair of parallel sides and two pairs of equal angles

c at least one pair of parallel sides, two pairs of equal angles and one line of symmetry.

Applying skills

Reasoning

1 This tile pattern is common in the area around Cairo.

A Cairo tile is a pentagon with

- five equal sides
- two right angles
- one line of symmetry.

a Draw two different Cairo tiles.

b Using one of your tiles as a template, draw and colour your own version of the pattern.

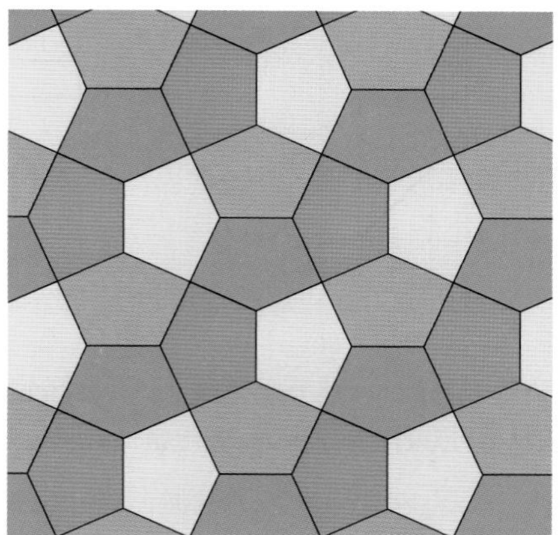

2 a Draw a 6 cm × 6 cm square.

Mark the four lines of symmetry with dotted lines as shown.

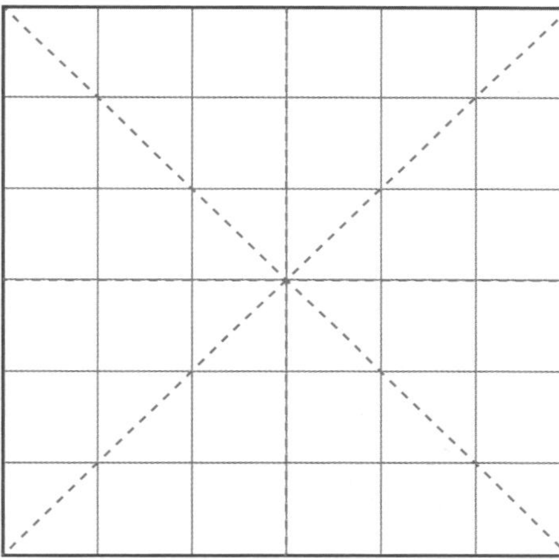

b Draw these four blue lines in one corner.

Reflect these lines in the horizontal line of symmetry, the vertical line of symmetry and the diagonal lines of symmetry, in order to produce a pattern with four lines of symmetry.

This is called a Rangok Pattern.

c Try one of your own.

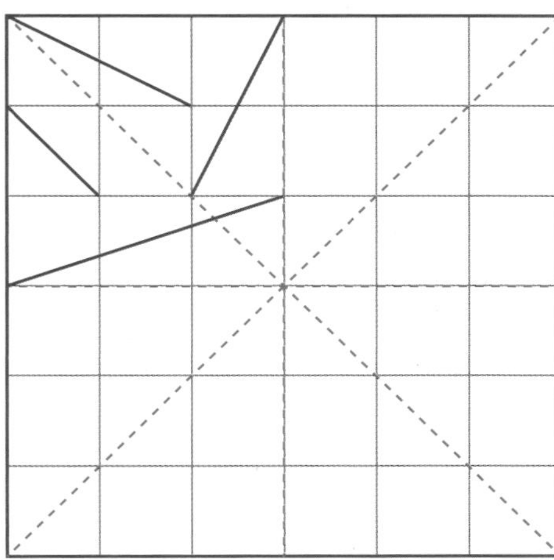

Reviewing skills

1 Copy each of these shapes on to squared paper.
Draw the reflections in the lines of symmetry.

a

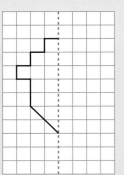

b

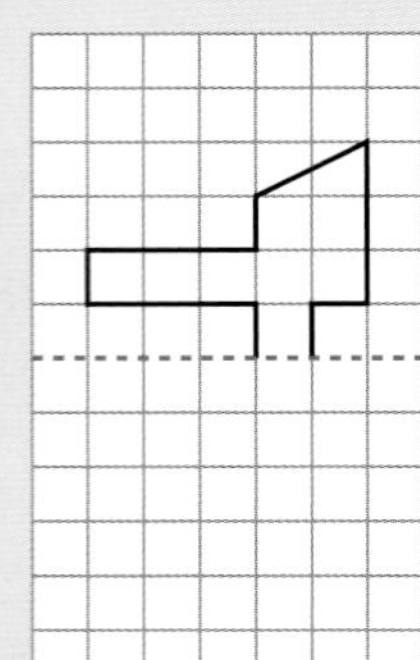

c

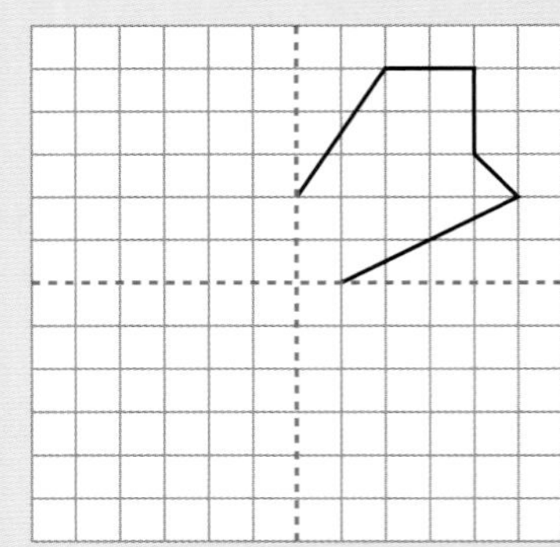

2 Which of these national flags have line symmetry?
Say how many lines of symmetry each flag has and their position.

a

Denmark

b

Jamaica

c

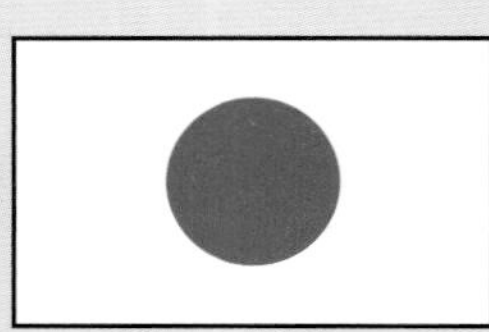

Japan

d

Norway

e

Bangladesh

f

Trinidad and Tobago

g

Botswana

3 Copy each of these part-shapes on to squared paper.
Complete them by reflecting in the lines of symmetry.

a

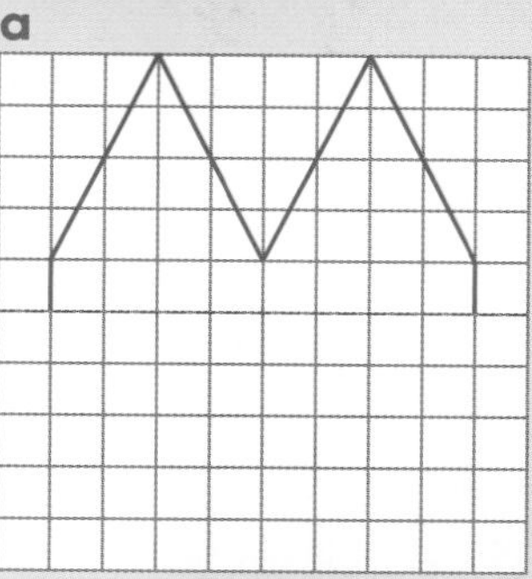

b

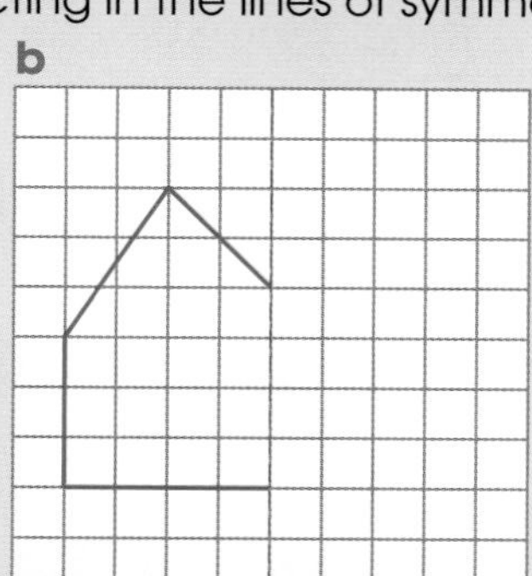

c

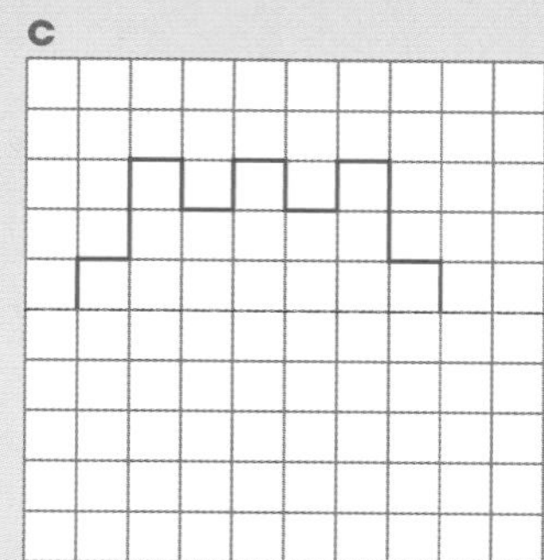

d

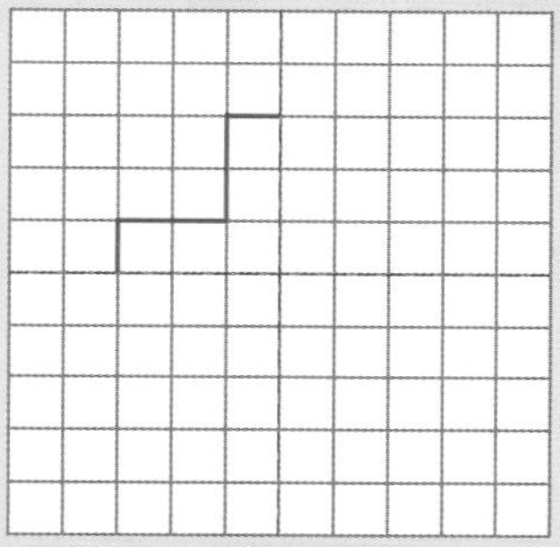

e

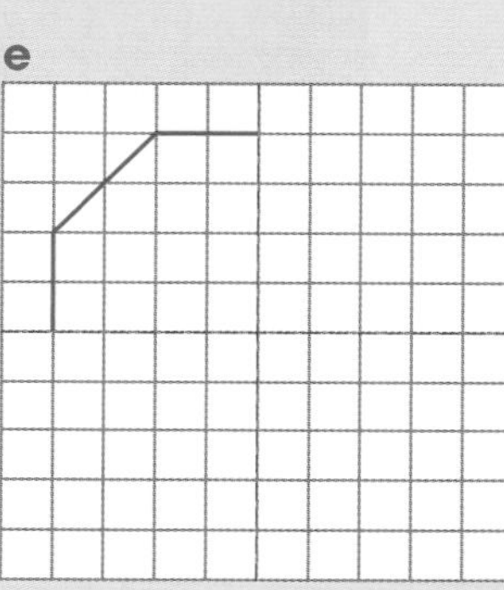

f

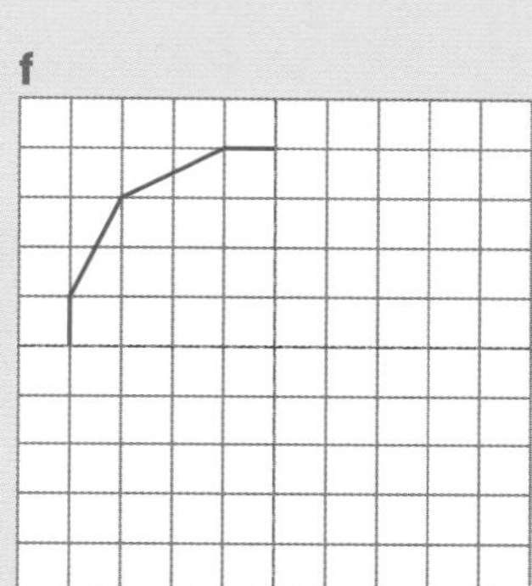

4 Copy the figures which have lines of symmetry.
Draw the lines of symmetry.

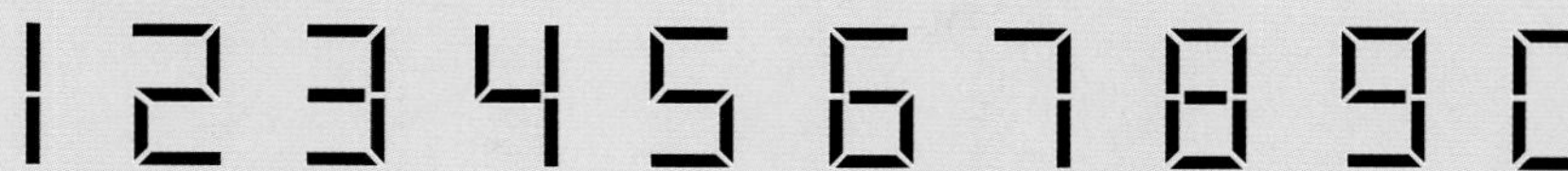

Building skills

Example outside the Maths classroom

Launch pads

Toolbox

An angle is a turn. It is usually measured in degrees.
Angles on a straight line add up to 180°.

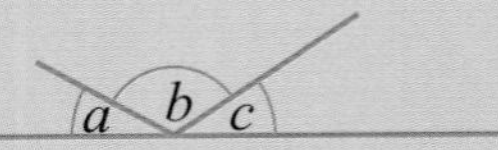

Angles around a point add up to 360°.

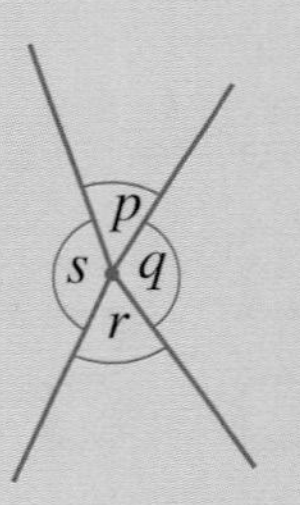

Vertically opposite angles are formed where two lines cross.
Vertically opposite angles are equal.

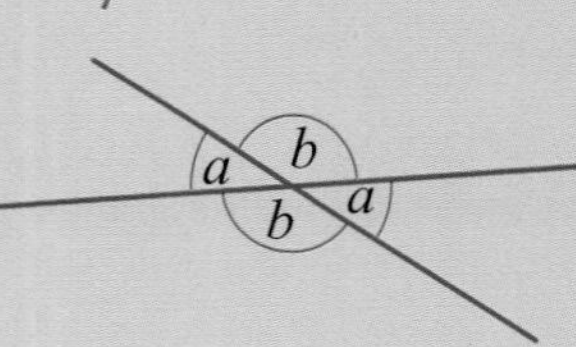

Example – Finding an angle on a straight line

Work out the size of angle b.

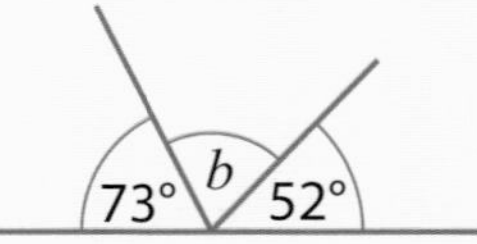

Solution

Angles on a straight line add up to 180°.

$$b + 73 + 52 = 180$$
$$b + 125 = 180$$
$$b = 180 - 125$$
$$b = 55°$$

Example – Finding an angle at a point

Work out the size of angle h.

105°
h
85°
110°

Solution

Angles at a point add up to 360°.

$$h + 105 + 110 + 85 = 360$$
$$h + 300 = 360$$
$$h = 360 - 300$$
$$h = 60°$$

Example – Finding angles when two lines cross

Find the size of the lettered angles in this diagram.

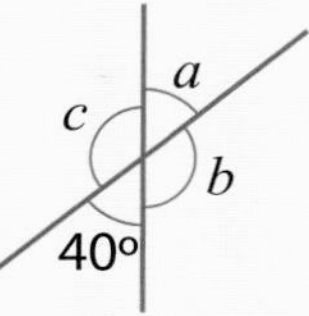

Solution

Vertically opposite angles are equal.

$a = 40°$

$b + 40° = 180°$ ← **Angles on a straight line add up to 180°**

$b = 180° - 40° = 140°$

$c + 40° = 180°$ ← **Angles on a straight line add up to 180°**

$c = 180° - 40° = 140°$ ← **Alternatively, since c is opposite b you can say that $c = 140°$ without doing the calculation twice.**

Remember:

- A full turn is 360°.
- Half a turn, a straight line, is 180°.
- Vertically opposite angles are equal.
- Missing angles can be worked out by thinking about how much more turn is needed.

Skills practice A

1 Calculate the size of each lettered angle.

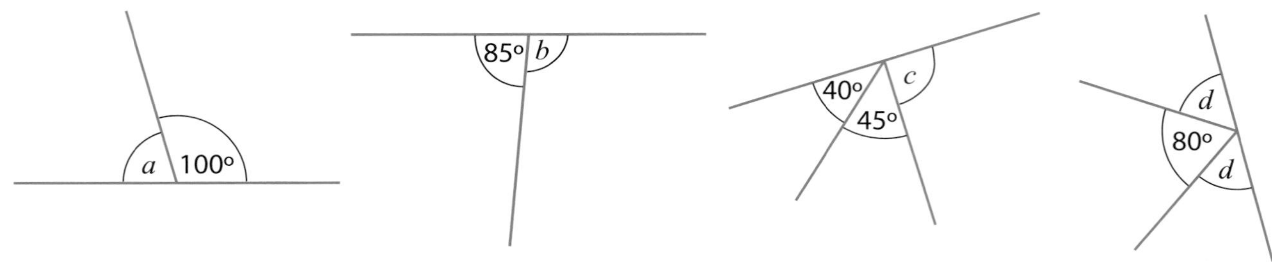

2 Calculate the angle between the spokes on this car wheel.

3 A pipe is bent through 45°.
What is angle x?

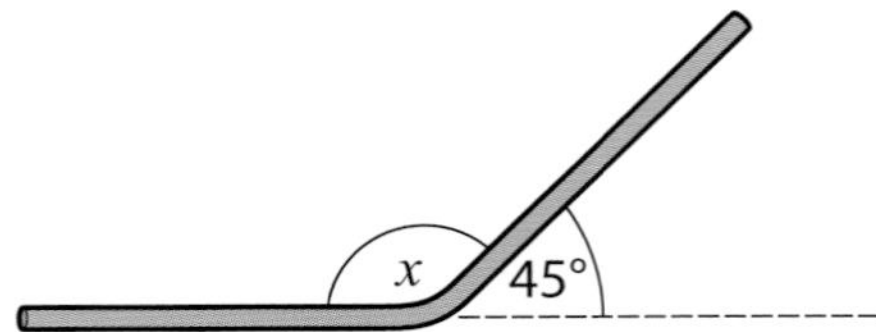

4 Copy these statements and fill in the missing words or numbers.

4 right angles make a ☐ ☐.
4 × 90° = ☐.
A whole turn is ☐ degrees.

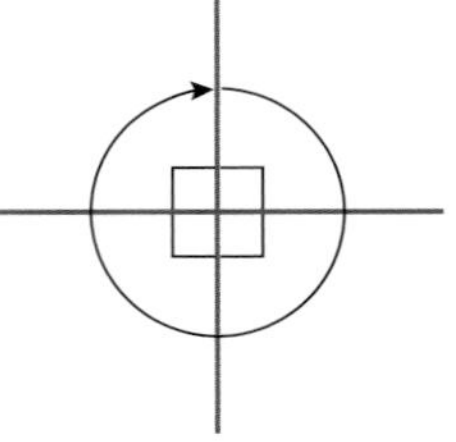

5 Calculate the size of the lettered angles.

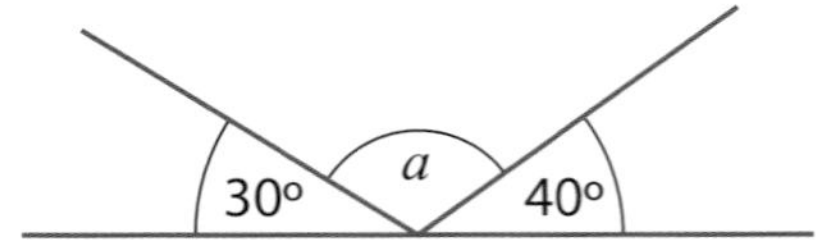

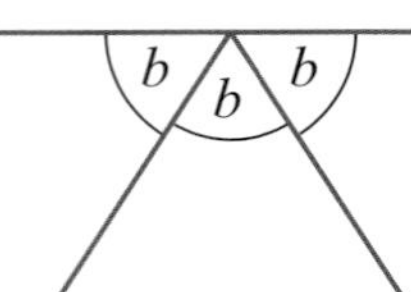

6 In this diagram, angle $p = 30°$.

a Use this to find angle q.

b Use your answer to part **a** to find angle r.

c Use your answer to part **b** to find angle s.

d Show that $p + s = 180°$.

e State which angles are equal.

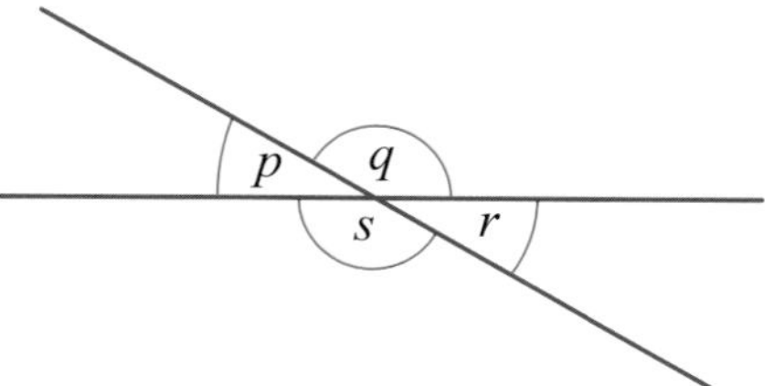

Skills practice B

1 Find the size of each lettered angle in these diagrams.

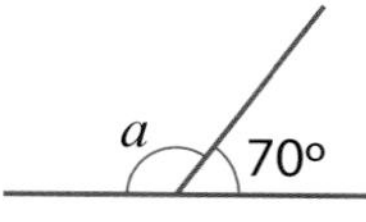

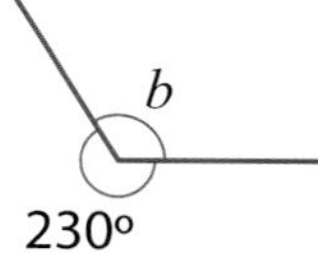

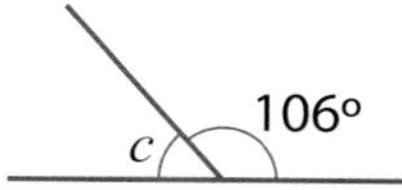

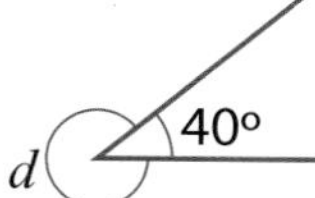

e 100° 145°

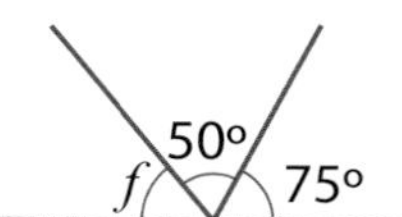

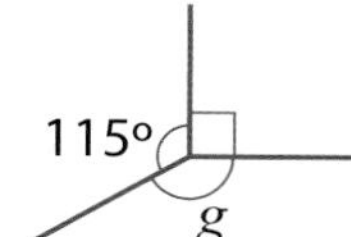

105° h 85° 110°

2 This water wheel has 20 blades.
Work out the angle between each blade.

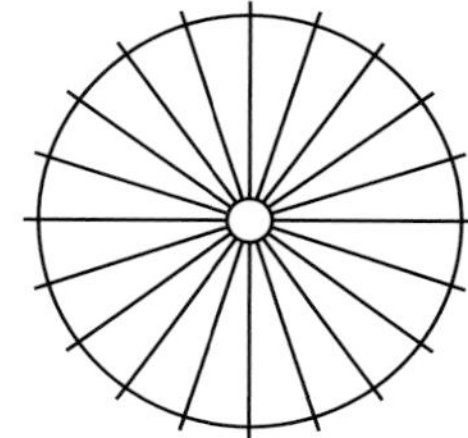

3 Work out the angle between the hands of a clock at

a 1 o'clock

b 8 o'clock.

4 Find the size of each lettered angle in these diagrams.

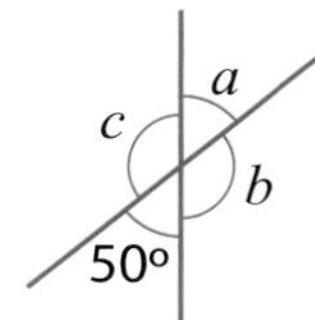

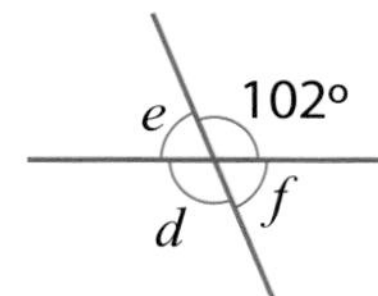

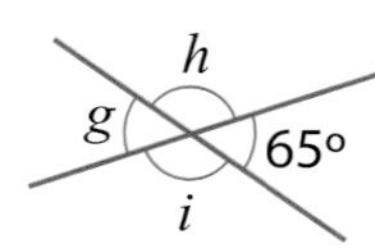

5 For each of these diagrams, calculate the size of the lettered angle.
Explain your answer.

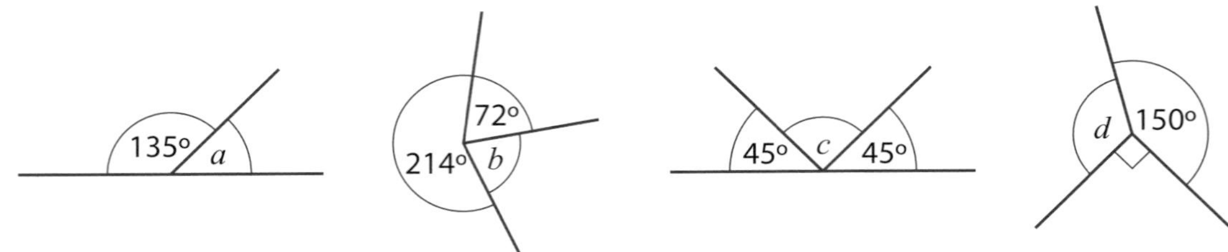

6 Calculate the size of each lettered angle.
Explain your answer.

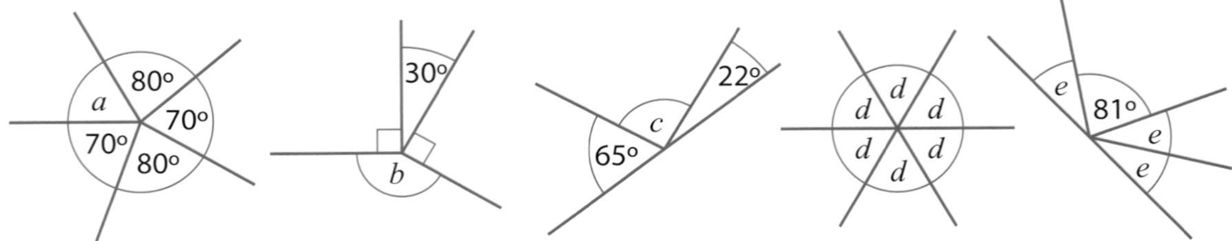

7 a, b, c, d, e and f are six angles.
Use these diagrams to work out their sizes.

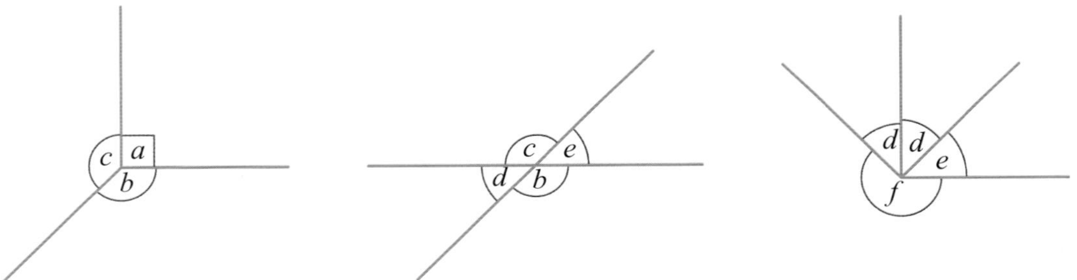

Wider skills practice

Reasoning

1 James is programming a robot. It can take steps forward and turn clockwise.
His instructions are

FWD 10
TURN 245°
FWD 8
TURN 100°
FWD 11
TURN 120°
FWD 9

James now wants the robot to face in the same direction as it was facing originally.
What should his last instruction be?

Reasoning

2 A dealer sells cars in three colours.
Look at this pie chart.
The number of green cars is twice the number of blue cars.
The number of red cars is three times the number of blue cars.
Calculate each lettered angle.

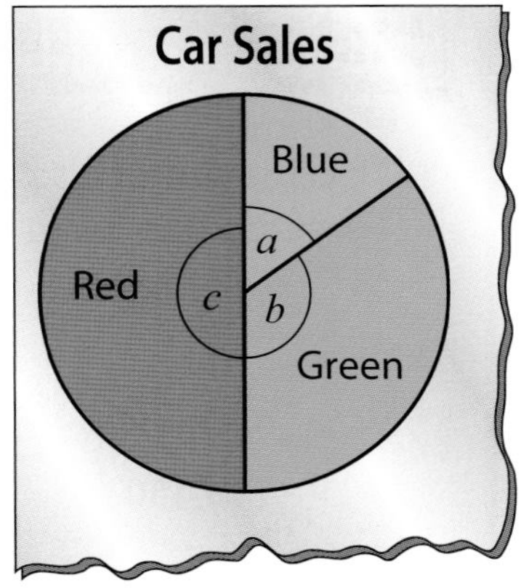

Applying skills

Problem solving

1 Here is a design for a new London Eye.
a It has 36 viewing capsules.
What is the angle made at the centre by two adjacent capsules?
b The wheel takes 30 minutes to complete a turn.
One capsule is at the highest point at exactly 12 noon.
How many seconds is it until the next capsule is at the highest point?

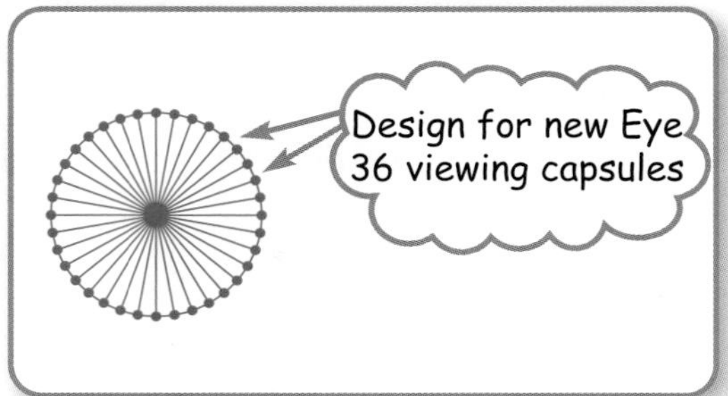

Problem solving

2 A metronome produces a steady beat for music.
These diagrams show the swings for different tempos (speeds).

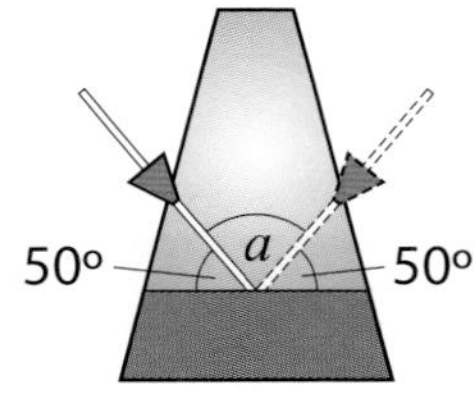

Moderato
116 beats per minute

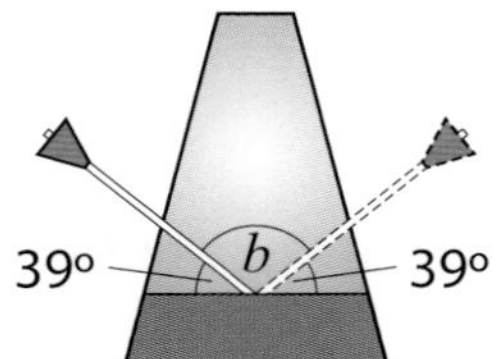

Largo
50 beats per minute

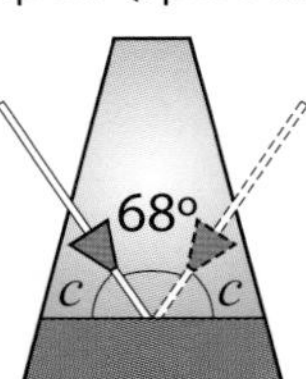

Presto
208 beats per minute

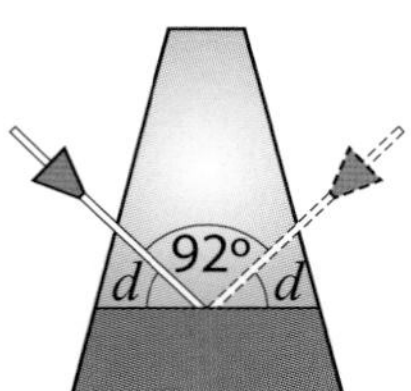

Adagio
72 beats per minute

a Calculate the size of each lettered angle.
b Write down the tempos in order of speed, slowest to fastest.

Reviewing skills

1 Find the size of each lettered angle in these diagrams.
For each one, write down the angle fact that you use.

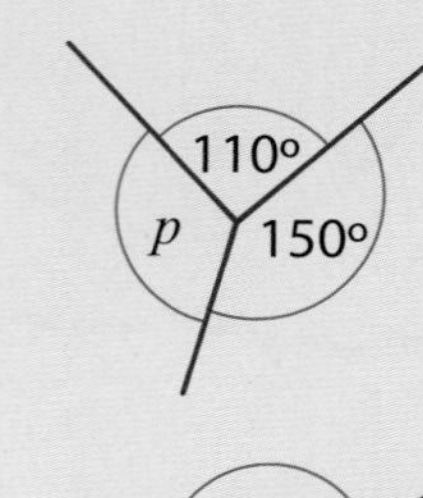

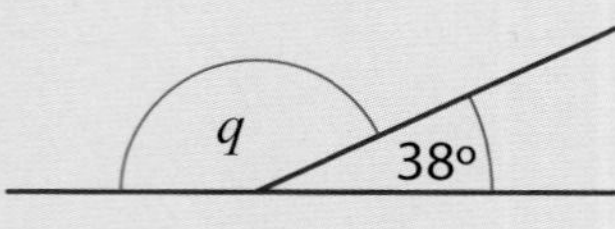

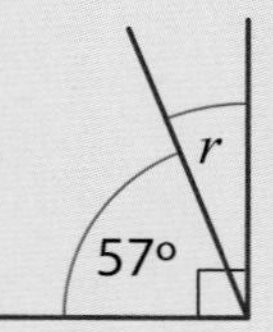

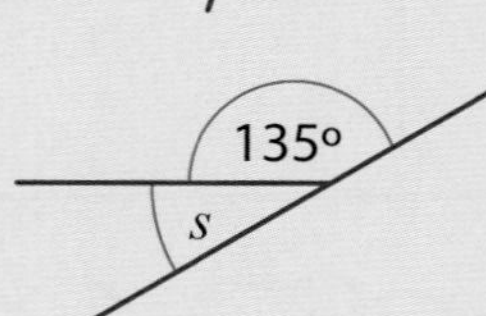

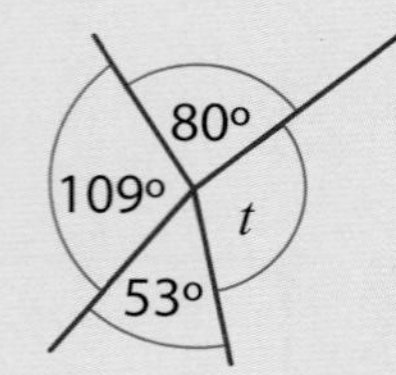

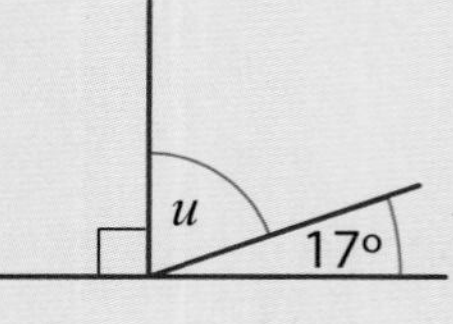

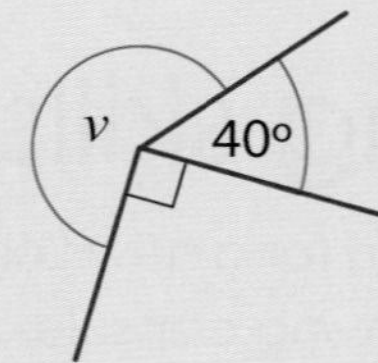

2 **a** Find angles a, b and c in this diagram.

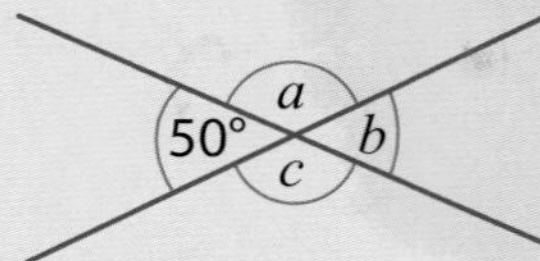

b Find angles q and r in this diagram.

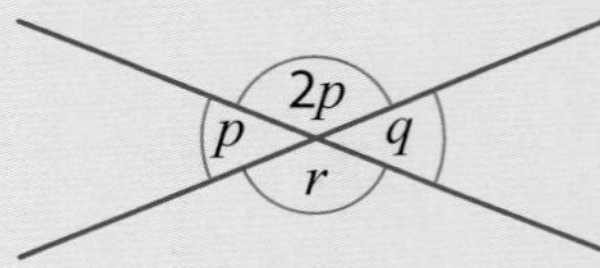

Unit 4 • Rotational symmetry • Band d

Building skills

Example outside the Maths classroom

Ferris wheel

Toolbox

The **order of rotational symmetry** is the number of times that a shape will fit on to itself in one complete turn.

The **centre of rotational symmetry** is the point about which the shape has to be rotated in order for it to fit on to itself.

This is usually just called the centre of rotation.

If a shape fits on to itself only once in one turn then it has rotational symmetry of **order 1**. This means it has **no** rotational symmetry.

When identifying the order of rotational symmetry it can be helpful to use tracing paper. Trace the outline of the shape and then rotate the tracing paper over the original image.

Example – Identifying rotational symmetry

What is the order of rotational symmetry of each of these shapes?

a

b

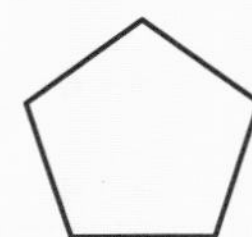

Solution

a The rectangle will fit on to itself two times in one turn so the order is 2.

b The pentagon will fit on to itself five times in one turn so the order is 5.

Example – Completing shapes with rotational symmetry

Look at the sign.

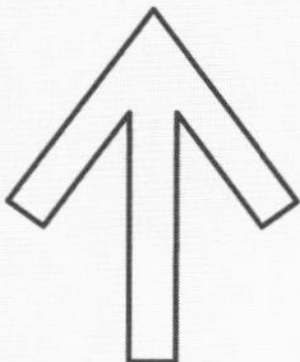

a What is the order of rotational symmetry of the sign?

b How can you change the sign so that it has rotational symmetry of order 2?

Solution

a It has no rotational symmetry so the order is 1.

b

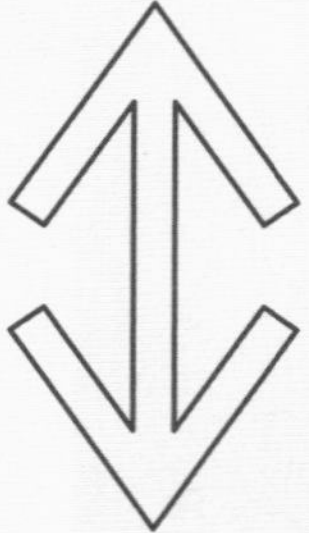

Remember:

- The order of rotational symmetry is how many times a shape will fit on to itself in one turn.
- A shape with rotational symmetry of order 1 has no rotational symmetry.

Skills practice A

1 Which of these dominoes has rotational symmetry?

a

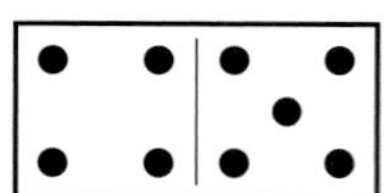

b

c

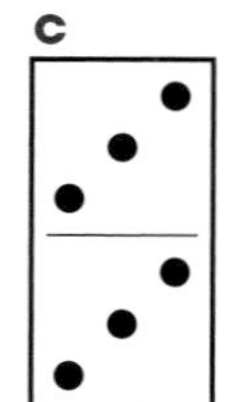

d

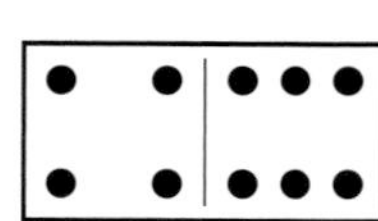

e

2 Write down the order of rotational symmetry of each of these shapes.

a  **b** 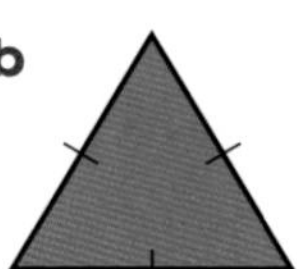 **c**  **d** 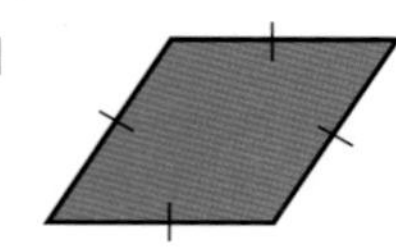

e **f**  **g** 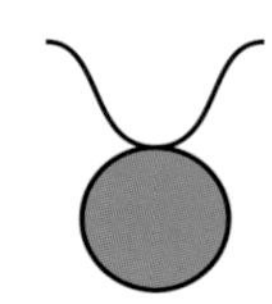 **h** 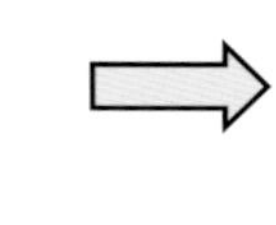

3 Write down the order of rotational symmetry of each of these shapes.

a **b**  **c**  **d**  **e** 

4 Copy each of these shapes carefully.

a  **b** 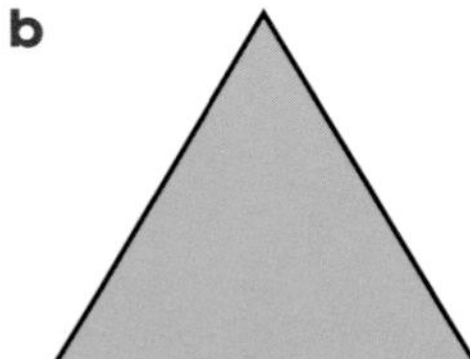 **c** 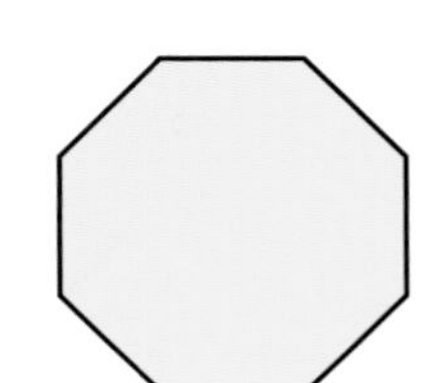 **d** 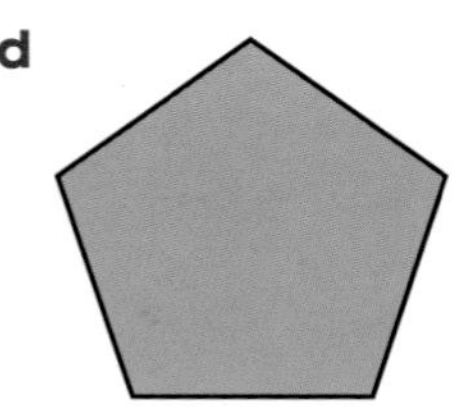

i Write down the name of each shape.

ii Write down the order of rotational symmetry of each shape.

iii Mark the centre of rotation of each shape with a cross.

5 These are car hub caps.

a  **b** 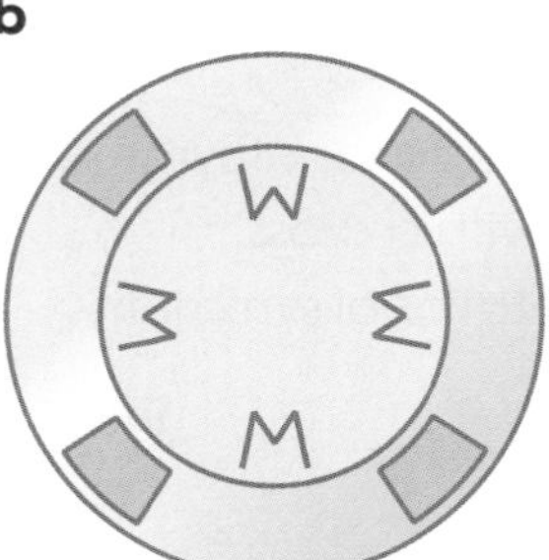 **c** 

i Write down the order of rotational symmetry of each hub cap.

ii Draw a sketch of each hubcap.
Mark the centre of rotational symmetry of each hub cap with a cross.

Reasoning

6 Is it always true, sometimes true or never true that a shape with rotational symmetry of order 3 will also have reflection symmetry of order 3?

Write a paragraph explaining your answer.

If you think *always*, explain how you can be so certain.

If you think *never*, explain how you can be so certain.

If you think *sometimes*, explain when it is and when it isn't true.

Skills practice B

1 Some of the shapes below have line symmetry, some have rotational symmetry and some have both.

Describe fully the symmetry of each shape.

a

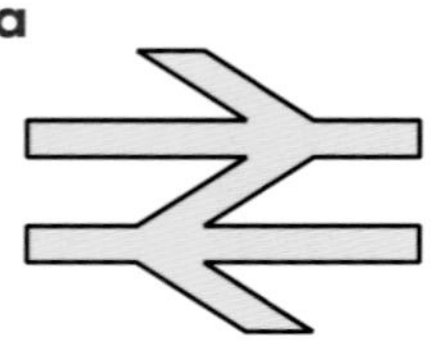

b

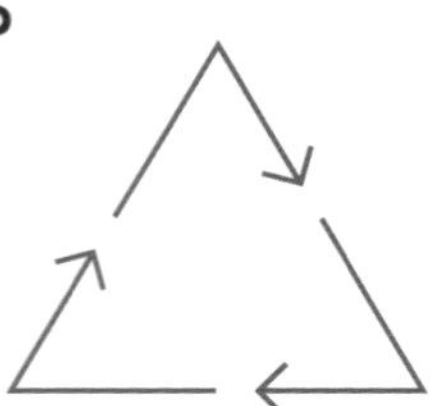

c

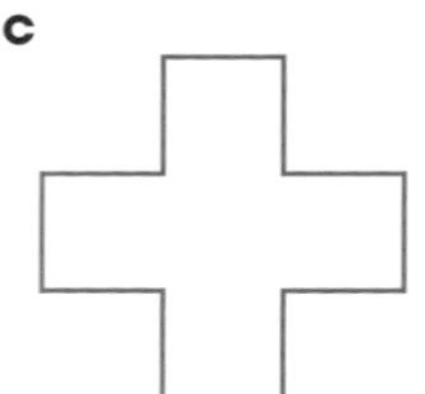

d

2 Here are parts of shapes which have two lines of symmetry.

i Copy them and draw in the rest of the shape.

ii Write down the order of rotational symmetry of each shape.

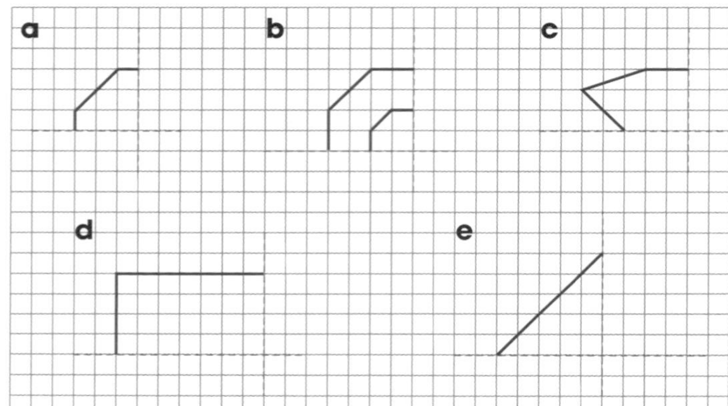

3 **a** For each shape, write down the order of rotational symmetry.

b Sketch the shape and mark the centre with a cross.

c Explain how you found the centres of rotational symmetry.

i

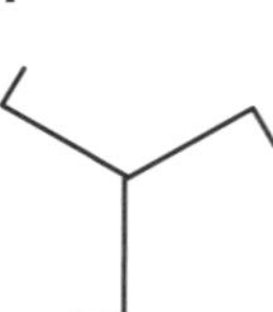

ii

iii

iv

4 i Copy and complete these shapes so that they have both line and rotational symmetry. Mark the centre of rotational symmetry with a cross.

ii For each one write down how many lines of symmetry it has and the order of rotational symmetry.

a

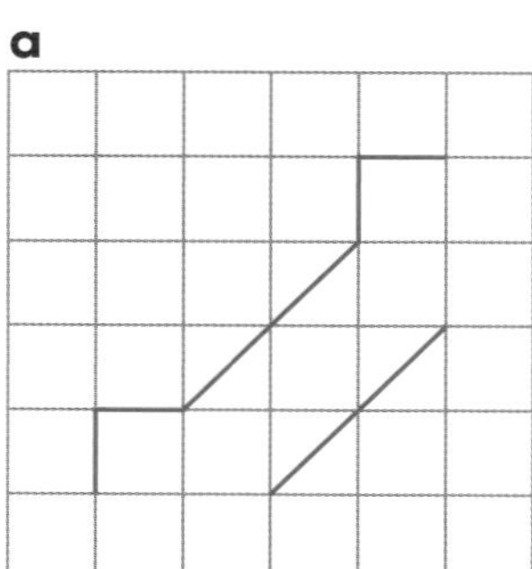

b

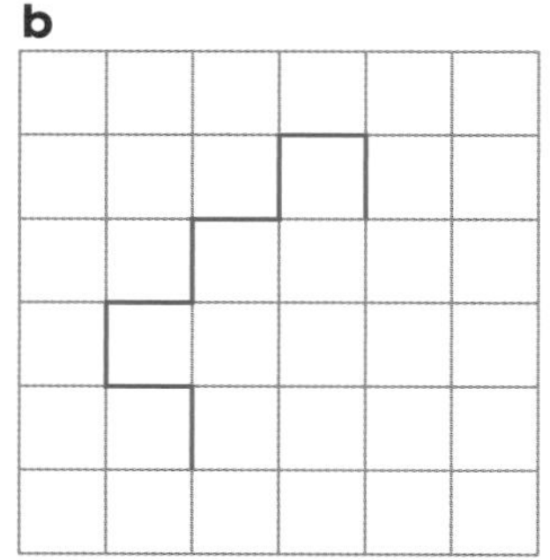

Reasoning

5 Copy this table, Draw a shape to fit into each section of this two-way table.

	Has rotational symmetry	Has no rotational symmetry
Has at least one line of symmetry		
Has no lines of symmetry		

Wider skills practice

1 a The diagram shows the face of a concrete block. Describe the symmetry of this shape.

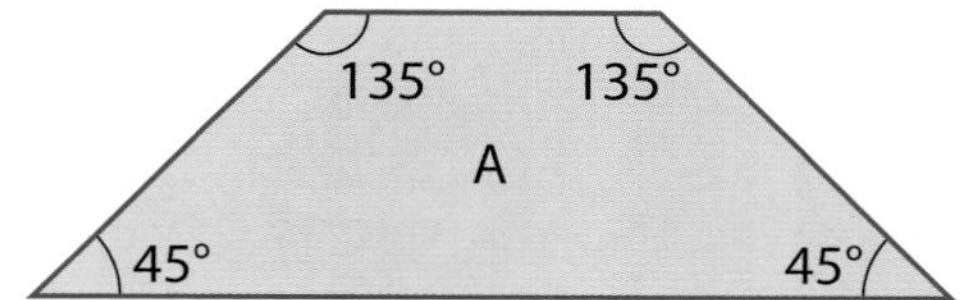

b Four of these concrete blocks are fitted together. They make this square shape with a square hole. Draw sketches of the three stages of the construction. In each case, give a full description of the symmetry.

i Blocks A and B only in place.

ii Blocks A, B and C are in place but not block D.

iii All four blocks, A, B, C and D, are in place.

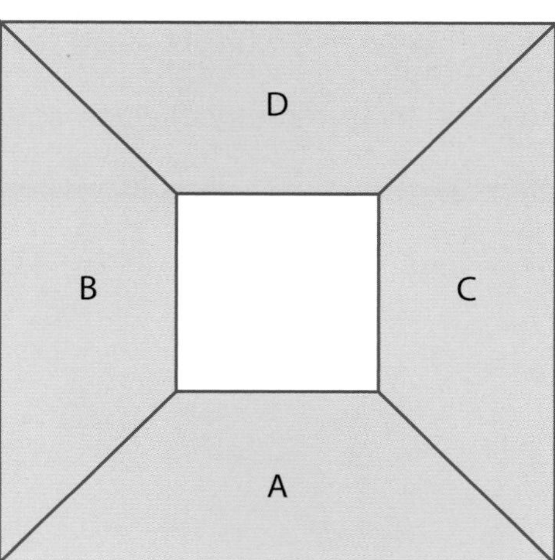

Applying skills

1 Snow is made when water vapour freezes to form crystals.
The crystals are usually shaped as regular hexagons.
They combine to form snowflakes with rotational symmetry of order 6.
Here is a snowflake drawn on isometric paper.
Design five more snowflakes.
Make them as large or as small as you like.
They can be very intricate, but remember that each one must have rotational symmetry of order 6.

Reviewing skills

1 Write down the order of rotational symmetry of these patterns.

a **b** **c** **d**

2 Look at these road signs.

a **b** **c** **d** **e** **f**

Some have line symmetry, some have rotational symmetry, some have both and some have no symmetry.
Describe the symmetry of each sign.

Unit 5 • Angles in triangles and quadrilaterals • Band e

Building skills

Example outside the Maths classroom

Chinese puzzles

Toolbox

- The three interior angles at the vertices in a triangle add up to 180°.

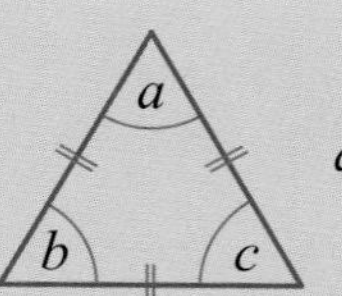

$a + b + c = 180°$

- The four interior angles at the vertices in a quadrilateral add up to 360°.

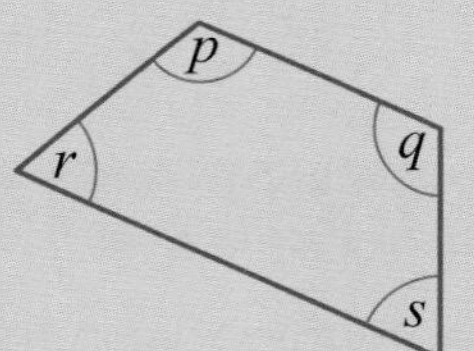

$p + q + r + s = 360°$

- The exterior angle of a triangle equals the sum of the two opposite interior angles.

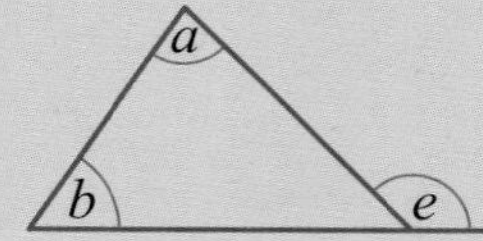

$e = a + b$

Example – Finding angles in triangles

Find the size of each unknown angle in these triangles.

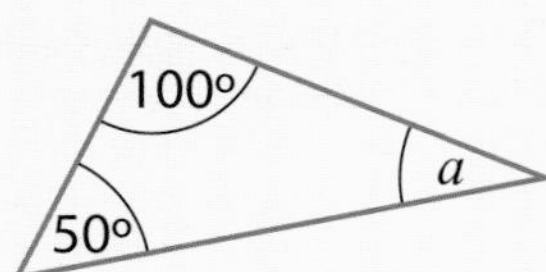

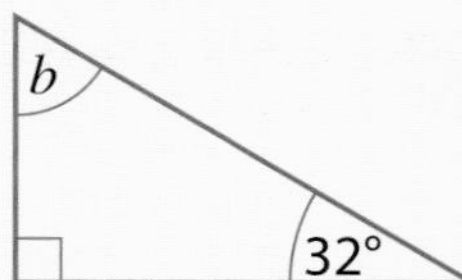

Solution

Angles in a triangle add up to 180°.

$a + 50 + 100 = 180$

$a + 150 = 180$

$a = 180 - 150$

$a = 30°$

Angles in a triangle add up to 180°.

$b + 90 + 32 = 180$

$b + 122 = 180$

$b = 180 - 122$

$b = 58°$

Example – Finding angles in quadrilaterals

Find the size of each unknown angle in these quadrilaterals.

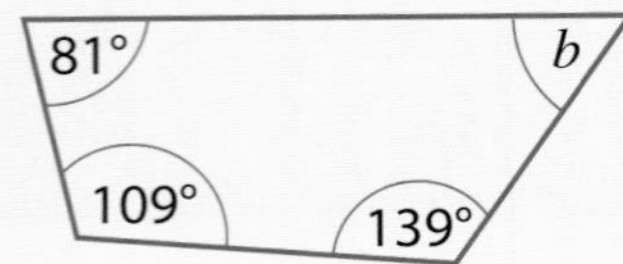

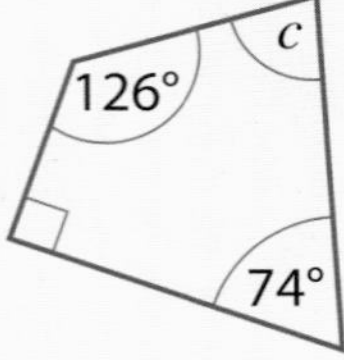

Solution

Angles in a quadrilateral add up to 360°.

$b + 81 + 109 + 139 = 360$

$b + 329 = 360$

$b = 360 - 329$

$b = 31°$

Angles in a quadrilateral add up to 360°.

$c + 74 + 90 + 126 = 360$

$c + 290 = 360$

$c = 360 - 290$

$c = 70°$

Example – Finding unknown angles

Find the size of each unknown angle in these diagrams.

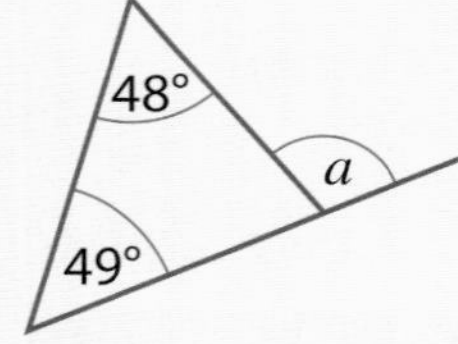

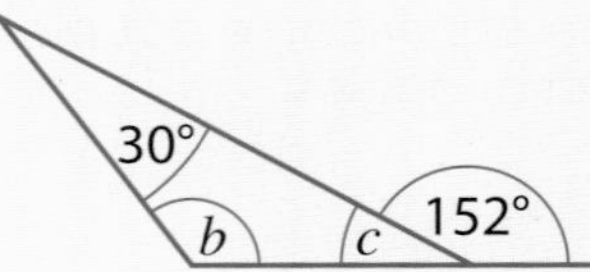

Solution

The exterior angle of a triangle equals the sum of the two opposite interior angles.

$a = 48 + 49$

$a = 97°$

The exterior angle of a triangle equals the sum of the two opposite interior angles.

$b + 30 = 152$

$b = 152 - 30$

$b = 122°$

Angles on a straight line add up to 180°.

$c + 152 = 180$

$c = 180 - 152$

$c = 28°$

Remember:

- Angles in a triangle add up to 180°.
- Angles is an equilateral triangle are all the same. They are always 60°.
- Isosceles triangles always have two angles the same.
- The exterior angle of a triangle equals the sum of the two opposite interior angles.

Skills practice A

1 Find the size of each lettered angle in these diagrams.

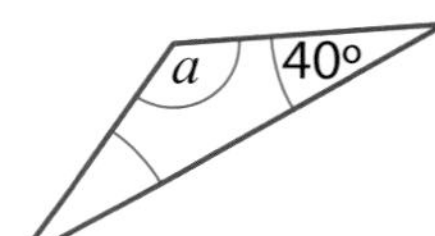

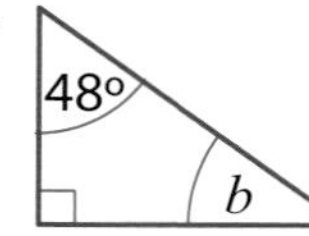

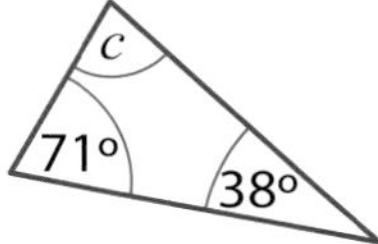

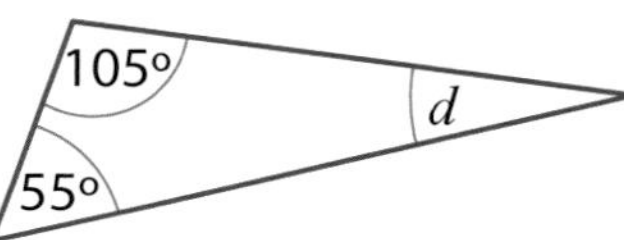

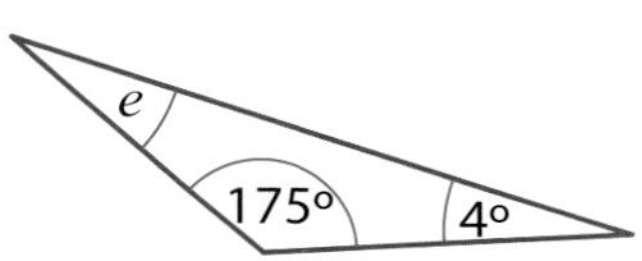

2 Find the size of each lettered angle in these diagrams.

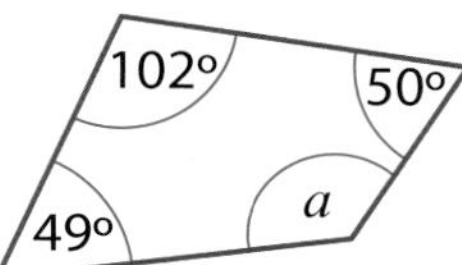

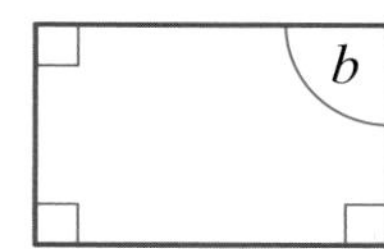

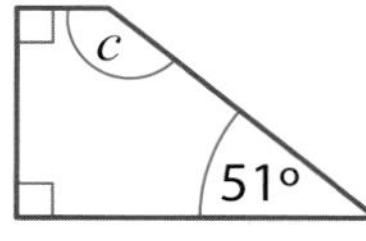

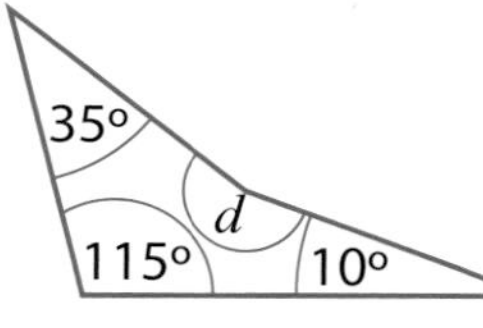

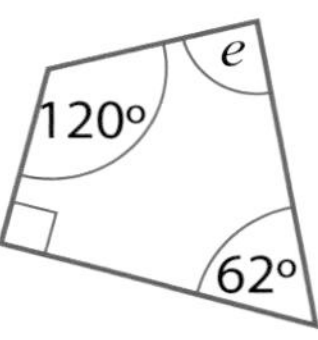

3 Find the size of each lettered angle in these diagrams.

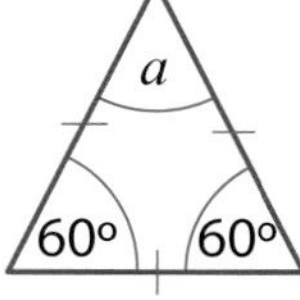

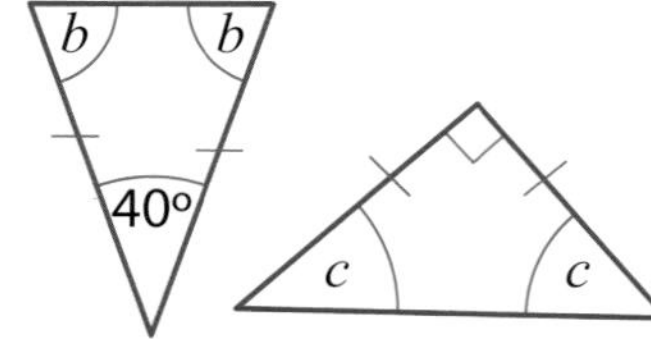

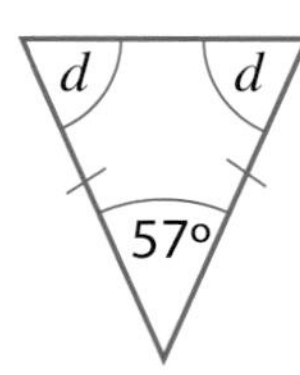

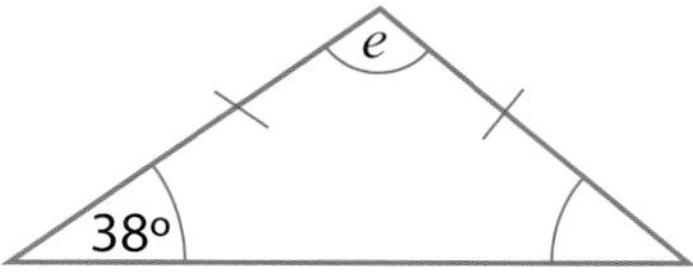

4 **a** Calculate the size of angle a.

b Calculate the size of angle b.

c Add angle b to 70°.
What do you notice?

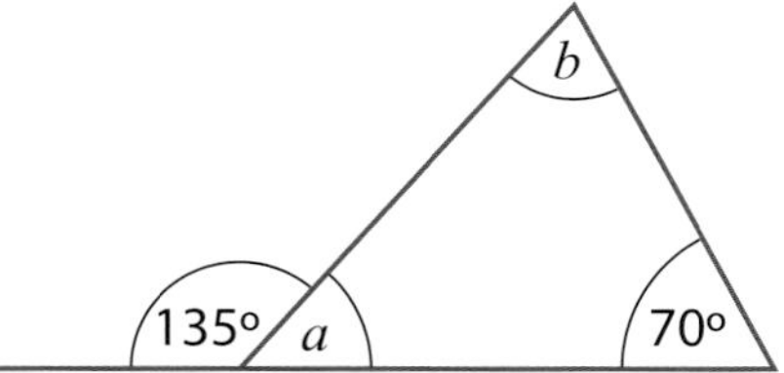

5 Find the size of each lettered angle in these diagrams.

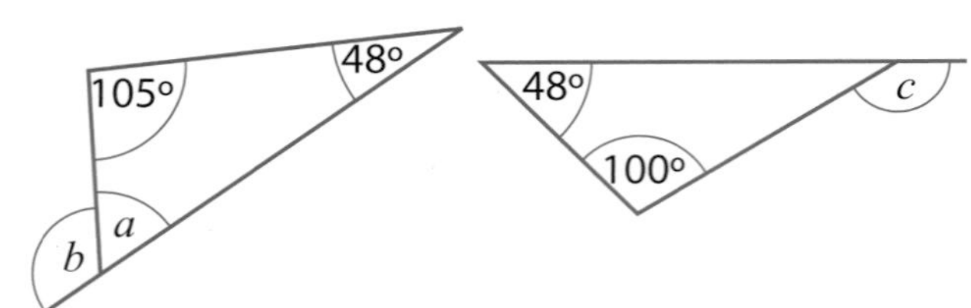

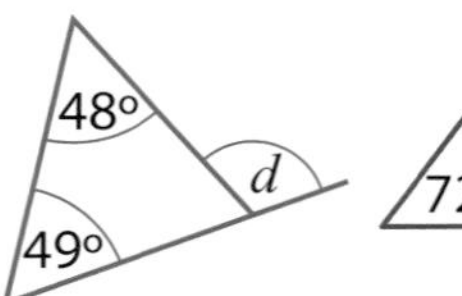

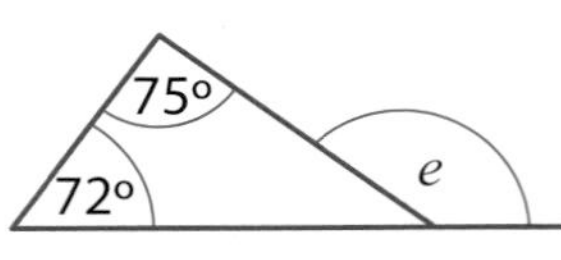

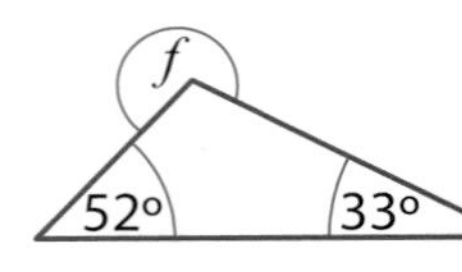

6 Find the size of each lettered angle in this diagram.

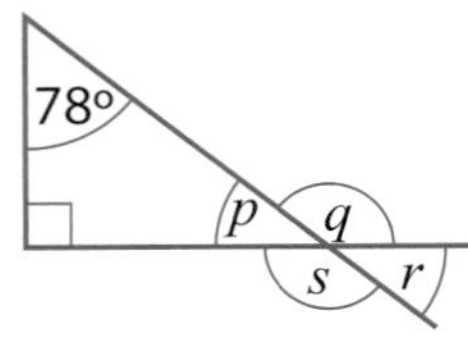

7 The diagram represents a ladder leaning against a wall. Calculate the size of the acute angle between the wall and the ladder.

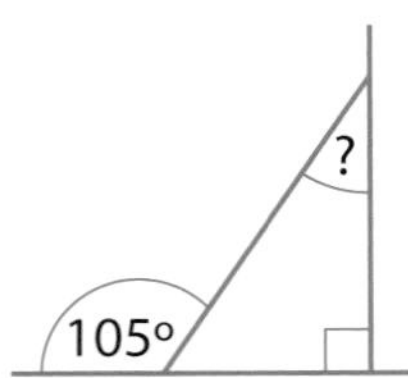

8 The diagram represents a tent. Calculate the size of the angle at the apex of the tent.

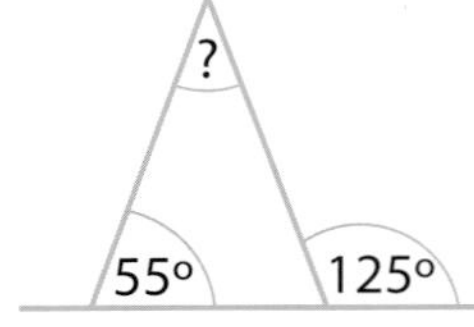

Reasoning

9 Is it always true, sometimes true or never true that a triangle has three acute angles?

Explain your answer.

If you think *always*, explain how you can be so certain.

If you think *never*, explain how you can be so certain.

If you think *sometimes*, explain when it is and when it isn't true.

Skills practice B

1 For safety, the angle this ladder makes with the horizontal ground must be between 65 and 75 degrees.

Between what values must the acute angle between the vertical wall and the ladder be?

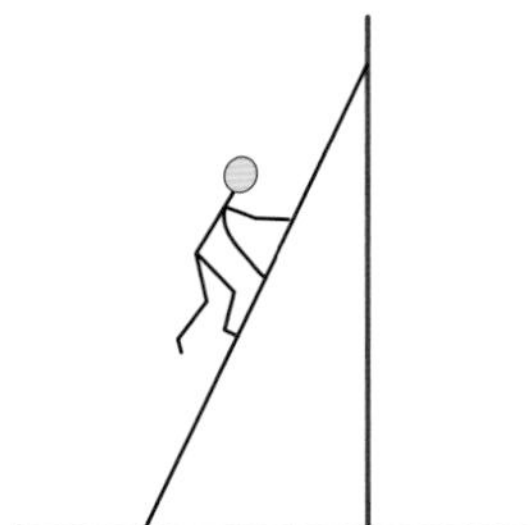

2 Here is the end wall of a house.

What is angle x?

What assumptions do you have to make to answer this question?

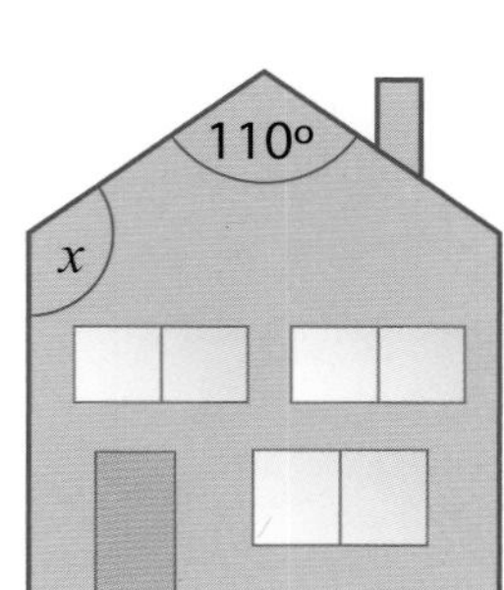

3 Look at the angles in this girder bridge.

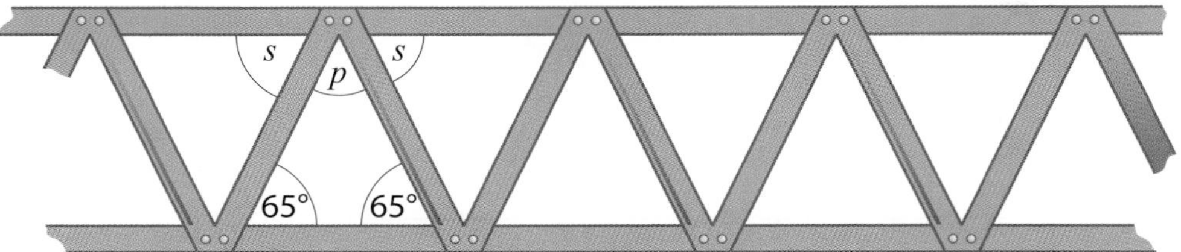

a Calculate the size of angle p.

b Calculate the size of angle s.

4 Calculate the size of angle x.

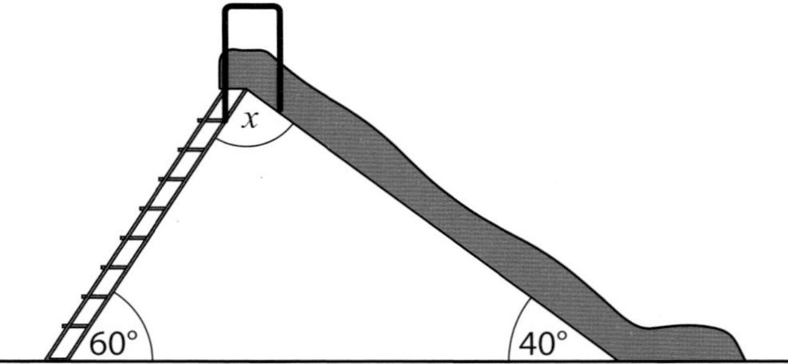

5 Each slope of the roof of this house makes an angle of 37° with the horizontal. What is the angle at the top of the roof?

Reasoning

6 This quadrilateral is split into two triangles.

a What is $a + b + c$?

b What is $p + q + r$?

c Explain why the sum of the interior angles of the quadrilateral is 360°.

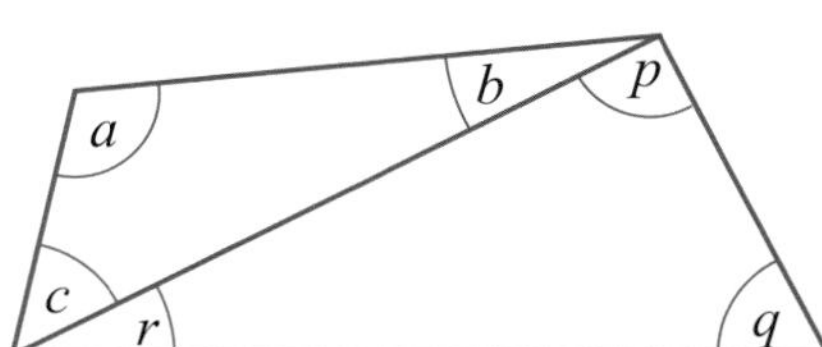

7 Each of the unknown angles in these triangles has a code letter in the table below.

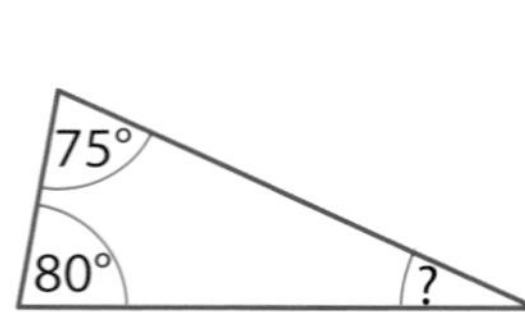

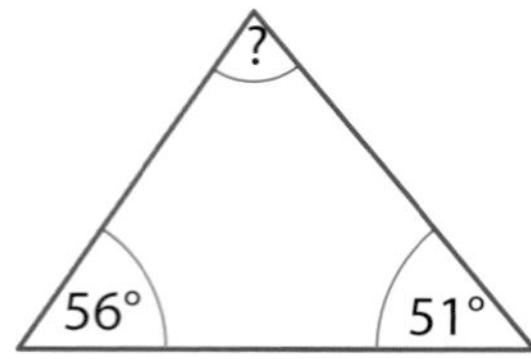

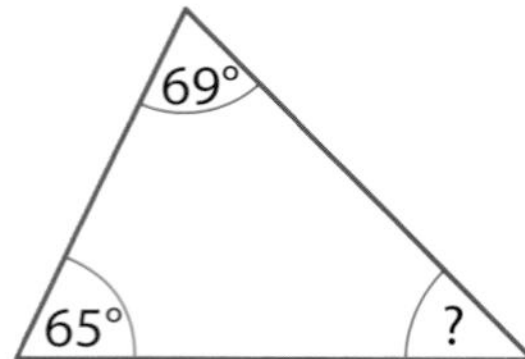

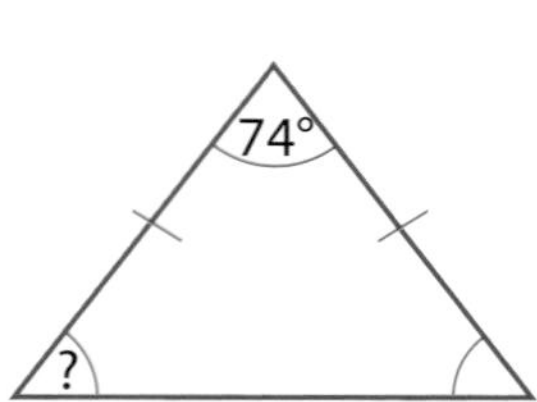

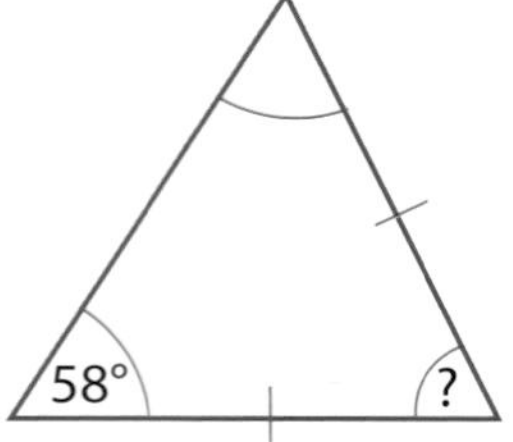

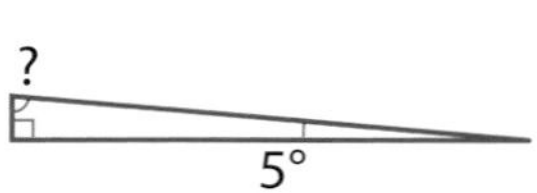

Find the six code letters.
Rearrange them to discover the code word.

n	e	g	i	s	m	a	r	o	l	q	y	w	u
82°	73°	66°	32°	53°	48°	25°	64°	21°	74°	46°	59°	33°	85°

8 The gable of a house is symmetrical.
Calculate angle x shown in the diagram

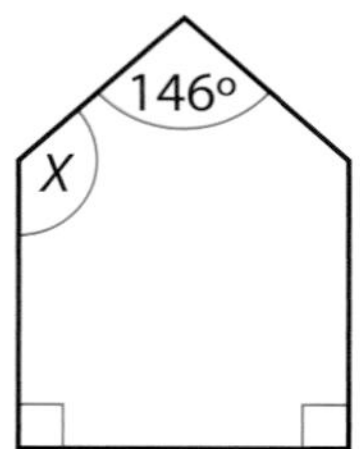

9 Calculate the lettered angles shown in the rectangle

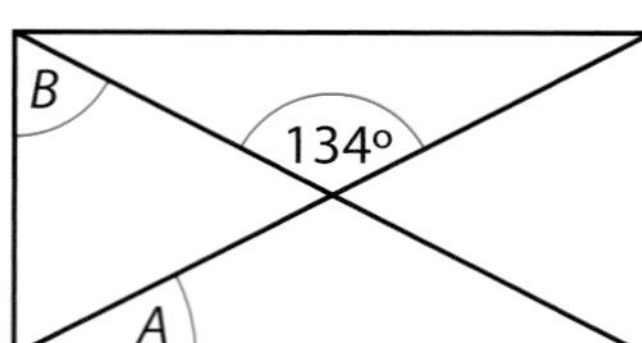

Wider skills practice

Reasoning

1 This question shows you why the angles of a triangle must add up to 180°.

Follow through the steps yourself.

a Start by drawing your own triangle ABC.
There should be nothing special about it.
Call the angles a, b and c.
You want to show that $a + b + c = 180°$.

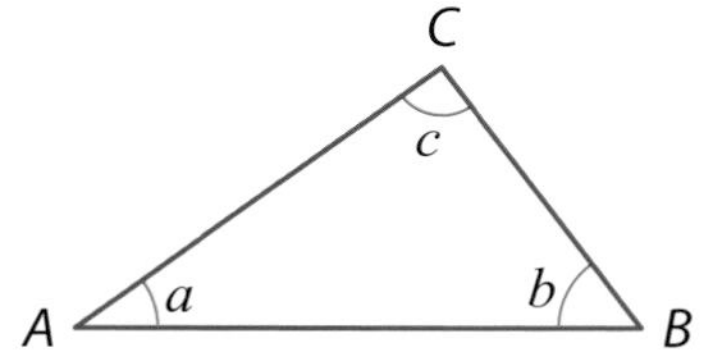

b Now draw a rectangle like this.
Call it AXYB.
Mark in all the right angles in your diagram.
The one at X is already done.

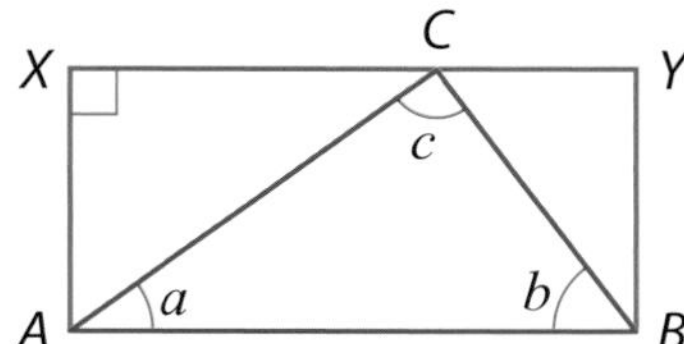

c Now draw in the line CZ.
It makes two more rectangles.
Mark in the new right angles.

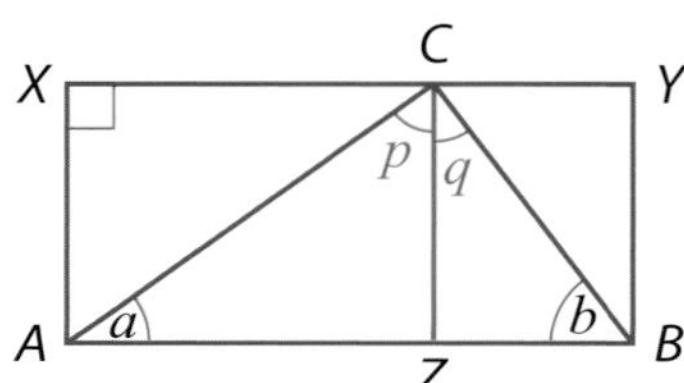

d The line CZ divides the angle c into two parts, p and q.
Copy and complete this statement.

$p + \square = c$

e Now look at rectangle AXCZ and complete this statement about its symmetry.
AXCZ has ☐ symmetry of order ☐.
So triangles AXC and CZA are ☐.
Mark in the missing angles on a copy of the diagram using one of the letters a or p.

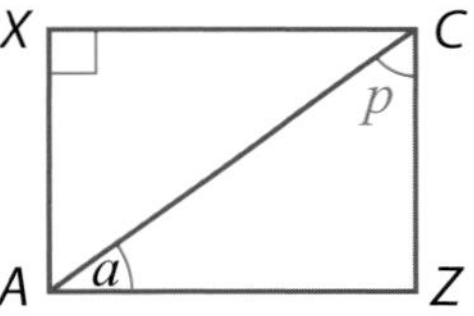

f Look at the angle at A and complete this statement.

$a + p = \square°$

g Now look at the rectangle ZCYB.
Describe its symmetry, mark the missing angles and complete this statement.

$b + q = \square°$

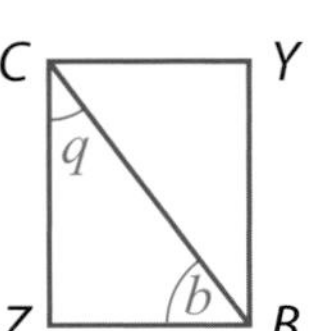

h Now complete these statements.

$a + p + b + q = \square° + \square° = \square°$

$a + b + p + q = \square°$

$a + b + c = \square°$

So the angles of triangle ABC add up to 180°.

2 **a** Draw a triangle and measure the angles inside.
You should find that they add up to 180°.

b Use one side of your triangle as the base and draw a second triangle.
You should now have drawn a quadrilateral.
What do all of the angles inside a quadrilateral add up to?

c Use one side of your quadrilateral as the base for another triangle.
You should now have a pentagon.
What do all of the angles inside a pentagon add up to?

d Keep adding triangles and measuring angles until you can confidently answer these questions.

i How many triangles do you need to draw to make a shape with 42 sides?

ii What do all of the angles inside a shape with 42 sides add up to?

Reasoning

Applying skills

Reasoning

1 This pattern is made by placing four equilateral triangles inside a rectangle.

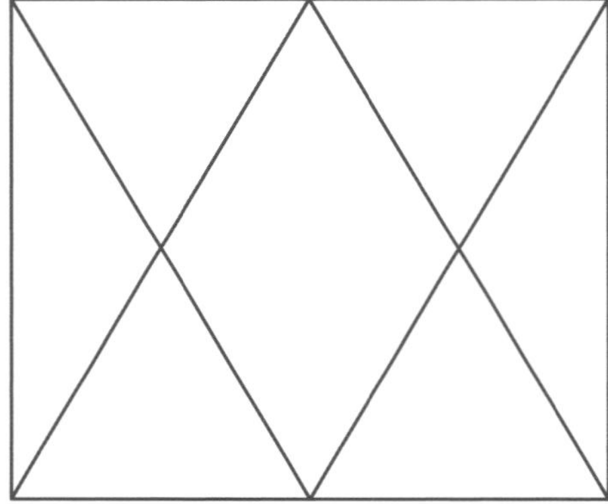

Work out the size of each angle in the pattern.

Reviewing skills

1 Find the size of each lettered angle in these diagrams.

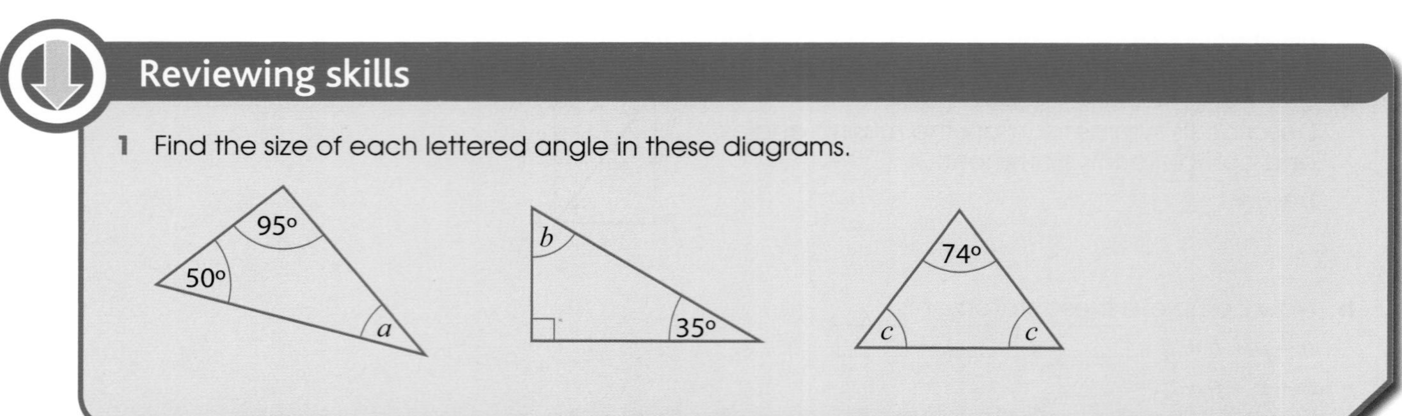

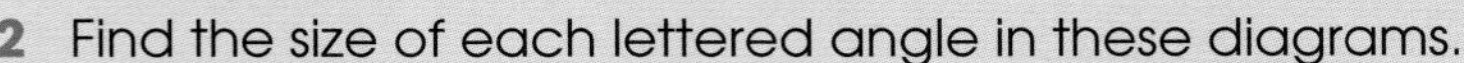

2 Find the size of each lettered angle in these diagrams.

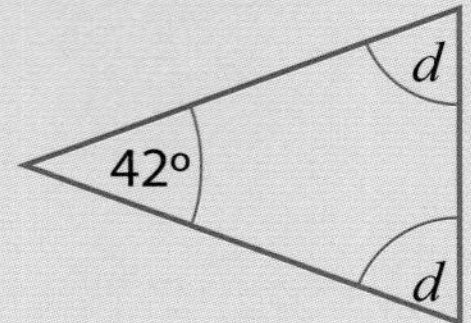

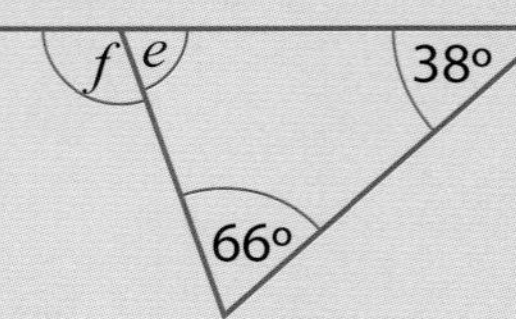

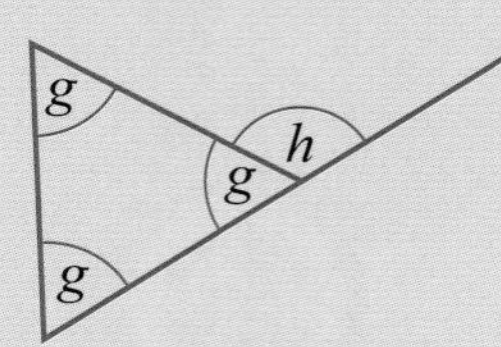

3 Find the size of each lettered angle in these diagrams.

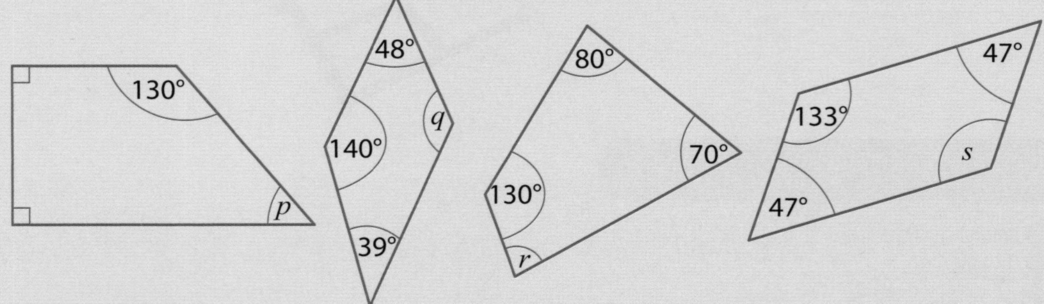

Unit 6 • Types of quadrilateral • Band e

Building skills

Example outside the Maths classroom

Pantograph

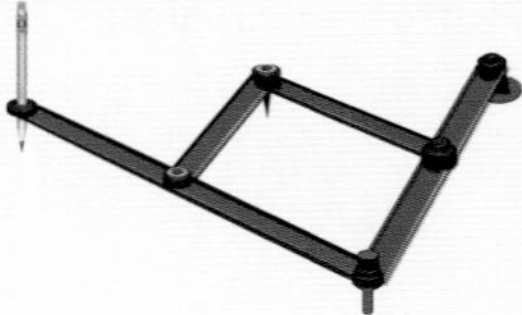

Toolbox

When classifying shapes the key things to look for are

- the number of sides
- the lengths of the sides
- the lines of symmetry
- whether opposite sides are parallel
- the sizes of the angles.

There are seven special types of quadrilateral to remember.

A **square**

A **rectangle**

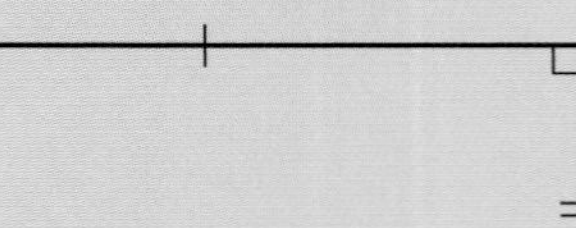

A **parallelogram**

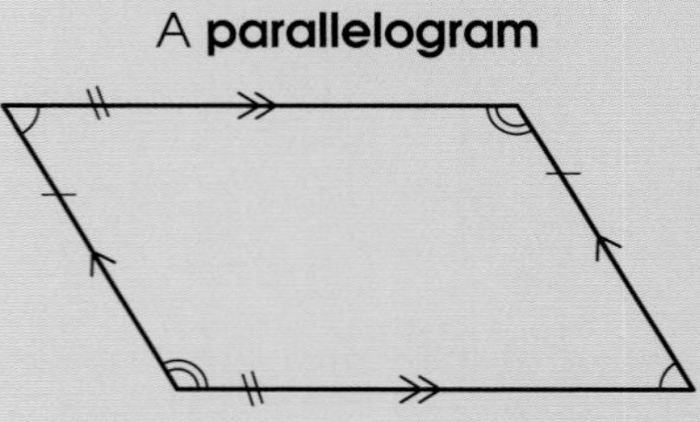

A **rhombus**

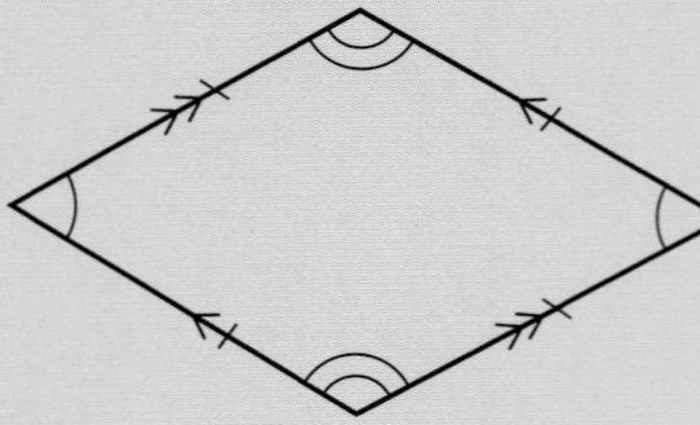

A **kite**

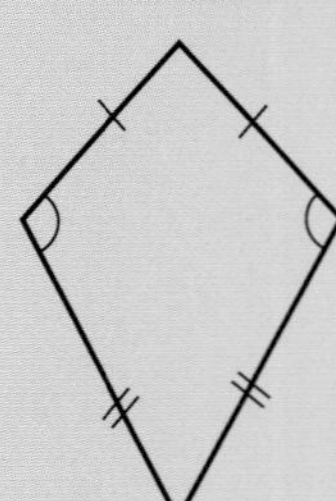

An **arrowhead**

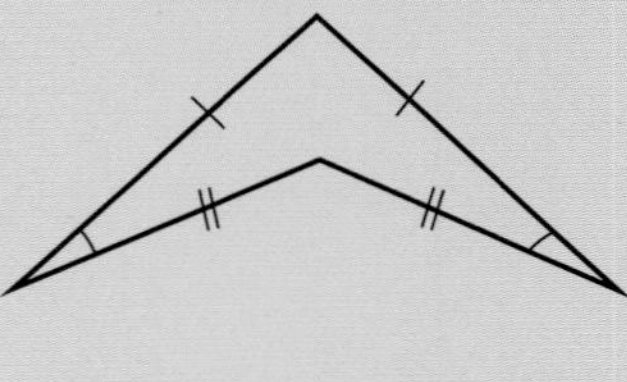

A **trapezium**

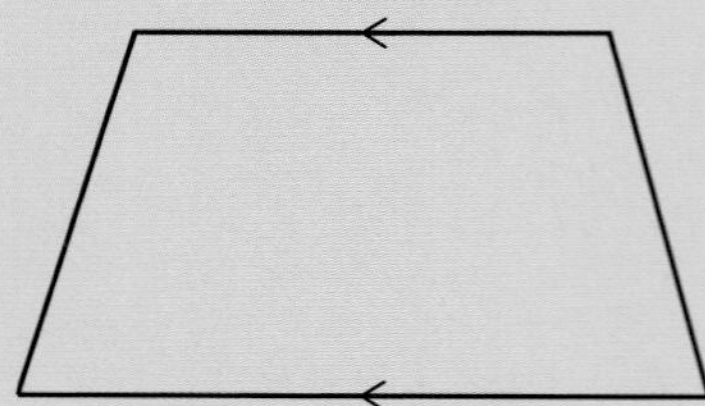

Equal sides are marked with the same number of dashes.
Parallel sides are marked with the same number of arrows.
Equal angles are marked with the same number of arcs.

Example – Classifying quadrilaterals

What type of quadrilateral is Meena thinking of?

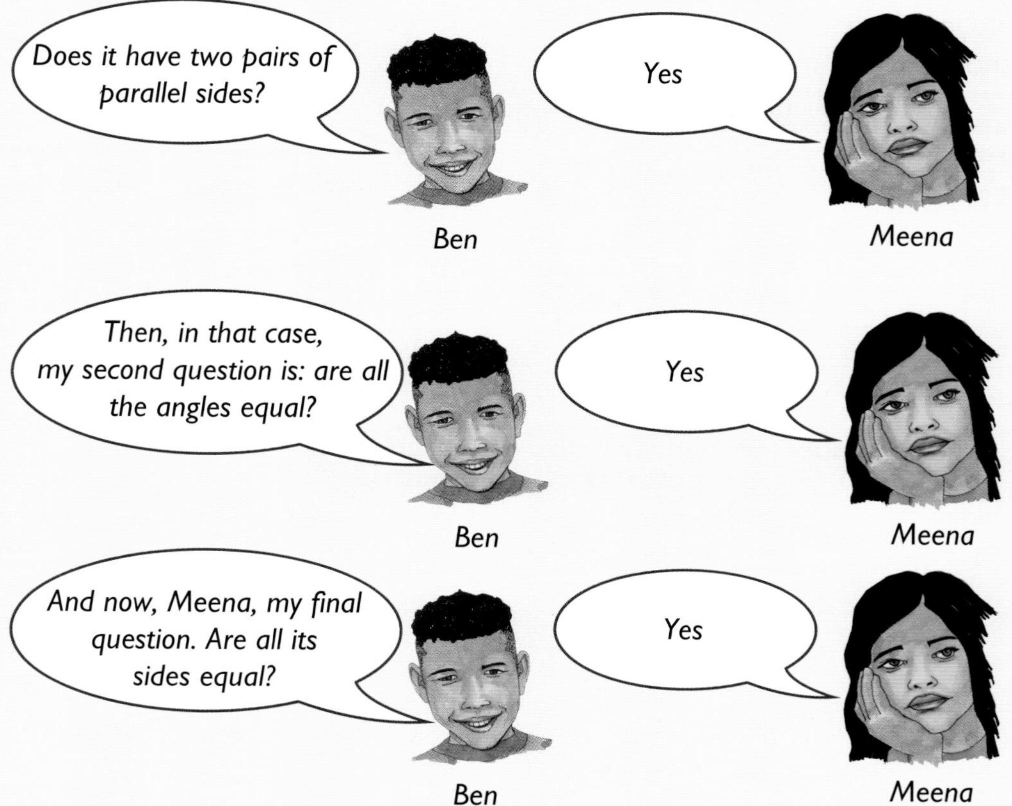

Does it have two pairs of parallel sides?
Yes
Ben
Meena
Then, in that case, my second question is: are all the angles equal?
Yes
Ben
Meena
And now, Meena, my final question. Are all its sides equal?
Yes
Ben
Meena

Solution

Squares, rectangles, parallelograms and rhombuses have two pairs of parallel lines.
Out of these, only squares and rectangles have all angles equal.
Out of these, only a square has equal sides.
Meena's quadrilateral must be a square.

Remember:

- When classifying quadrilaterals, consider equal sides, equal angles and parallel sides.
- Parallel sides are marked with arrows and equal sides with dashes.

Skills practice A

1 Look at this pattern.
It is made of triangles and quadrilaterals.
Copy and complete this table.
Only count each shape once.

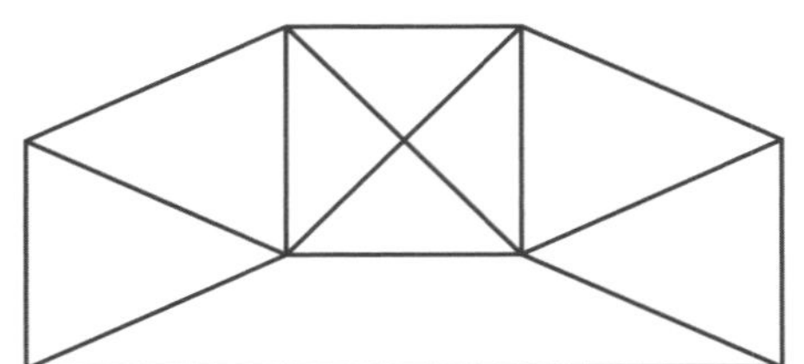

Shape	How many in pattern?
Right-angled triangle	
Other isosceles triangle	
Square	1
Other parallelogram	
Other trapezium	
Kite	

2 Copy and complete this table.

	Name of shape	How many pairs of parallel sides?	How many pairs of equal sides?
	Square	Two pairs	
		None	

3 Pete draws a quadrilateral with four right angles.
What types of quadrilateral can it be?

4 Dolores draws a quadrilateral with opposite sides parallel.
What types of quadrilateral can it be?

5 The diagram shows two sides of a quadrilateral.
Make four copies of the diagram on squared paper.
Use them for parts **a** to **d**.

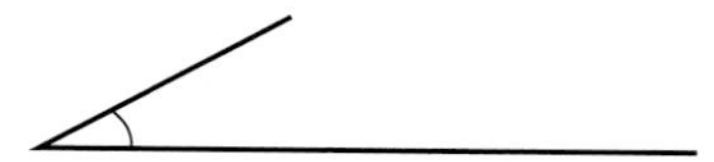

- **a** Add two sides to make a parallelogram.
- **b** Add two sides to make a symmetrical trapezium.
- **c** Add two sides to make a trapezium with a right angle.
- **d** Add two sides to make a different trapezium which is not symmetrical.

Reasoning

6 Which of the special quadrilaterals
- **a** can contain a right angle
- **b** can include a reflex angle?

7 Say whether each of these statements is true or false.
Explain your answer, using a diagram if it helps.
- **a** A parallelogram with a right angle is a rectangle.
- **b** A trapezium with a right angle is a rectangle.
- **c** A rectangle with equal sides is a square.
- **d** Every kite is also a rhombus.
- **e** An arrowhead cannot have an obtuse angle.

Skills practice B

1 Draw x and y axes from 0 to 6.
Plot the points A(1, 6), B(5, 4) and C(5, 2).
Plot the point D, and write down its co-ordinates, when ABCD is
- **a** a parallelogram
- **b** a kite
- **c** a trapezium containing a right angle.

2 Look at this pattern.
Find a square, a trapezium, a rhombus, a rectangle, a kite and a parallelogram.
For example, HOSM is a square.
No letter is used more than once.
Which letter is not used?

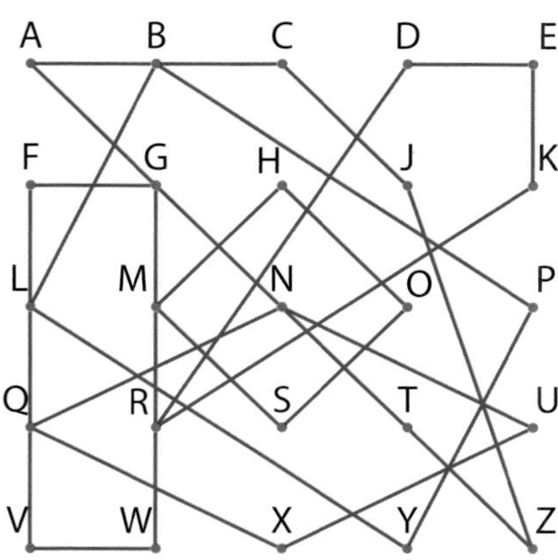

3 Describe each of these shapes as fully and as accurately as you can.

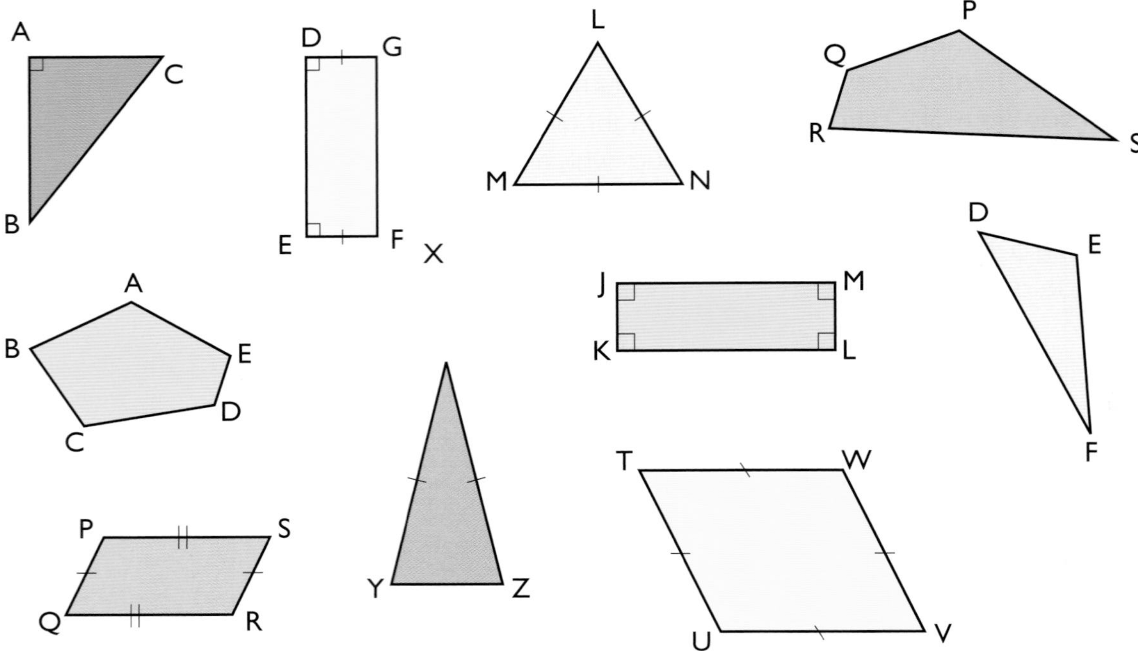

4 This shape is a regular octagon.

Find ways in which you can use the points and lines on it to make these quadrilaterals:

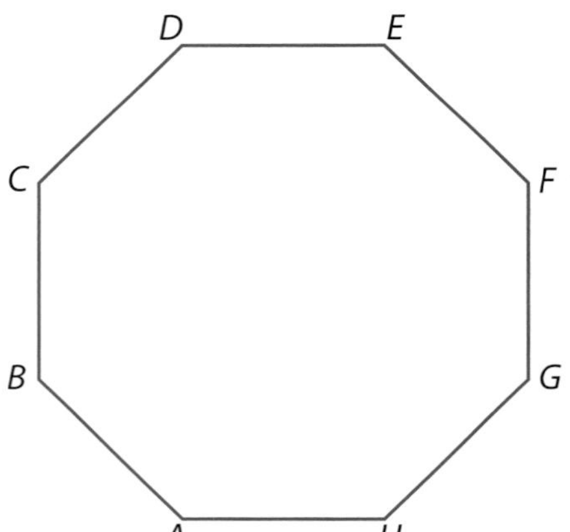

a a square

b a rhombus

c a trapezium

d a kite

e an arrowhead

f a rectangle that is not a square

g a parallelogram that is not a rectangle or a rhombus

h any other quadrilateral.

You may extend the lines that are on the diagram.

You may also make new lines by joining the vertices and you may extend these lines.

5 List all the types of quadrilateral that have these features.

a Four equal angles

b Four equal sides

c At least one pair of parallel sides

d Two pairs of parallel sides

e One but not two pairs of equal angles

Reasoning

6 Samir, Jo, Angus and Lucy are talking about quadrilaterals.

I think a rectangle is a type of square.

Jo

I think a square is a type of rectangle.

Samir

I think a rectangle is a type of parallelogram.

Angus

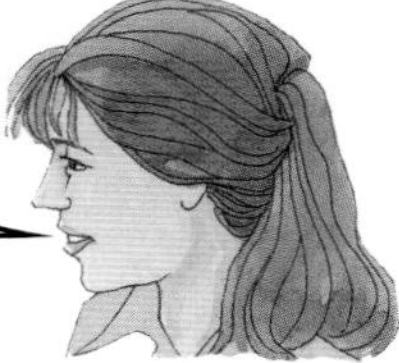
I think a parallelogram is a type of rectangle.

Lucy

Who, if anybody, do you think is right?
Explain why you think this.

Wider skills practice

1 Copy and complete this table.

	Name of shape	How many lines of symmetry?
	Rectangle	2

2 **a** Place these various types of quadrilateral in this two-way table.

Square Rectangle Rhombus Parallelogram Kite Arrowhead Trapezium

Angles / Parallel sides	None equal	One pair equal	Two pairs equal	All angles equal
None				
One pair parallel				
Two pairs parallel				

b You should find that three of the boxes contain two quadrilaterals each. In each of these cases, give a further test to tell them apart.

Applying skills

1 Draw a square and then draw two lines on it.
What different shapes can you make?
Here is one example showing four squares.

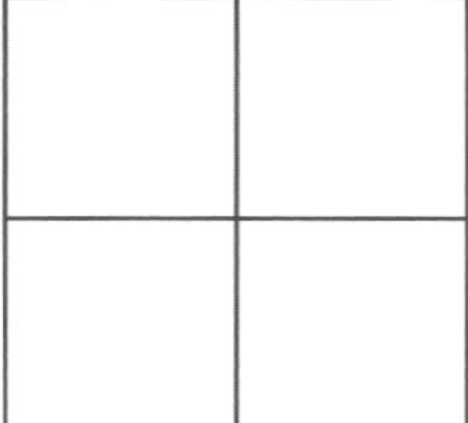

2

I'm thinking of a quadrilateral.
You have three questions to find it, Ben.

Meena

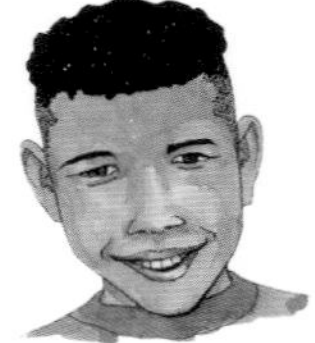

Ben

Draw a decision diagram to show what questions Ben could ask to discover any type of quadrilateral.

Reviewing skills

1 Describe these shapes as fully and accurately as you can.

a

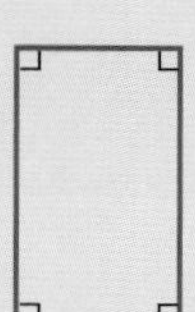

b

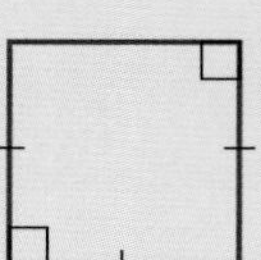

c

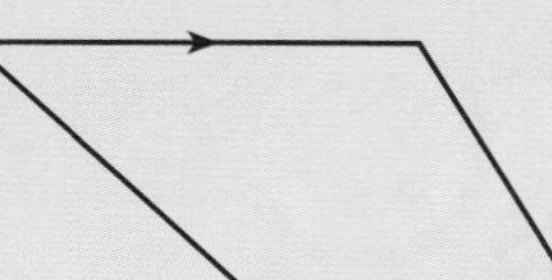

d

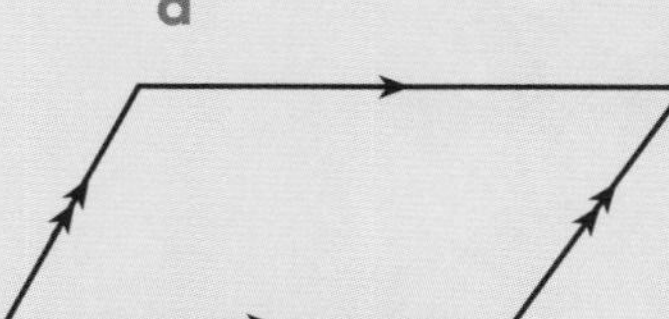

e

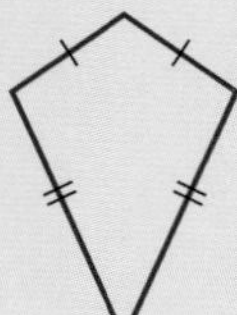

f

2 Look at these diagrams.
In each case one vertex of the quadrilateral is missing.
Write down the co-ordinates of the missing vertex to make

a a square **b** a kite **c** a rhombus

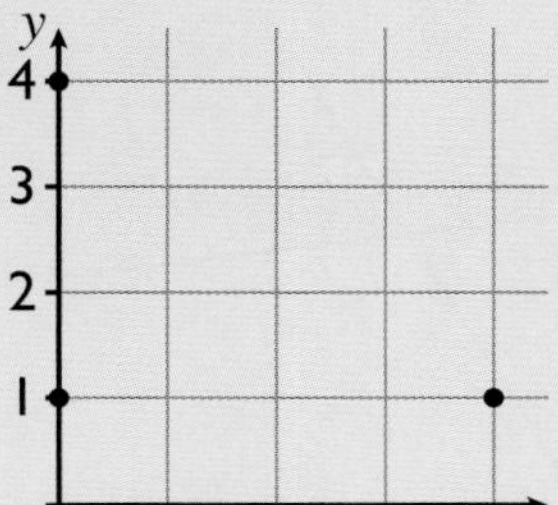

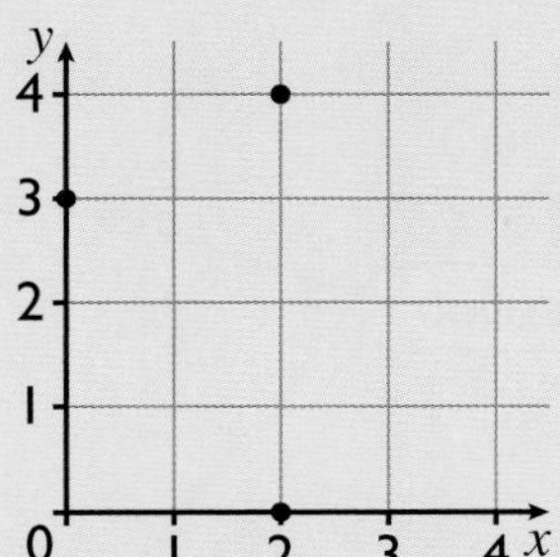

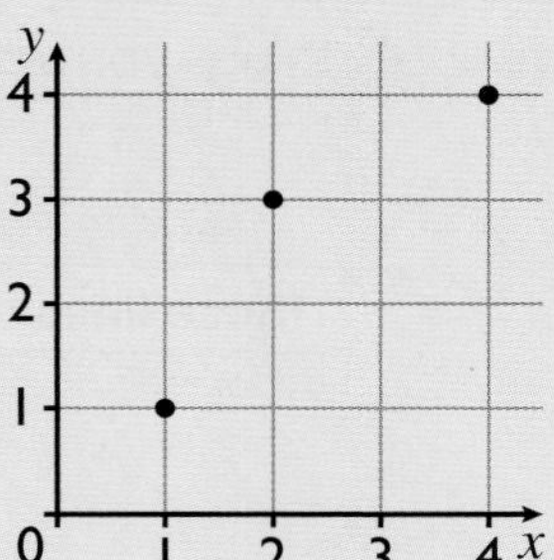

Building skills

Example outside the Maths classroom

Chart navigation

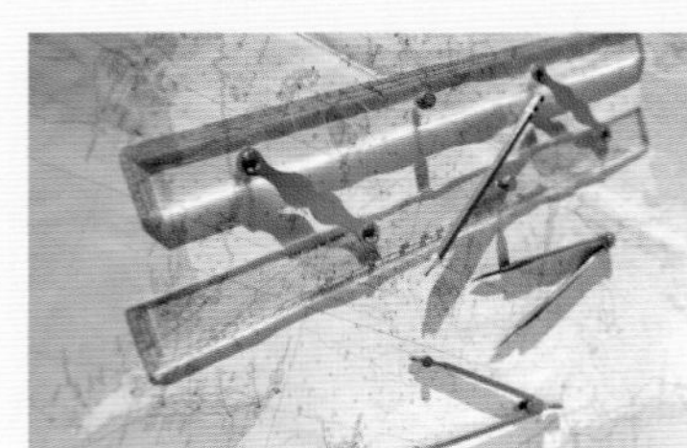

Toolbox

Parallel lines go in the same direction. They never meet, no matter how far they are extended.
When a third line crosses a pair of parallel lines it creates a number of angles.
Some of these are equal; others add up to 180°.
The three diagrams show things you will see.

These lines are parallel.

The line that crosses them is sometimes called the transversal.

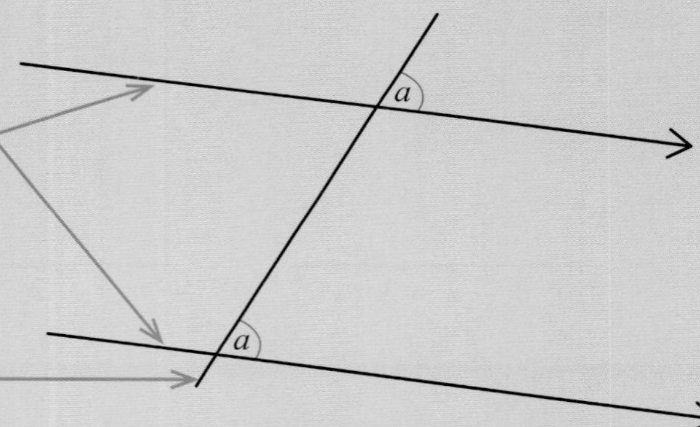

The two angles marked a are exactly the same.
They are called **corresponding** angles.

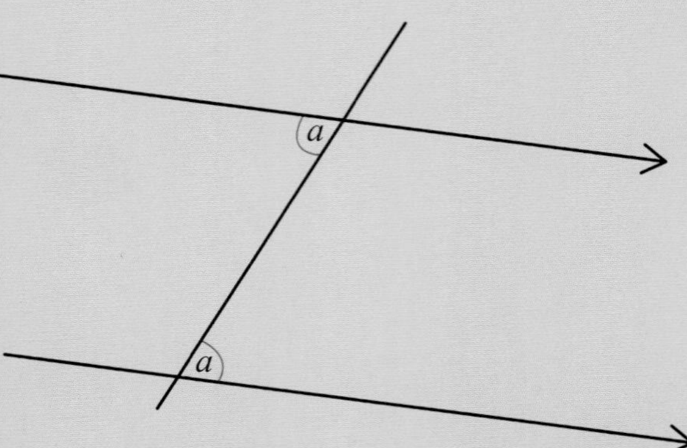

The two angles marked a are also the same.
Angles in this position are called **alternate** angles.

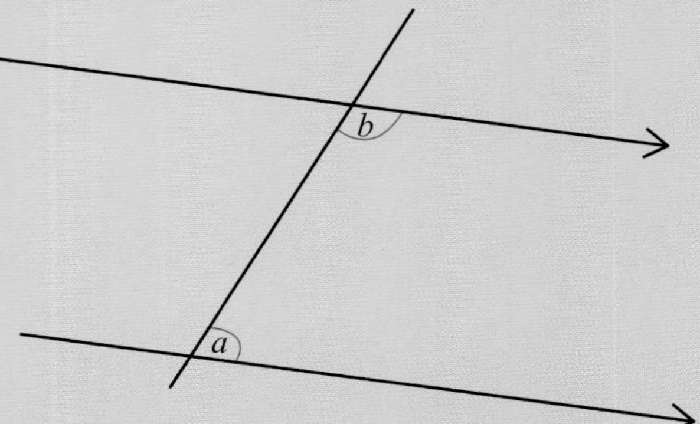

Angles a and b add up to 180°.
They are **supplementary** angles.

Example – Using alternate angles

The diagram shows a pair of parallel lines and an intersecting line.
Work out the size of the lettered angles.
Give a reason for each of your answers.

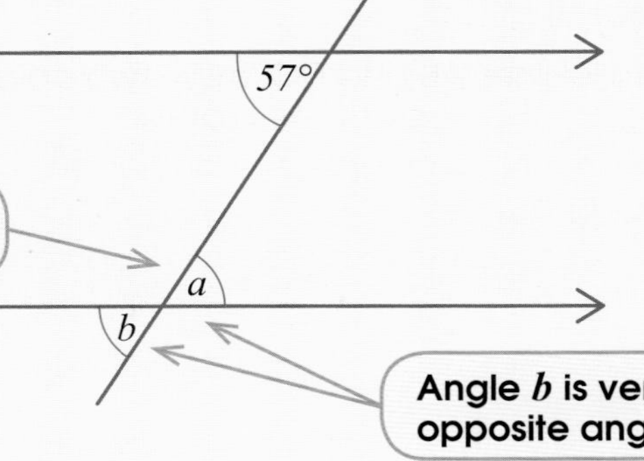
57°
a
b
Angle *a* and the given angle are alternate angles.
Angle *b* is vertically opposite angle *a*.

Solution

Alternate angles are equal.

$a = 57°$

Opposite angles are equal.

$b = 57°$

Example – Using supplementary and corresponding angles

The diagram shows a pair of parallel lines and an intersecting line.
Work out the size of the lettered angles.
Give a reason for each of your answers.

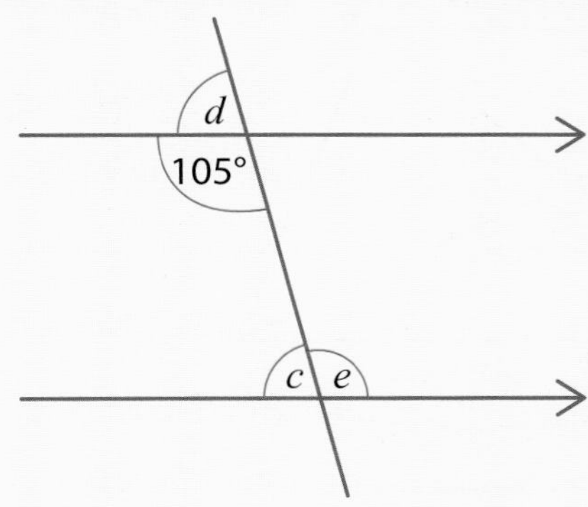
d
105°
c
e

Solution

Supplementary angles add upto 180°

$c = 180 - 105 = 75°$

c and d are corresponding angles.

Corresponding angles are equal.

$d = 75°$

Alternate angles are equal.

e is an alternate angle to the given angle.

$e = 105°$

Example – Finding angles in a parallelogram

ABCD is a parallelogram.

Its diagonals meet at M.

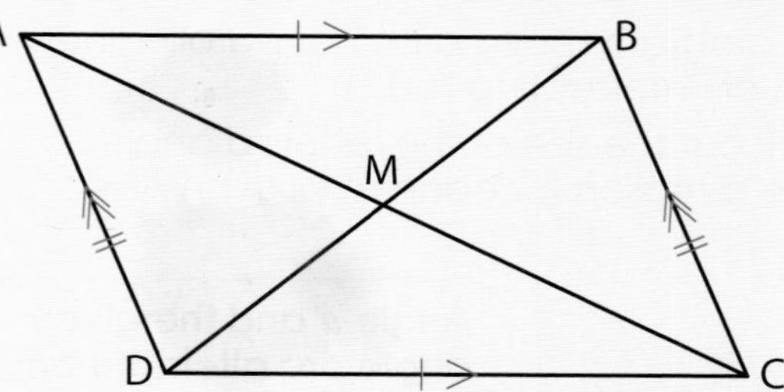

a Using the parallel lines AB and DC, mark two pairs of alternate angles on a copy of the diagram.

b Mark two pairs of vertically opposite angles on a copy of the diagram.

c Show a pair of congruent triangles on a copy of the diagram.

Solution

a Alternate angles

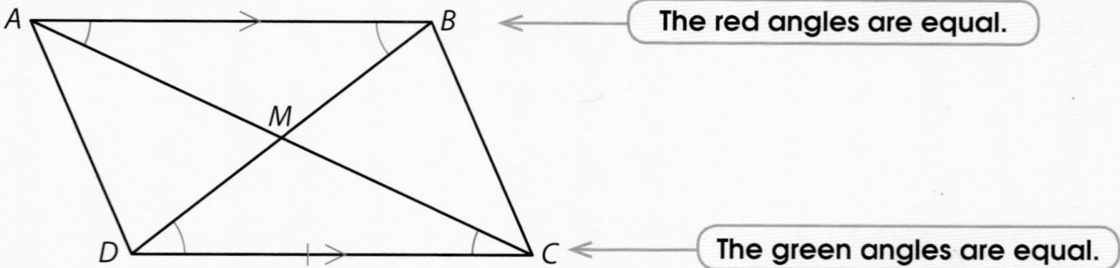

b Opposite angles

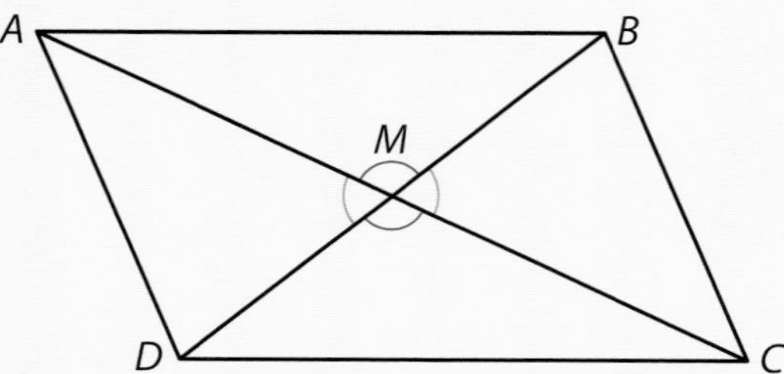

c Congruent triangles

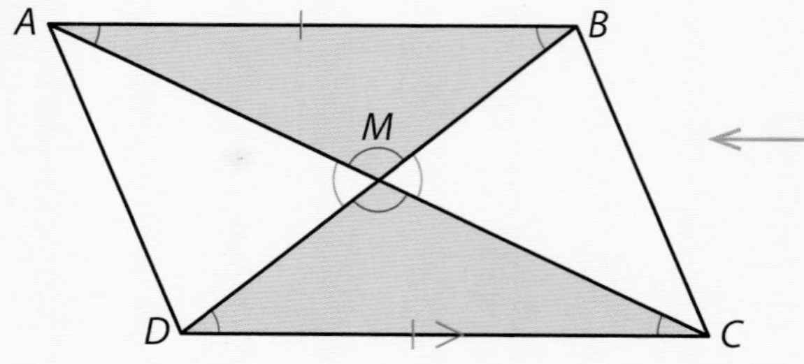

The three angles of these triangles are the same so they are the same shape.

Also AB and CD are the same length so they are the same size.

Remember:

When you have a pair of parallel lines with a third line crossing them:

- a and c are corresponding angles. They are equal.
- c and b are alternate angles. They are equal.
- c and d are supplementary angles. They add up to 180°.

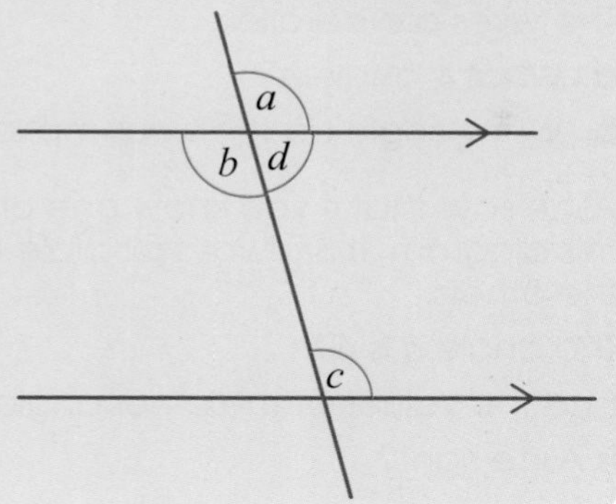

Skills practice A

1 Find the size of each lettered angle in these diagrams.

a

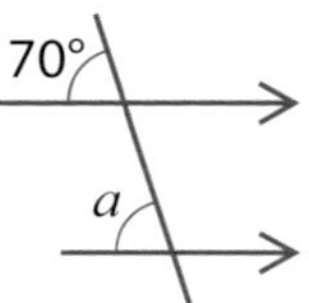

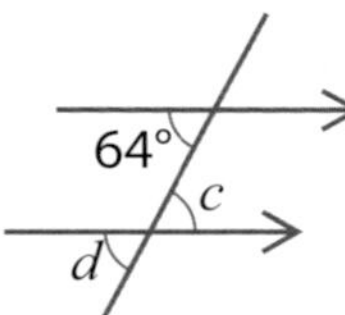

b

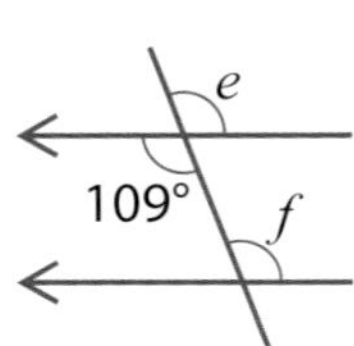

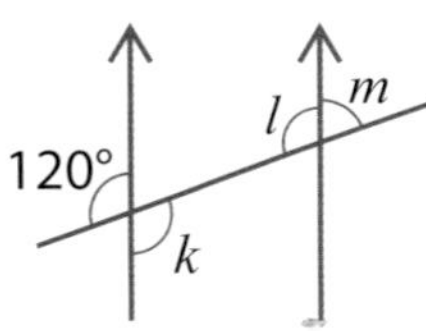

c

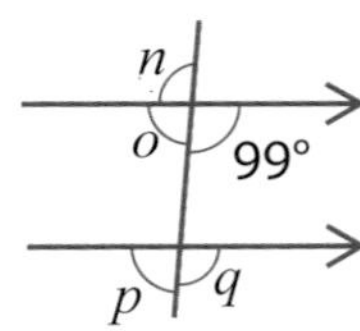

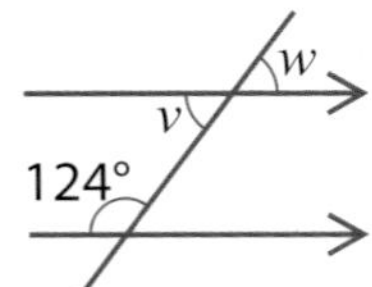

2 Ameet makes a wire fence.
The posts are vertical.
The wires are parallel.
Work out the size of the angles a, b and c.

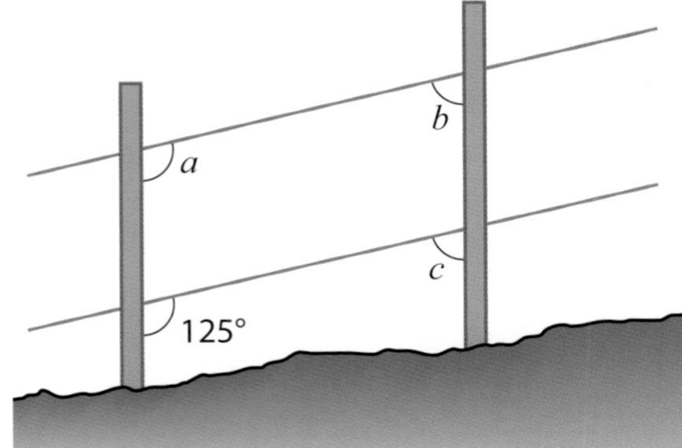

3 Here is a fence going down a hill.
The posts are vertical.
The wires are parallel.
a What is angle a?
b What angle do the wires make with the horizontal?

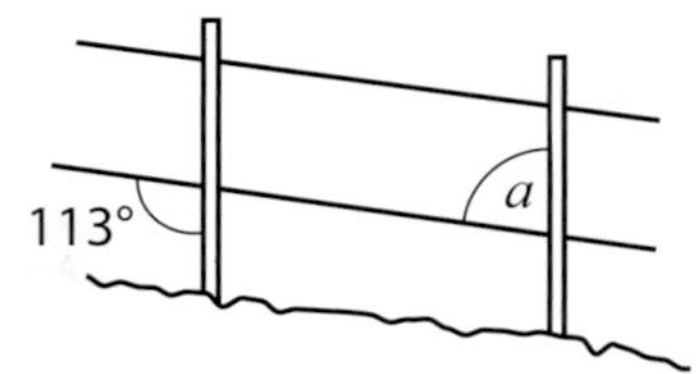

Reasoning

4 Alice says that if you know one of the angles in this diagram, it is always possible to find all of the others.
The angle a is 40°.
Find the values of the other angles in the diagram.
Is Alice right?

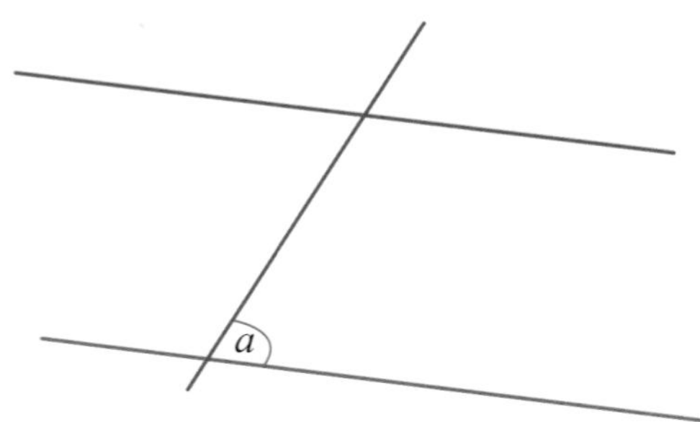

Skills practice B

1 Find the size of each lettered angle in these diagrams.
For each one write down the angle fact(s) that you use.

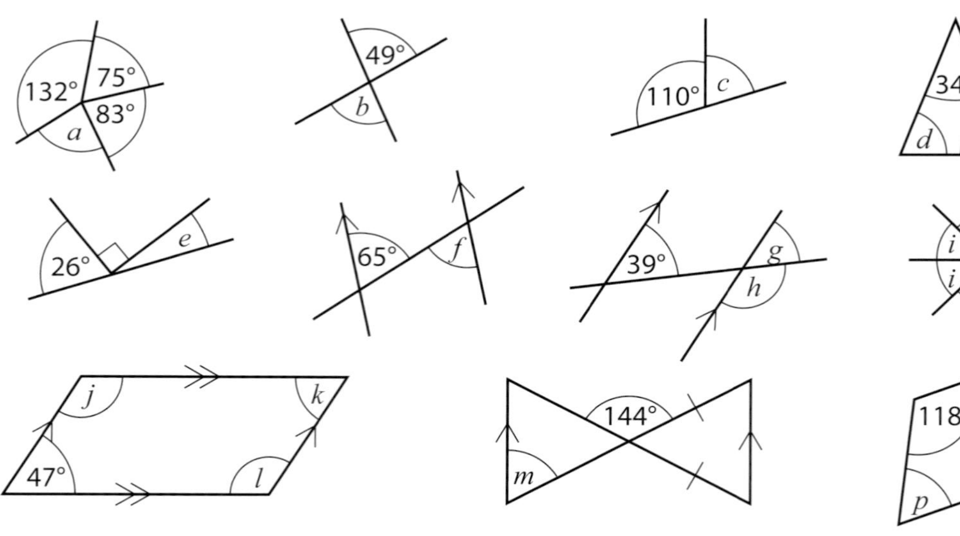

2 **a** Find the size of each lettered angle.
b What is $a + b + c + d$?

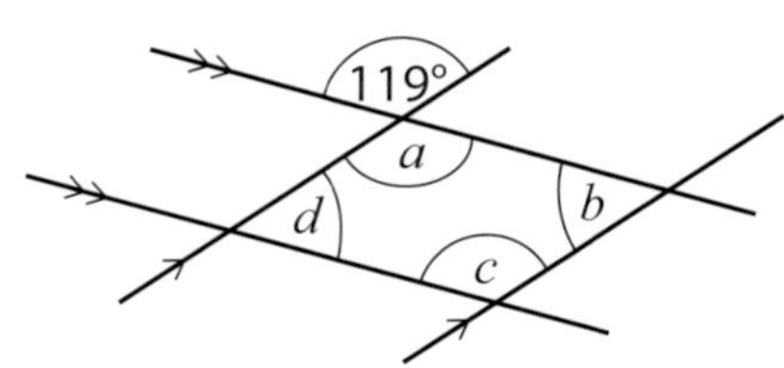

3 **a** Write down the size of each lettered angle.
For each one give your reason.

b What is $f + d + e$?

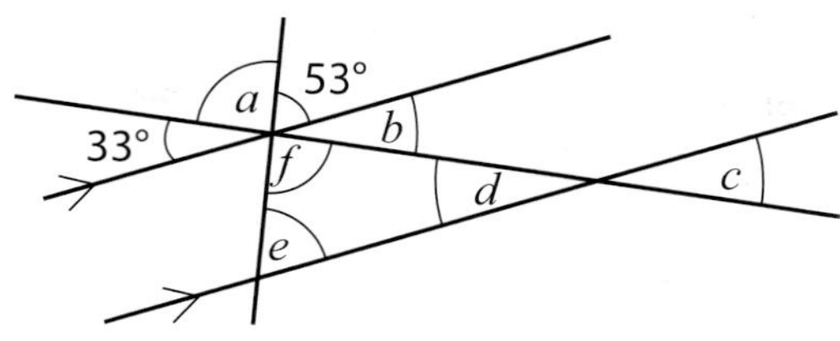

4 Find the size of each lettered angle in these diagrams.
For each one write down the angle fact(s) that you use.

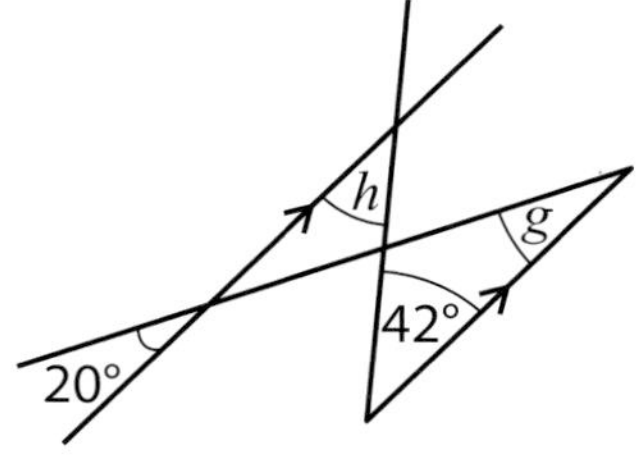

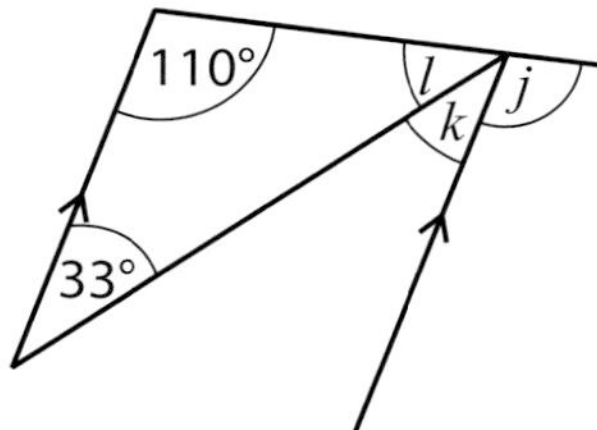

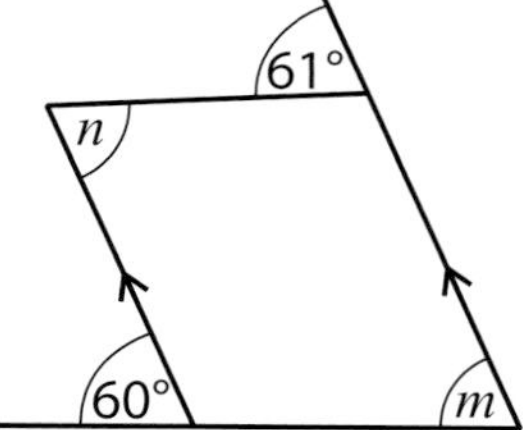

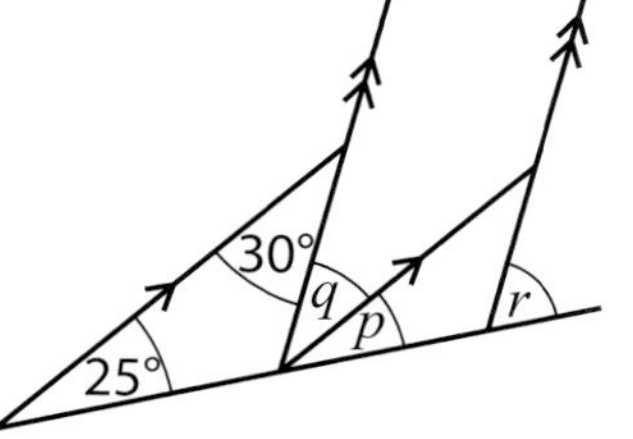

Reasoning

5 Sam says that the triangle ABC has the same angles as triangle ADE in each of these diagrams.

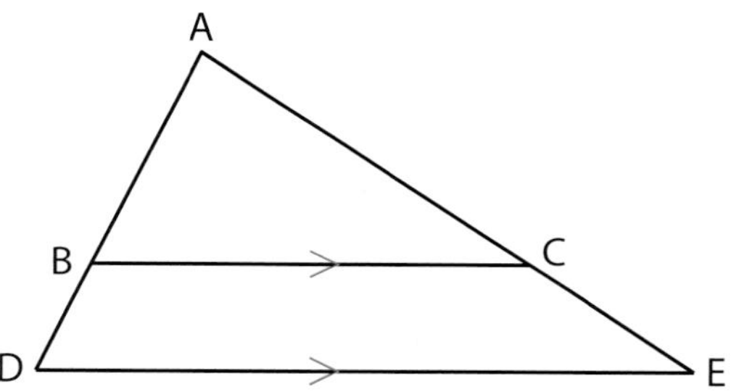

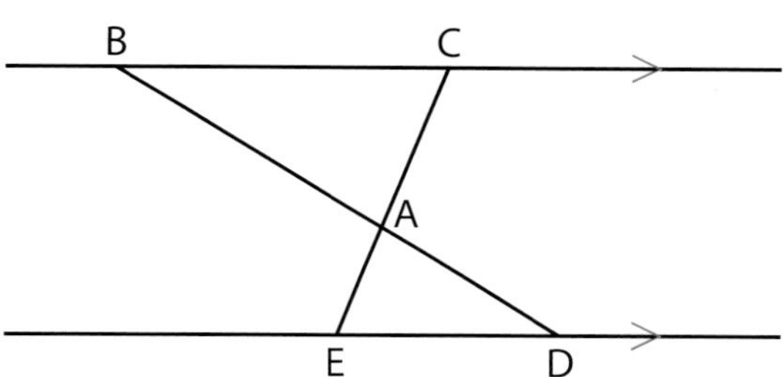

Is Sam right in each case?
How do you know?

6 Calculate angles *a* and *b* shown in the diagram representing an electricity pylon

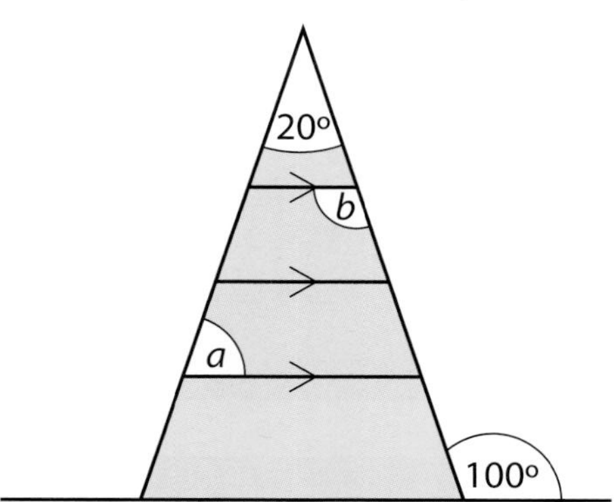

Wider skills practice

Reasoning

1 Use the following method to prove that the angles of a triangle add up to 180°.

  a Draw your own triangle PQR. It should be nothing special.
  Call its angles x, y and z.

  b Now draw rectangle PSTR.
  The point Q is on the line ST.

  c Explain how you know that ST is parallel to PR.

  d Mark on your diagram another angle that is equal to x and explain how you know they are equal.

  e Mark on your diagram another angle that is equal to z and explain how you know they are equal.

  f Use your diagram to prove that the three angles of triangle PQR add up to 180°.

2 ABCD is a trapezium.
Its diagonals meet at E.

  a Show that the three angles of triangle AEB are the same size as the three angles in triangle CED.

  b Say what is the same and what is different about the triangles AEB and CED.

  c Are triangles AEB and DEC congruent?

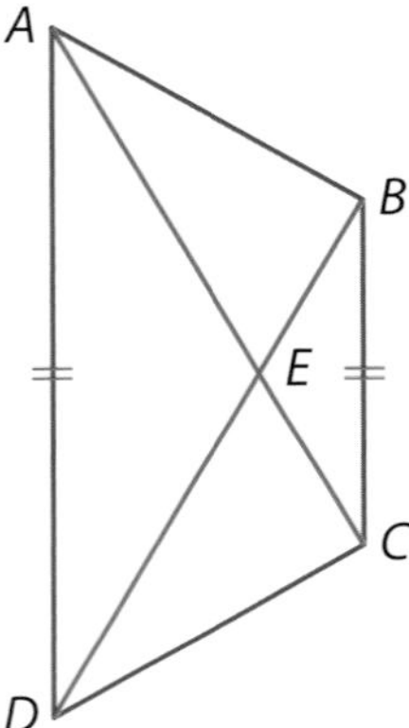

Applying skills

1 Kelly is making a rabbit hutch.
This is a scale drawing of one end.
It is made from three pieces of wood cut from 3 m lengths that are 40 cm wide.

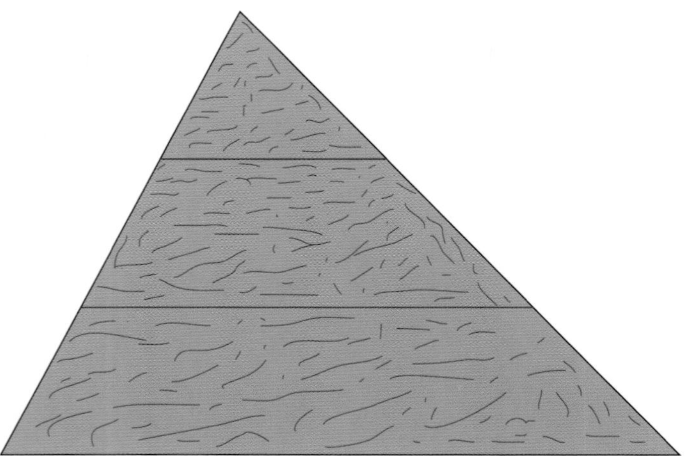

Make a scale drawing of a piece of wood.
Show where Kelly should cut it to make the end from the least amount of wood.

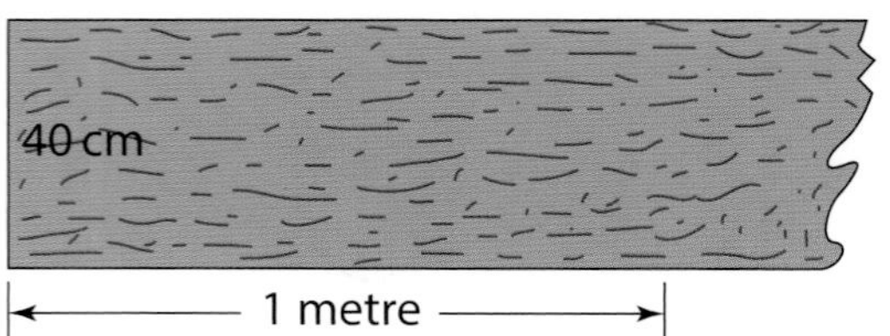

Reviewing skills

1 Calculate the size of each lettered angle and write down your reason.

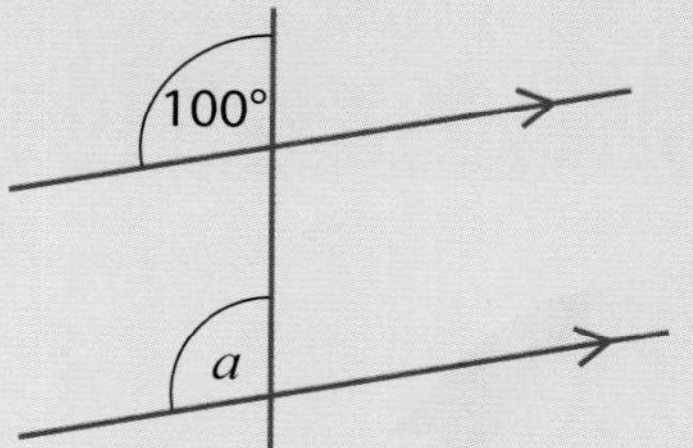

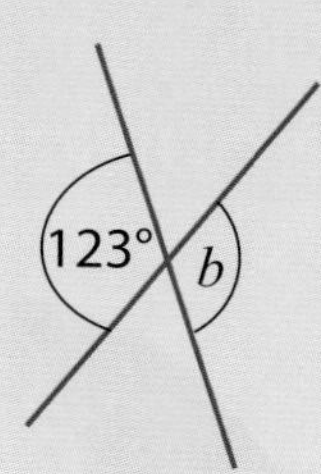

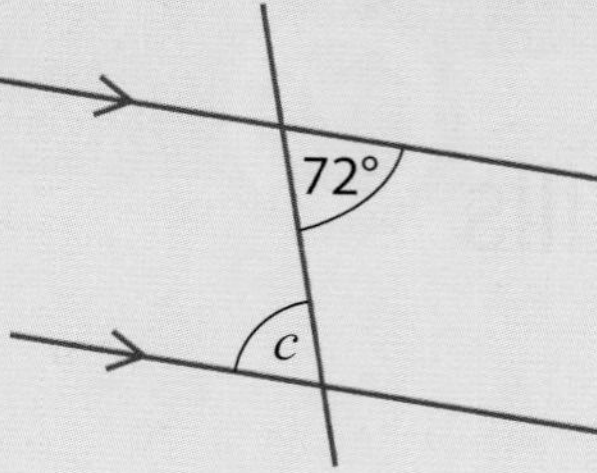

2 **a** Calculate the size of each lettered angle and write down your reason.

b What is $w + x + y + z$?

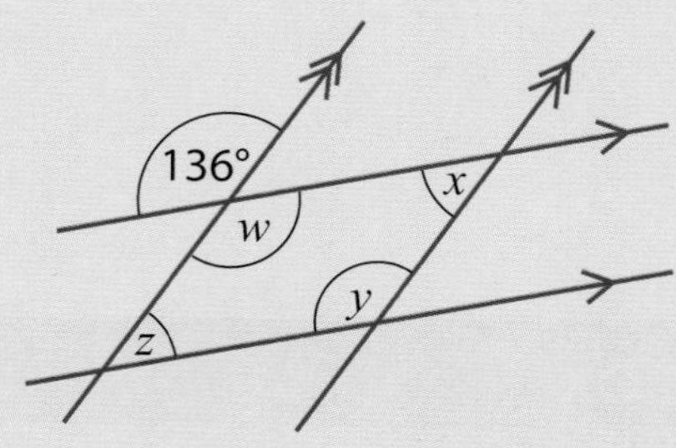

3 Find the size of each lettered angle in these diagrams.

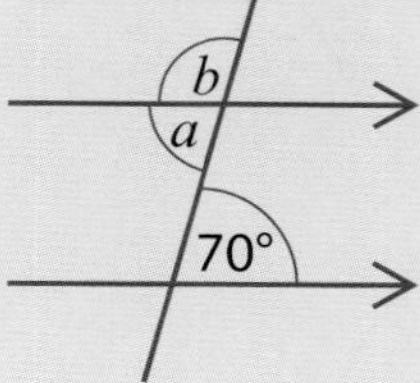

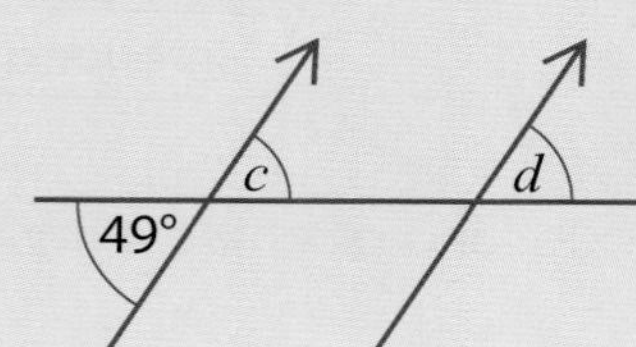

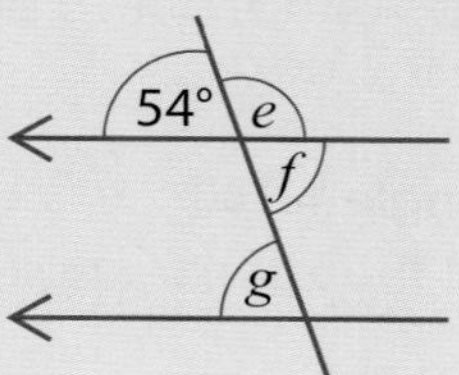

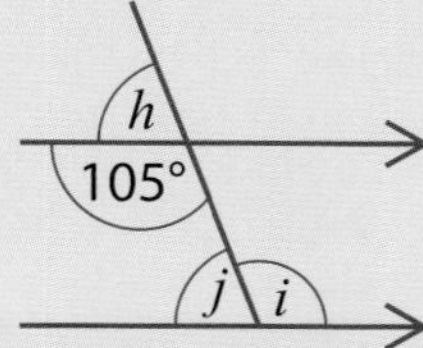

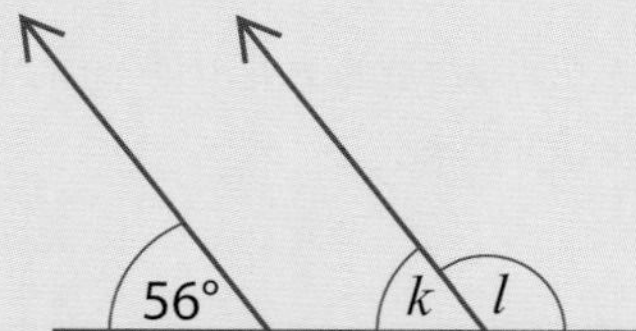

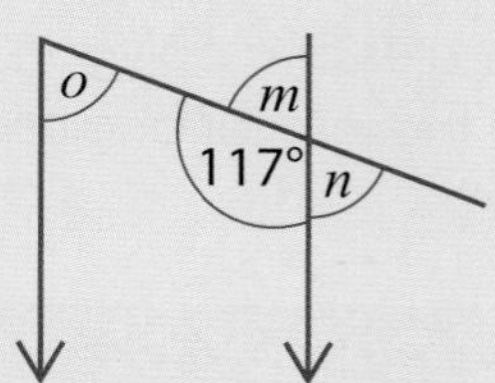

Unit 8 • Angles in a polygon • Band g

Building skills

Example outside the Maths classroom

Umbrellas

Toolbox

An **interior angle** is the angle inside the corner of a shape.

An **exterior angle** is the angle that has to be turned through to move from one side to the next.

At each vertex,
interior angle + exterior angle = 180°.

In a **regular polygon** all of the interior angles are equal and all of the exterior angles are equal.

The exterior angles of a polygon always make one complete turn and so **add up to 360°**.

Exterior angle

Interior angle

The sum of the interior angles of a polygon $=180(n-2)°$ where n is the number of sides.

For example, the sum of the interior angles of a hexagon is
$180 \times (6-2) = 720°$.

Example – Finding angles of regular polygons

The diagram shows a regular octagon.

a Work out the size of an exterior angle of a regular octagon.

b Hence find the size of an interior angle of a regular octagon.

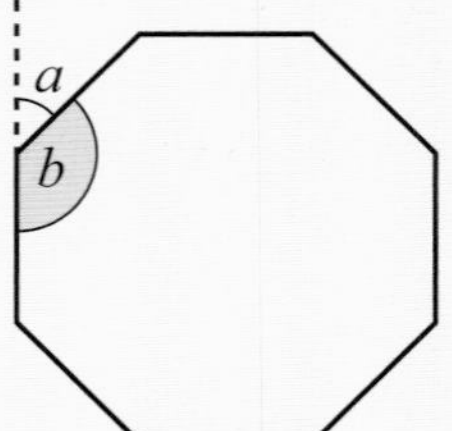

Solution

a The exterior angles of a polygon add up to 360°.
For a regular octagon,
each exterior angle = 360° ÷ 8 = 45° ← **This is angle a in the diagram.**

b Interior angle + exterior angle = 180°
For a regular octagon,
each interior angle = 180° – 45° ← **This is angle b in the diagram.**
= 135°

Example – Finding the number of sides of a regular polygon

A regular polygon has an interior angle of 162°.
How many sides does the polygon have?

Solution

First find the exterior angle.

Interior angle + exterior angle = 180°

162° + exterior angle = 180°

Exterior angle = 180° − 162° = 18°

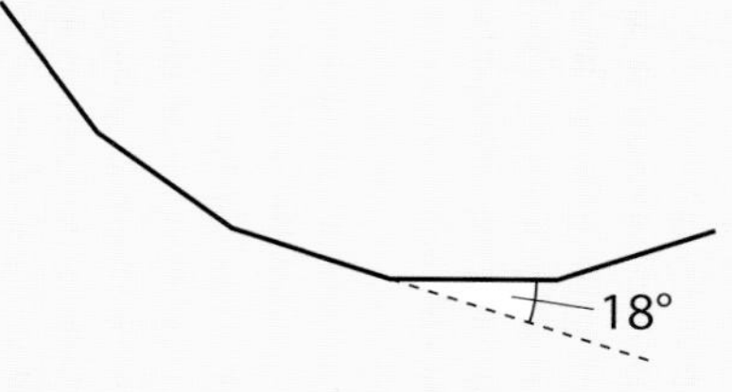

The exterior angles add up to 360°.

Number of exterior angles = 360 ÷ 18 = 20

The polygon has 20 sides. ← **The number of sides is the same as the number of exterior angles.**

Remember:

- The exterior angles of a polygon add up to 360°.
- The sum of the interior angles can be found using the expression $180(n-2)°$.
- At each vertex, interior angle + exterior angle = 180°.

Skills practice A

1 a Calculate the exterior angle a of a regular pentagon.

b Calculate the interior angle b of a regular pentagon.

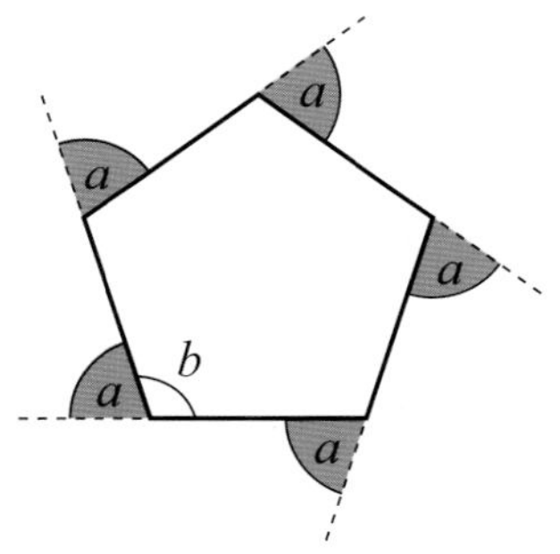

2 a Calculate the exterior angle a of a regular hexagon.

b Calculate the interior angle b of a regular hexagon.

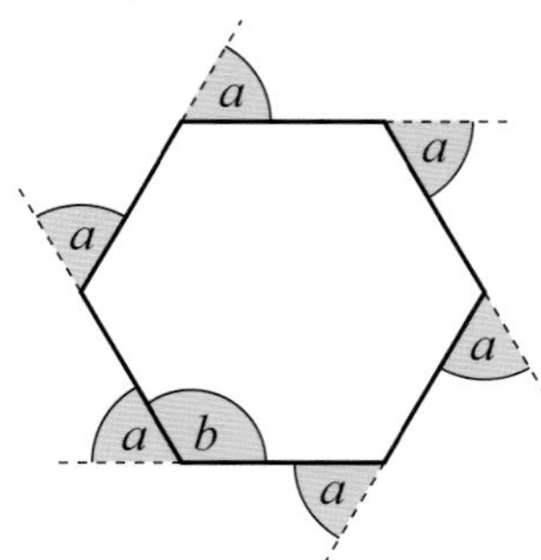

3 Calculate the size of each lettered angle.

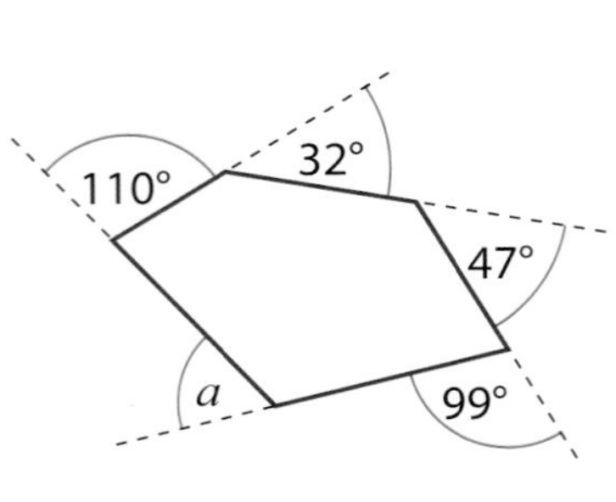

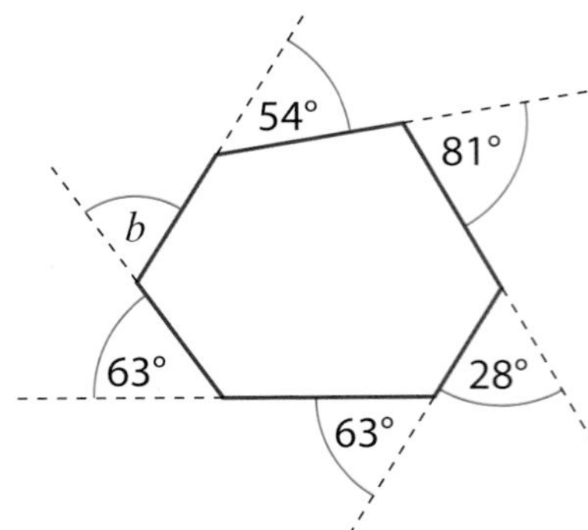

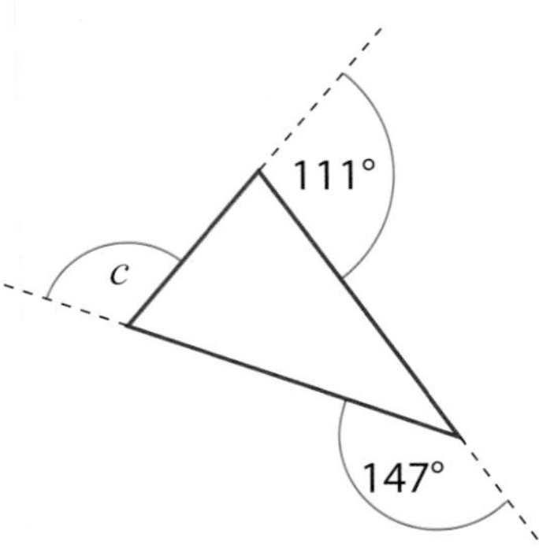

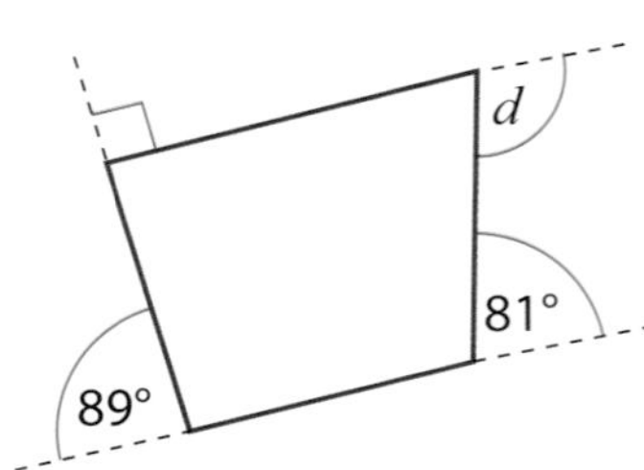

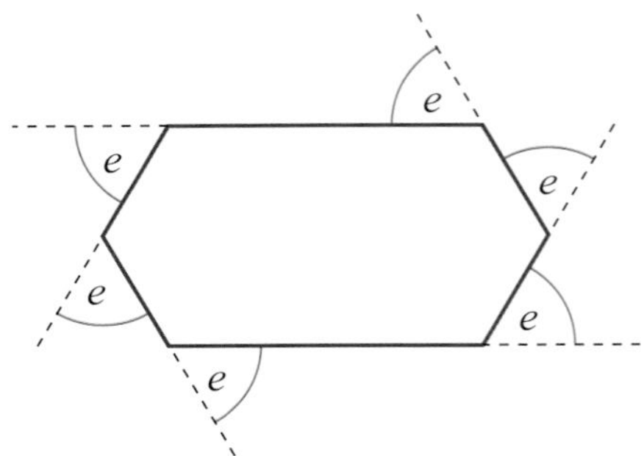

4 The diagram shows a pentagon split into five triangles.
What is the sum of

- **a** all the angles in all the triangles
- **b** the angles at the centre
- **c** the interior angles of the pentagon?

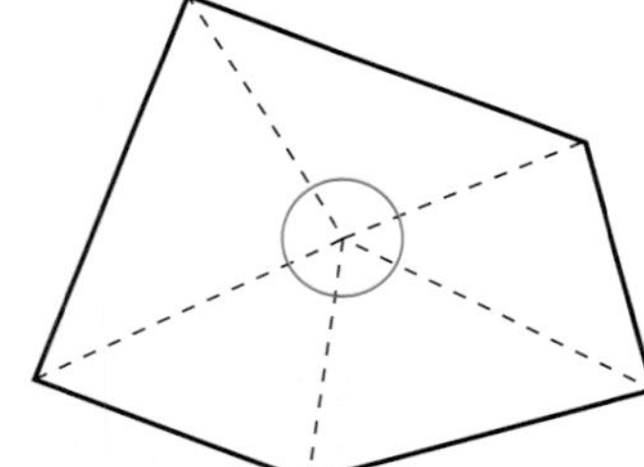

5 Draw a large pentagon.
Extend each side as shown.
Label each exterior angle with a letter.

- **a** Use a protractor to measure each angle and write it down.
  For example $a = 82°$, $b = 43°$, …
- **b** Add your exterior angles together.
  What should the sum be?
  How accurate is your drawing?
  How accurate are your angle measurements?

Repeat this activity with different polygons.

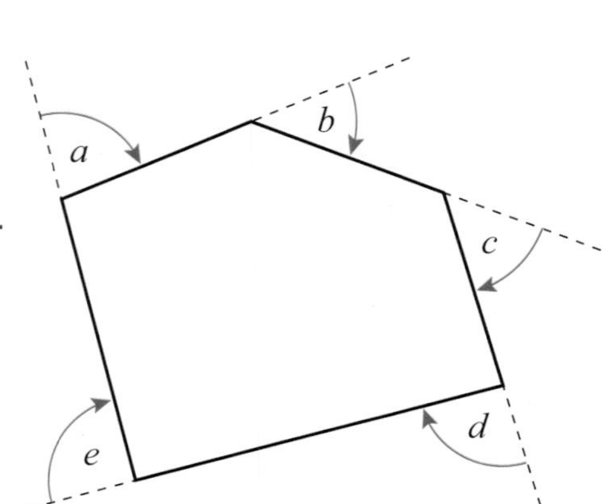

Skills practice B

1 The diagram shows an octagon.

The vertices are all joined to a point O in the middle, making eight triangles.

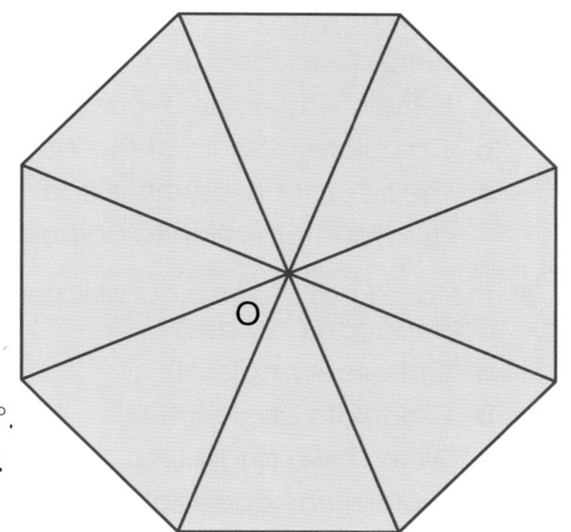

- **a** Work out the total of all the angles in the eight triangles.
- **b** What is the total of the angles at the point O?
- **c** Use your answers to parts **a** and **b** to find the total of the interior angles of the octagon.
- **d** Use the diagram to explain why the sum of the interior angles of a polygon with n sides is $n \times 180° - 360°$.
- **e** Show that the formula in part **b** is the same as $(n - 2) \times 180°$.
- **f** Find the size of each interior angle if the octagon is regular.

2 This diagram shows an octagon.

One of the vertices, D, is joined to all the other vertices.

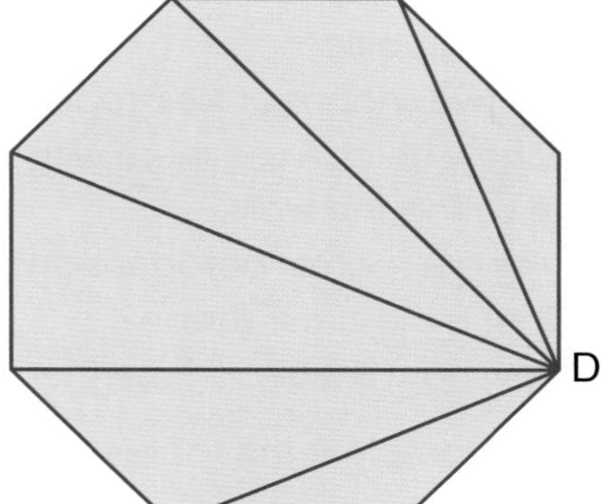

- **a** The lines from D divide the octagon into triangles.
  How many triangles are there?
- **b** What is the total of all the angles in the triangles.
- **c** What is the total of all the interior angles of the octagon.
- **d** Find the size of each interior angle if the octagon is regular.

3 A regular polygon has exterior angles of 20°.

- **a** What is the sum of the exterior angles?
- **b** How many sides does the polygon have?
- **c** What is the size of each interior angle?
- **d** What is the sum of the interior angles of the polygon?

4 Calculate the sum of the interior angles of

- **a** a hexagon
- **b** a nonagon (9 sides)
- **c** a polygon with 23 sides
- **d** a quadrilateral
- **e** a triangle
- **f** a polygon with 501 sides.

5 Look at the diagram of a heptagon.
Some of the sides are parallel.

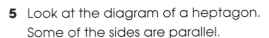

a How can you tell it is not regular?

b Write down the size of angle a.

c Explain why angle a = angle b.

b Calculate the sum of the interior angles of the heptagon.

e The heptagon is symmetrical.
Use this information to calculate the size of angle c.

Reasoning

6 The diagram shows a quadrilateral inside a regular dodecagon (12 sides).

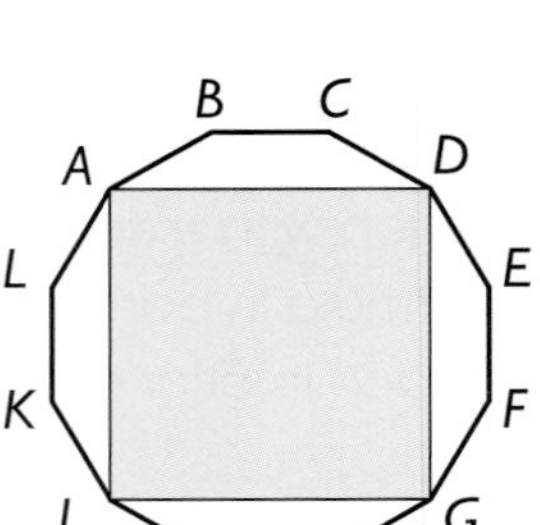

a Calculate angle ABC.

b Calculate angle BAD.

c Write down angle LAJ.

d Explain why quadrilateral ADGJ must be a square.

7 a Mark is tiling a floor.
He has two shapes of tiles: regular hexagons and equilateral triangles.
Show how he can use them without leaving any gaps.

b Sheuli is tiling the walls.
Her tiles are regular octagons and one other regular shape.

i What is the other shape?

ii Show how Sheuli can use them without leaving any gaps.

8 Copy and complete this table for regular polygons.

Number of sides n	Name of polygon	Total of interior angles $180(n-2)°$	Each interior angle $\frac{180(n-2)°}{n}$
3	Equilateral triangle		
4	Square	360°	90°
5			
6			
8			
10			
12	Dodecagon		
20	Icosagon		

Wider skills practice

Reasoning

1 The angles on a straight line add up to 180°.
The interior angles of a triangle add up to 180°.
Use these facts to prove that the exterior angle of a triangle equals the sum of the opposite two interior angles.

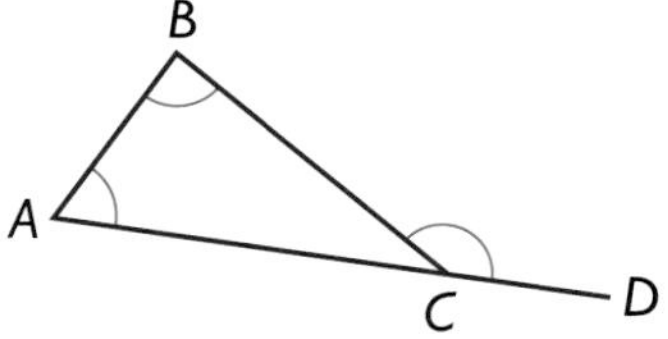

Angle BCD = angle BAC + angle ABC

Reasoning

2 The diagram shows an irregular hexagon.
Two of its interior angles are right angles.

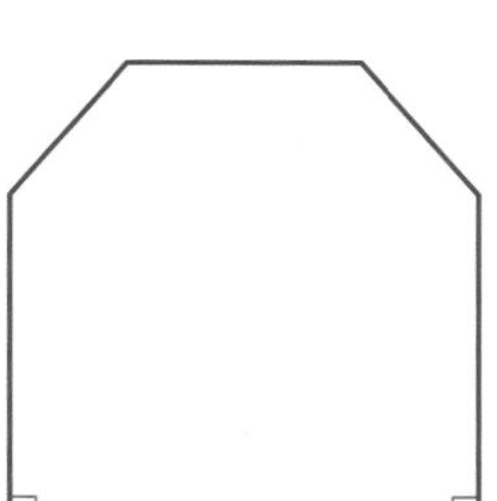

a Draw irregular hexagons with
- **i** one right angle
- **ii** three right angles
- **iii** four right angles
- **iv** five right angles

b Prove that the interior angles of a hexagon cannot be six right angles.

Applying skills

1 The diagram shows a gutter around a bay window on a house.
Calculate the size of angle x.

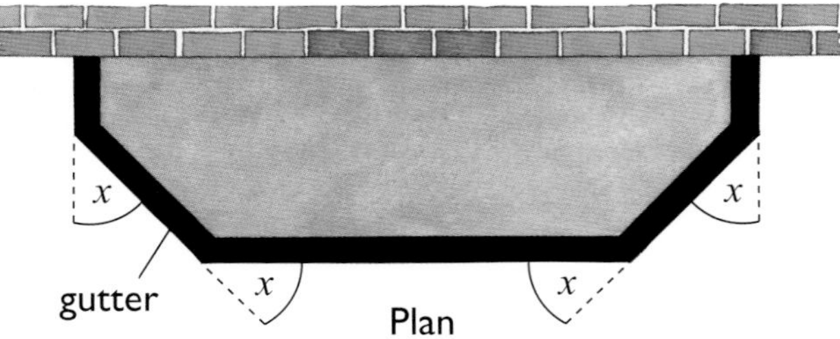

2 Oliver is a plumber.
He uses pipe connectors like this one.
Why does he call it a 20° connector?

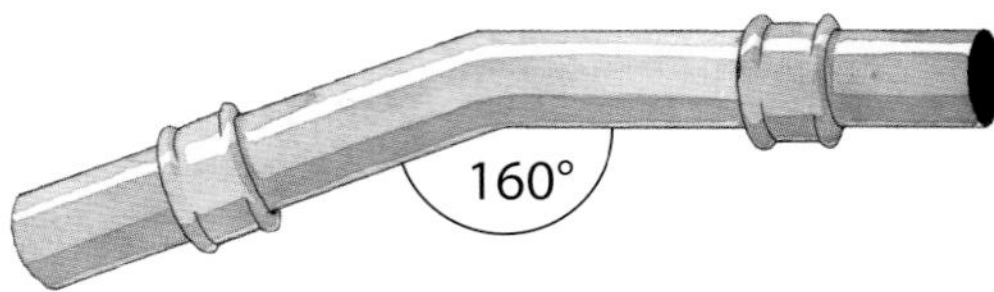

Reviewing skills

1 Calculate the size of each lettered angle.

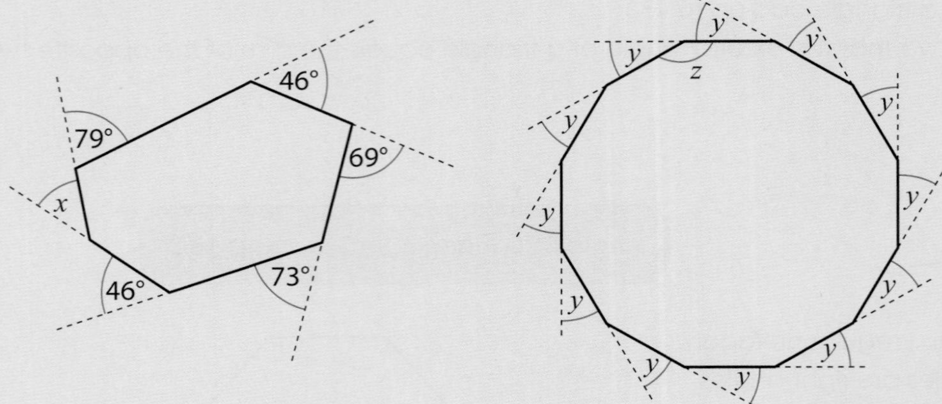

2 Calculate the size of the interior angle of a regular polygon with 120 sides.

3 A regular polygon has an interior angle of 144°.
How many sides does the polygon have?

Strand 3 • Measuring shapes

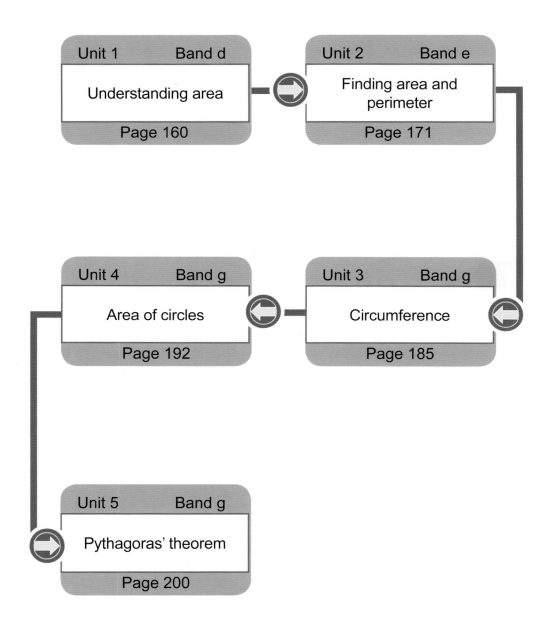

Unit 1	Band d	Understanding area	Page 160
Unit 2	Band e	Finding area and perimeter	Page 171
Unit 3	Band g	Circumference	Page 185
Unit 4	Band g	Area of circles	Page 192
Unit 5	Band g	Pythagoras' theorem	Page 200

Unit 1 • Understanding area • Band d

Building skills

Example outside the Maths classroom

Varnishing a floor

Toolbox

Area is the amount of space inside a two-dimensional shape.

It is measured in square units such as cm^2 or m^2.

One way to find the area of a shape is to count the number of squares inside it.

$1\,m^2$ is the space inside a square of side 1 m.

An area of $4\,m^2$ means an area equivalent to $4 \times 1\,m^2$.

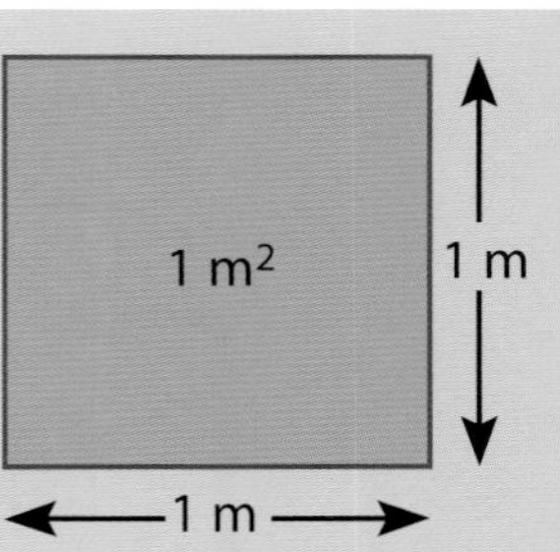

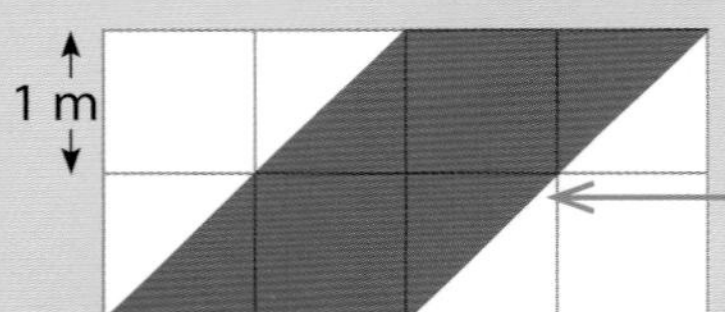

This parallelogram covers two full squares and four half squares. So in total it covers 4 squares. Its area is $4m^2$.

Example – Finding the area of a rectangle

Joe covers a flat roof on his house with decking.

The roof is 9 metres long and 4 metres wide.

What is the area of the roof?

Solution

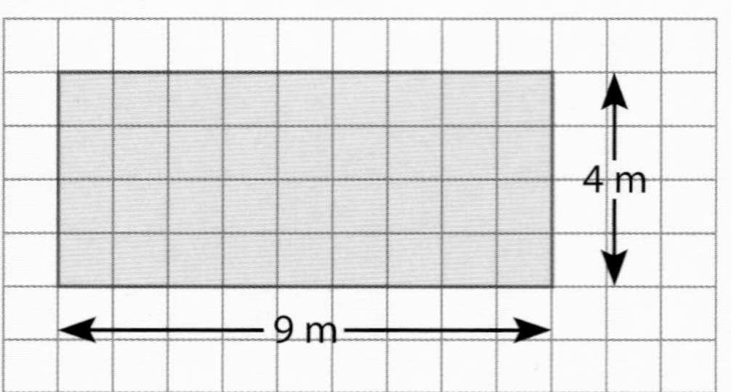

This rectangle is 9 squares long and 4 squares wide. So each square represents $1\,m^2$. Say this as 'one square metre'.

The roof is made up of 4 rows of 9 squares.

So the area is $4 \times 9 = 36\,m^2$.

Example – Finding the area of irregular shapes

Find the area of this shape.

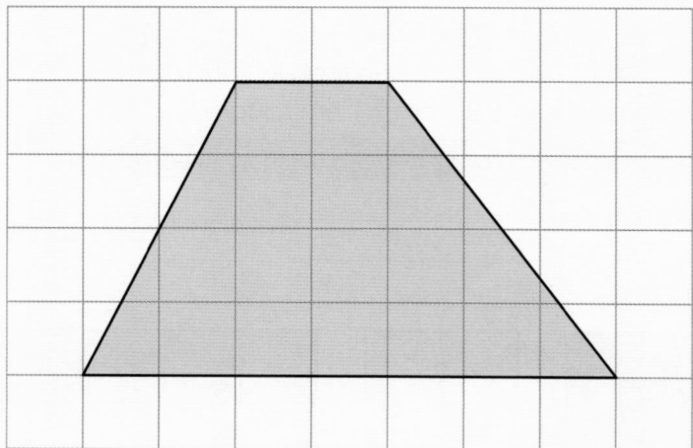

Solution

Each square is 1 cm × 1 cm.

1 cm
1 cm

So its area is 1 cm^2. ← Say this as 'square centimetre'.

There are 10 whole squares.

P, Q and R are a half square each, so $1\frac{1}{2}$ squares in total.

A and B add to one whole square.

C and D add to one whole square.

Total area = $10 + 1\frac{1}{2} + 1 + 1 = 13\frac{1}{2}$ squares

The area is $13\frac{1}{2}$ cm^2.

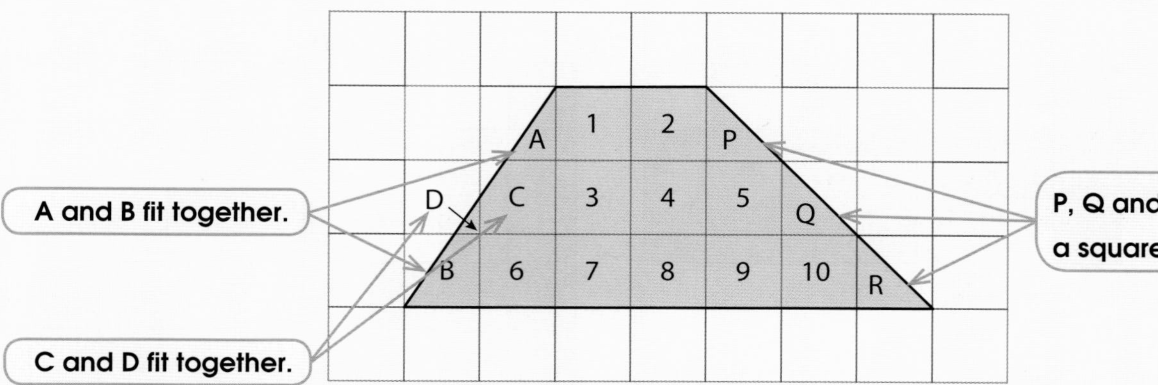

Remember:

- More complicated areas can be found by reallocating part squares to make full squares and then counting.
- Remember to include square units when you find areas.

Skills practice A

Reasoning

1 a Copy and complete this table for rectangles A to F.
The first row of the table has been filled in for rectangle A.
Each square represents a 1 cm × 1 cm square.

Rectangle	Length (cm)	Width (cm)	Area (cm^2)
A	3	2	6
B			
C			

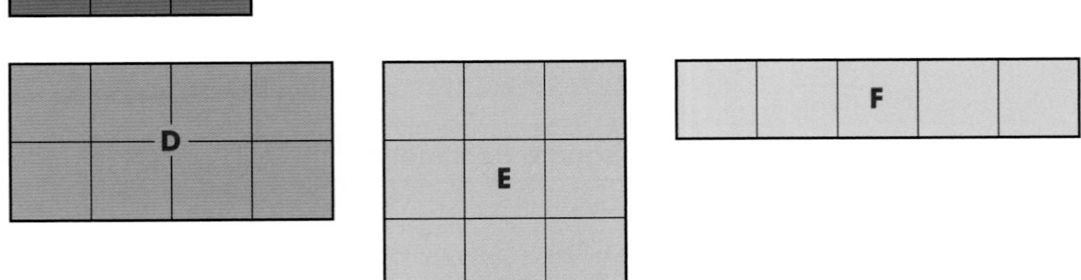

b Can you see a way to check that the figures in the table are correct without looking at the rectangles?

2 a Copy these shapes. Find the area of each shape. Each small square represents a 1cm × 1cm square.

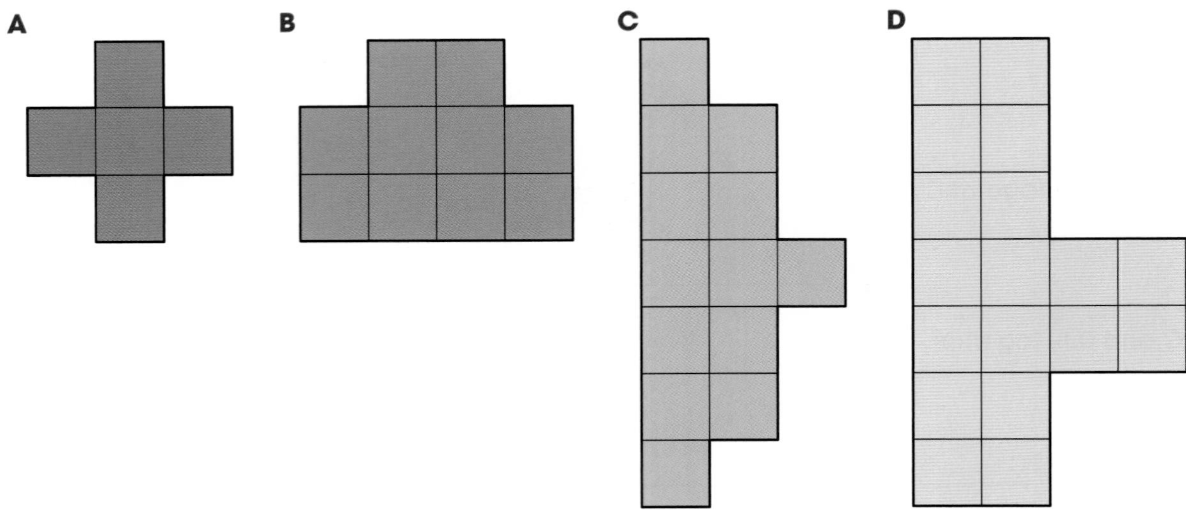

b Draw two different shapes with an area of 12 cm^2.

3 Copy these pairs of shapes.
Explain which has the larger area.

a

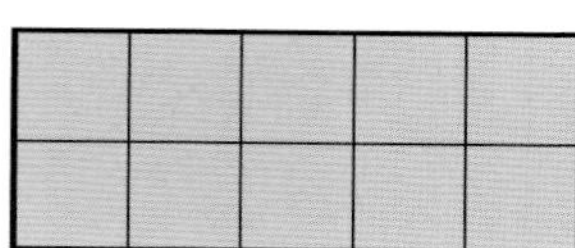

b

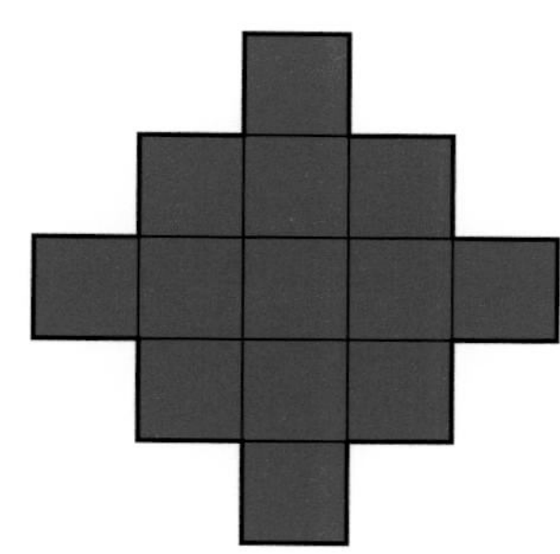

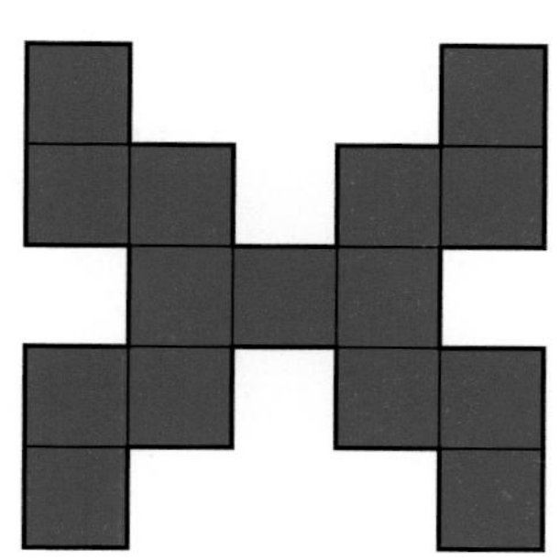

4 Tim wants a sign on his bedroom door.
He writes his name on squared paper.
Each square represents 1 square centimetre.

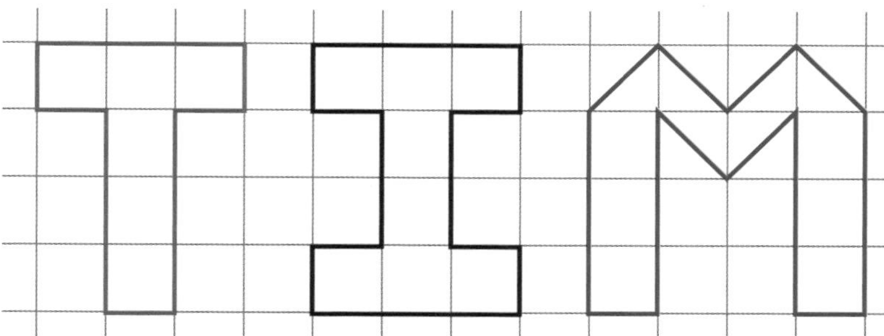

a What is the area of the letter T in Tim's sign?
b What is the area of the M in Tim's sign?
c What is the total area of the letters in Tim's sign?

My name's bigger than yours, Tim.

Jane also writes her name on squared paper.

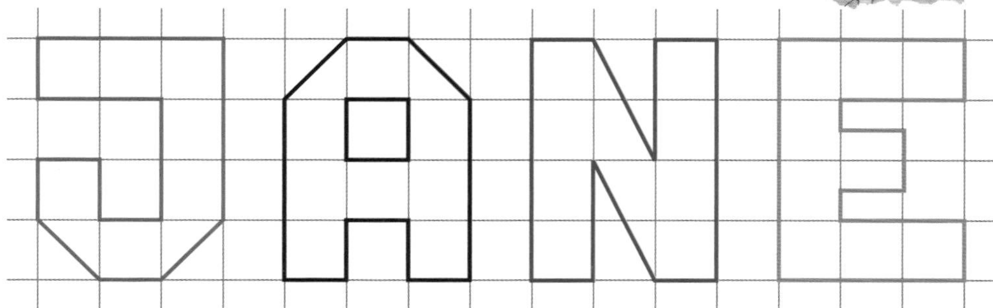

d What is the area of Jane's name?
e Is Jane right?
f Now write your own name on centimetre-squared paper.
Find the area of your name.

If your name is very long, just write your initials.

5 Find the area of each of these triangles.

a

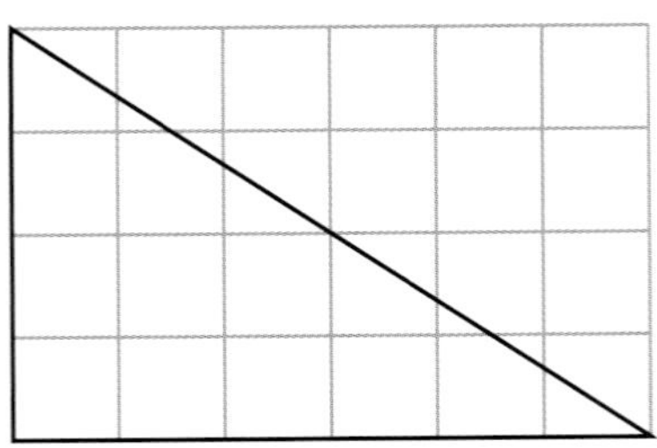

b

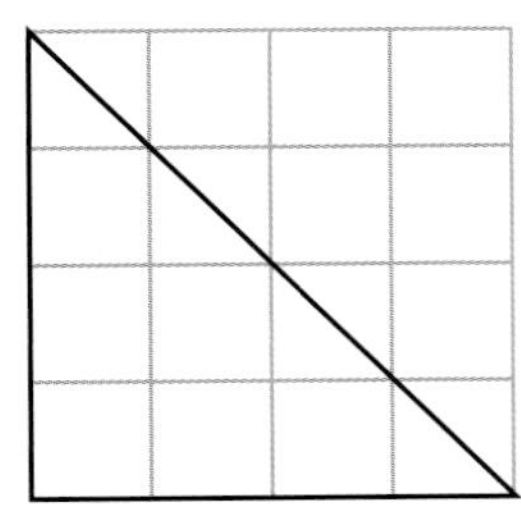

c

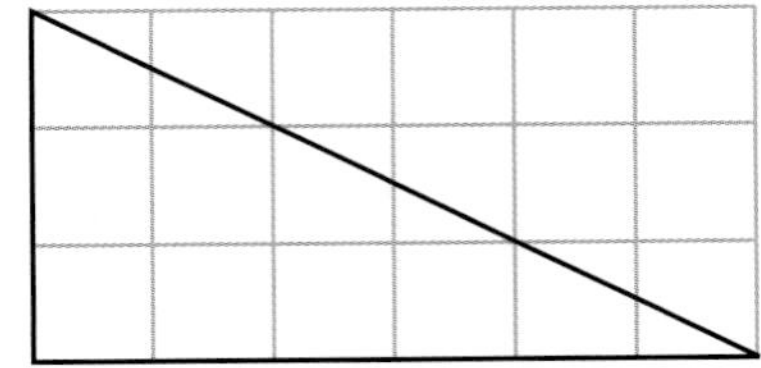

d

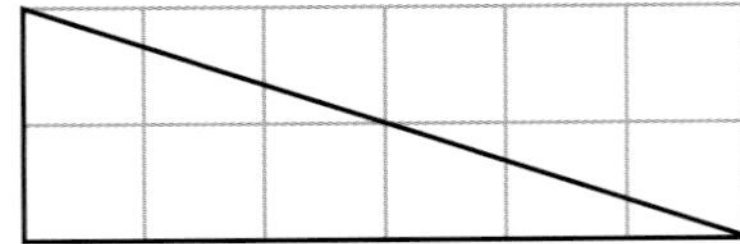

e

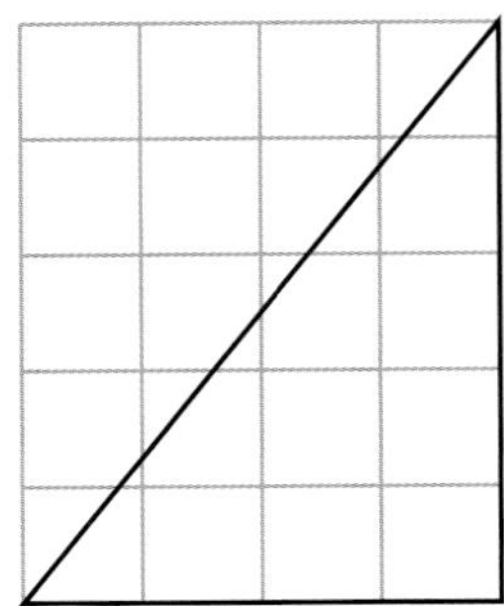

6 Copy each of these shapes and find its area.
Each square is 1 cm^2.

A

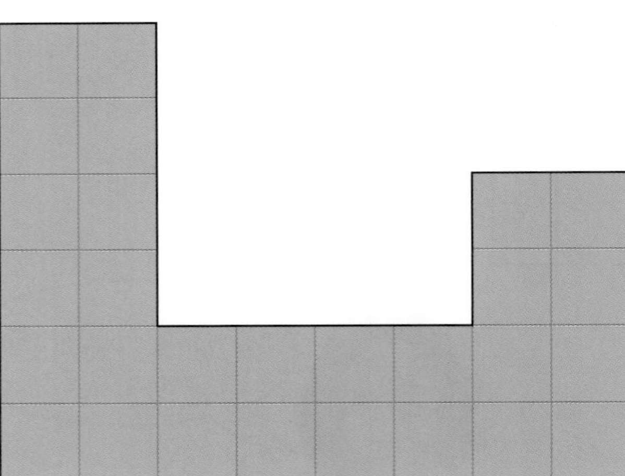

B

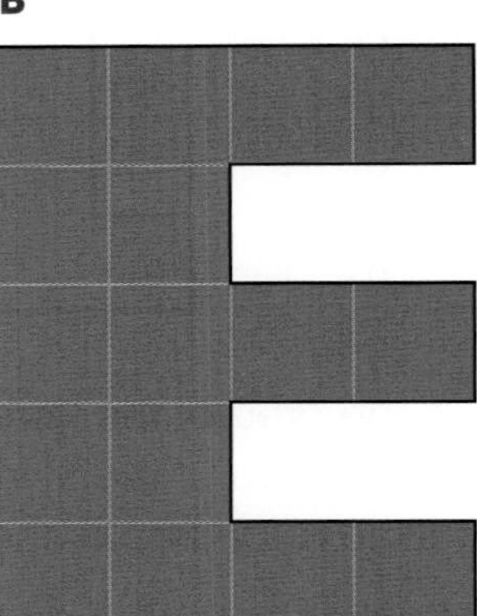

7 **a** Find the area of each of these shapes.

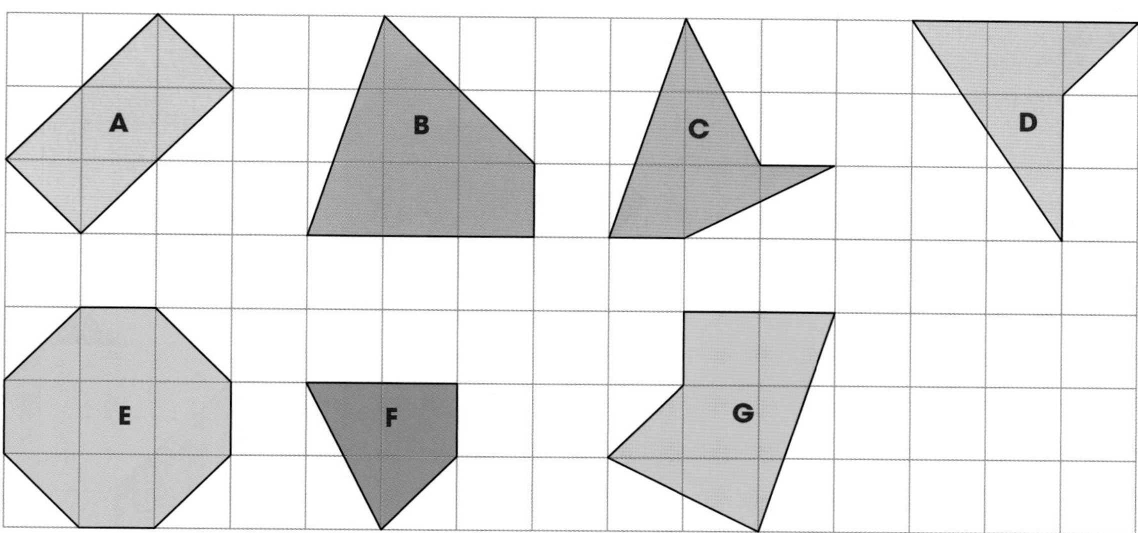

b Sort the shapes in order of area, largest first.

8 Find the area in square units of each of the shaded pieces.

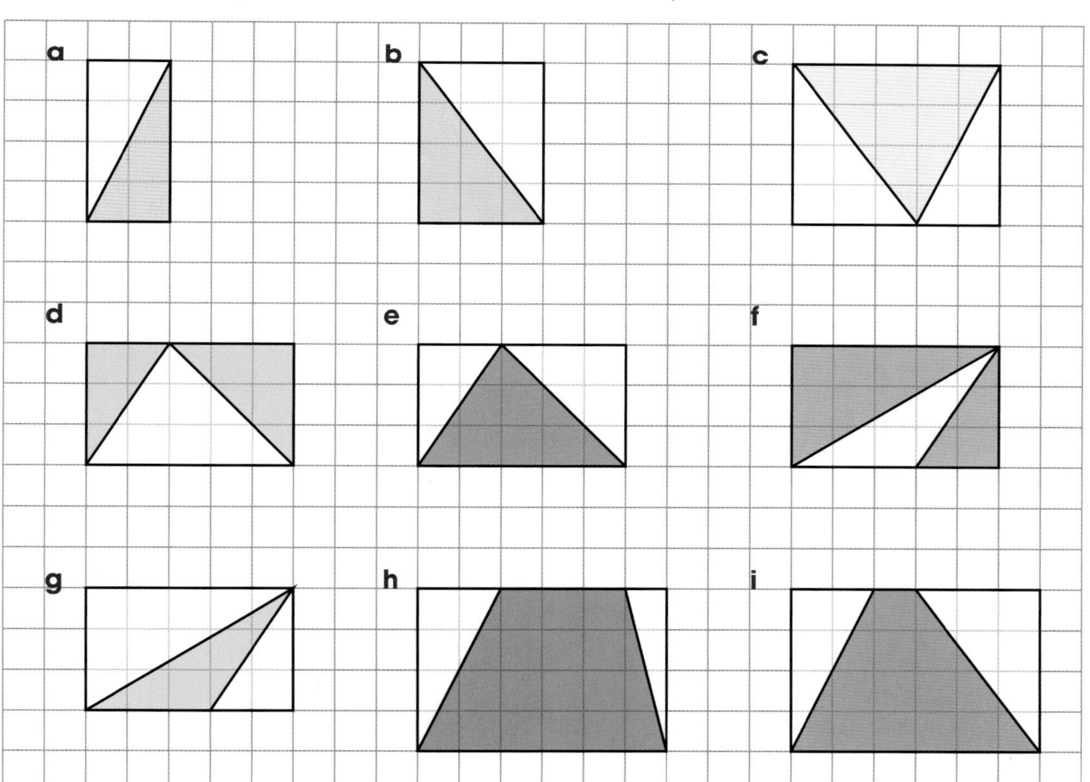

Skills practice B

1 Look at this picture of a dog.

a How many whole squares are there in this shape?

b How many half squares are there?

Each square measures 1 cm^2.

c What is the total area of the shape?

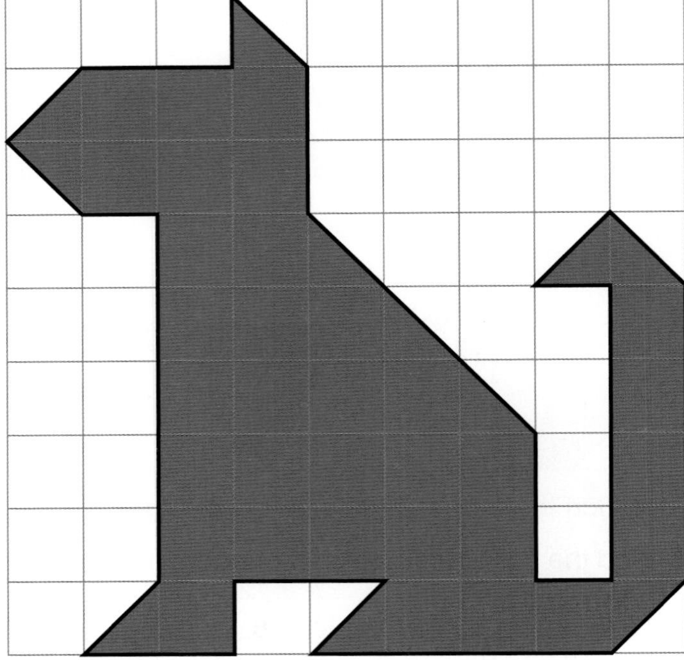

2 Work out the area of each of these shapes.

a

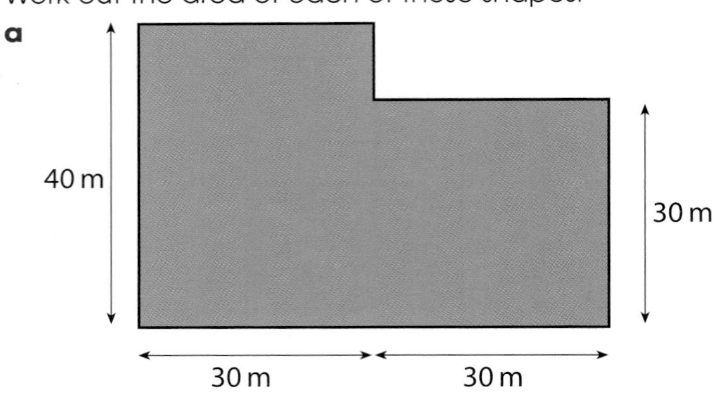

b

c

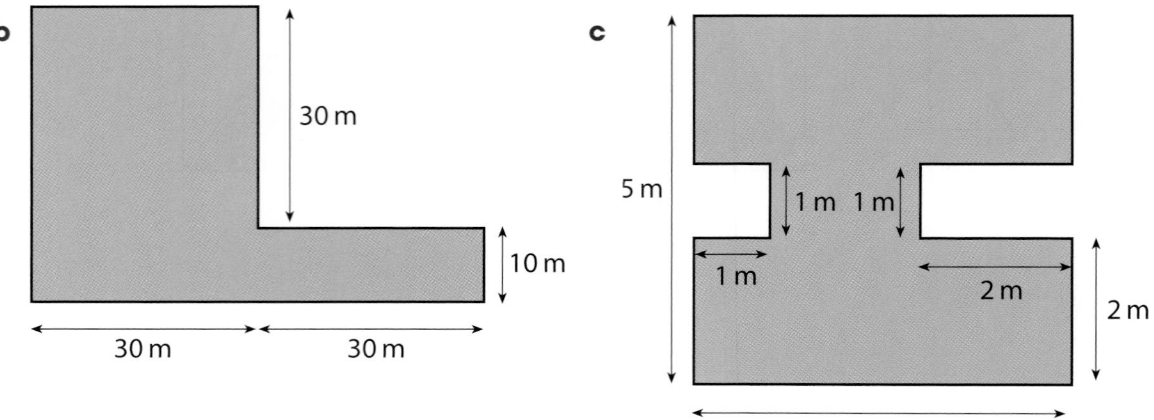

Reasoning

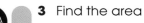

3 Find the area of each of these shapes.

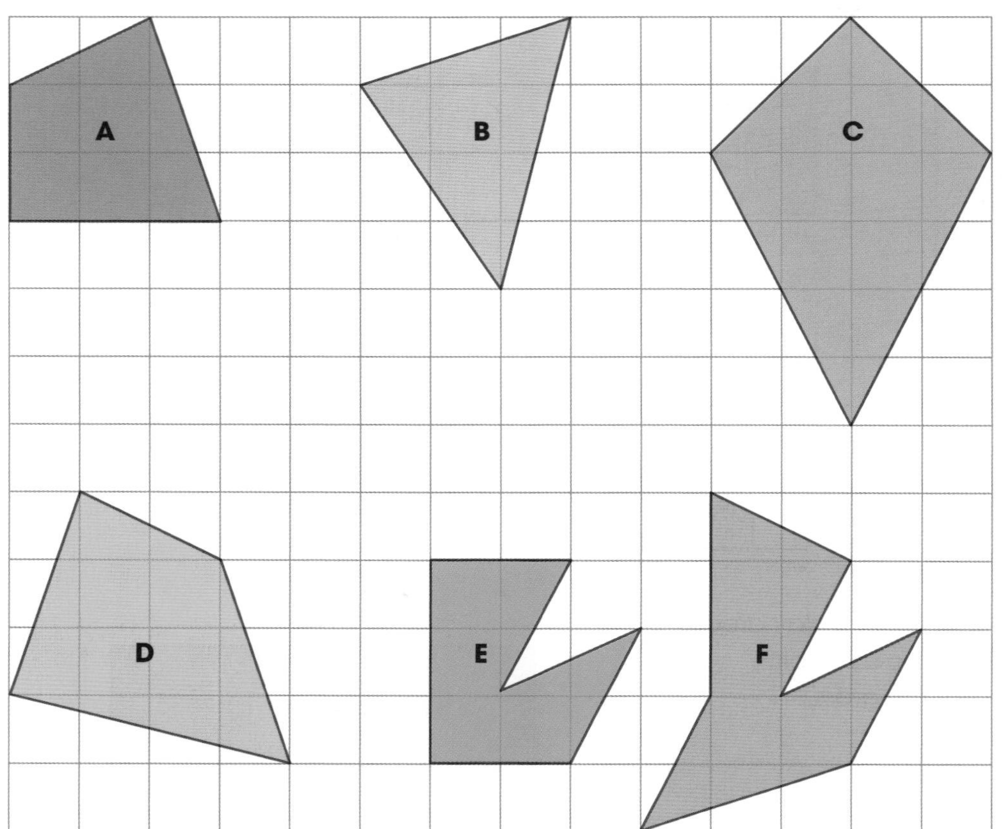

4 This is the plan of an ornamental lake in Avonford Park.

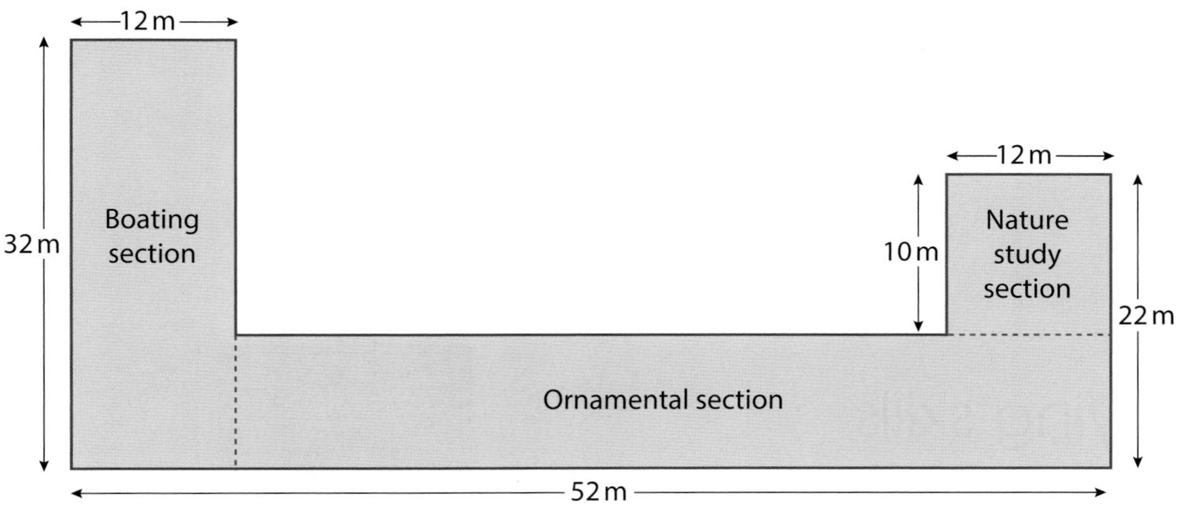

a Work out the area of the boating section.
b Work out the area of the nature study section.
c Work out the lengths of the ornamental section.
d Work out the area of the ornamental section.
e What is the total area of the lake?

5 Calculate the area of each of these objects.

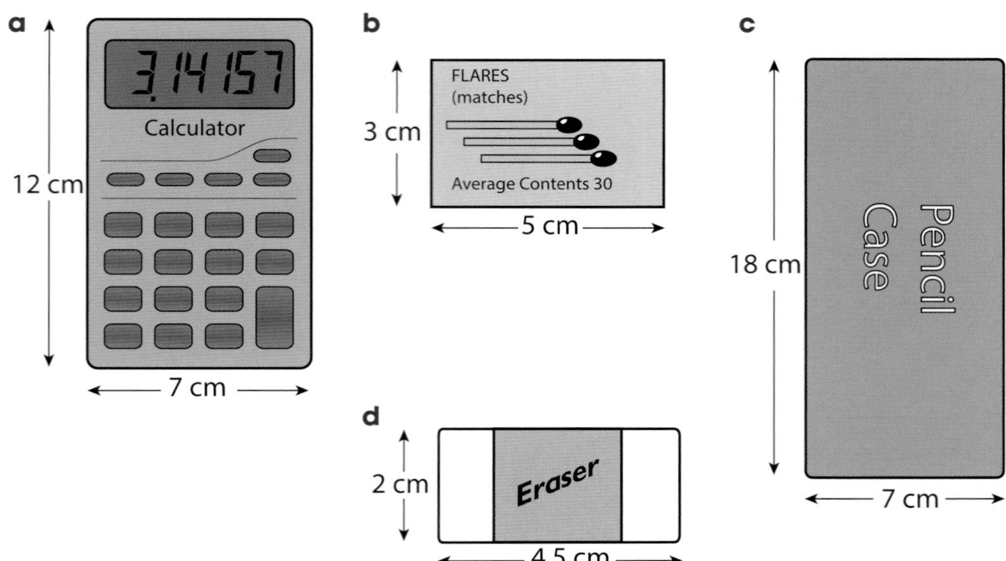

6 The diagram shows a children's play area.

a Work out the area of the sand-pit.

b Work out the area of the grass.

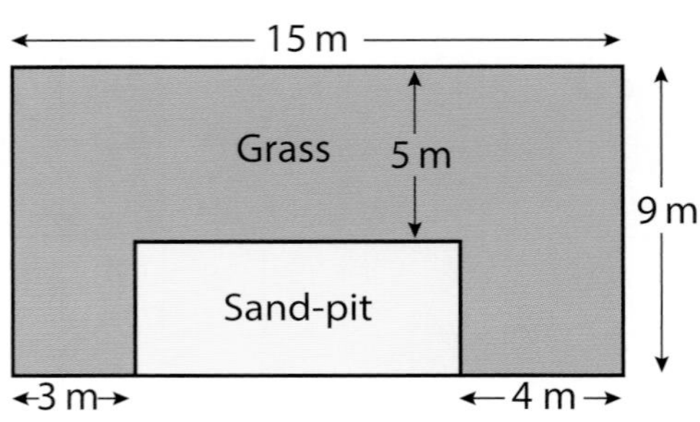

Wider skills practice

Problem solving

1 Plot each of these sets of points on graph paper, with both x and y axes labelled from 0 to 8.
Join them to make shapes.
Work out the area of each shape.
Your answers will be in square units.

a (0, 1), (3, 8), (6, 1)

b (0, 5), (4, 5), (6, 0), (2, 0)

c (0, 0), (2, 6), (5, 6), (8, 0)

d (0, 3), (1, 3), (4, 1), (7, 1), (8, 3), (7, 6), (4, 6), (1, 5), (0, 5)

Applying skills

Problem solving

1 A pad of writing paper has 30 sheets.

a What is the area of one side of one sheet of paper?

b How many sides of paper are there in the pad?

c What is the total writing area in the pad?

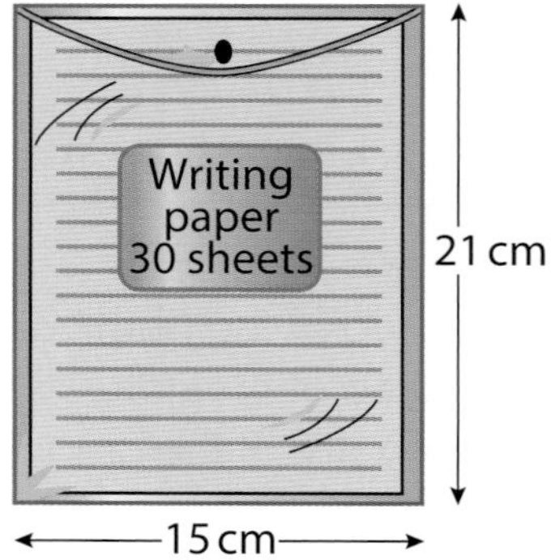

Problem solving

2 Find a way to show that these four shapes have the same area.

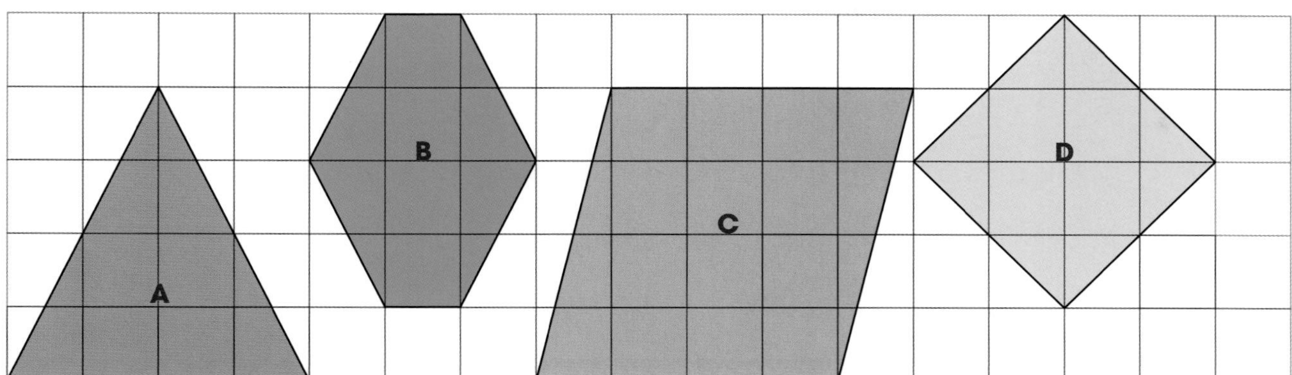

Reviewing skills

1 Look at these pairs of shapes.
In each case, say which has the larger area.

a

b

2 For each rectangle

i measure the length and width in centimetres

ii work out the area in cm^2.

a

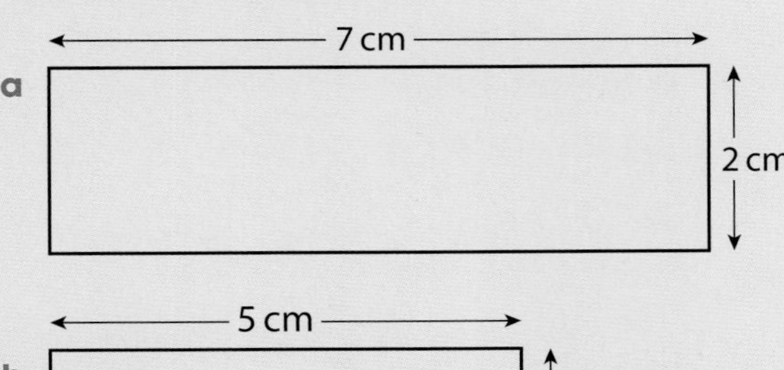

b

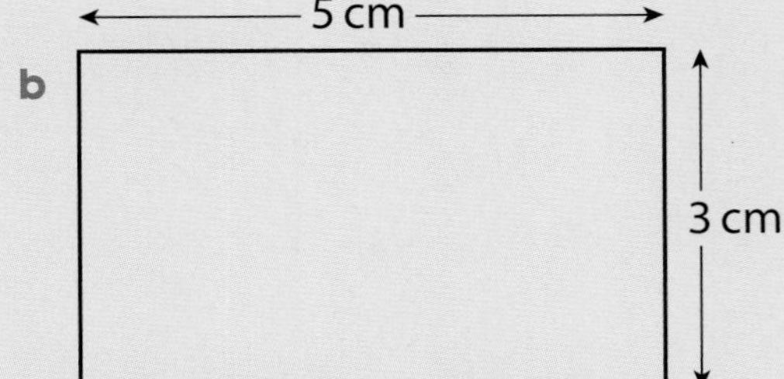

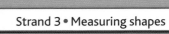

3 This shape is drawn on centimetre-squared paper.
Calculate its area.

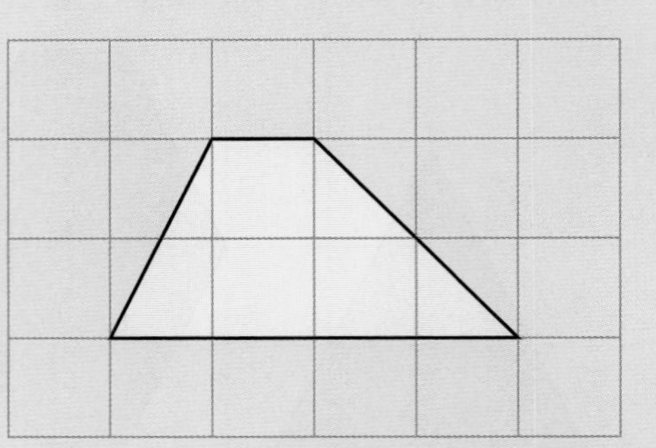

Building skills

Example outside the Maths classroom

Fencing

Toolbox

Perimeter is the distance all the way round the outside of a shape.
Area is the amount of space inside a two-dimensional shape.
Area is measured in square units such as cm^2 and m^2.
One way to find the area of a shape is to count the number of squares inside it.
However, it is not always practical to count the area inside the shape and when this is the case, a formula is used.
Rectangle: Area = length × width
Triangle: Area = $\frac{1}{2}$ × base × height
Parallelogram: Area = base × height
Choose one of the sides as the base.
Height is the distance between the base and its parallel side.

Example – Using the formulae for area

Find the area of each of these shapes.

a

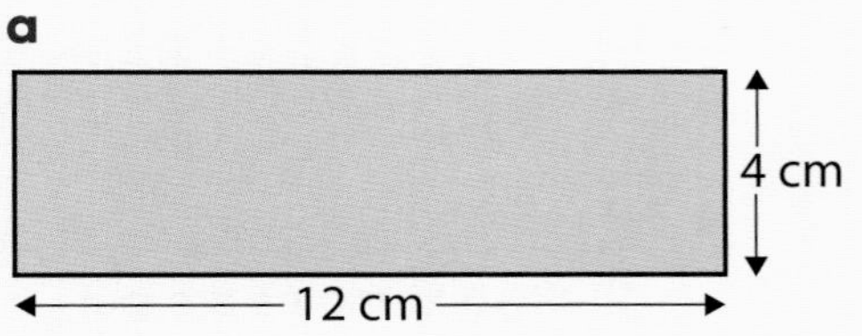

b

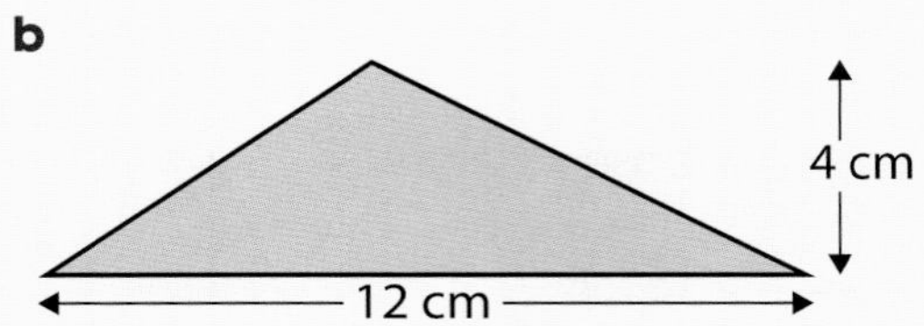

c

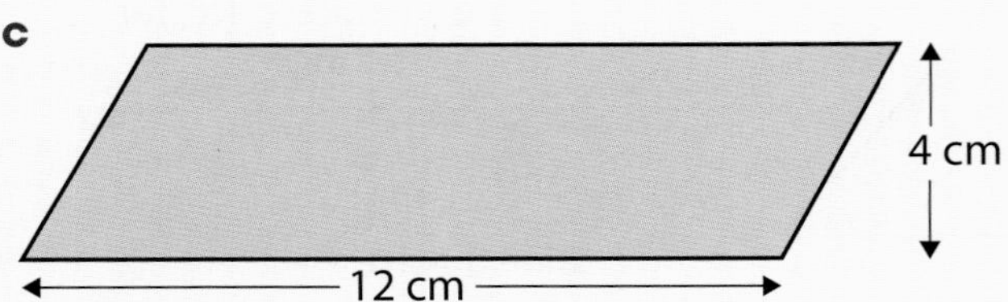

Solution

a Area of a rectangle = length × width

$= 12 \times 4$

$= 48\,\text{cm}^2$

b Area of a triangle = $\frac{1}{2}$ × base × height

$= \frac{1}{2} \times 12 \times 4$

$= 24\,\text{cm}^2$

c Area of a parallelogram = base × height

$= 12 \times 4$

$= 48\,\text{cm}^2$

Example – Finding the area of a trapezium

Find the area of this shape.

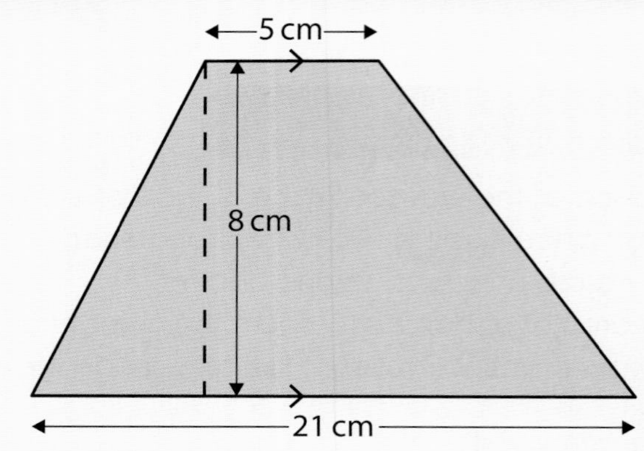
5 cm
8 cm
21 cm

Solution

Area of rectangle = length × width

$= 8 \times 5$

$= 40\,\text{cm}^2$

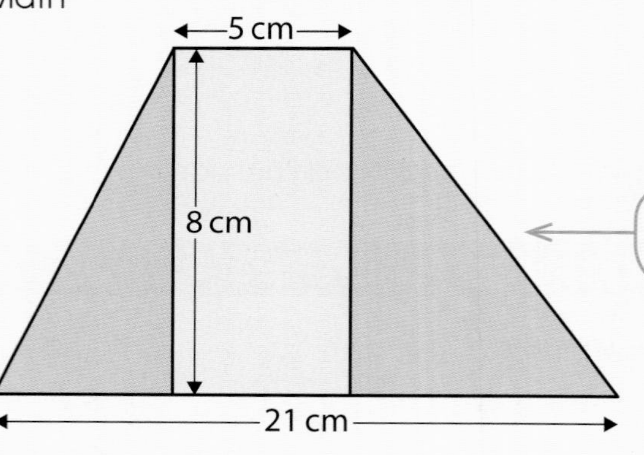
5 cm
8 cm
21 cm
The area of the rectangle is 8 × 5 = 40 cm².

Area of triangle = $\frac{1}{2}$ × base × height

$= \frac{1}{2} \times 16 \times 8$

$= 64\,\text{cm}^2$

Total area = $104\,\text{cm}^2$

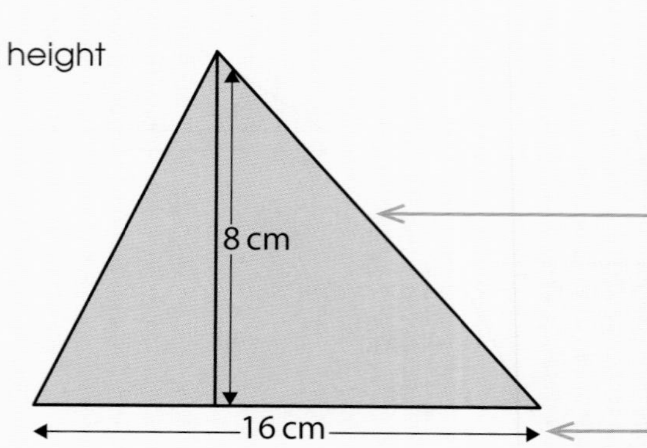
8 cm
16 cm
The two triangles are put together.
They make a larger triangle.
21 – 5 = 16

Example – Finding the perimeter and area of a compound shape

Find the perimeter and area of each of these shapes.

a

3 m
4 m
3 m
2 m

b

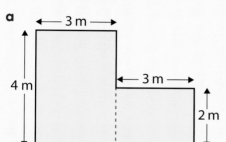

3 cm
3.5 cm
7 cm
11.3 cm

Solution

a Perimeter = 4 m + 3 m + 2 m + 3 m + 2 m + 6 m
= 20 m

Be systematic.
Start at one corner and work your way about the shape.

Area of whole shape = Area A + Area B
= (3 m × 4 m) + (3 m × 2 m)
= 12 m² + 6 m²
= 18 m²

The shape can be divided into two rectangles.

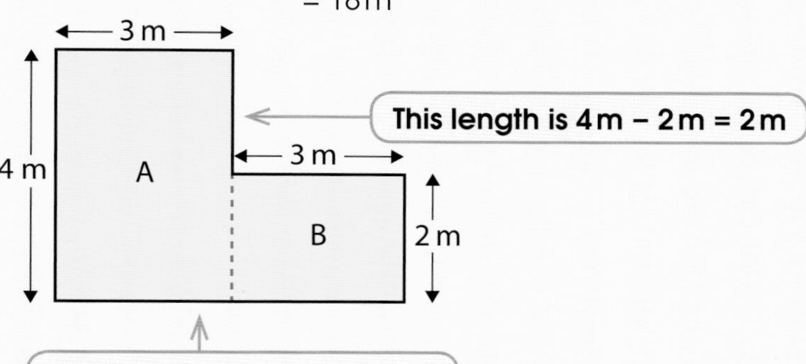

The base is 3 m + 3 m = 6 m

b Perimeter = 7 cm + 11.3 cm + 3 cm + 3 cm + 7 cm + 11.3 cm
= 42.6 cm

Area of whole shape = (7 cm × 11.3 cm) − (3 cm × 3.5 cm)
= 79.1 cm² − 10.5 cm²
= 68.6 cm²

You could think of this shape as a large rectangle with a small rectangle cut out.

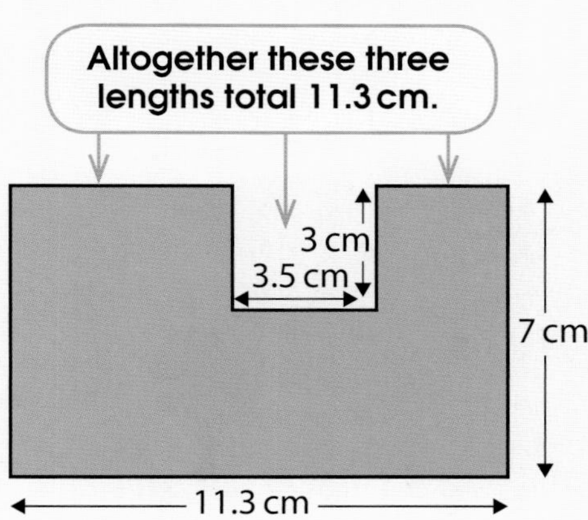

Remember:

- Make sure you include lengths for all the sections of the perimeter. They may not all be given.
- In addition to the base, you need the height of a shape for calculating area. This may be different from the side length.
- Perimeter is a length, but remember to use square units for area.

Skills practice A

1 For each rectangle

i measure the length and width in centimetres

ii work out the area in cm^2.

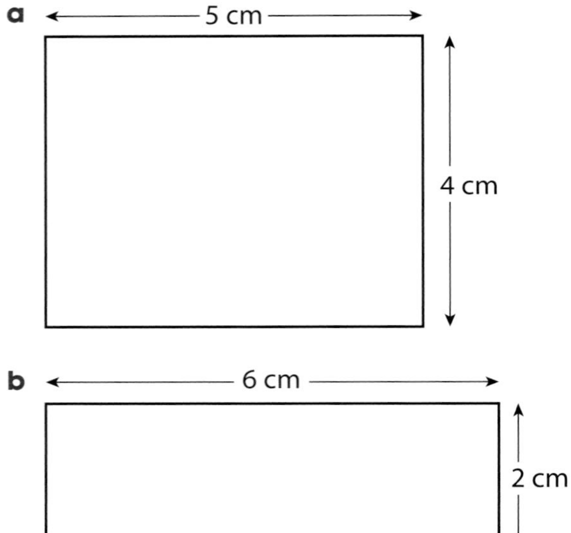

2 This is a sports field at Ali's school.

Before a football match, the team runs all the way round the field to warm up.

How far do the players run?

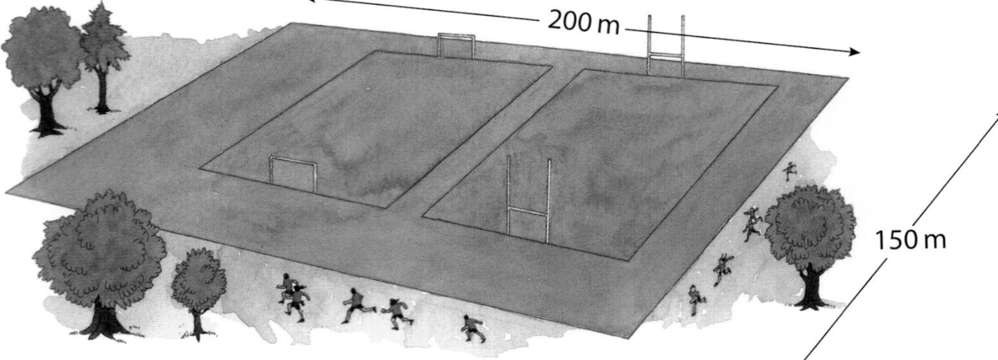

3 Carmen is a keen gardener.
She is making a herb bed in the shape of a rectangle.
The length is 8 feet and the width is 6 feet.
She is going to mark the edges with wire edging.
Each piece is 1 foot long.
How many pieces will she need to go all the way round the herb bed?

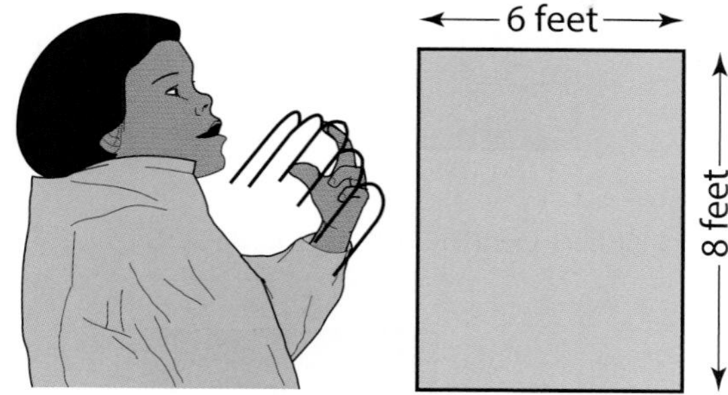

4 Look at this shape.
- **a** Find the area of part A.
- **b** Find the area of part B.
- **c** What is the area of the whole shape?

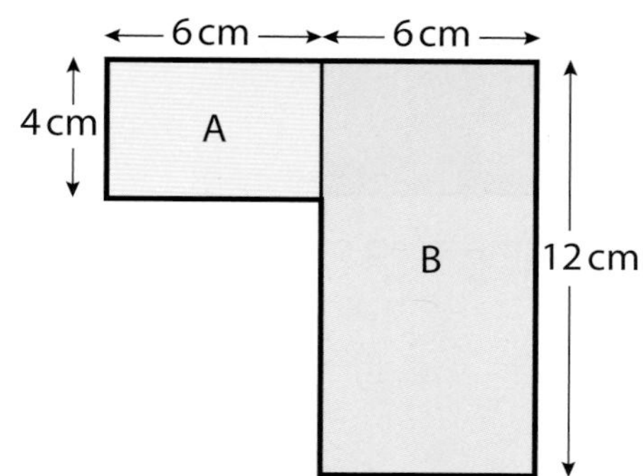

5 This is Karl's garden.
- **a** Find the area of the whole garden.
- **b** Find the area of the patio.
- **c** What is the area of the grass?

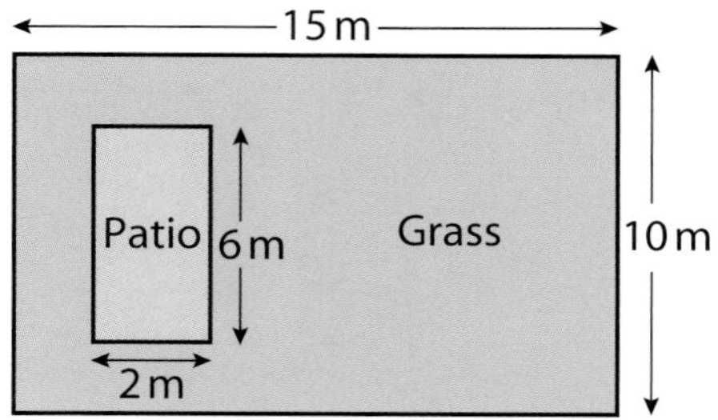

6 Work out the area of each of these shapes.

a

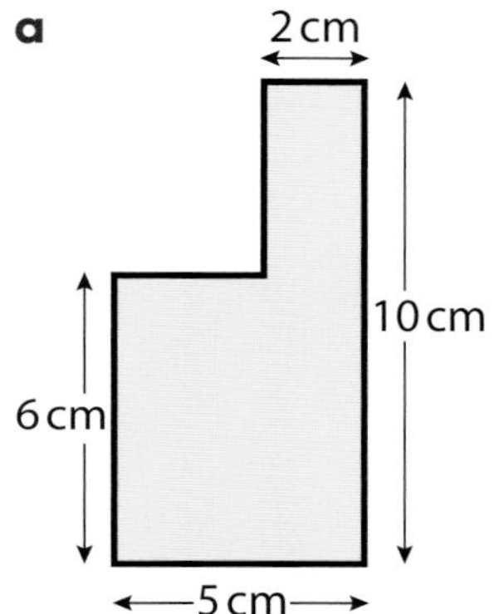

b

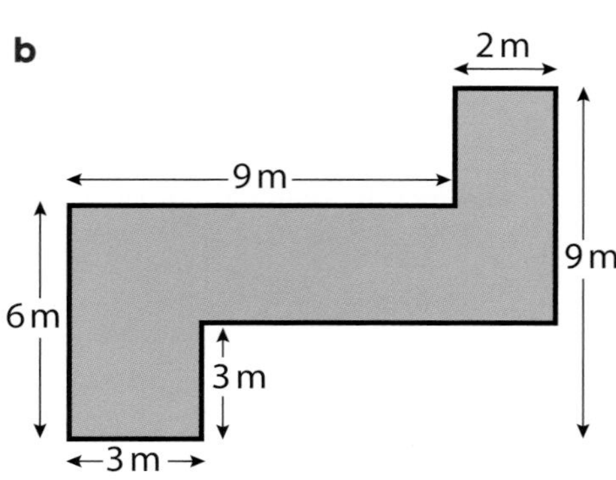

7 Find the area of each of these shapes.

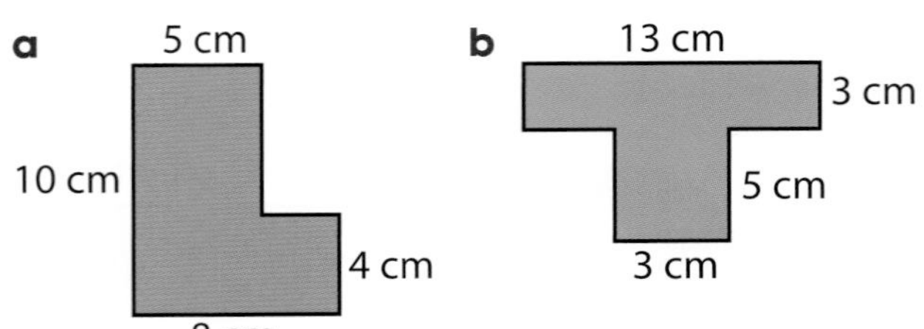

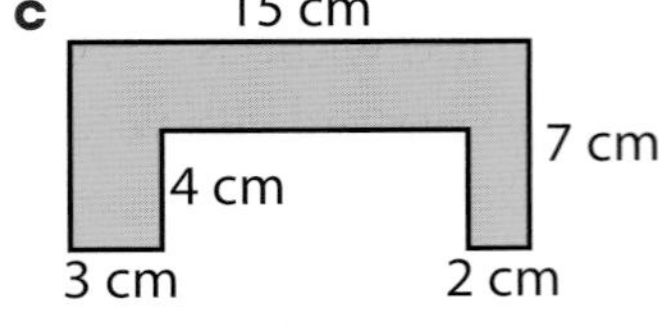

8 Find the perimeter of this shape.

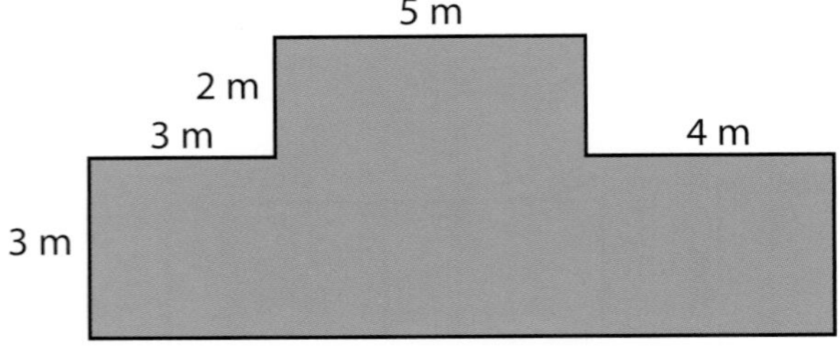

9 Work out the area of each of these triangles.

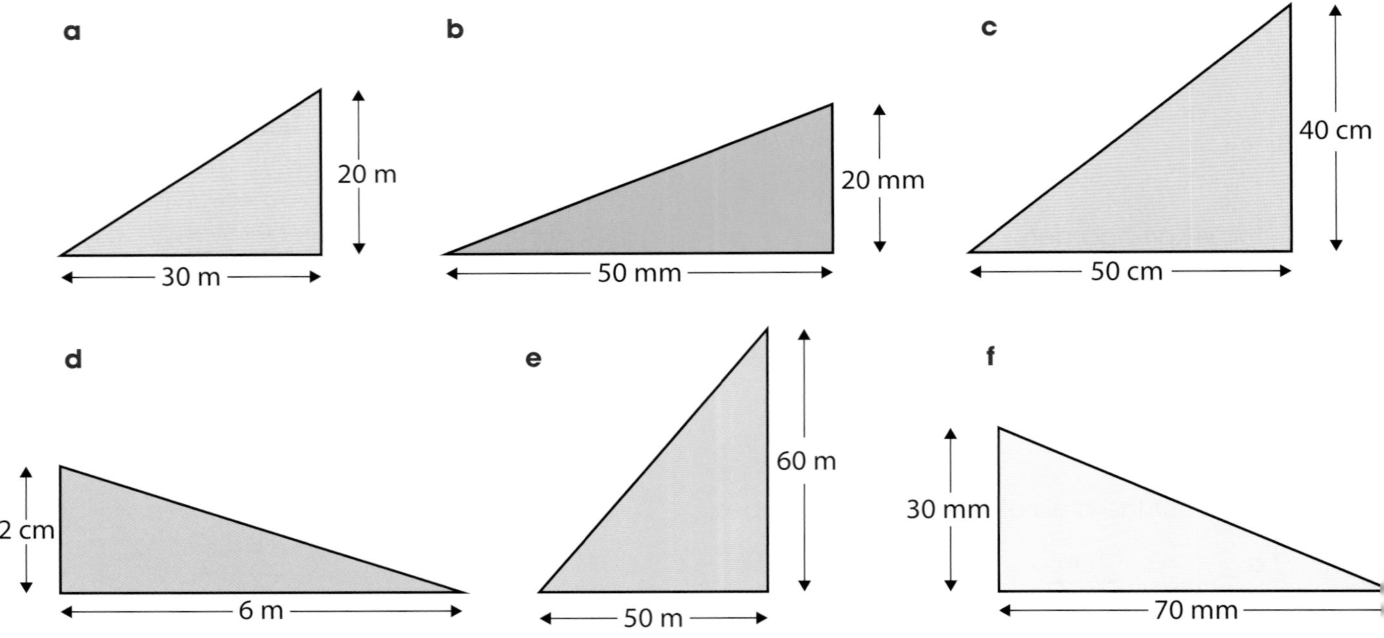

10 Choose the appropriate formula and use it to calculate the area of this shape.

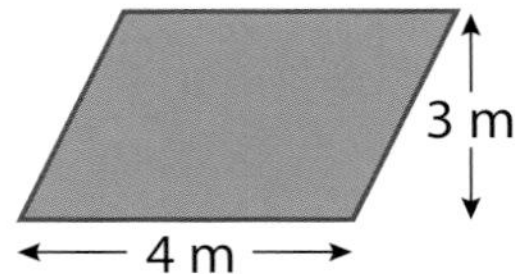

11 Find the area of each of these parallelograms.

a

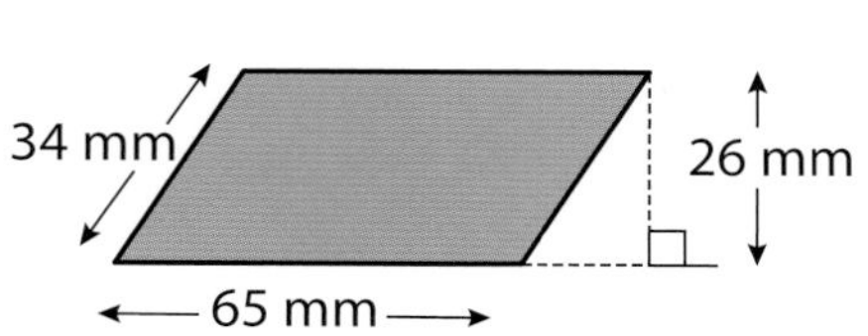

b

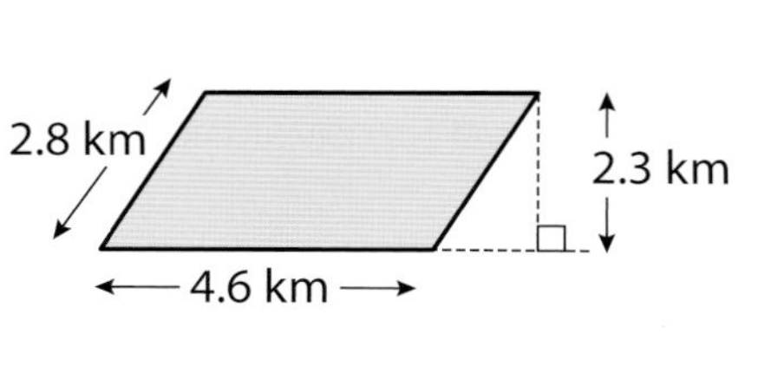

c

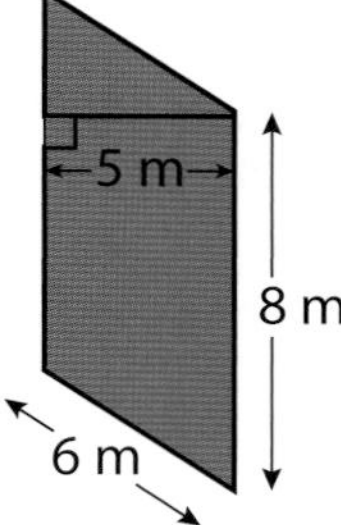

d

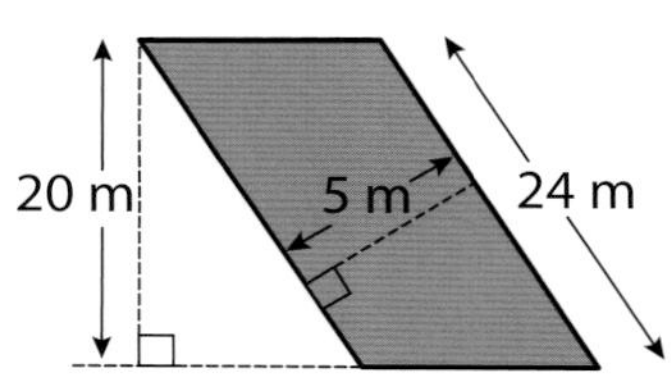

e

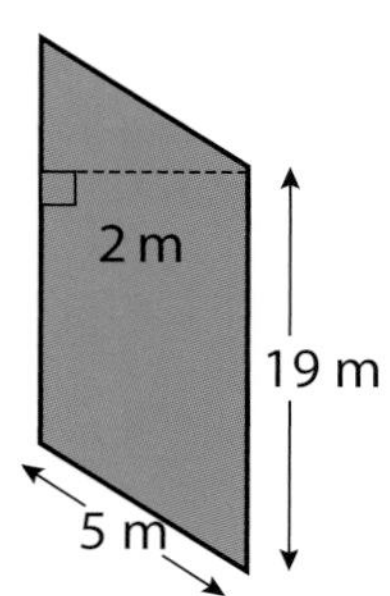

12 a Find the area of this parallelogram.

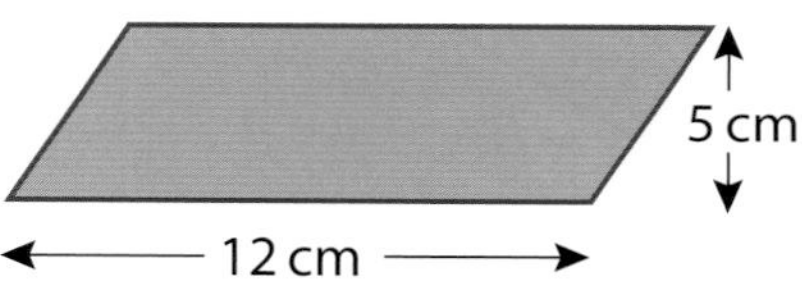

b What is the area of each trapezium in this diagram?

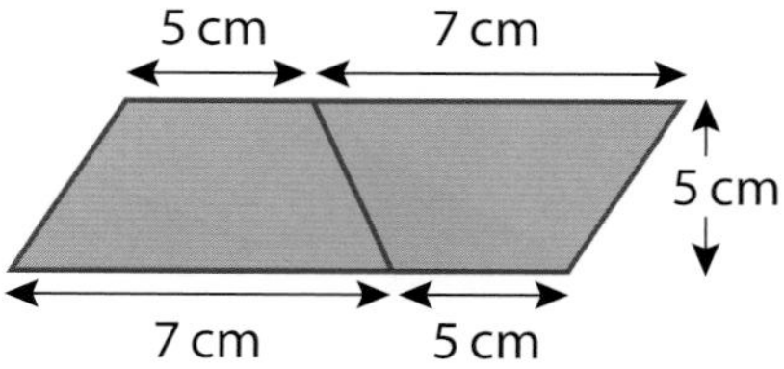

13 Calculate the area of this trapezium.

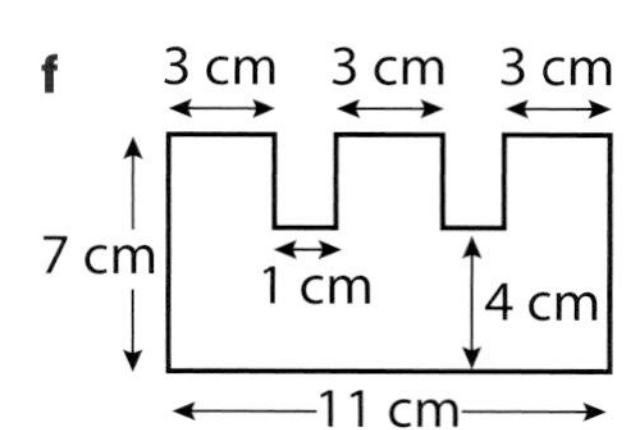

Skills practice B

1 Find the area of each of these shapes.

a

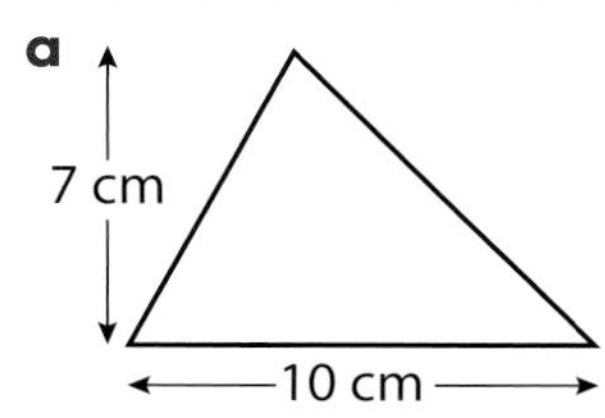

b

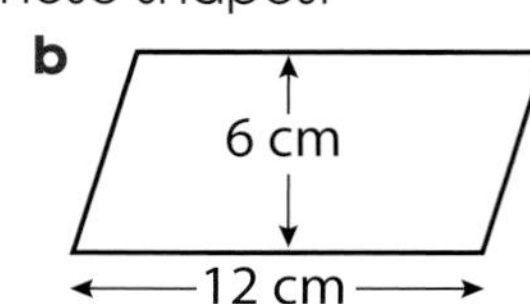

c

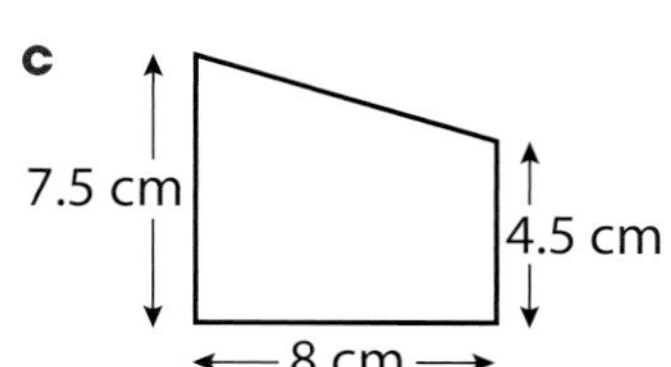

d

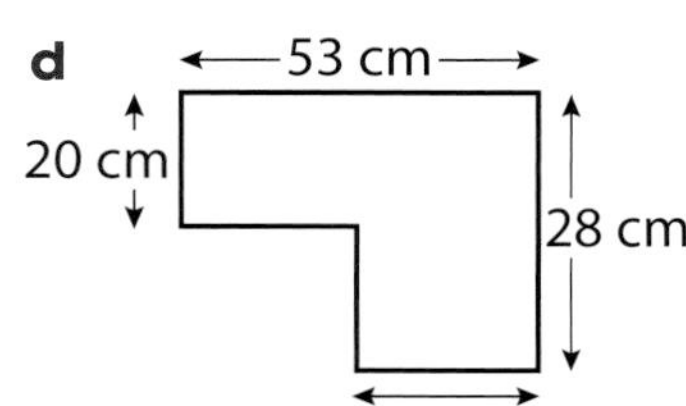

e

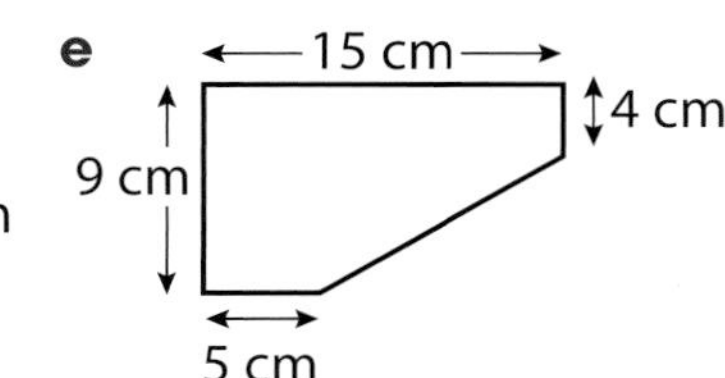

f

3 cm 3 cm 3 cm

7 cm 1 cm 4 cm

11 cm

2 Phoebe is going to gravel an area of her garden as shown in the diagram.

a What is the area of ground to be covered?

b One bag of gravel covers 0.75 m^2.
How many bags will Phoebe need to cover the plot?

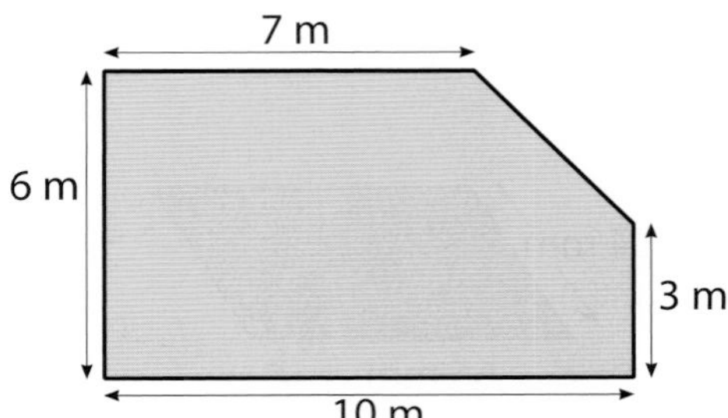

3 Find the area of each of this trapezium.

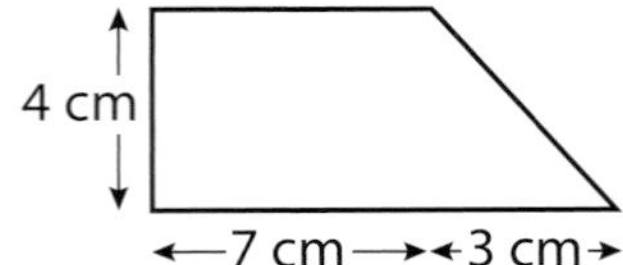

4 Calculate the area of each of these triangles by measuring.

a

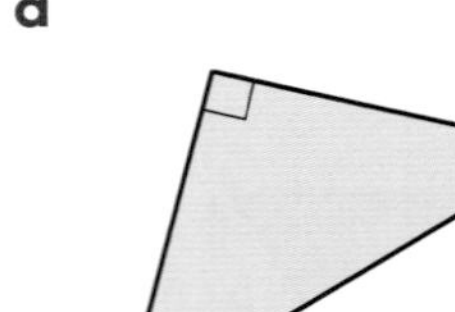

b

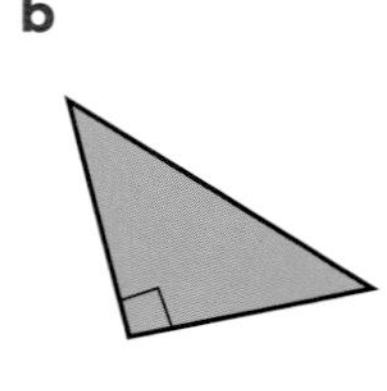

c

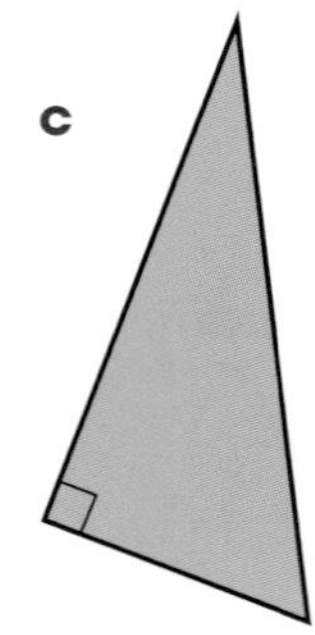

d

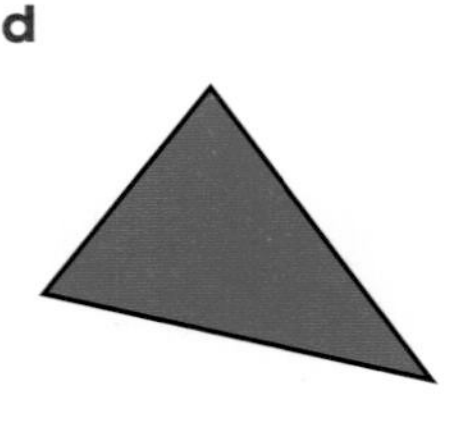

e

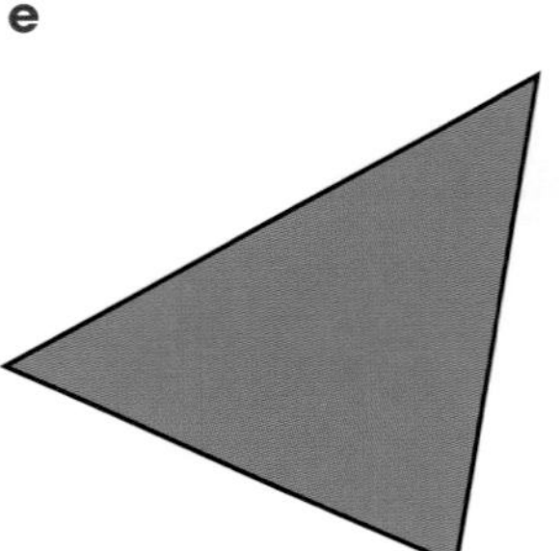

f

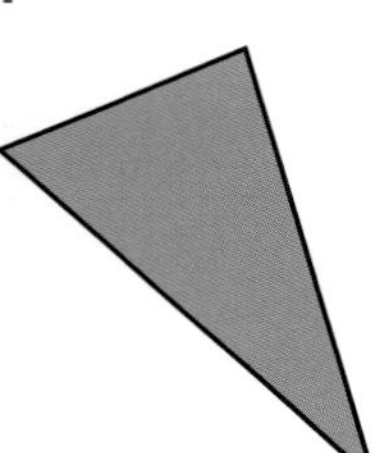

g

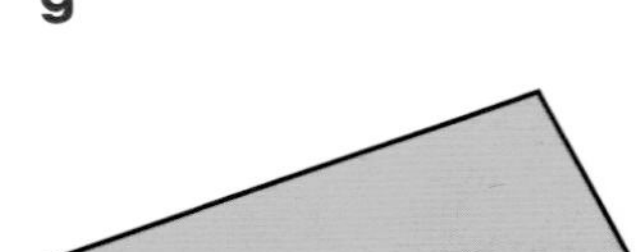

h

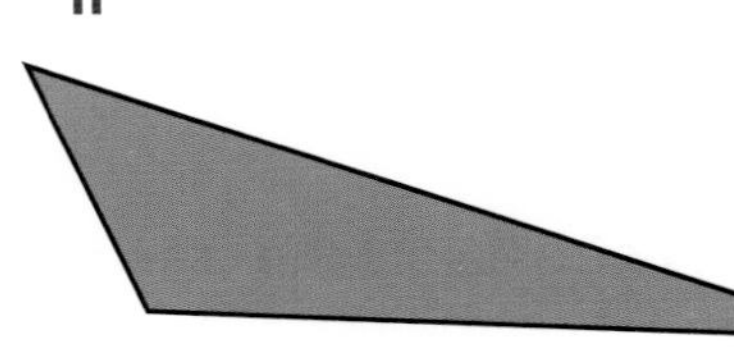

5 The diagram shows the plan of a bungalow.
Calculate the area of each of these rooms.

a Kitchen

b Bedroom

c Lounge

d Bathroom

e Hall

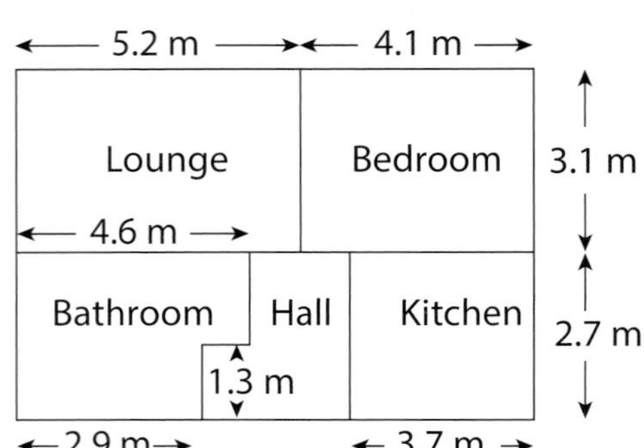

Reasoning

6 Find the area of each of the shaded shapes.

a

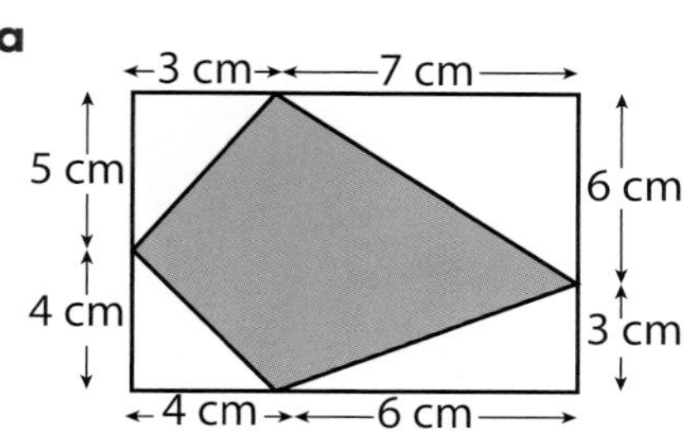

b

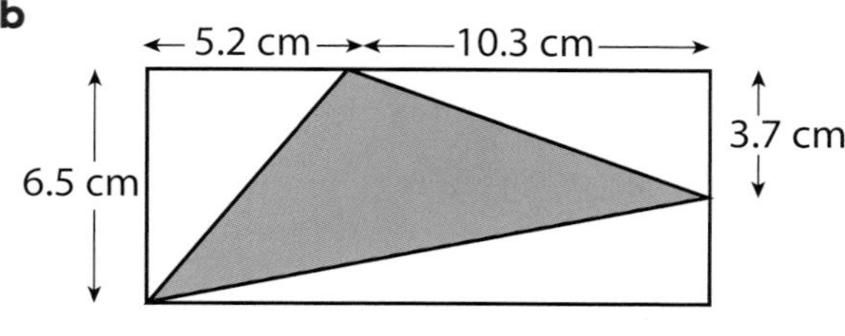

Reasoning

7 This letter E is painted on the side of a van.
What is the area of the painted letter?

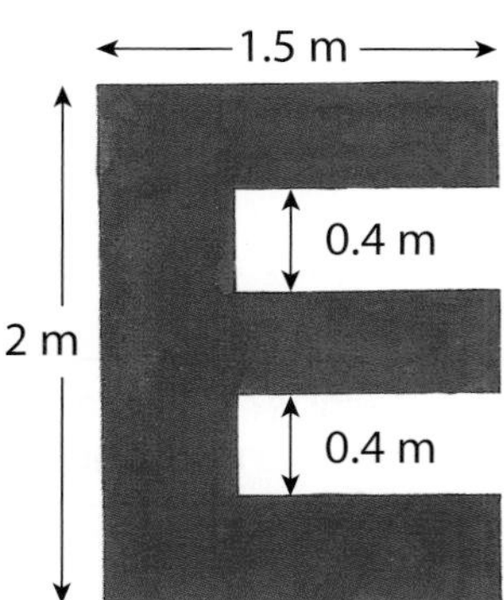

8 The perimeter of a square photograph is 40 cm.

a What is the length of one side of the photograph?

b What is the area of the photograph?

Reasoning

9 An underpass is 12 m long and 2.5 m high.
Jake is painting both walls.
A tin of paint covers 12.5 m^2.
How many tins does Jake need?

10 For his school design project, Samir makes a T-shirt.

a What shapes can you see in Samir's design?

b What is the area of one sleeve?
What about two sleeves?

c What is the area of the body?

d What is the total area of Samir's T-shirt?

Remember a T-shirt has a front and a back!

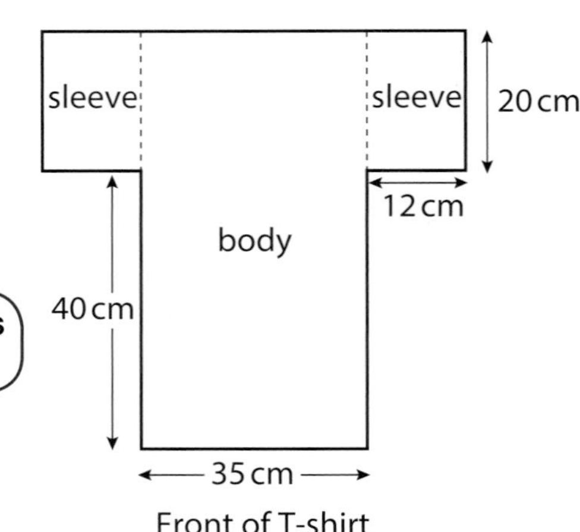

Front of T-shirt

e Samir is also making a green and white tablecloth.
How much green fabric does Samir need to make his tablecloth?

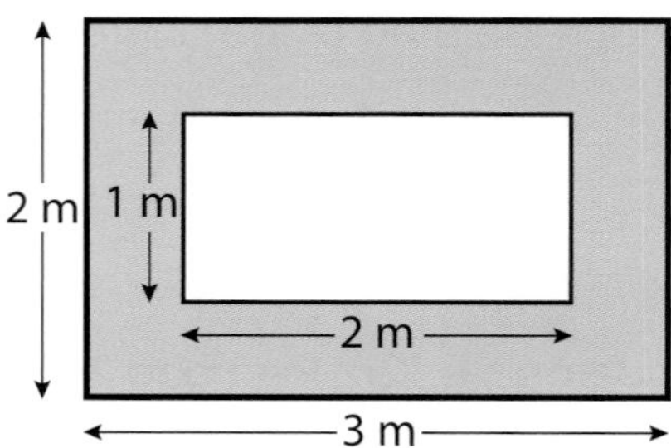

11 Work out the area of each of these parallelograms.

a

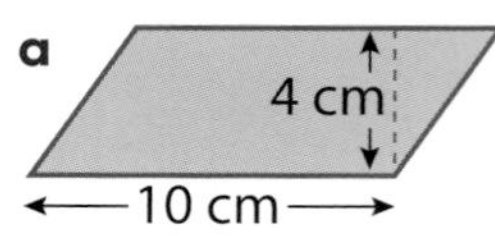

b

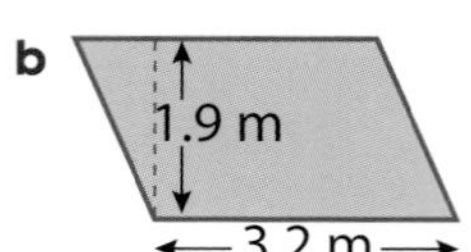

c Write the formula for the area of this parallelogram using b and h.

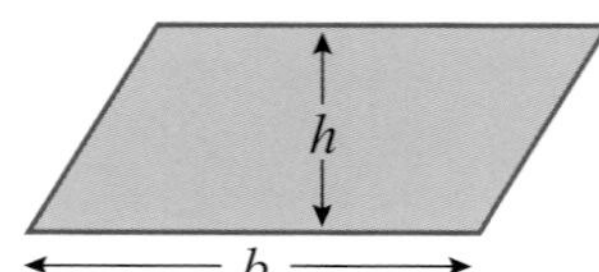

Reasoning

12 This diagram shows two squares.
The measurements are in metres.
Think about the areas of the squares.
Work out the length l.
Give your answer correct to the nearest centimetre.

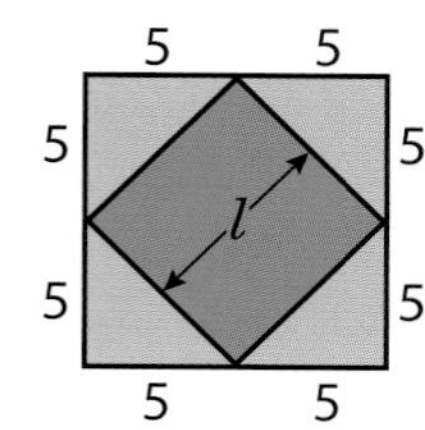

Reasoning

13 This parallelogram has an area of 144 cm².
Find the lengths of AD and BC.

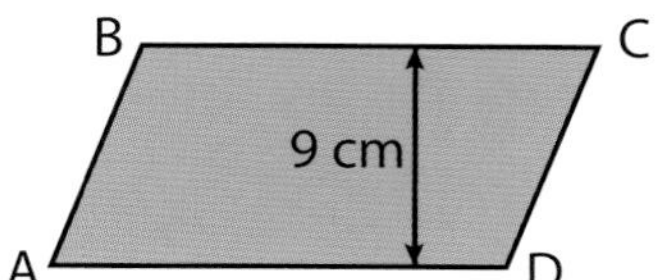

14 Kim and her dad are decorating the living room.

Kim measures the walls of the living room.
She draws a plan.

a How can they find the perimeter?
b What are the missing lengths?
c What is the perimeter of the room?

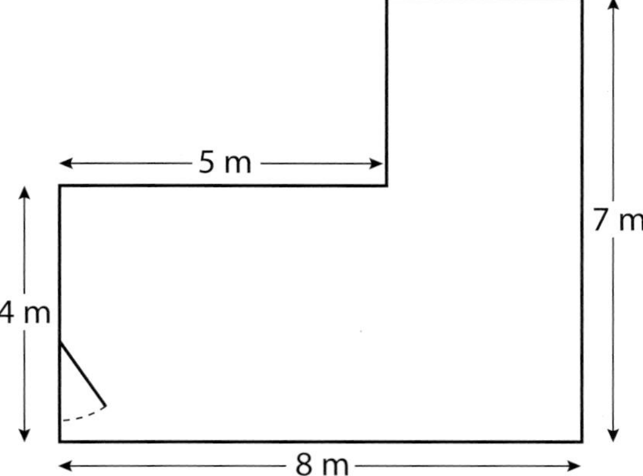

This is a sketch plan of the upstairs of Kim's house.

d Work out the missing lengths.

e Find the perimeter of each room.

f A square room has a perimeter of 16 m. What is the length of one wall of the room?

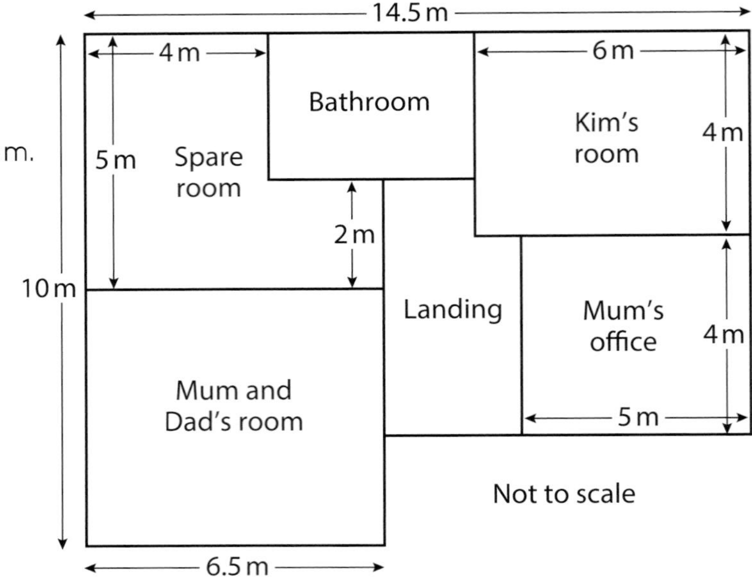

Wider skills practice

Reasoning

1 Here is a plan of Megan's garden.

a Fencing costs £1.20 per metre. Calculate the cost to fence around the garden.

b Megan decides to sow the garden with lawn seed. One packet of seed covers 12.5 m^2. How many packets will she need?

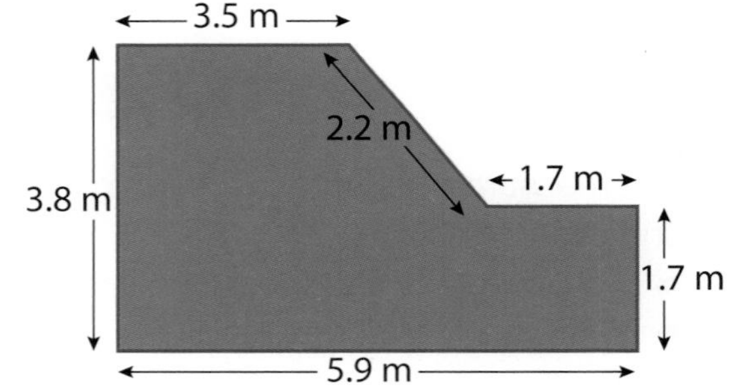

2 Find the area of the parallelogram ABCD with vertices at A(0, 1), B(0, 5), C(6, 7) and D(6, 3).

Reasoning

3 John has a new rug for his bedroom.

a What is the area of the rug in square metres?

b What are the measurements of the rug in centimetres?

c What is the area of the rug in square centimetres?

d How many square centimetres make 1 square metre?

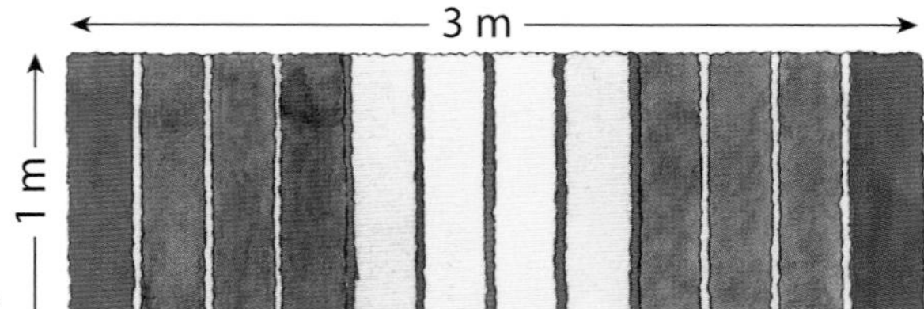

Reasoning

4 The full formula for the area of a triangle is

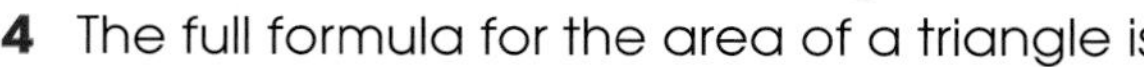

$\frac{1}{2}$ × base × perpendicular height.

How does the diagram explain the formula?

Why is it called perpendicular height?

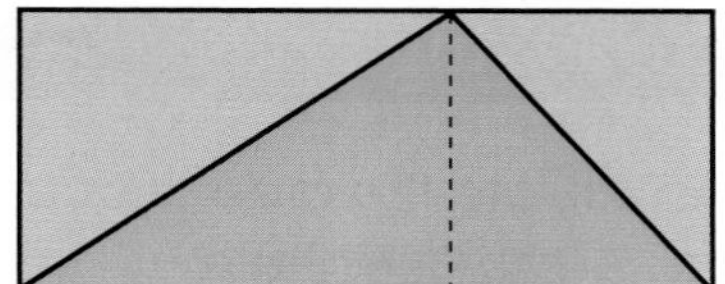

Applying skills

Problem solving

1 Liban is painting the end of his house.

Each tin of paint covers 25 m^2.

How much paint will he need to buy to cover the wall?

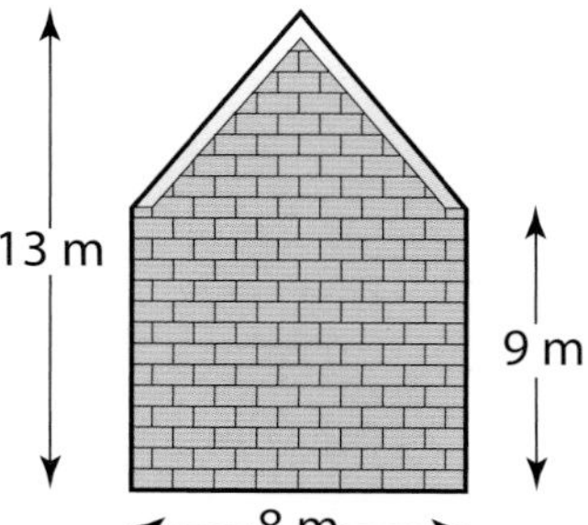

Problem solving

2 The diagram shows the floor and four walls of Tom's bedroom.

It has one door and two identical windows.

The other two walls are plain rectangles.

a Tom is putting a new skirting board round the room.

What is the total length of wood he will need?

b He plans to put new wallpaper on the wall opposite the door.

The wallpaper is sold in rolls which are 55 cm wide and 11 m long.

How many rolls will he need to buy?

c The other three walls are going to be painted.

Tom needs to know the area to be painted.

Find the total area of the three walls he is going to paint.

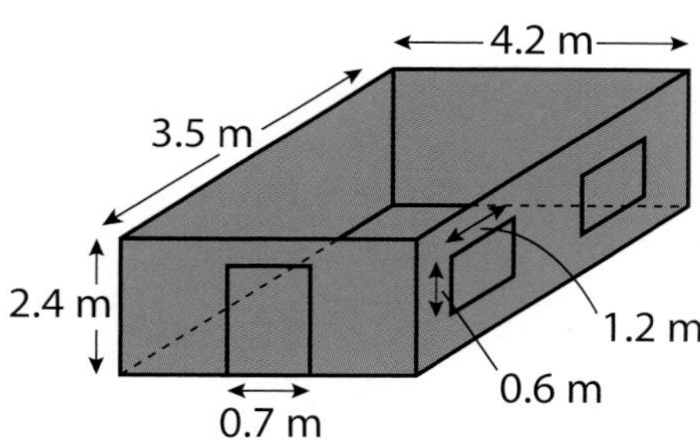

Problem solving

3 Mercy is buying carpet for her bedroom.

Carpet is sold by the whole square metre.

How much carpet does Mercy buy?

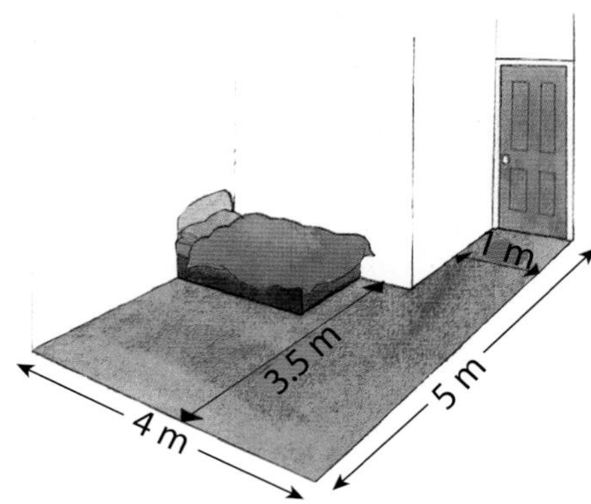

Reviewing skills

1 Calculate the area of each of these shapes.

a

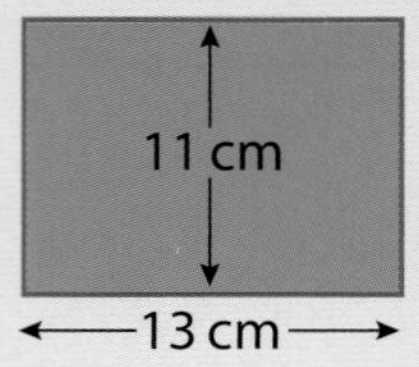

b **c**

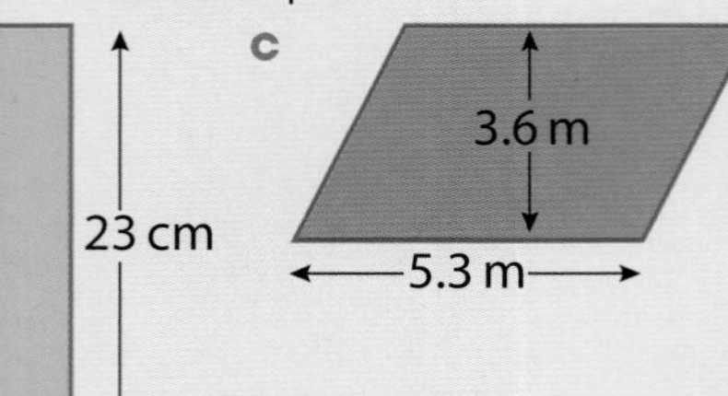

d

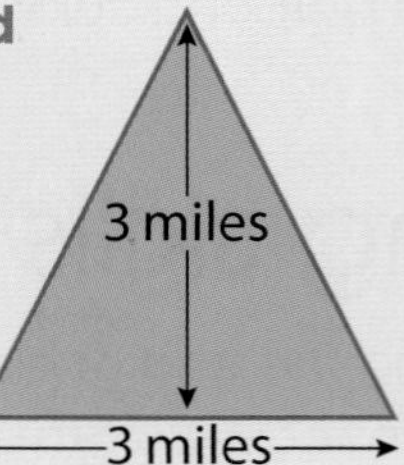

e

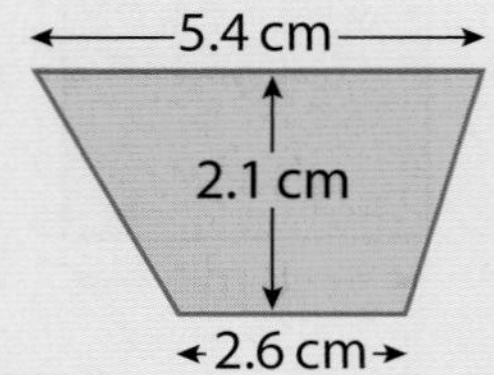

f

5 cm

5 cm

2 Find the perimeter and area of each of these shapes.

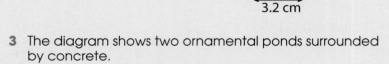

a **b** **c**

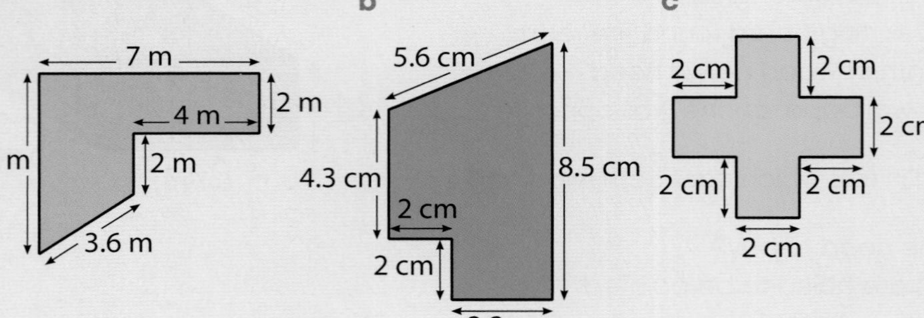

3 The diagram shows two ornamental ponds surrounded by concrete.

a Calculate the area of concrete.

Concrete costs £2.70 per square metre.

b Calculate the cost of the concrete surrounding the ponds.

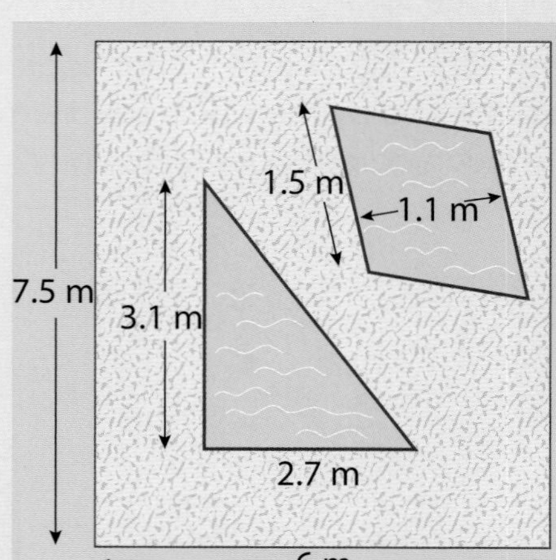

Unit 3 • Circumference • Band g

Building skills

Example outside the Maths classroom

Cogs and wheels

Toolbox

The **circumference** is the distance around the edge of a circle.

The **radius** of a circle is the distance from the centre to the circumference.

The **diameter** is the distance across the circle through the centre.

The circumference, C, of a circle of diameter d is

$$C = \pi d.$$

It is also given by

$$C = 2\pi r$$

where r is the radius.

π (pi) is a Greek letter which represents the number value 3.141 592 654… .

An approximate value of π is 3.14.

To be more accurate use π on your calculator.

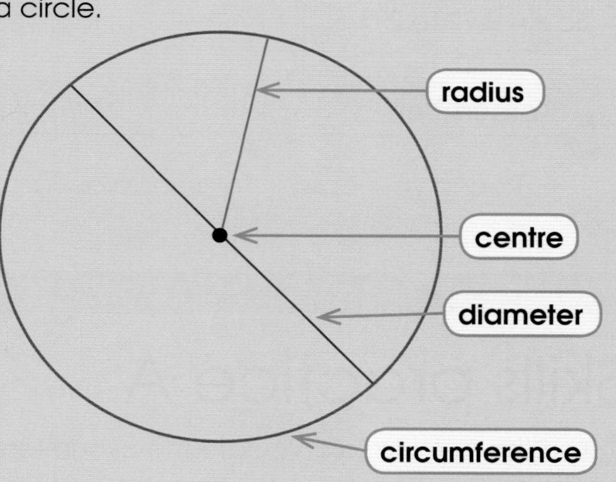

Example – Finding the circumference given the radius

Find the circumference of this circle.

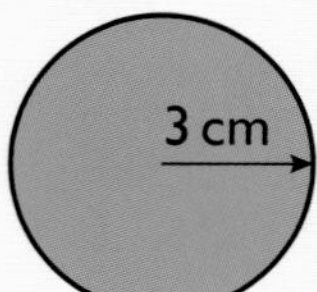

Solution

$C = 2\pi r$ ← **The radius is given so use the formula $C = 2\pi r$.**

$= 2 \times \pi \times 3$ ← **r = 3 Use the value of π stored in your calculator or 3.14**

$= 18.8$ to 1 d.p. ← **Round your answer.**

Example – Finding the diameter and radius given the circumference

A circle has a circumference of 5.76 m.

a Calculate the diameter.

b Calculate the radius.

5.76 m

Solution

a

$$C = \pi d$$

$$5.76 = \pi d$$

$$\frac{5.76}{\pi} = d$$ ← **Divide both sides by π.**

$$1.833... = d$$

So d = 1.83 to 2 d.p.

b $r = \frac{d}{2}$

$= \frac{1.833...}{2}$ ← **Use the unrounded value of d from your calculator.**

$= 0.916...$

So r = 0.92 to 2 d.p.

Remember:

✦ Round your answer to an appropriate degree of accuracy.

Skills practice A

1 Use this formula to calculate the circumference of each of these circles.

$C = \pi d$

a

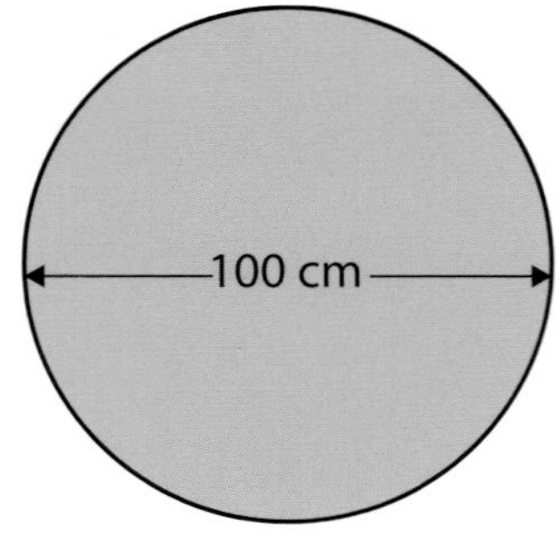

b

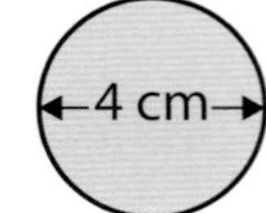

c

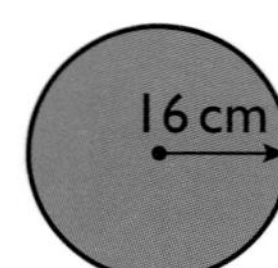

d

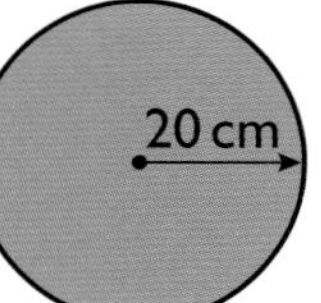

e

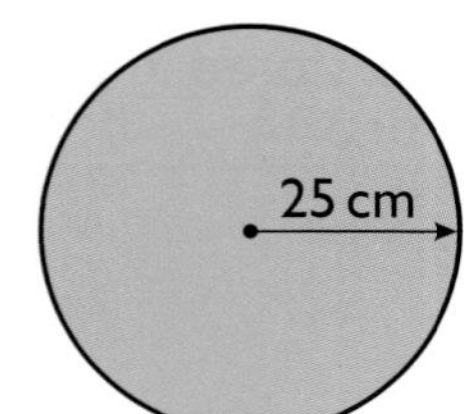

f

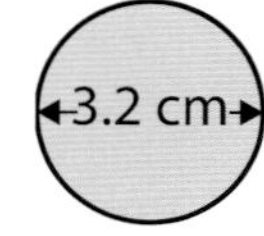

2 Use $\pi = 3.14$ to calculate the circumference of the circles with these dimensions.

a Diameter = 10 cm

b Diameter = 2 m

c Radius = 2.5 cm

3 Use your calculator to calculate the circumference of the circles with these dimensions.

a Diameter = 8 cm

b Diameter = 23.6 cm

c Radius = 3.8 m

4 A circle has a circumference of 6.28 m.
Calculate

a the diameter of the circle

b the radius of the circle.

5 Copy and complete this table.

Radius	Diameter	Circumference (to 2 d.p.)
4 cm	8 cm	
	12 cm	
		21.98 cm
	15 cm	
12.5 cm		
21.6 cm		
		1.26 m
		31.40 km

6 On this clock face, the hour hand is 5 cm long and the minute hand is 8 cm long.

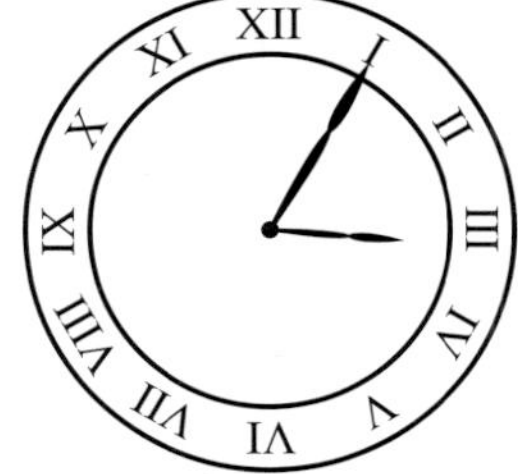

a How far does the tip of the minute hand travel in one hour?

b How far does the tip of the minute hand travel in 12 hours?

c How far does the tip of the hour hand travel in 12 hours?

d How far does the tip of the hour hand travel in one hour?

Reasoning

7 In this diagram, there are three shapes:

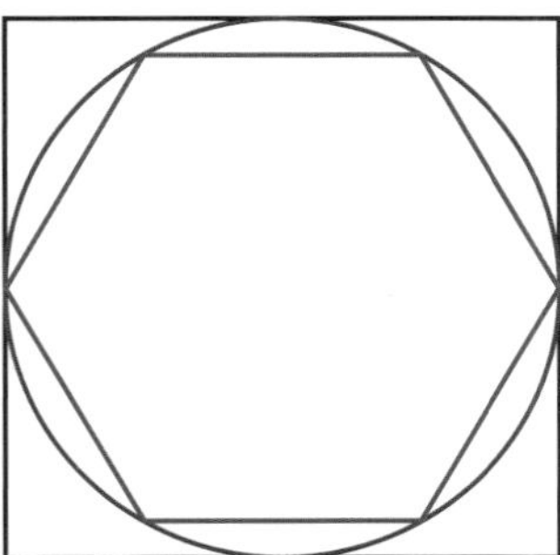

- a circle of radius 2 cm
- a square outside the circle but touching it in four places
- a hexagon with its vertices on the circle.

a Measure the sides of the square and the hexagon.

b Copy and complete this statement.
The circumference of the circle is greater than ☐ centimetres and less than ☐ centimetres.

c Explain how this tells you that the value of π is between 3 and 4.

Skills practice B

1 The large wheel on Andy's wheelchair has a diameter of 60 cm.
Andy pushes the wheel round exactly once.

a How far has Andy moved?

Andy crosses a busy road, 20 m wide.

b How many times does he have to rotate the large wheel to do this?

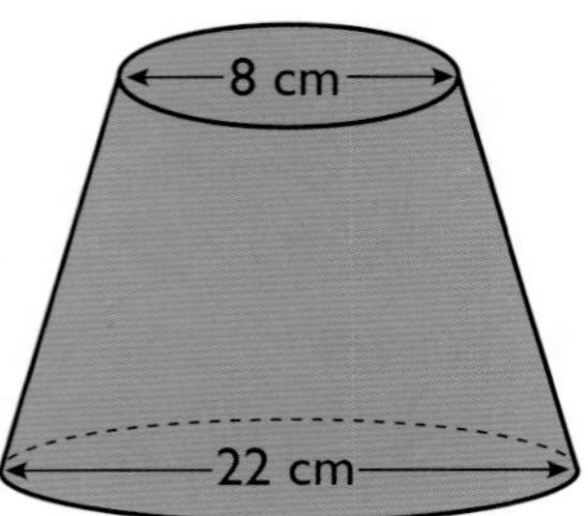

2 Christine is trimming a lampshade with ribbon at the top and the bottom.
How much ribbon does she need

a to decorate the top of the lampshade

b to decorate the bottom of the lampshade

c in total?

Reasoning

3 This is a penny-farthing.
The large wheel of the penny-farthing has a diameter of 56 inches.

a Calculate the distance travelled in one complete turn by a point on the circumference of the large wheel.

The small wheel has a diameter of 12 inches.

b Calculate the distance travelled in one complete turn by a point on the circumference of the small wheel.

c How many times does the small wheel turn for one complete turn of the large wheel?

The pedal goes round in a circle with a radius of 6 inches.

d How far does the cyclist's right foot travel in one revolution of the pedal?

4 A cotton reel has a diameter of 2 cm.
There are 500 turns of thread on the reel.
How long is the thread?

5 The wheels on a bicycle both have a diameter of 50 cm.
Calculate, to one decimal place

a how many metres the bicycle travels when the wheels rotate 100 times

b how many turns of the wheels are needed to cover a distance of 5 km.

Reasoning

6 A 400 m racetrack consists of two 100 m straights and two 100 m semicircles, one at each end, as shown in the diagram.
Find distance x.

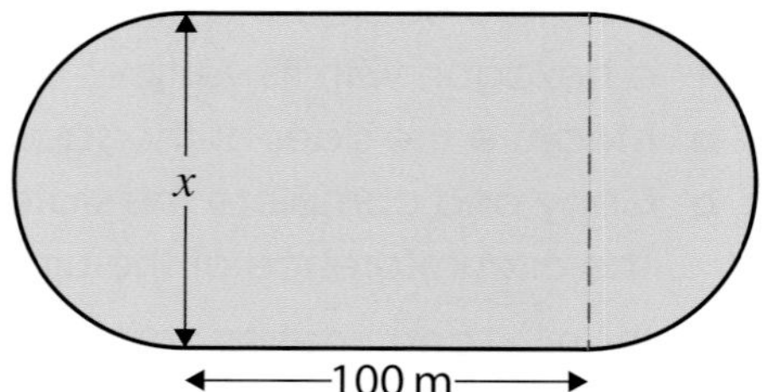

7 The large wheel on Maisie's bicycle has a diameter of 50 cm.

a What is the circumference of this wheel?

Maisie cycles 48 m to the end of her road.

b How many times does the large wheel go round?

The small wheel has a radius of 20 cm.

c How many times does the small wheel go round as Maisie cycles to the end of the road?

Reasoning

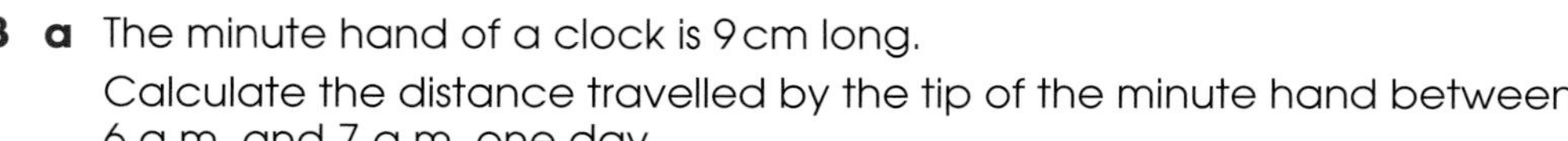

8 **a** The minute hand of a clock is 9 cm long.

Calculate the distance travelled by the tip of the minute hand between 6 a.m. and 7 a.m. one day.

b The hour hand of the clock is 6 cm long.

Calculate the distance travelled by the tip of the hour hand between 6 a.m. and 7 a.m. one day.

c Calculate the distance travelled by the two hands between 7 a.m. and 7.15 a.m.

9 A satellite orbiting in space travels so that it is always 1120 km above the surface of the Earth.

The Earth is approximately a sphere of radius 6380 km.

Calculate the distance travelled by the satellite in one complete orbit.

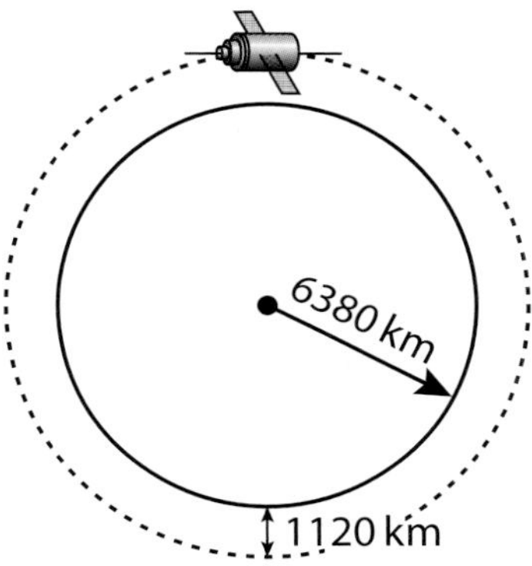

10 Look at this picture of a wrought iron gate.

The curve at the top is a semicircle.

Calculate the total length of iron rod used to make the gate.

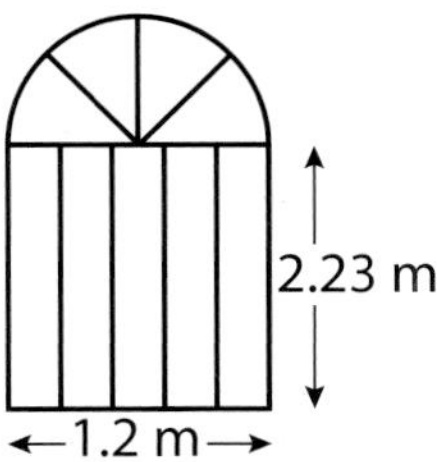

11 A lawn is laid in a garden.

The lawn is a circle of radius 3.5 m.

a What is the diameter of the lawn?

b What is the circumference of the lawn?

Edging is fitted round the circumference of the lawn to keep it neat.

The edging comes in flexible 1.5 m lengths and costs £2.99 per length.

c How many lengths of edging are needed?

d How much does the edging cost?

12 A circular table has a radius of 1.8 m.

George has a circular tablecloth with circumference 12 m.

a Show that the tablecloth is large enough for the table.

b By how much will it overhang the table?

Wider skills practice

1 Mark measured the radius of this cart wheel to be 80 cm.

a Calculate its circumference.

b How far does the cart travel when the wheel turns 100 times? Give your answer in

i centimetres **ii** metres **iii** kilometres.

2 π does not have an exact numerical value.
A rough approximation for π is 3.
Some calculators have a button for π.
A ten-digit display calculator gives π as 3.141 592 654.

a Phil says 'π is 3.142'.
Is Phil right?

Other approximations for π are $\frac{22}{7}$ and $\sqrt{10}$.

b **i** Use a calculator to change $\frac{22}{7}$ to a decimal.

ii What sort of decimal is your answer?

iii Compare $\frac{22}{7}$ and the value of π on your calculator.
How many decimal places are the same?

c **i** Use a calculator to change $\sqrt{10}$ to a decimal.

ii Compare your answer with the value given by the π button.

Applying skills

1 A circular table has a diameter of 10 metres.
Twenty politicians are invited to an international conference.
Each politician needs 1.5 m around the circumference of the table.
Can they all sit around the table together?

2 The picture shows a flower pot of radius 12 cm.
There is a leaf design pattern around the top.
The leaf design fits exactly 16 times.
What is the length of the leaf design?

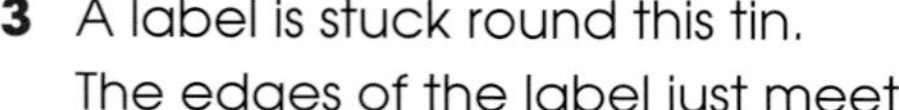

Problem solving

3 A label is stuck round this tin.
The edges of the label just meet.

a Calculate the circumference of the tin.

b The label is 10 cm high.
What area of paper is needed to make one label?

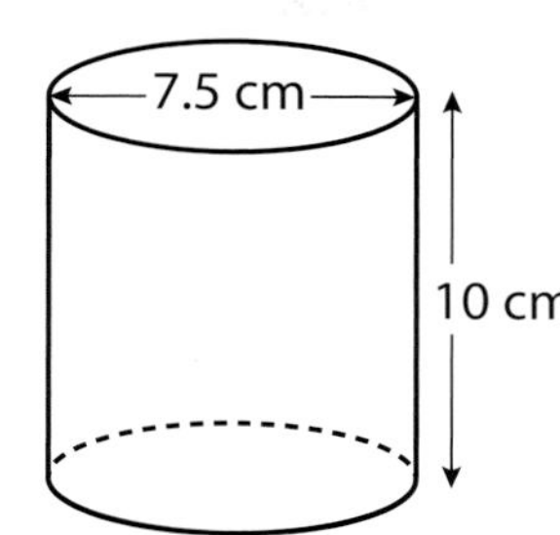

Reviewing skills

1 Find the circumference of each of these circles.

a

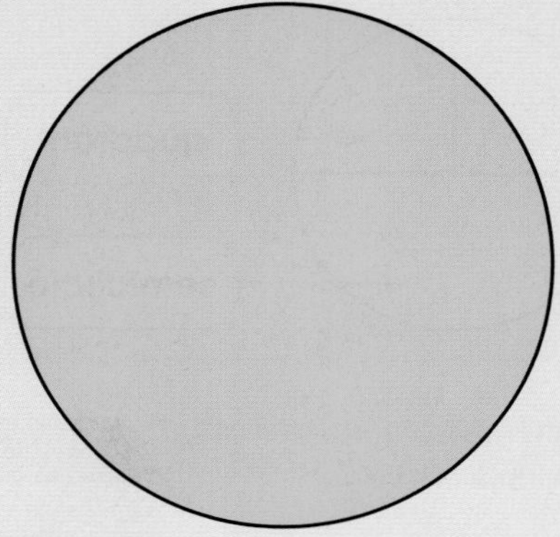

b A circle of diameter 120 m.

2 Here are three timpani.
For each timpani find the circumference of the rim.

a

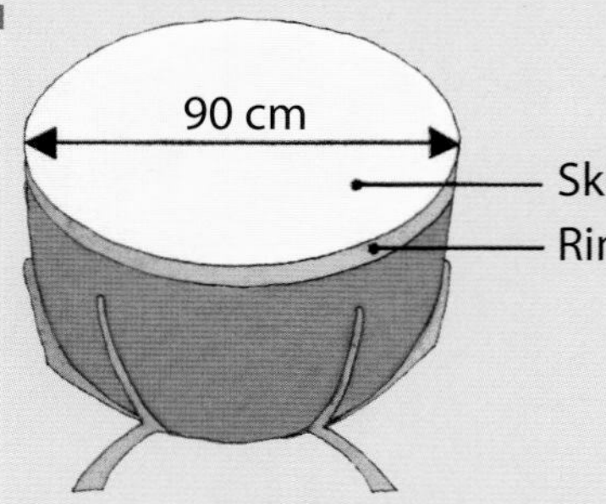

F to C

b

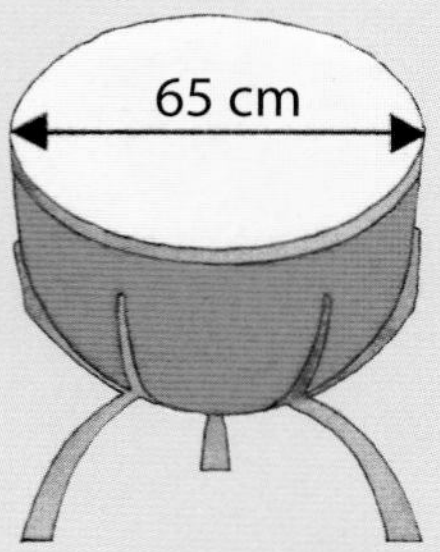

B to F

c

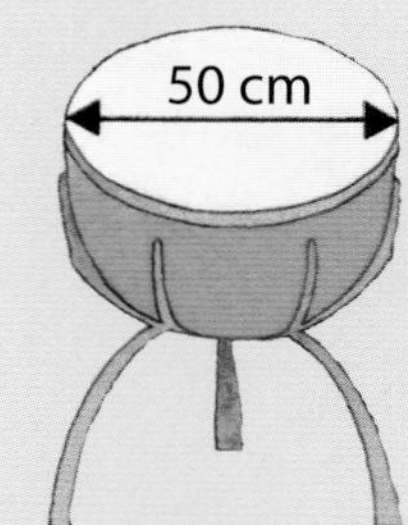

E to G

3 A bin has a circumference of 100 cm.
What is its diameter?

Unit 4 • Area of circles • Band g

Building skills

Example outside the Maths classroom

Amphitheatres

Toolbox

The area, A, of a circle of radius r is given by

$A = \pi r^2$.

A **semicircle** is half a circle.

Its area is $\frac{1}{2}\pi r^2$.

A **quadrant** is a quarter of a circle.

Its area is $\frac{1}{4}\pi r^2$.

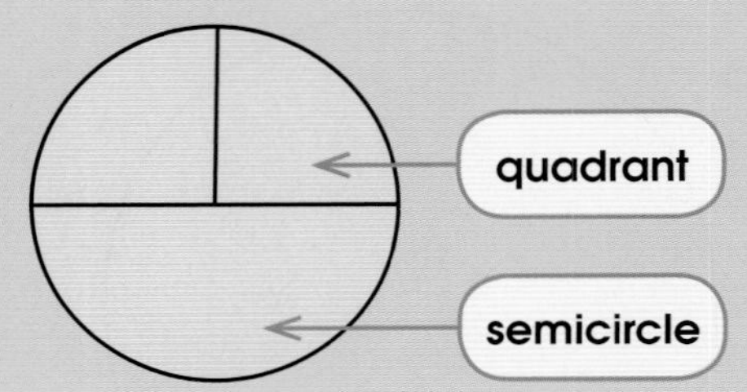

Example – Finding the area of a circle given the diameter

Find the area of this circle.

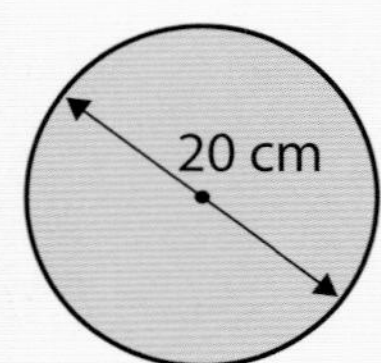

Solution

First find the radius.

$$r = \frac{d}{2} = \frac{20}{2} = 10\,\text{cm}$$

$$\begin{aligned} A &= \pi r^2 \\ &= \pi \times 10^2 \\ &= \pi \times 100 \\ &= 314.2\,\text{cm}^2 \text{ (to 1 d.p.)} \end{aligned}$$

Example – Finding the radius and diameter of a circle given the area

The area of this circle is 50 cm^2.

Calculate

a the radius

b the diameter.

Give your answers to the nearest centimetre.

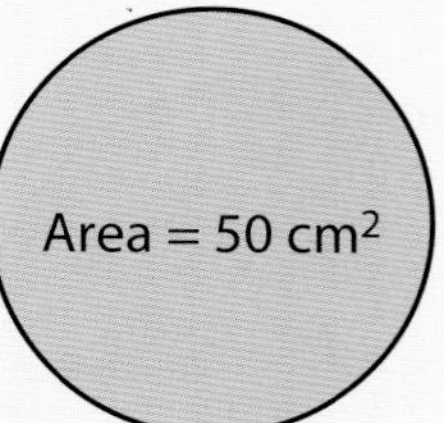

Solution

a $A = \pi r^2$

$50 = \pi r^2$

$\frac{50}{\pi} = r^2$ ← **Divide both sides by π**

$\sqrt{\frac{50}{\pi}} = r$ ← **Take the square root of both sides.**

$r = \sqrt{\frac{50}{\pi}}$

$r = 3.989\ldots$ ← **Your calculator often gives you a long number. Write it like this, using … for the later digits. That is better than writing them all down.**

The radius is 4 cm
(to the nearest centimetre).

b Diameter = 2 × radius

= 2 × 3.989… ← **Use the most accurate value.**

= 7.979…

The diameter is 8 cm
(to the nearest centimetre).

Keep the number on your calculator until you get your final answer.

Remember:

- Area is measured in square units such as cm^2 or m^2.

Skills practice A

1 Calculate the area of each of these circles.

a

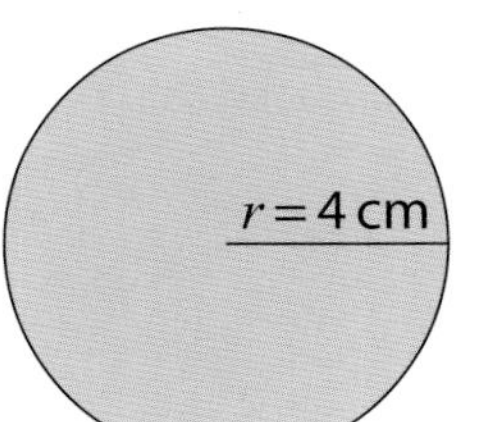

b

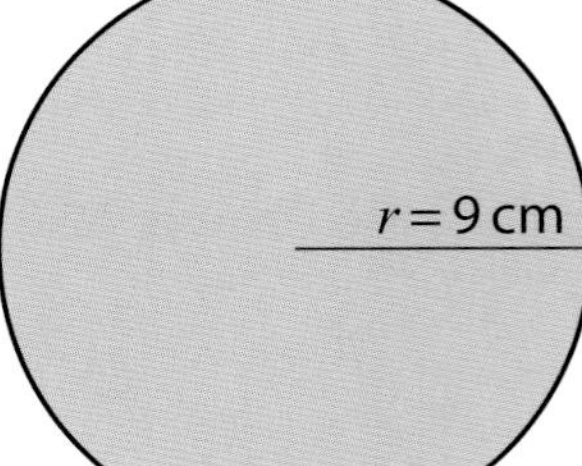

c

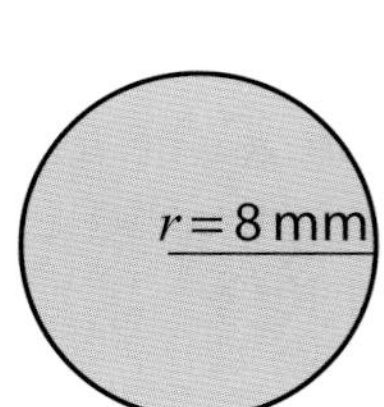

d

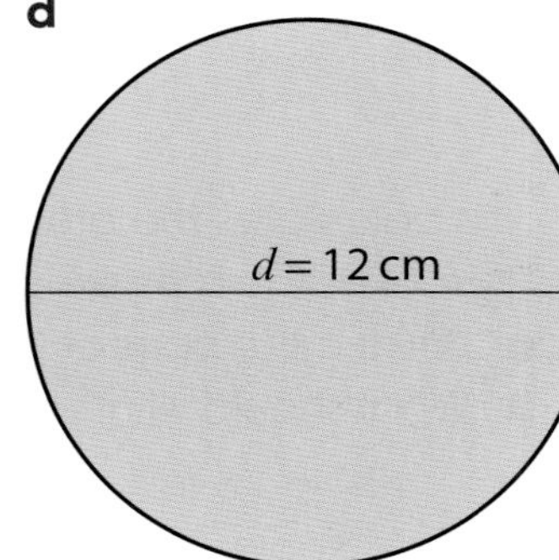

2 Calculate the area of each of these circles.

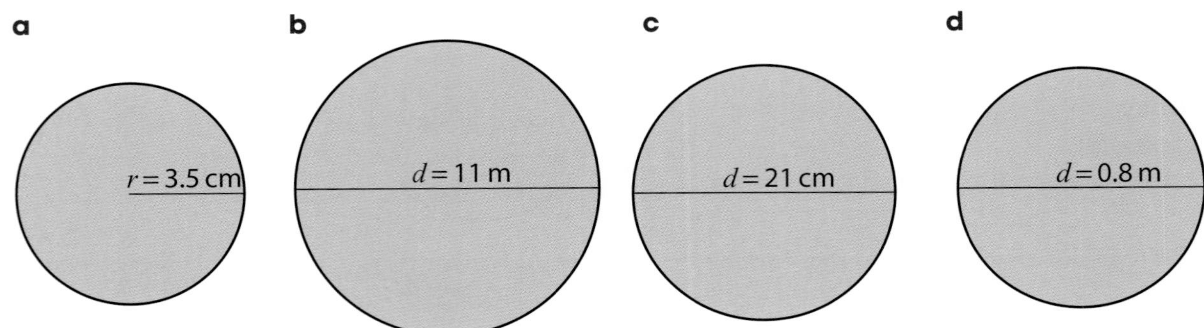

3 Calculate the areas of the circles with these dimensions.

a Radius = 10 cm
b Radius = 5 cm
c Radius = 2.6 m
d Diameter = 2 km
e Diameter = 30 feet

4 Copy and complete this table.

Radius	Diameter	Area (to 2 d.p.)
4 cm	8 cm	
	12 cm	
		12.56 m^2
	18 inches	
		1256.00 cm^2
2.7 cm		
		907.46 mm^2

5 Find the area and perimeter of each of these shapes.

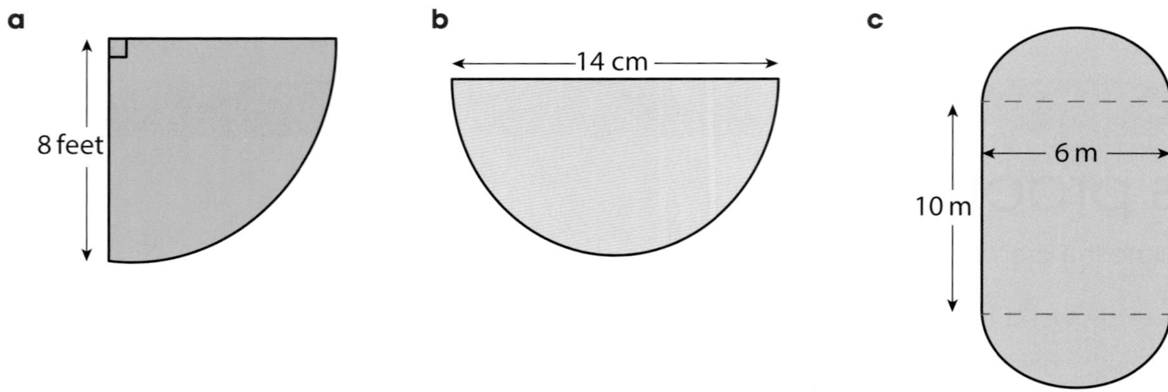

6 Jitin calculates the area of a circle with radius 10 cm.
He makes the mistake of thinking that 10^2 is 20 and gets the answer 62.8 cm^2.

a What is the correct value of 10^2?
b What should Jitin's answer be?

Reasoning

7 Zoe calculates the area of a circle with diameter 10 cm.
She forgets that radius = diameter ÷ 2.
She gets the answer 314 cm².
What is the correct answer?

8 A sumo wrestling ring has a diameter of 24 m.
Find the radius and area of the ring.

9 Calculate the area of the playing surface of this CD.

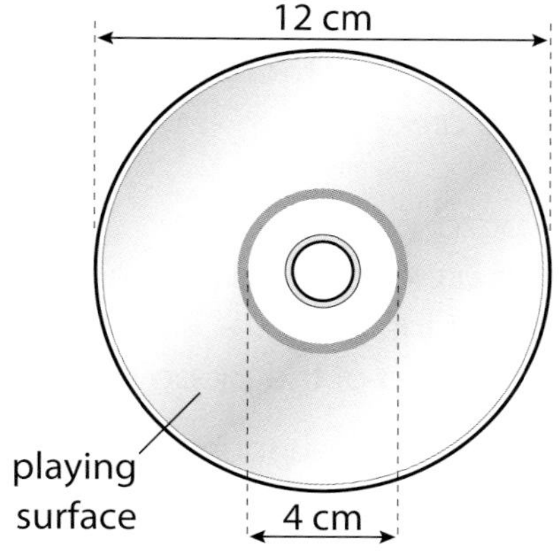

Skills practice B

Reasoning

1 Find the area and perimeter of each of these shapes.

a

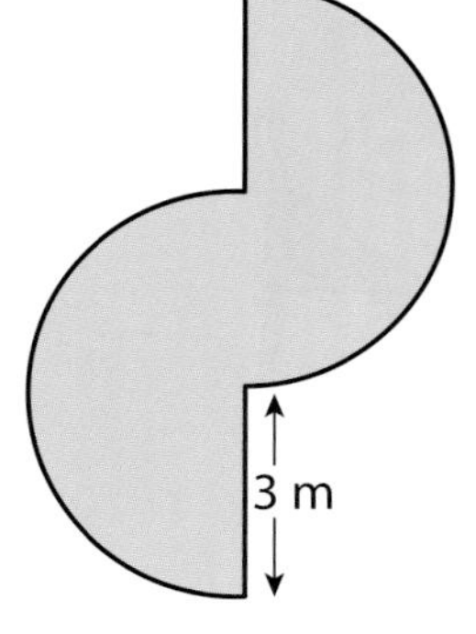

b

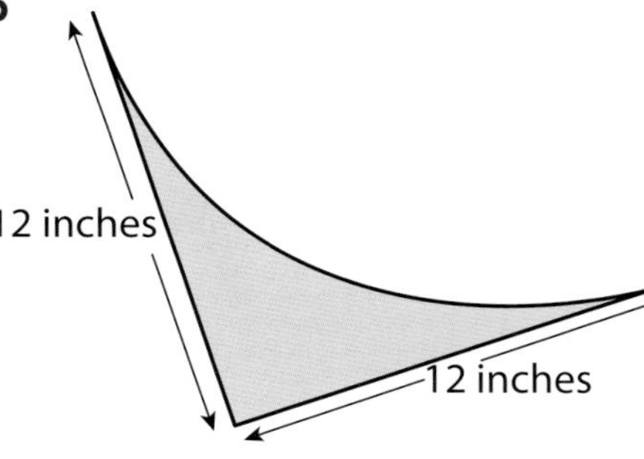

c

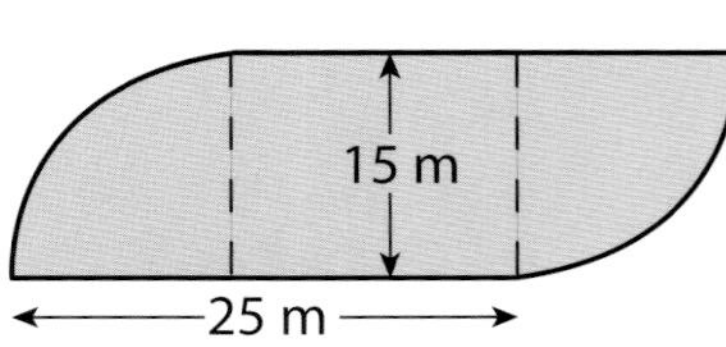

d

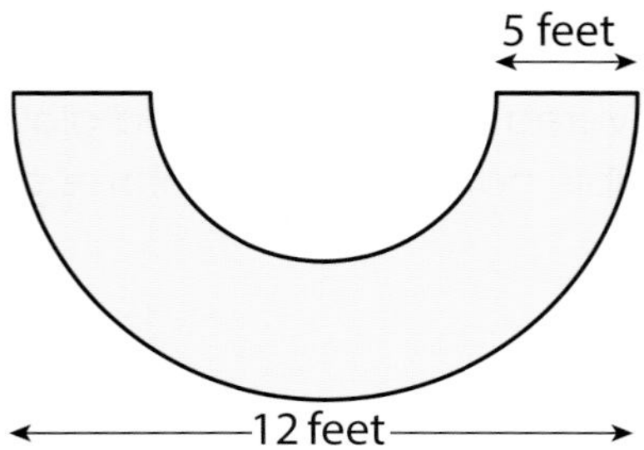

2 Here is a diagram of a castle door.
The arch at the top is a semicircle.
Calculate the area of the door.

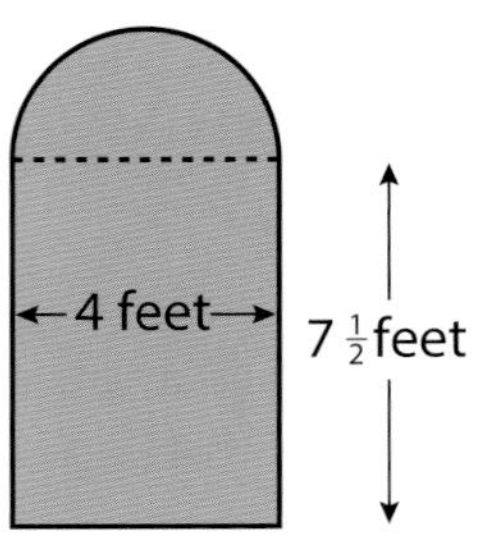

3 The radius of a pizza is 15 cm.
Calculate the area of

a the pizza

b half the pizza.

Spencer cuts a piece of pizza for himself.
It is a quarter of the pizza.

c What is the area of the piece he cuts?

15 cm

Reasoning

4 This is a circular pond, surrounded by a path.
The path is 50 cm wide.
Calculate

a the area of the pond

b the area of the path.

Give your answers in square metres, correct to two decimal places.

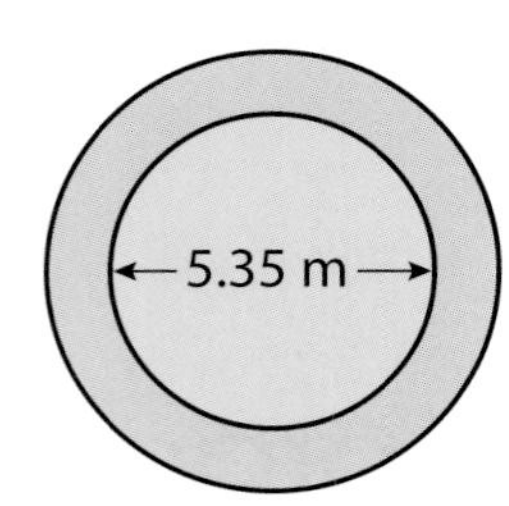

5 At a family barbecue the patio table is laid ready for the meal.
The table is circular and has a radius of 90 cm.

a What is the area of the top of the table?

Plates of radius 10 cm are placed on the table along with glasses of base radius 3 cm.

b What is the area of one plate?

c What is the area of the base of one glass?

Four places are set with a plate and a glass.

d What is the total area covered by the plates and glasses?

e What is the area of the table not covered by plates and glasses?

Reasoning

6 The top of this birthday cake is a circle of radius 12 cm.

a What is the area of the iced top of the cake?

The cake is displayed on a circular board of radius 14 cm.

b What is the area of the cake board?

c What is the area of board that is visible when the cake is placed on it?

The design on the cake includes a fish of area 80 cm^2 and three circular bubbles with radii 1 cm, 1.5 cm and 2 cm.

d What is the total area of the design?

e What area of icing is not covered by the design on the top of the cake?

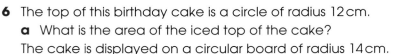

Reasoning

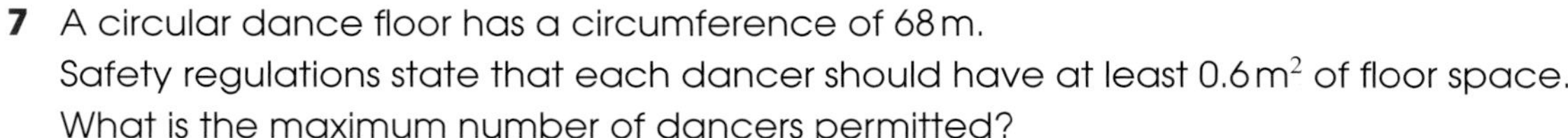

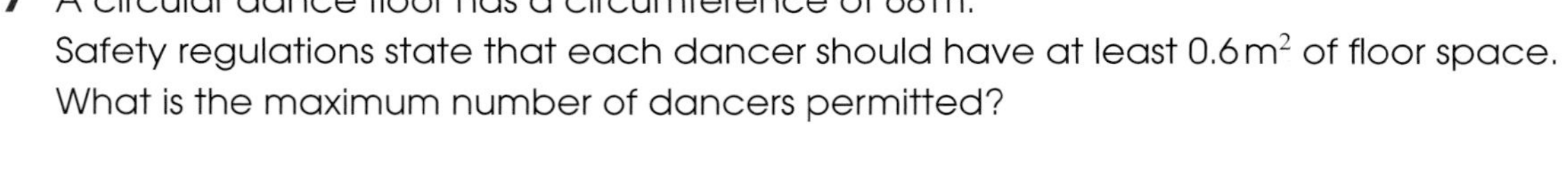

7 A circular dance floor has a circumference of 68 m.

Safety regulations state that each dancer should have at least 0.6 m^2 of floor space.

What is the maximum number of dancers permitted?

Reasoning

8 **a** Calculate the area of the concrete surround of this circular swimming pool.

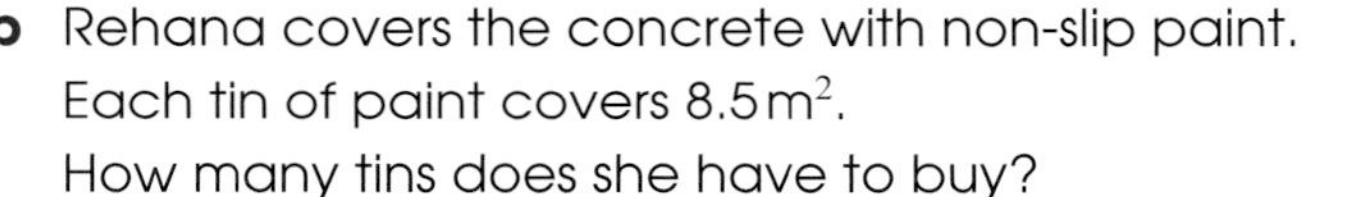

b Rehana covers the concrete with non-slip paint.

Each tin of paint covers 8.5 m^2.

How many tins does she have to buy?

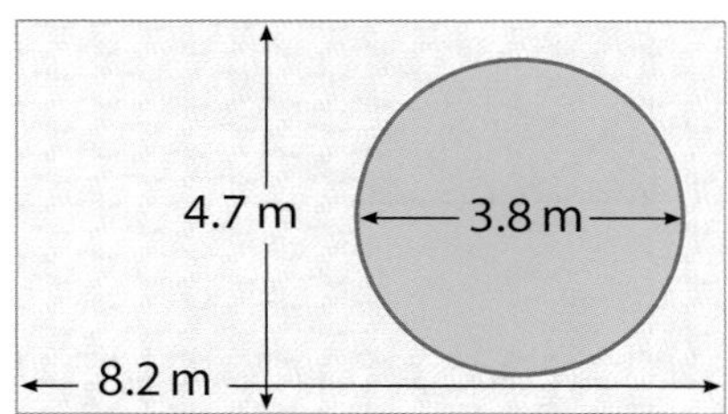

Wider skills practice

1 This stained-glass window is made from three large red circles and two small blue circles.

The surrounding glass is clear.

a Calculate the area of one of the red circles.

b Calculate the area of one of the blue circles.

c Calculate the area of the clear glass.

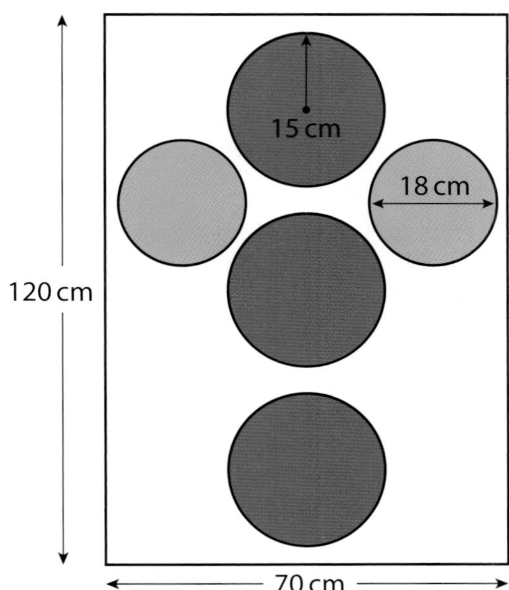

Reasoning

2 A race track consists of two semicircles of diameter 58 metres connected by two straight lines of length 70 metres.
How many laps are required for a 10 000 metre race?

3 Put these shapes in order of size by area from smallest to biggest.

a Circle, radius 12 cm
b Quadrant, radius 25 cm
c Rectangle, length 31 cm, width 15 cm
d Semicircle, diameter 36 cm
e Square, perimeter 88 cm
f Triangle, base 20 cm, height 48 cm

Reasoning

4 What percentage of paper is left when a circle is cut out from a square as shown?

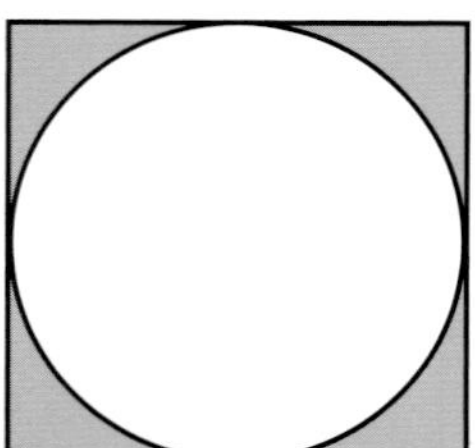

Applying skills

Problem solving

1 A cricket club has a circular field with a circumference of 110 metres.
Turf costs £9 per square metre and is only available in 1 m × 1 m square pieces.
How much would it cost the club to turf the whole field?

Problem solving

2 Here is a plan of a garden.

a What is the total area of the garden?
b What is the area of the deck?
c What is the area of the surface of the pond?
d What is the area covered by the stepping stones?
e What is the feeding area on the bird table?
f What is the area of the summer house?
g What is the total area of the flower beds?
h What is the total area of the lawn?

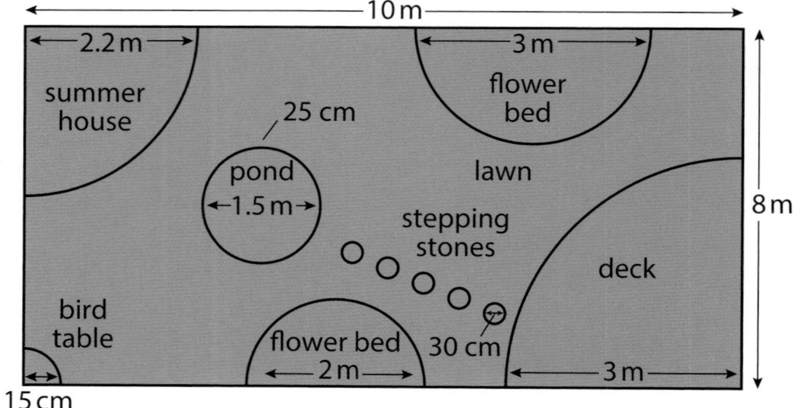

3 The small pizza serves two people.

a How many people would a medium pizza serve?

b What about a large pizza?

c A village in Italy is having a pizza festival.
They want to make one giant pizza for 100 people.
What diameter does the pizza need to have?

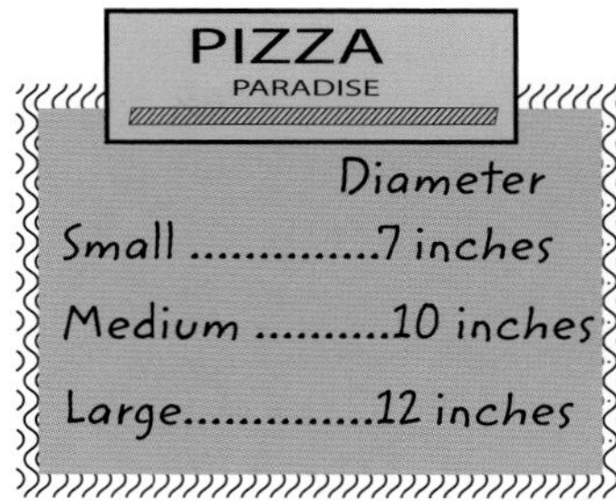

Reviewing skills

1 For each of these circles, find the area.
Give your answers to one decimal place.

a

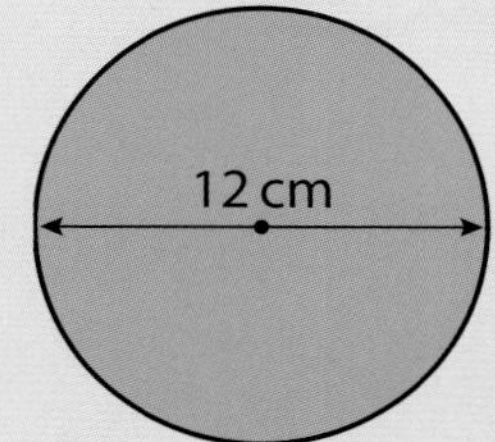

b

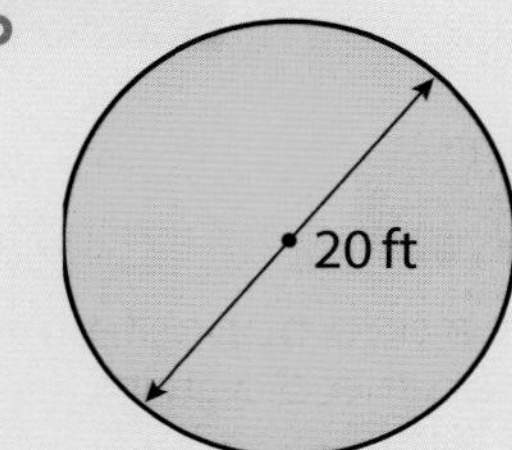

c

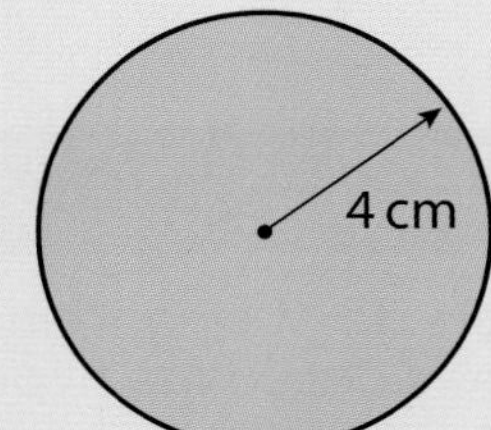

2 Calculate the area of the green region.

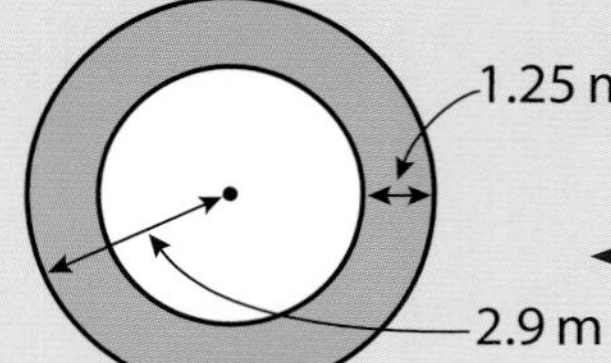

These circles are concentric; they have the same centre.

3 Find the perimeter and area of this shape.

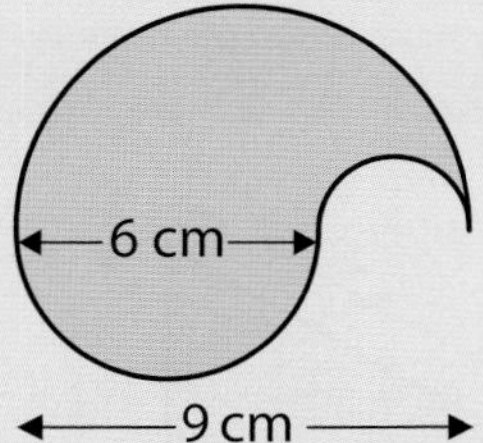

Building skills

Example outside the Maths classroom

Buildings

Toolbox

The longest side of a right-angled triangle (the side opposite the right angle) is called the **hypotenuse**, side h on the diagram.

Pythagoras' theorem says that for a right-angled triangle, the square on the hypotenuse is equal to the sum of the squares on the other two sides.

Pythagoras' theorem allows the third side of a right-angled triangle to be calculated when the other two sides are known.

Pythagoras' theorem: $a^2 + b^2 = h^2$

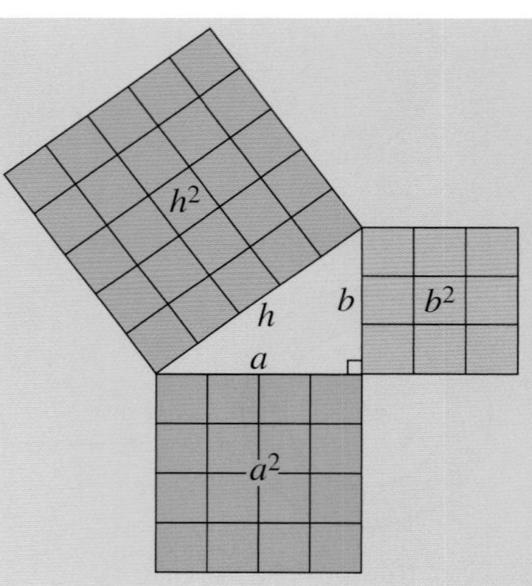

Example – Finding the length of the hypotenuse

Find the length of side h in this triangle.

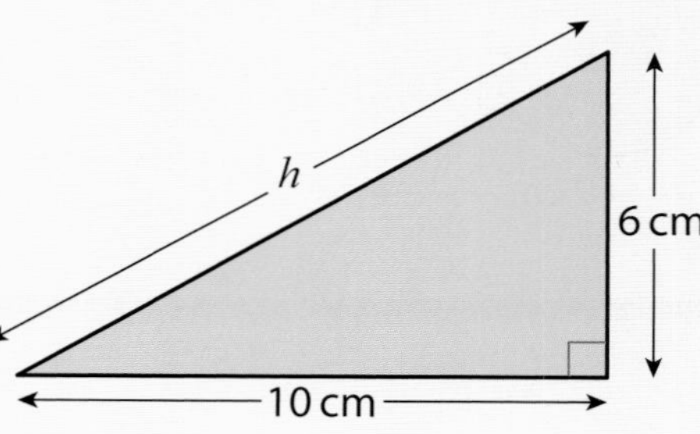

Solution

Let $a = 10$ and $b = 6$ ← **It could be $a = 6$ and $b = 10$**

$a^2 + b^2 = h^2$ ← **Pythagoras' theorem**

$10^2 + 6^2 = h^2$ ← **Substitute in the lengths.**

$100 + 36 = h^2$

$136 = h^2$

$h = 11.661...$

$h = 11.7$ cm (to 1 d.p.)

Example – Solving a problem involving a right-angled triangle

Gill is a decorator.

She uses a ladder 6 m long.

The safety instructions on the ladder say that the foot of the ladder must be at least 1.75 m from the wall on horizontal ground.

What is the maximum height her ladder will reach up a vertical wall?

Solution

$1.75^2 + y^2 = 6^2$ ← **Pythagoras' theorem rewritten for this problem.**

$3.0625 + y^2 = 36$

$y^2 = 36 - 3.0625$ **Draw a sketch and label it. y stands for the height reached by the ladder.** →

$y^2 = 32.9375$

$y = \sqrt{32.9375}$ ← **Take the square root of both sides.**

$y = 5.739...$

Maximum height of ladder = 5.74 m (to the nearest centimetre)

6 m

y

1.75 m

Remember:

- Pythagoras' theorem only works for right-angled triangles.
- You need to know the length of two of the sides in order to use Pythagoras' theorem.

Skills practice A

1 Grace is trying to find the length of the hypotenuse of this triangle.
Copy and complete her calculation.

$h^2 = 24^2 + 5^2$

$h^2 = 576 + 25$

$h^2 =$

$h =$

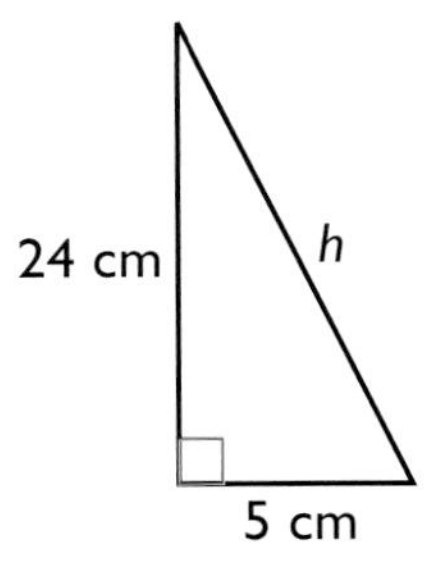

2 Find the length of the hypotenuse of each of these triangles.

a 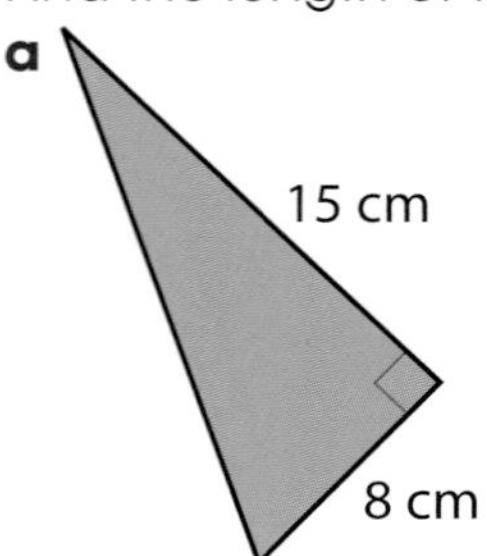

b 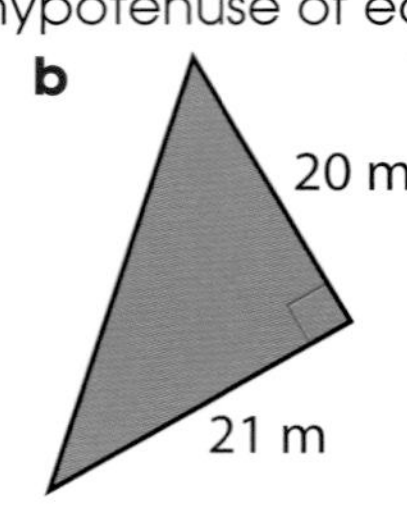

c 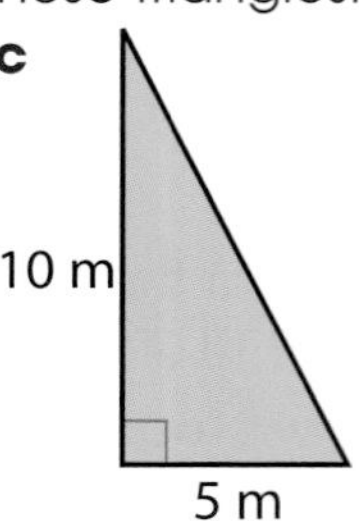

d 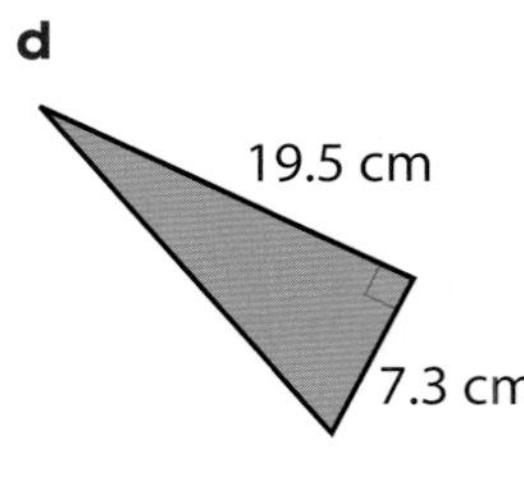

3 Find the length of the hypotenuse of the right-angled triangles with these dimensions.

a Base = 21 cm, height = 28 cm

b Base = 20 cm, height = 15 cm

c Base = 60 cm, height = 63 cm

d Base = 27 cm, height = 120 cm

e Base = 65 cm, height = 72 cm

4 Find the lengths of the sides marked with a letter in each of these triangles.
Give your answers correct to one decimal place.

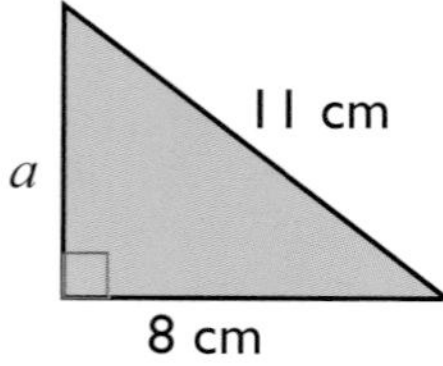

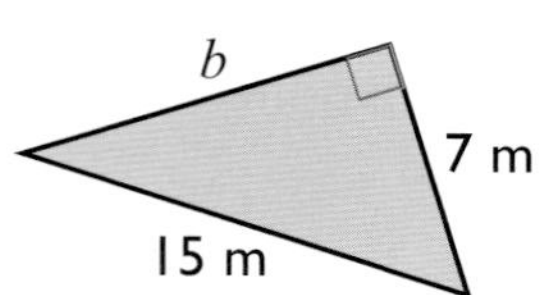

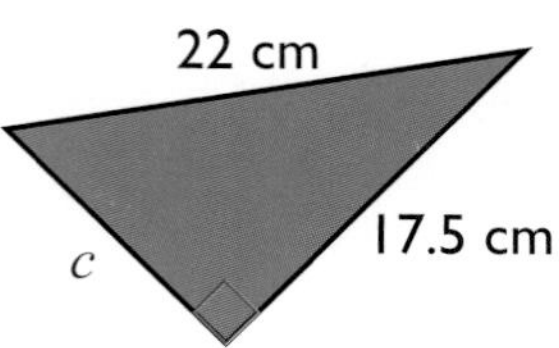

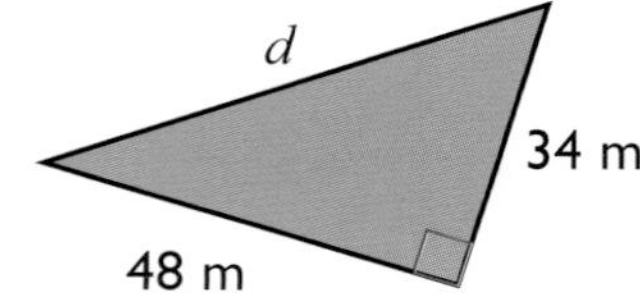

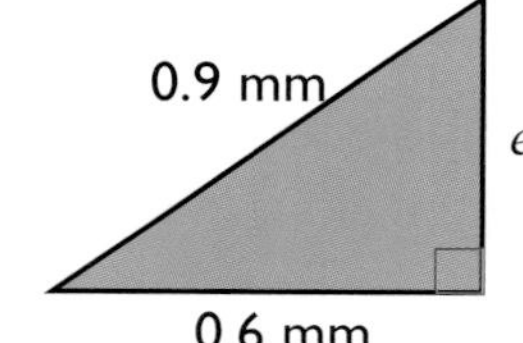

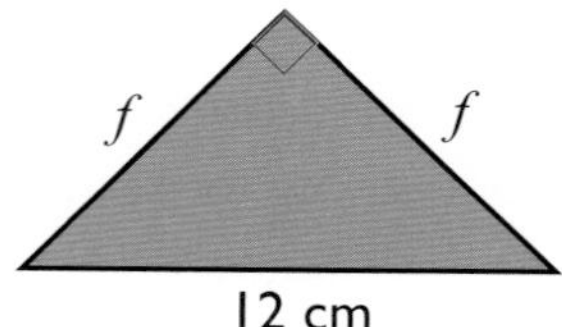

5 Calculate the length of the diagonal of this rectangle.
Give your answer in centimetres correct to one decimal place.

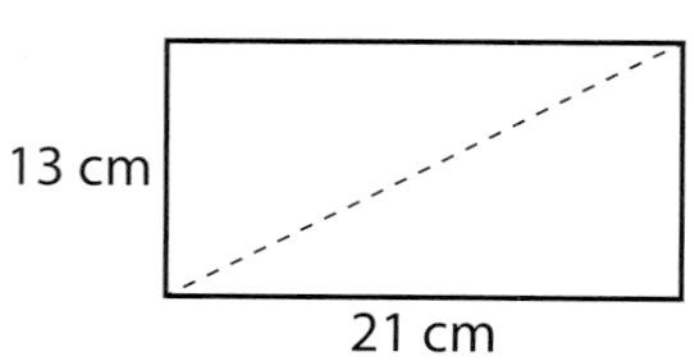

6 Calculate the length of the unknown sides in these triangles.

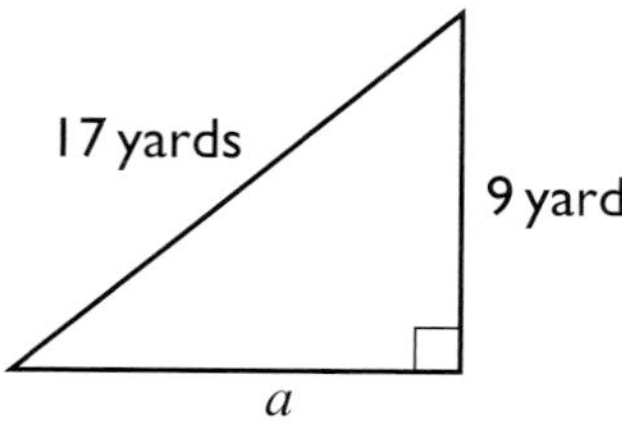

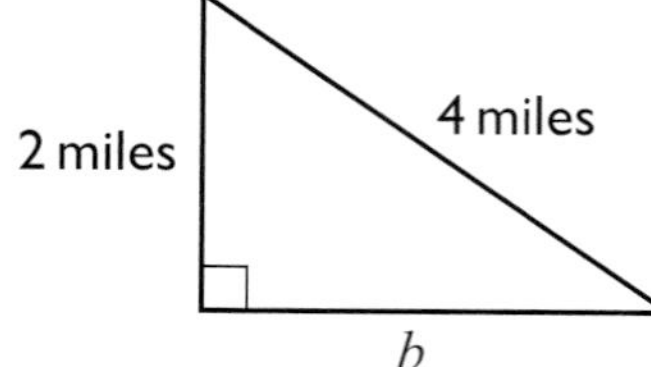

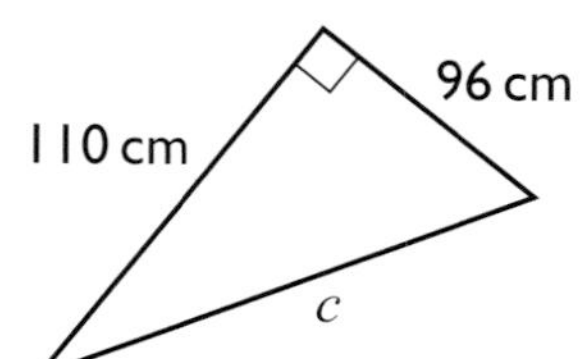

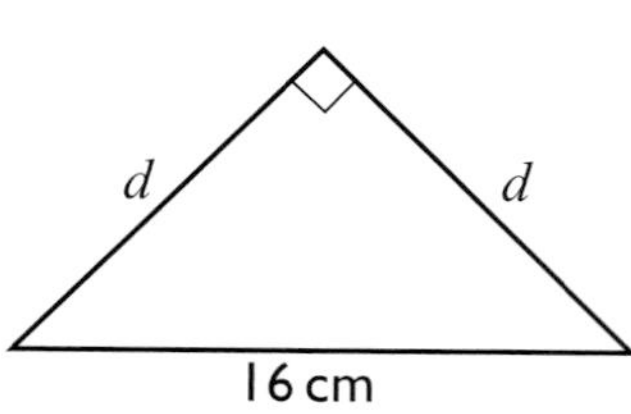

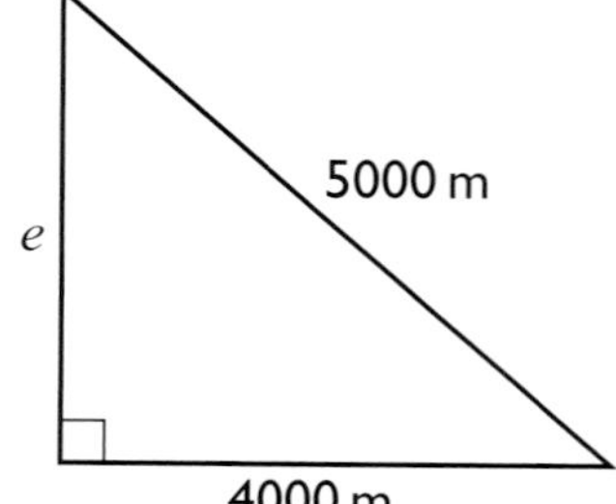

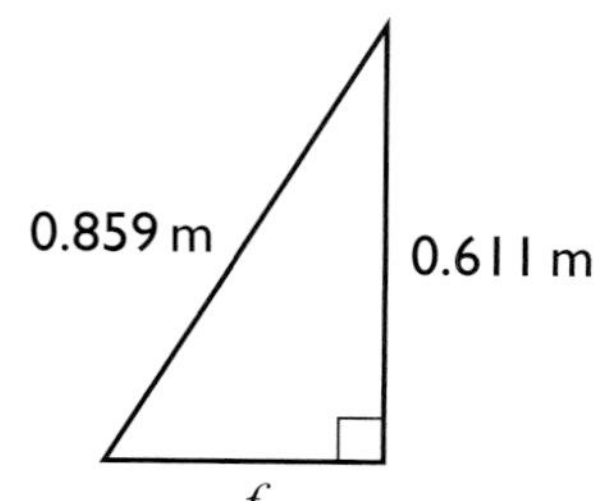

7 A right-angled triangle DEF has a right angle at F.
DF = 32 cm. EF = 24 cm.
Find the length of DE.

8 **a** Triangle ABC has a right angle at C.
AC is 3 cm and CB = 4 cm.
Make an accurate drawing of triangle ABC on squared paper. Label it.

b Draw squares ACDE and CBFG as shown.

c Measure length AB.
Draw square ABHI on a separate piece of squared paper.

d Glue square ABHI onto the hypotenuse AB.

e Shade the two smaller squares in one colour.
Shade the largest square in a different colour.

f Repeat steps **a** to **e** for right-angled triangles with sides AC and CB as shown in the table.

g Copy and complete the table.

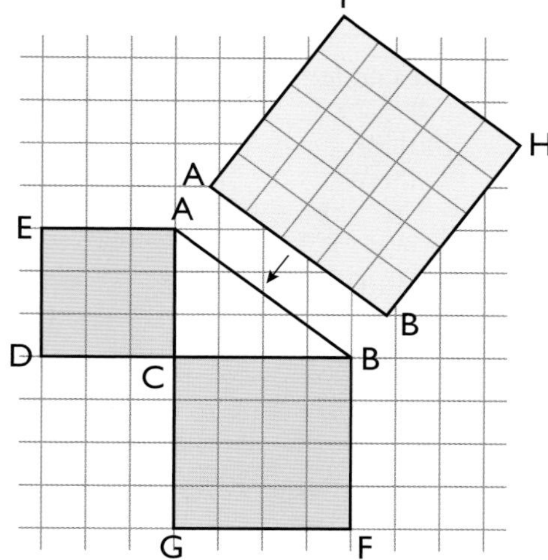

Shorter sides		Hypotenuse	Sum of areas of smaller squares	Area of largest square
AC	CB	AB		
3 cm	4 cm		$9\text{cm}^2 + 16\text{cm}^2 = 25\text{cm}^2$	
5 cm	12 cm			
8 cm	6 cm			

Skills practice B

Reasoning

1 Here is one of Callum's homework problems.

A right-angled triangle GHI has a right angle at G.

GH = 28 cm and HI = 100 cm.

Find the length of GI.

Solution

$a^2 + b^2 = h^2$

$28^2 + b^2 = 100^2$

$3600 + b^2 = 10\,000$

$b^2 = 10\,000 - 3600 = 6400$

$h = \sqrt{6400}$

$h = 80\text{ cm}$

There is a mistake in Callum's solution.
Find it and correct his answer.

Reasoning

2 Here is another of Callum's homework problems.

> The triangle ABC has a right angle at B.
>
> AC = 22 cm. BC = 17 cm.
>
> Find the length of AB.
>
> Solution
>
> $a^2 + b^2 = h^2$
>
> $22^2 + 17^2 = h^2$
>
> $484 + 289 = h^2$
>
> $773 = h^2$
>
> $h = \sqrt{773} = 27.8028\ldots$
>
> $h = 27.8$ cm (to 1 d.p.)

This solution is wrong.
Find the mistake and write a correct solution.

3 A ship leaves harbour and sails North for 17 km, then West for 11 km.
It then sails straight back to the harbour.
How far does it sail in total?

4 The diagram shows the flight path of an aircraft.
It has travelled 600 miles.
By this time it has gone 300 miles North.
Find how far East it has travelled.

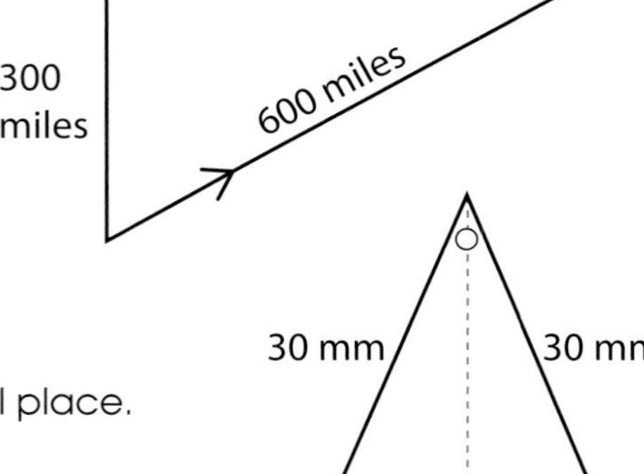

5 Sarah makes earrings.
Here is one of her designs, called 'Isosceles'.

a Calculate the perpendicular height of the triangle.
Give your answer in millimetres correct to one decimal place.

b Calculate the area of the earring.

30 mm 30 mm
24 mm

6 An aircraft leaves an airport and flies South-East for 200 km.
The diagram shows the flight path.

a What is the bearing of South-East?

b What type of triangle can you see in the diagram?

c Calculate how far East the aircraft has flown.

d How far South has the aircraft flown?

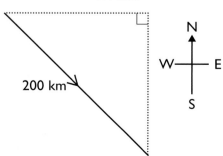

200 km
N W E S

Reasoning

7 The diagram shows a ball touching a floor and a wall.
Calculate the distance SR.

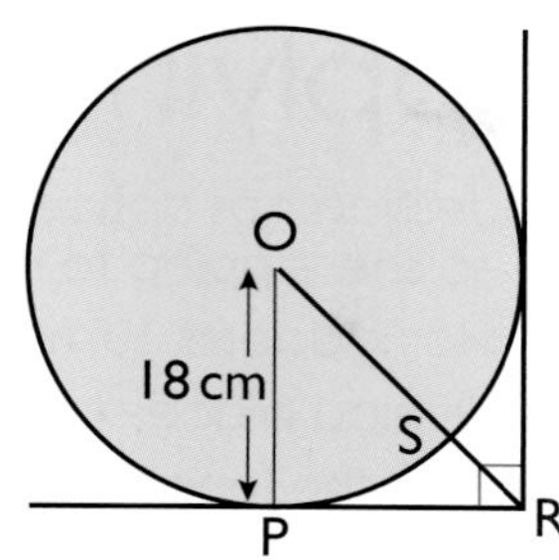

Reasoning

8 Calculate the area of this triangle.

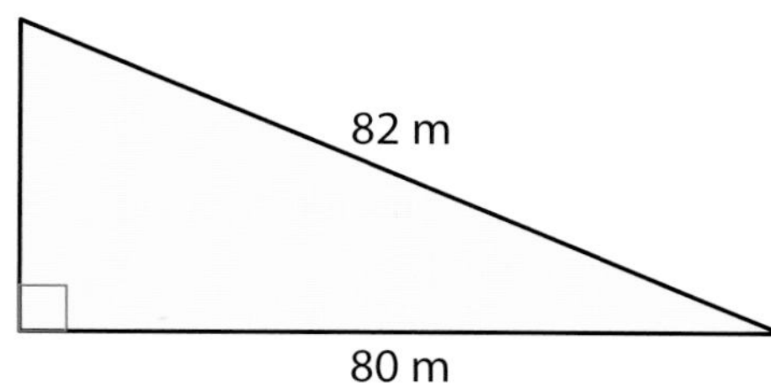

Wider skills practice

Reasoning

1 The number set 3, 4, 5 is called a Pythagorean triple because $3^2 + 4^2 = 5^2$.

a Sort these 18 numbers into six sets of Pythagorean triples.

8	21	20	48	75	4	60	63	41
17	9	3	52	72	87	15	40	5

b Look for more sets of Pythagorean triples.
Find as many different ones as you can.

Reasoning

2 Look at this triangle.
The longest side is l cm and the largest angle is $x°$.
The two shorter sides are a cm and b cm.

a l $x°$ b

a Draw a triangle with $a = 6$, $b = 8$ and $l = 10$.
Look at angle x.
Is it acute, a right angle or obtuse?
Copy this table and enter your results in the first row.

a	b	l	Compare l^2 to $a^2 + b^2$ Is l^2 smaller (<), equal (=) or bigger (>)?	Is x acute, a right angle or obtuse?
6	8	10		
6	8	12		
6	8	9		

b Draw triangles with

i $a = 6$, $b = 8$ and $c = 12$

ii $a = 6$, $b = 8$ and $c = 9$.

Complete the next two rows of your table.

c Draw two more triangles with $a = 6$ and $b = 8$.
Enter your results in your table.

d What do you notice?

Applying skills

Problem solving

1 Jessica has noticed a fault in the brickwork of her house, so she is going to have a closer inspection.

Her ladder is 10 m long and the house is 8 m high.

Jessica places the ladder with the foot 2 m from the wall of the house.

Find the length, x m, of the ladder that is above the house.

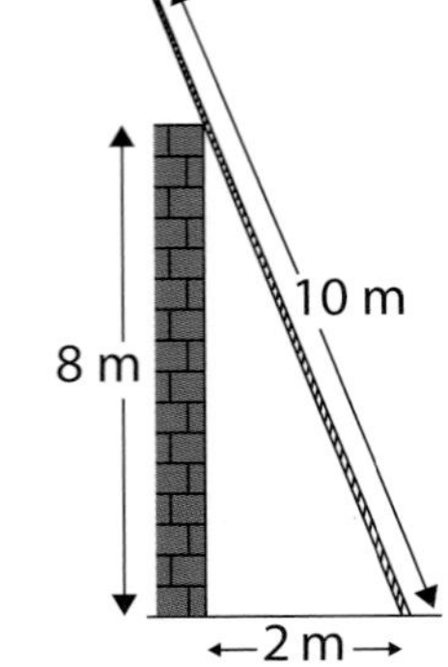

Problem solving

2 In the diagram, an object B is hanging between two walls.

It is suspended by two strings, AB and CB.

A and C are at the same level.

B is 12 m below them.

B is 5 m from the nearer wall.

The total length of the strings is 33 m.

a Find the distance between the two walls.

b Prove that the angle ABC is not 90°.

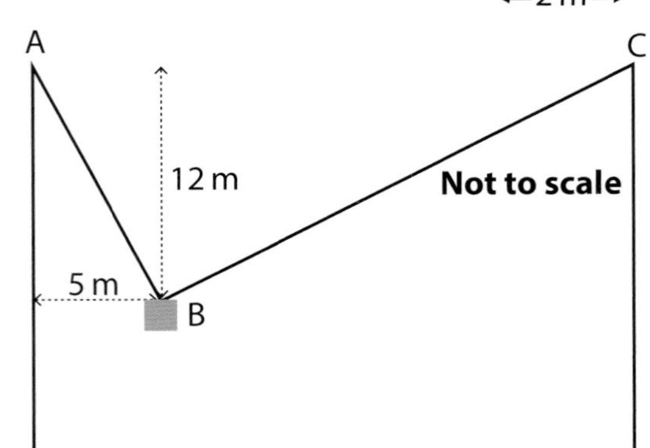

Reviewing skills

1 Find the lengths of the lettered sides in these right-angled triangles.

Give your answers correct to one decimal place.

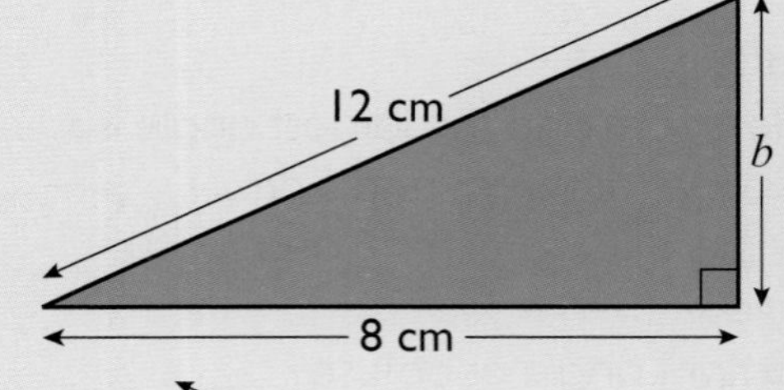

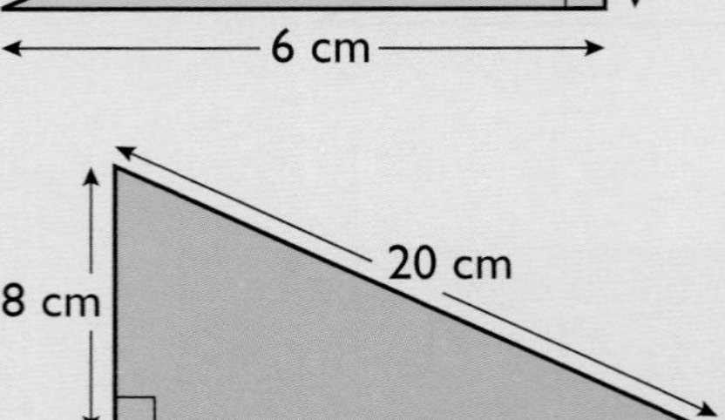

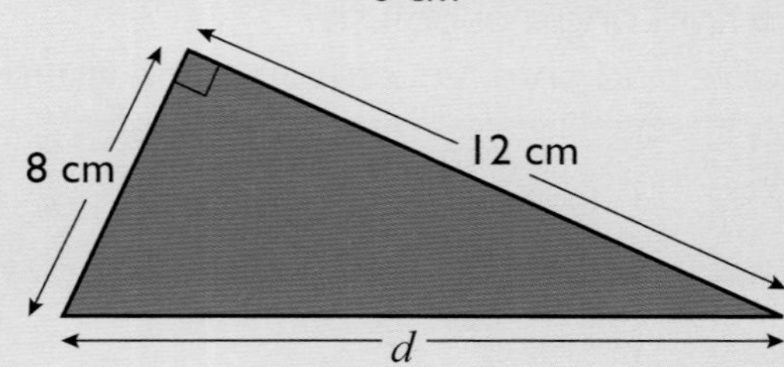

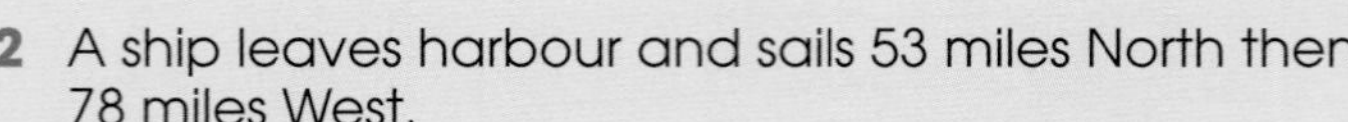

2 A ship leaves harbour and sails 53 miles North then 78 miles West.

Calculate the shortest distance back to the harbour.

Give your answer in miles correct to one decimal place.

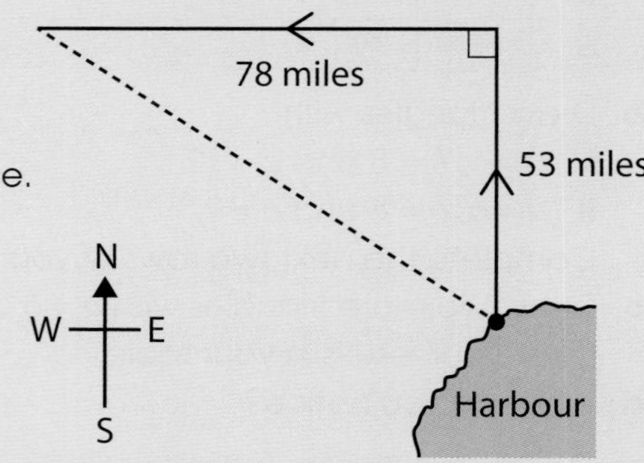

Strand 4 • Geometric construction

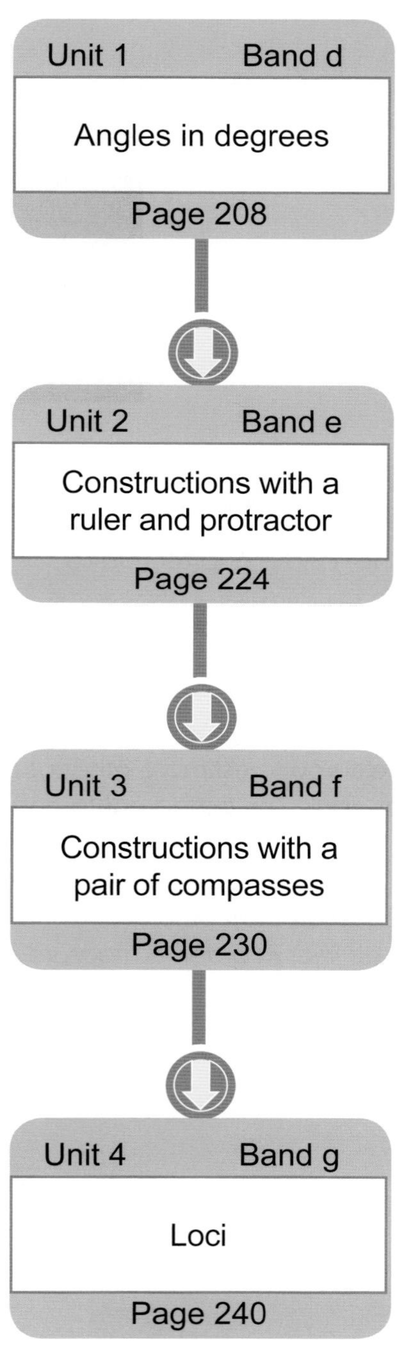

Unit 1 • Angles in degrees • Band d

Building skills

Example outside the Maths classroom

Refraction

Toolbox

A whole turn, a half turn (a straight line) and a quarter turn (a right angle) are easy to recognise.

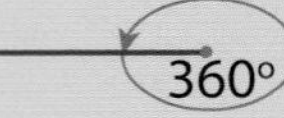

180°

Mental images of these angles will help you to estimate other angles.
Angles which are smaller than a right angle are called **acute** angles.
Angles larger than a right angle but smaller than a straight line are called **obtuse**.
Angles which are larger than a straight line are called **reflex** angles.
Use a protractor to draw or measure the amount of turn.
Sometimes it is necessary to extend the lines in order to measure accurately from the scale.

Naming an angle

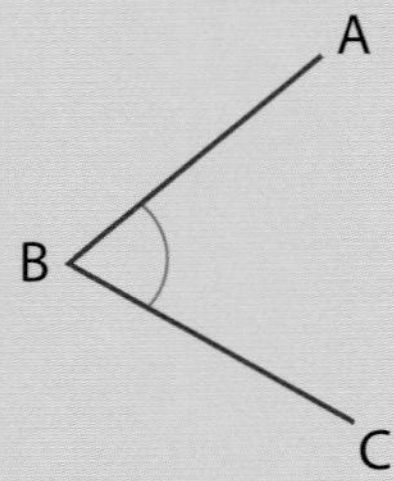

This angle can be referred to as

< B
angle B
< ABC
angle ABC
or $A\hat{B}C$.

Example – Recognising angles

What is the angle between the hands of a clock at

a 3 o'clock

b 1 o'clock

c 5 oclock?

Draw diagrams to explain your reasoning.

Solution

a At 3 o'clock, the angle between the hands is the same as a quarter turn. The angle is 90°.

b The angle between the hands at 1 o'clock is a third of the angle at 3 o'clock. So the angle is 30°.

c The angle between 12 and 3 is 90° (a quarter turn).

The angle between 3 and 4 and 4 and 5 is the same as between 12 and 1, so they are both 30°.

Total angle between the 12 and 5 = 90 + 30 + 30 = 150°.

Example – Measuring an angle

a Measure these three related angles.

i

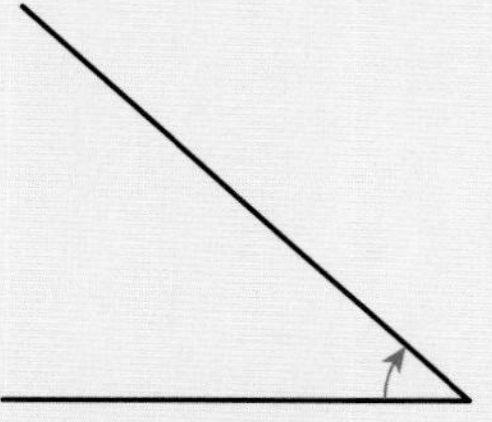

ii

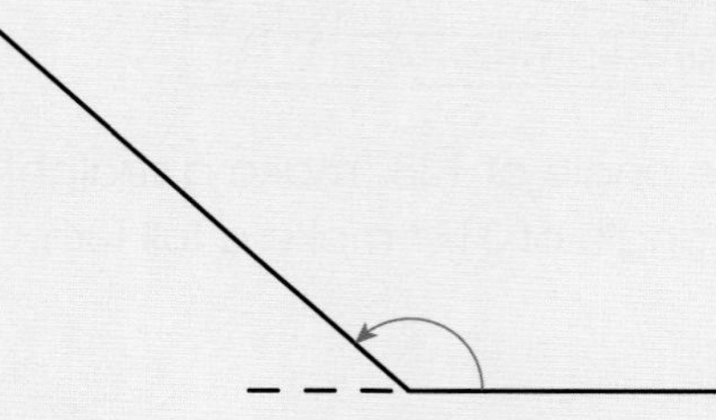

iii

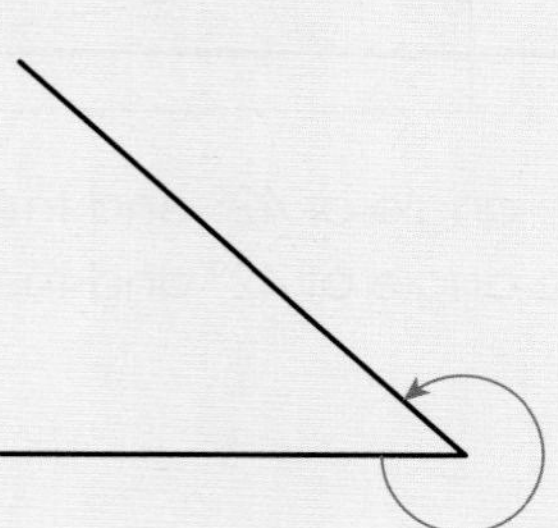

b Explain why they are related.

Solution

a

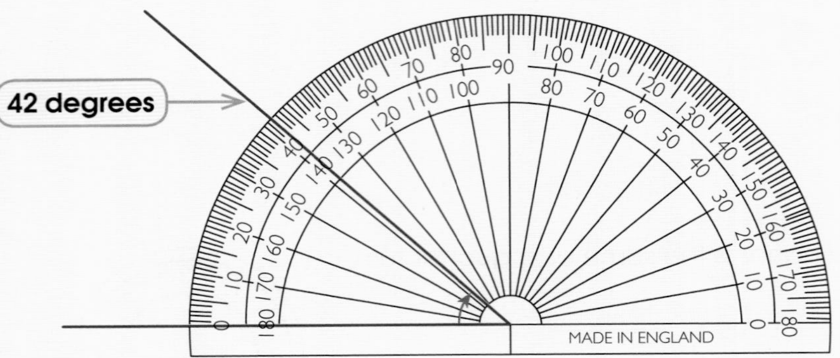

i The angle is 42°.

ii For the second angle, the zero line is the one on the right and so this is 138°.

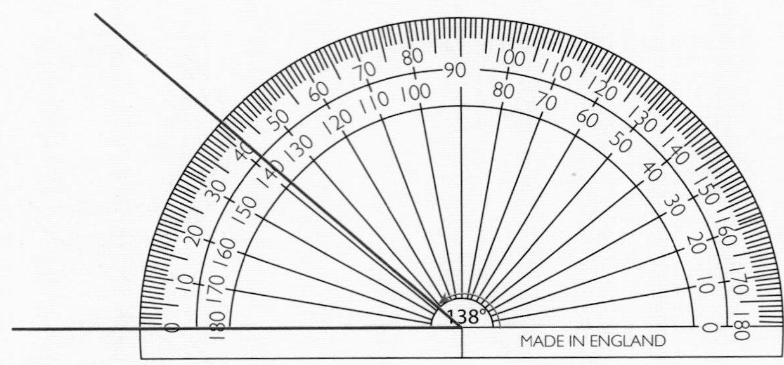

iii The angle is 180° + 138° = 318°.

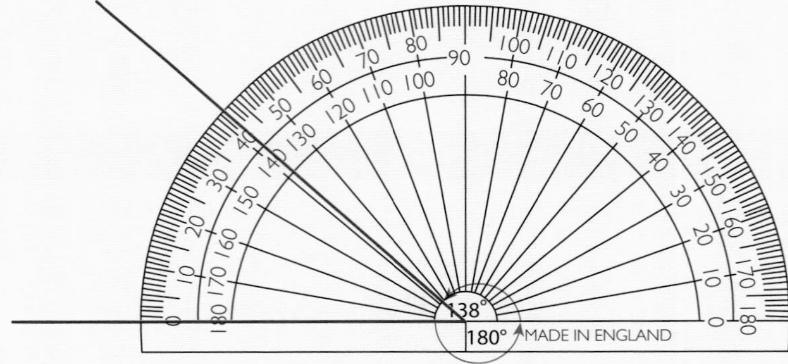

b The acute angle of 42° and the obtuse angle of 138° make a straight line.
The acute angle of 42° and the reflex angle of 318° make a full turn.

Example – Estimating angles

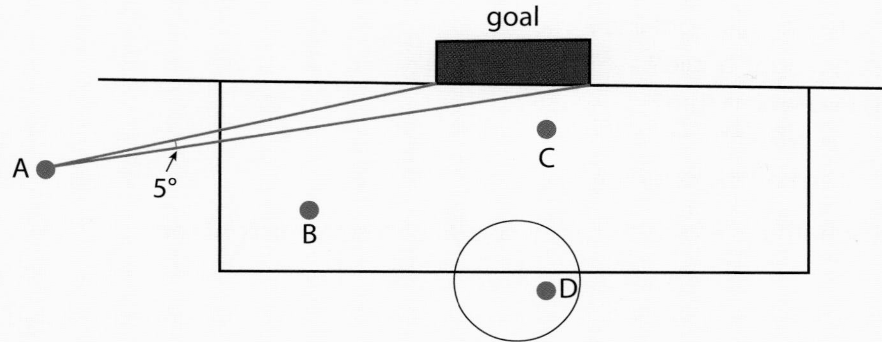

The diagram shows the position of four football players A, B, C and D during a game.

Player A has a very small angle to kick for goal – only about 5°.

Estimate the angles for the other three players.

Who has the best angle?

Solution

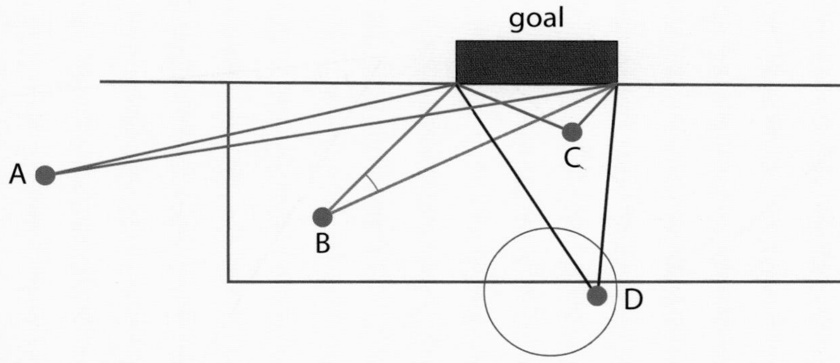

Draw lines from each player to the sides of the goal and estimate the angle.

The angle at B is about a third of a right angle.

So B is about 30°.

The angle at C is just over 90°

So C is about 100°.

The angle at D is just under half a right angle.

So D is about 40°.

Player C, closest to the goal, has the best angle.

Remember:

- A protractor has two scales, one from the left and one from the right. Make sure you are reading the correct one.
- Estimate the angle before you measure it – then you will know roughly what to expect.
- You need a mental picture of angles like 60°, 90° and 45°.

Skills practice A

1 **a** How many right angles are there on a straight line?

b How many degrees are there in a right angle?

c How many degrees are there in a whole turn?

d How many right angles are there in 270°?

e How many right angles are there in a whole turn?

2 **a** For each of these angles, say whether it is acute, a right angle, obtuse, a straight line, reflex or a full turn.

b Now measure the angles.

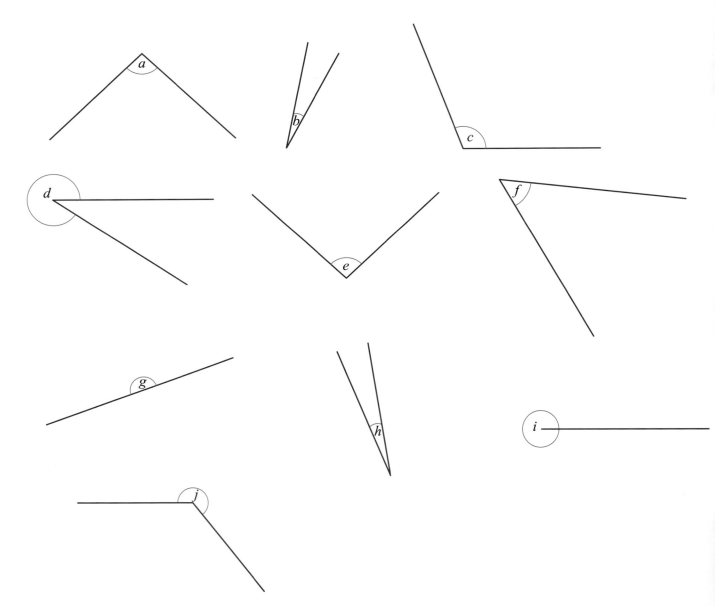

a b c d e f g h i j

3 Without measuring, match these angles to the sizes given.

- about 130°–135°
- about 60°–65°
- about 260°–265°
- about 90°–95°
- about 320°–325°
- about 30°–35°
- about 80°–85°
- about 10°–15°

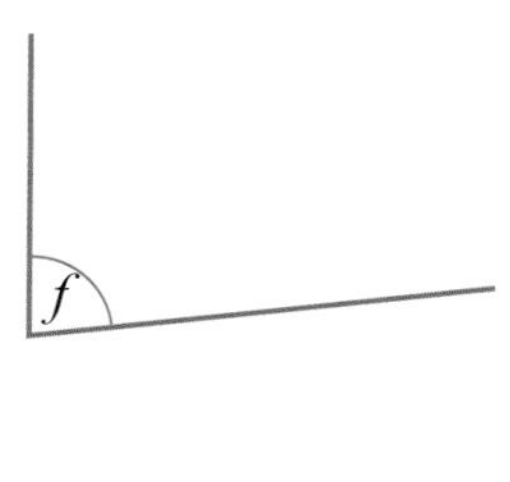

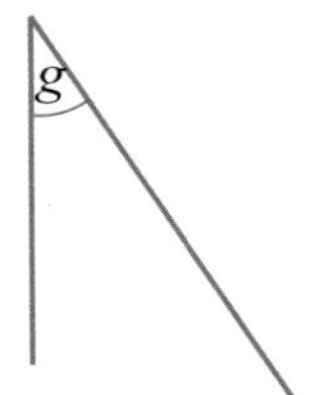

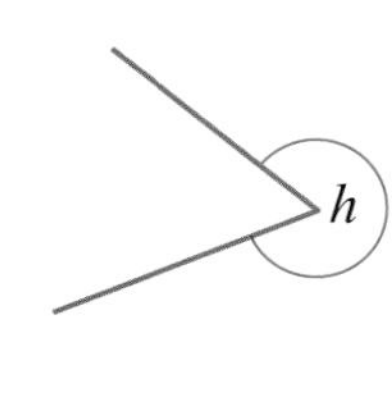

4 Measure these angles.

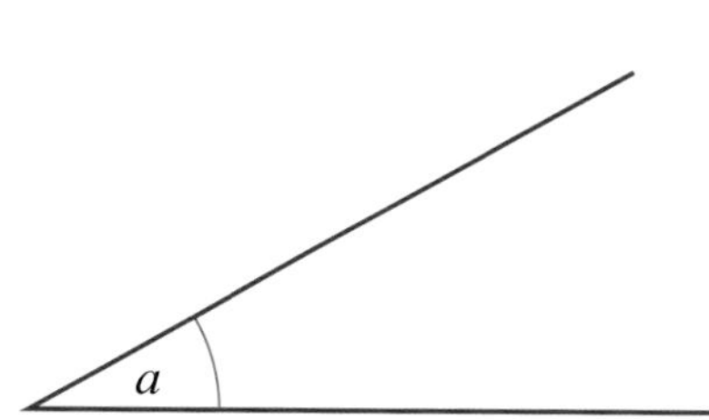

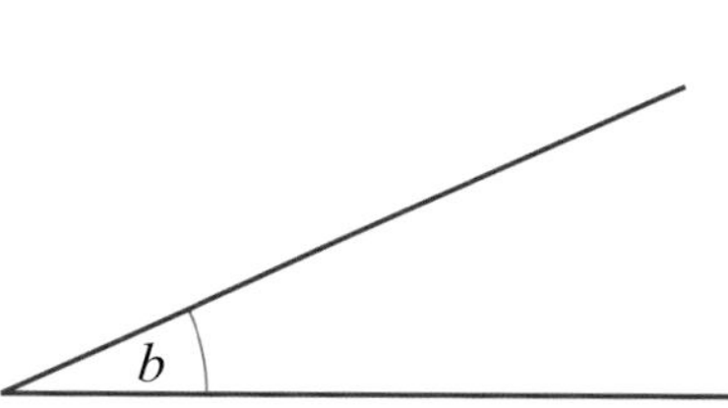

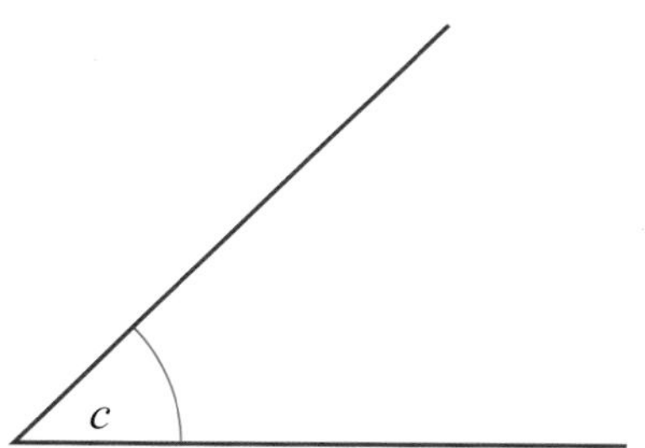

5 Draw these angles on a plain piece of paper.

a 65° **b** 80° **c** 90° **d** 95°
e 110° **f** 160° **g** 155°

6 Measure both angles on this straight line.

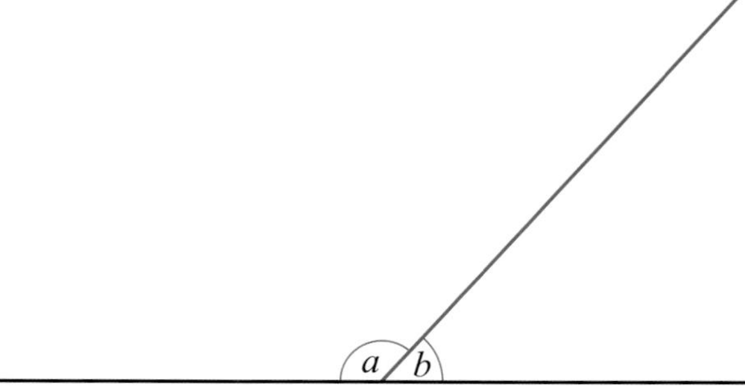

7 **a** Measure these angles.

b Now draw them on a plain piece of paper.

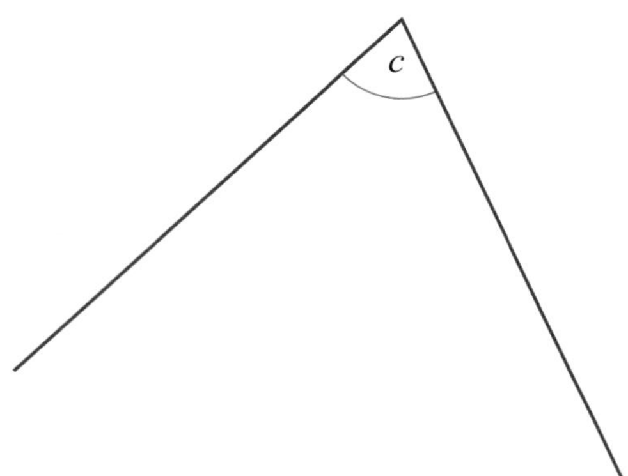

8 **a** Make an accurate drawing of a reflex angle of 207°.

You have to do a calculation to draw the correct smaller angle with your protractor.

i Calculate the smaller angle.

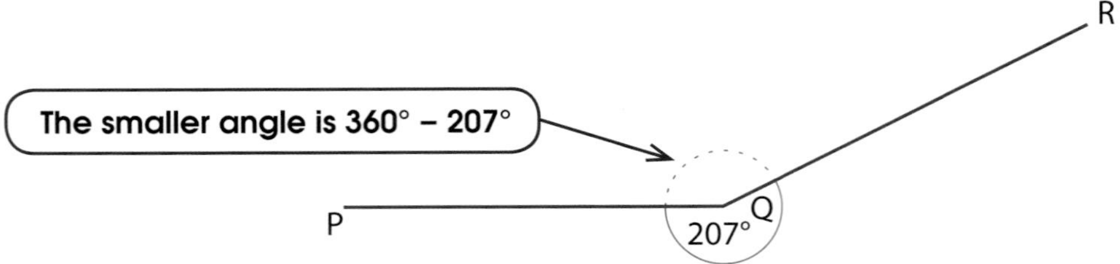

ii Use your protractor to draw the smaller angle.

iii Label the reflex angle.

b Use this method to draw and label these reflex angles.

Write down the calculation next to each one.

i 200° **ii** 243°

iii 300° **iv** 270°

v 181° **vi** 315°

9 Draw these angles.

Ask your partner to check your angles for neatness and accuracy.

a

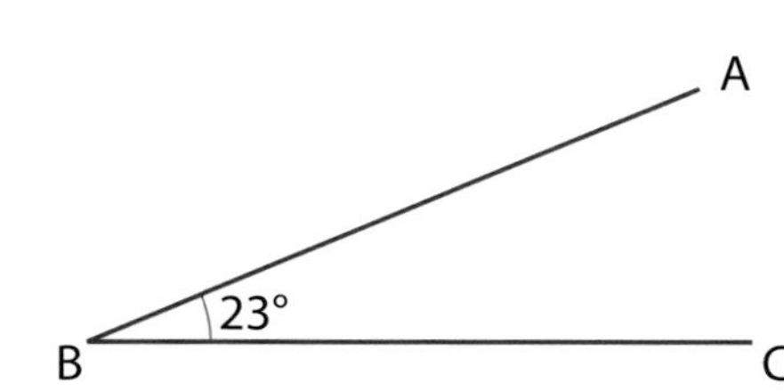

b

c

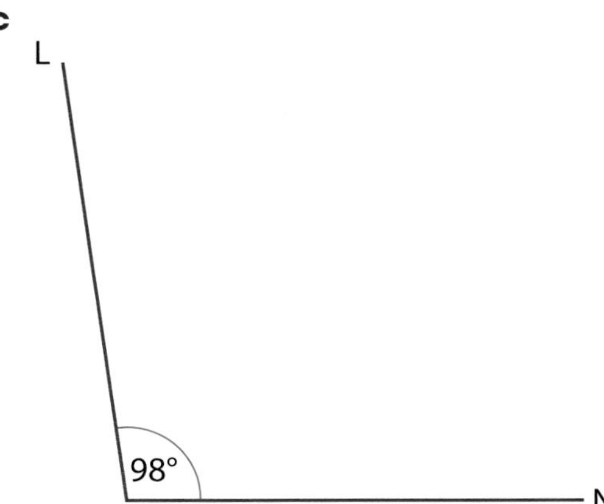

d

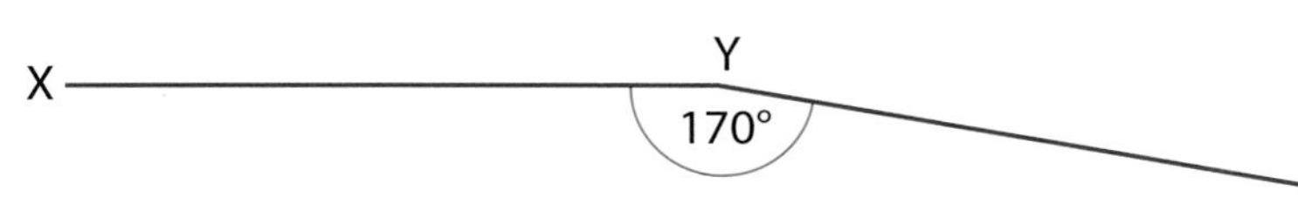

10 The yellow triangle is equilateral.

Its angles are all 60°. It is folded along a line of symmetry to give the green triangle.

a What are the sizes of the angles in the green triangle?

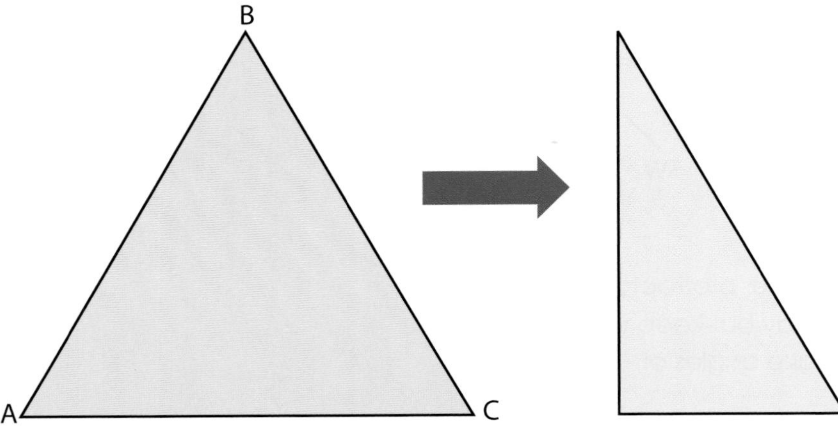

b Use your answers to part **a** to estimate the sizes of these angles.

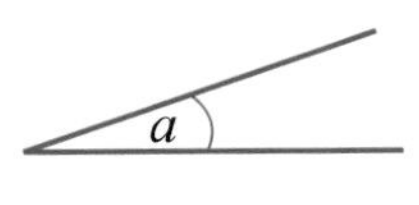

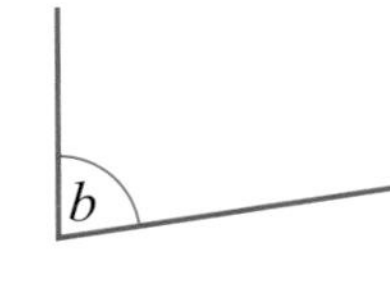

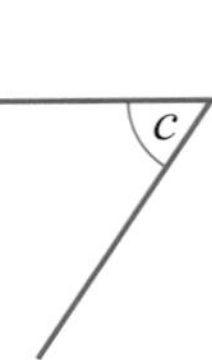

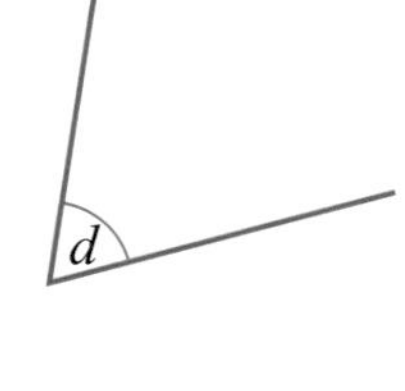

Skills practice B

1 Match the angle between the minute hand and the hour hand and the time.

9:00 a.m.	30°
4:00 p.m.	90°
11:00 p.m.	120°

2 These ten angles are five pairs.

Each pair can be fitted together to make a full turn, but first you will have to turn one of the angles in the pair.

Find the five pairs.

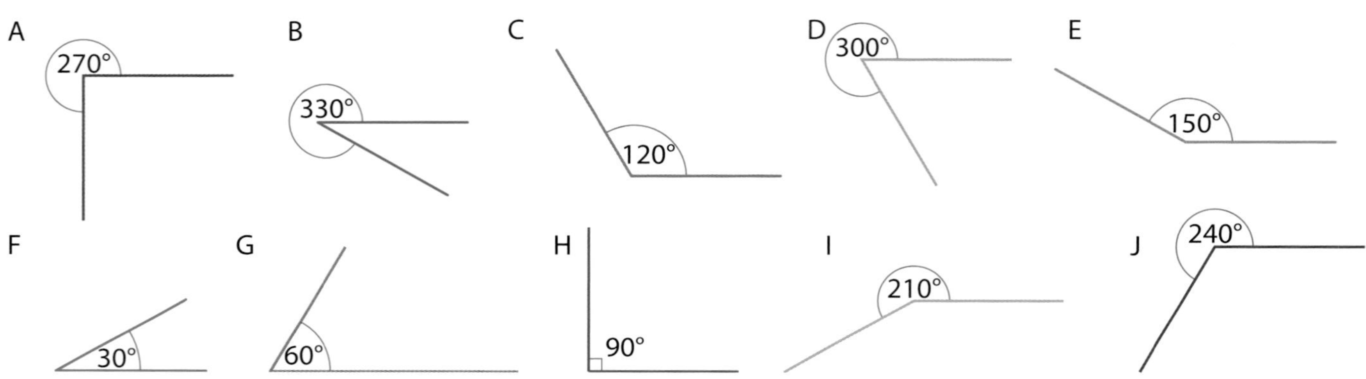

3 Look at this diagram of a compass.

Measure the angles between

a NE and E

b NE and SE

c NE and S

d NE and SW

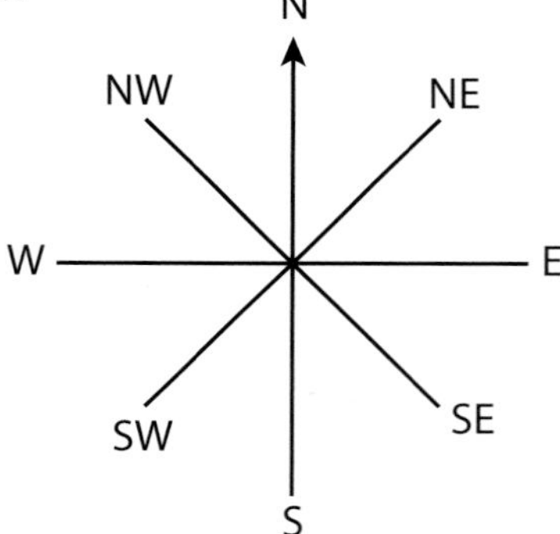

Reasoning

4 a Draw an angle of 50° using your protractor and a ruler.

b Now put your protractor away but keep your ruler out.

Use your angle of 50° to make angles of

i 130°

ii 310°

iii 230°.

5 **a** Make a rough estimate of the size of each of these angles.

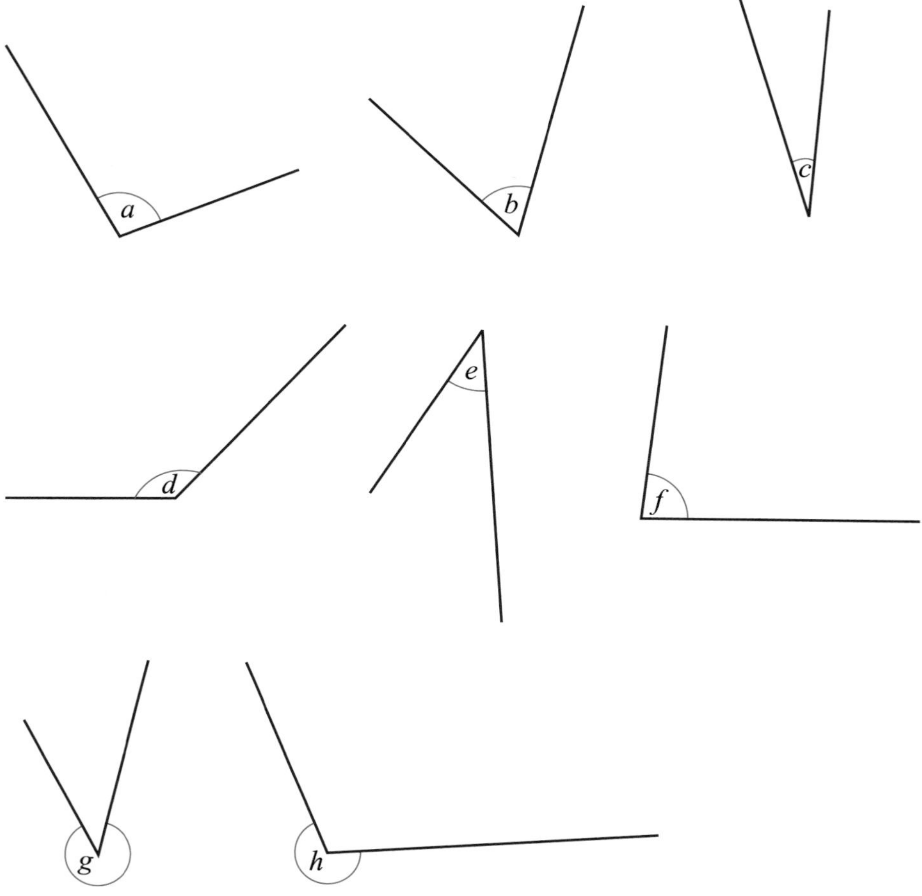

b Measure the angles with a protractor.
Check that your measurements agree roughly with your estimates in part **a**.

6 Fold a piece of filter paper in half.
Fold it in half again, and again and once again.
What angles are created by folding a piece of filter paper like this?

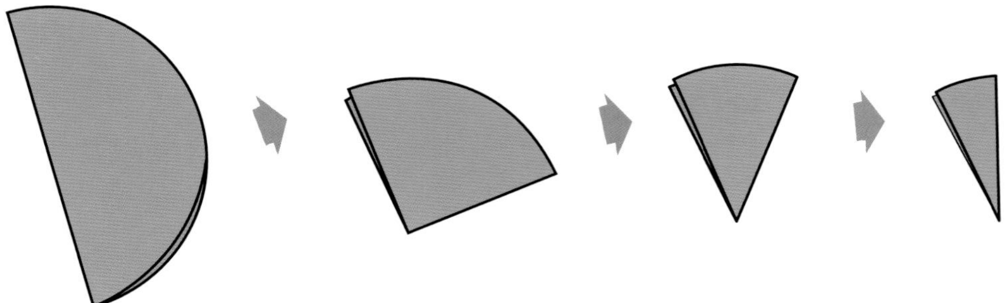

Reasoning

7 The diagram shows a regular octagon.
Some extra lines have been added in red.

a Measure the angles a, b, c and d.

b What do your answers show about the angles a, b, c and d?

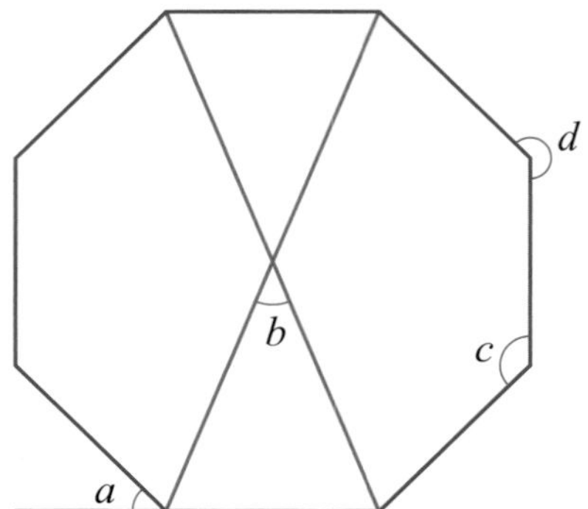

Reasoning

8 The diagram shows part of a honeycomb (enlarged).

a Measure the angles a, b and c.

b Explain why there are no gaps in a honeycomb.

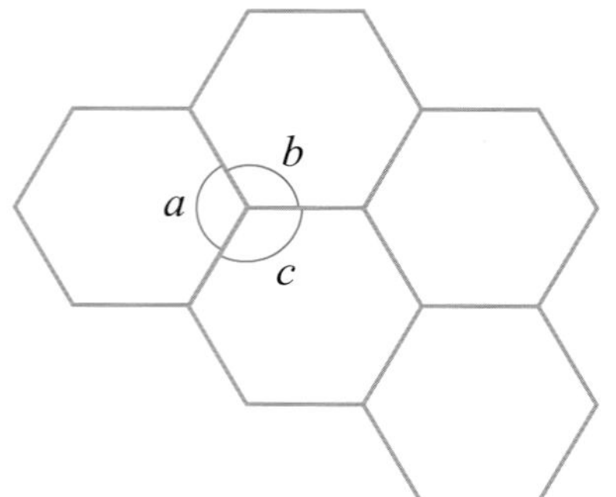

Reasoning

9 Look at this five-pointed star.
Measure suitable angles.
What angles can you see in the diagram?

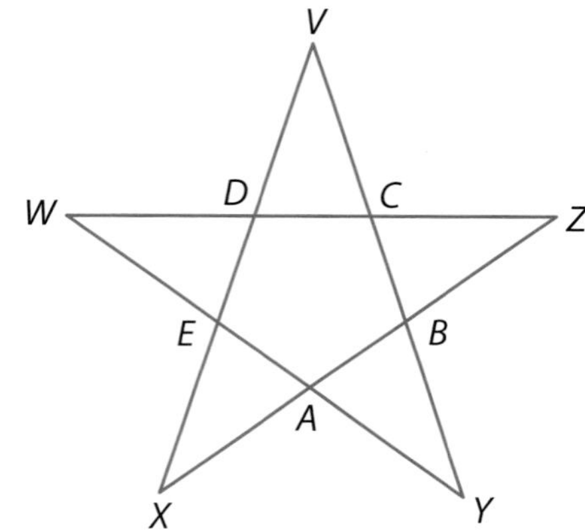

10 See how accurately you can draw angles:

a Write down an angle you think you can draw.

b Draw it.

c Measure it.

d Score 1 if you are within 10°; score 2 if you are within 5°.

e Repeat this process for 4 more angles.

f What do you think would help you improve your score?

11 The radio telescope receives signals from satellite A then turns to each of the other satellites in alphabetical order.

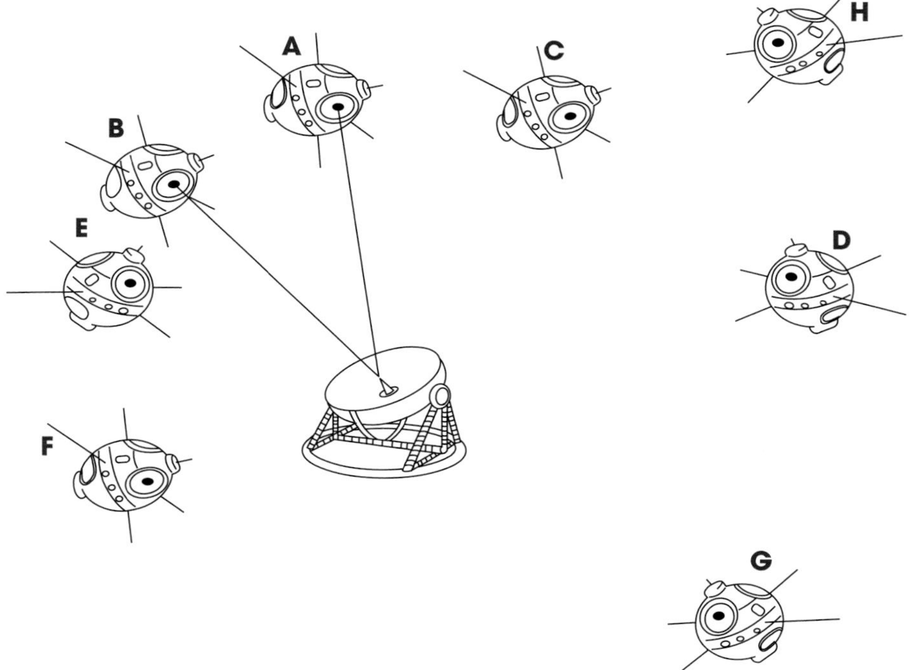

a Copy the table below.

Turn	Estimate	Clockwise or anticlockwise	Measure
A to B			
B to C			
C to D			
D to E			
E to F			
F to G			
G TO H			

b Estimate the angle, clockwise or anticlockwise, that the telescope must be turned to aim at the next satellite.

Add your estimates to the table.

c Measure the angles to check the accuracy of your estimates.

Wider skills practice

Reasoning

1 a Measure the angles in these diagrams.

i

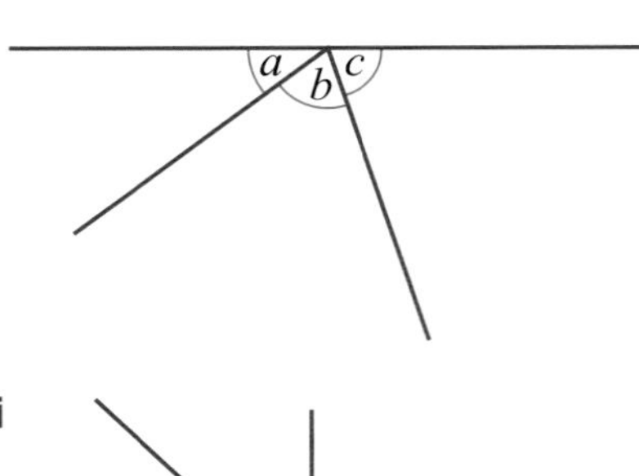

ii

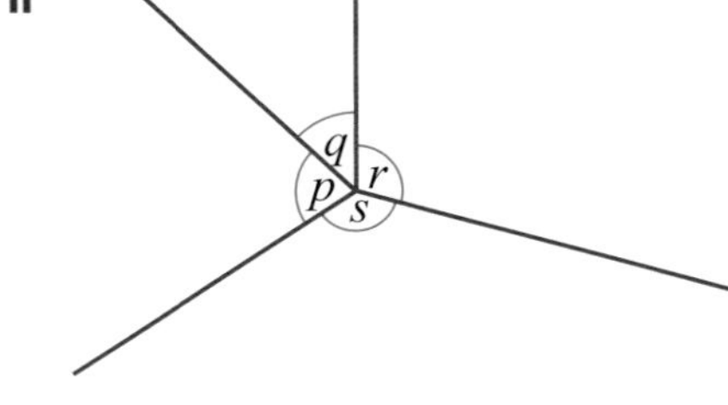

b In each case, add the angles together.
Explain your answer.

2 Draw a triangle.
Measure its three angles and add them together.
What do you notice?

Applying skills

Problem solving

1 A whole turn can be made up of an acute angle and a reflex angle.

a Copy and complete these sums.

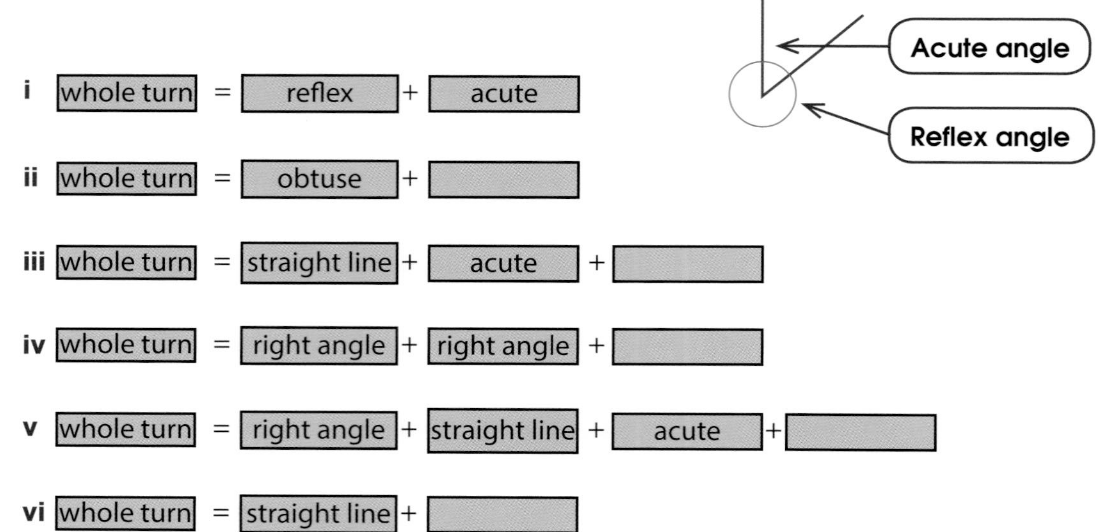

b For each sum in part **a**, give possible angles in degrees for each of the boxes.

2 Estimate the size of the angle in each of these situations.

a The angle you open a closed door to walk through.

b The angle you turn a tap when you wash your hands.

c Your angle of vision when you are looking at your computer.

d The minimum opening angle for the lid of a pedal-operated rubbish bin.

Reviewing skills

1 **a** Write these angles in order of size, with the largest first.

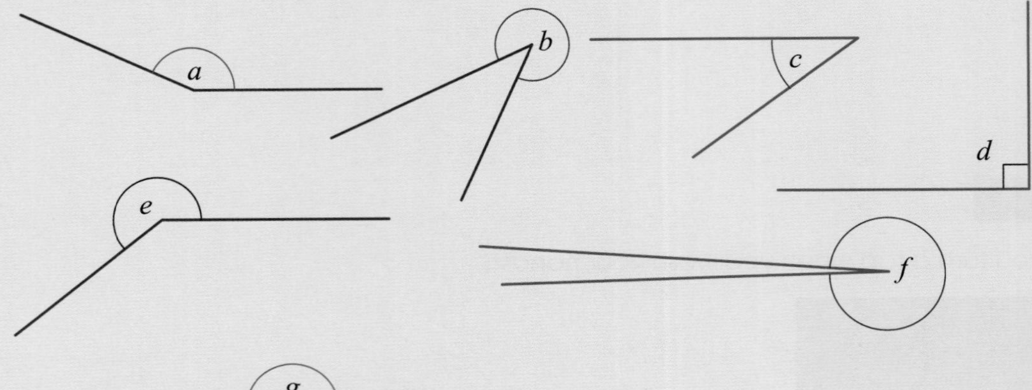

Describe each angle as acute, a right angle, obtuse, a straight line, reflex or a full turn.

b Now measure the angles in part **a**.

2 Match these angles with the diagrams below.

135° 330° 90° 30° 210°

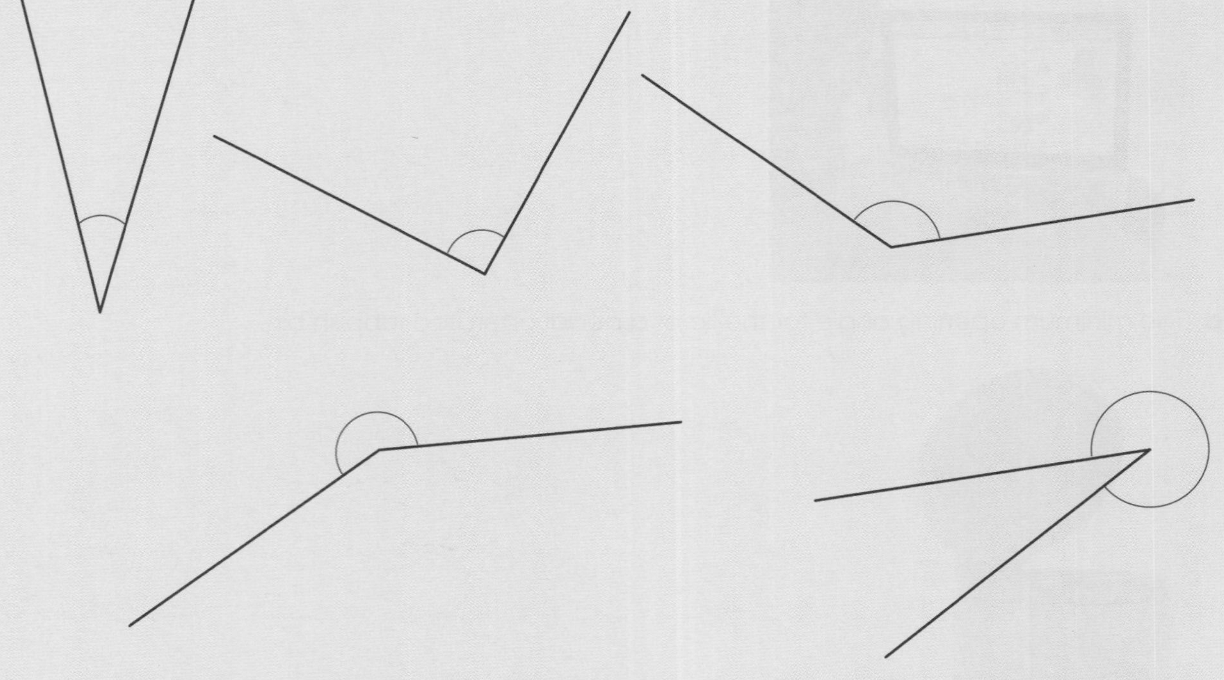

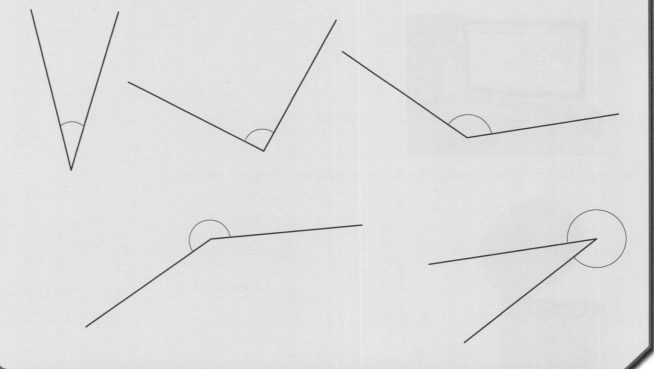

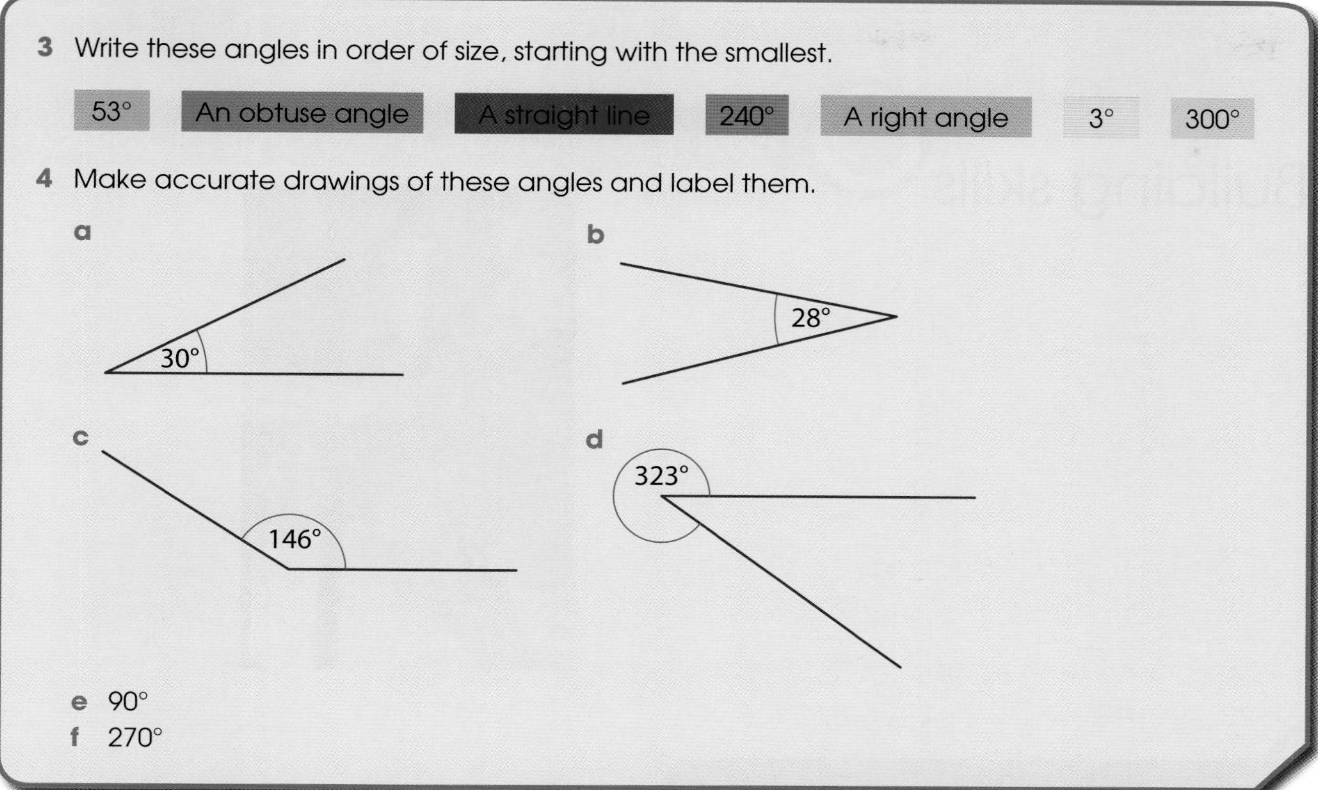

3 Write these angles in order of size, starting with the smallest.

53° | An obtuse angle | A straight line | 240° | A right angle | 3° | 300°

4 Make accurate drawings of these angles and label them.

a 30°

b 28°

c 146°

d 323°

e 90°

f 270°

Unit 2 • Constructions with a ruler and protractor • Band e

Building skills

Example outside the Maths classroom

Carpenters

Toolbox

Constructing a triangle when two sides and the included angle are given

- Use a ruler to draw one side of the correct length.
- Use a protractor to draw the correct angle at one end of the side.

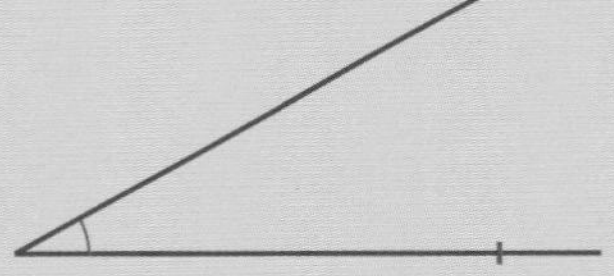

- Use a ruler to mark the correct length on the second side.

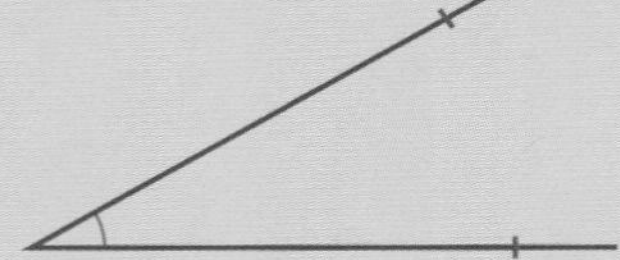

- Draw the third side to form the triangle.

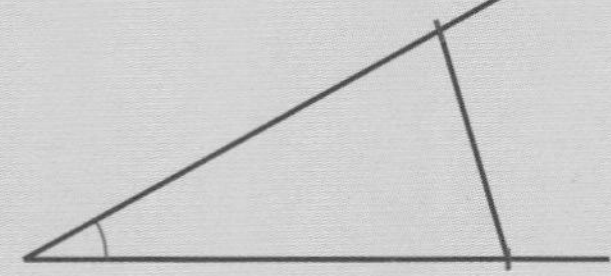

Constructing a triangle when two angles and the included side are given

- Use a ruler to draw one side of the correct length.

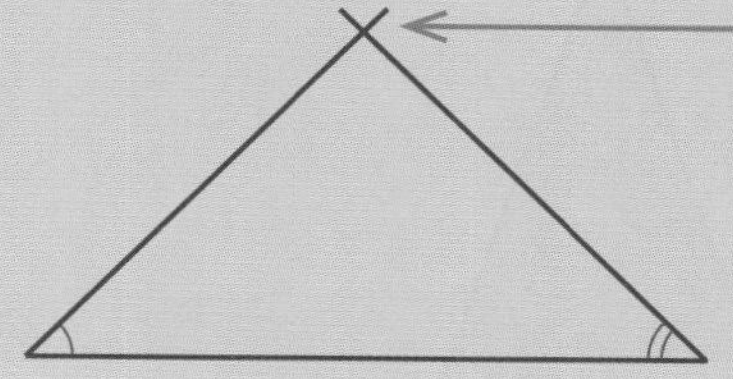

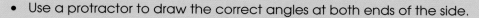

- Use a protractor to draw the correct angles at both ends of the side.

The lines cross at the vertex of the triangle.

Example – Drawing a construction to solve a problem

A snow hut has a roof with base angles of 65° so the snow will slide off the roof.

The hut is 8 m wide.

How high is the roof?

Solution

Use a scale of 1 cm for 1 metre.

Draw a base line of 8 cm to represent the width of 8 m.

Use a protractor to measure angles of 65° at each end of the line.

Complete the triangle.

Measure the height of the top above the base line.

It should be 8.5 cm.

The roof is 8.5 m high.

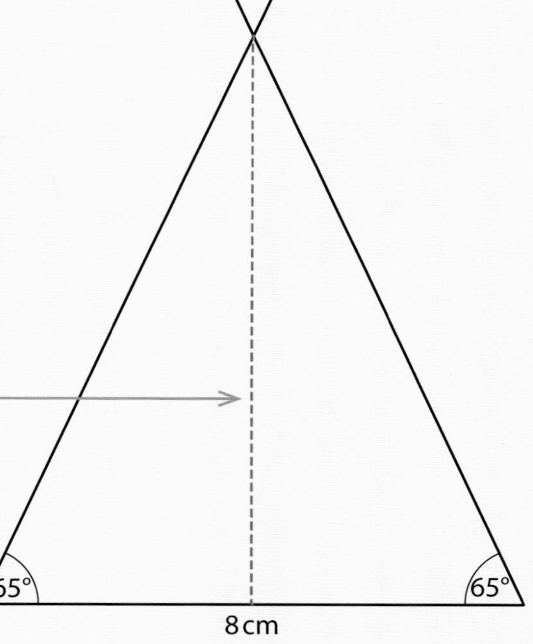

Remember:

- Always start with the base line of the triangle. Measure the angles and other lines from this starting point.

Skills practice A

1 Use a ruler and protractor to make accurate drawings of these triangles. In each case, measure the remaining sides and angles.

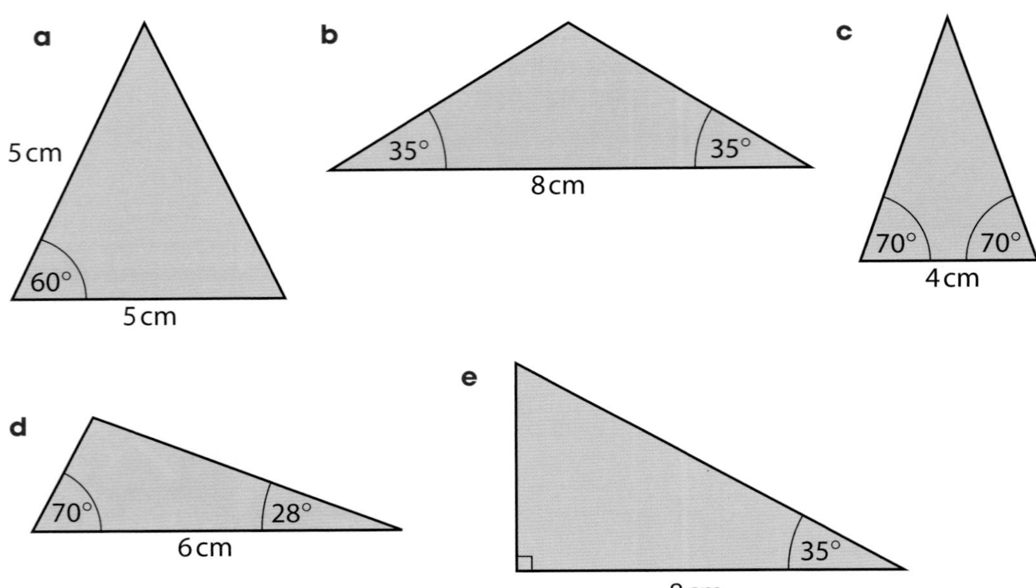

2 Use a ruler and a protractor to make accurate drawings of these triangles. In each case, measure the remaining sides and angles.

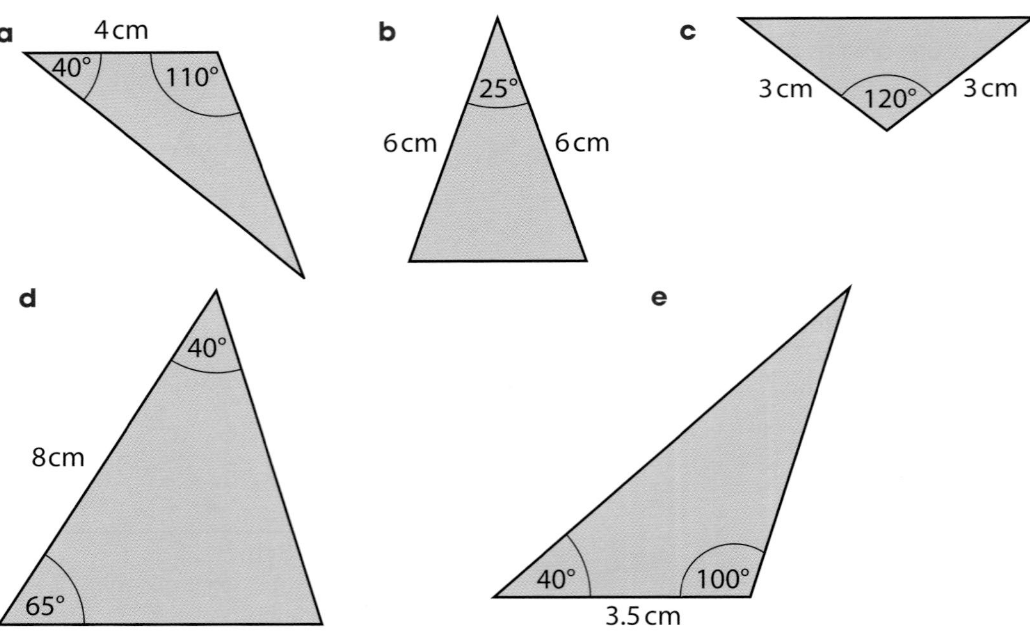

Reasoning

3 a Use a ruler and protractor to construct this triangle.
b Measure the lengths of the other two sides.
Why are these equal?

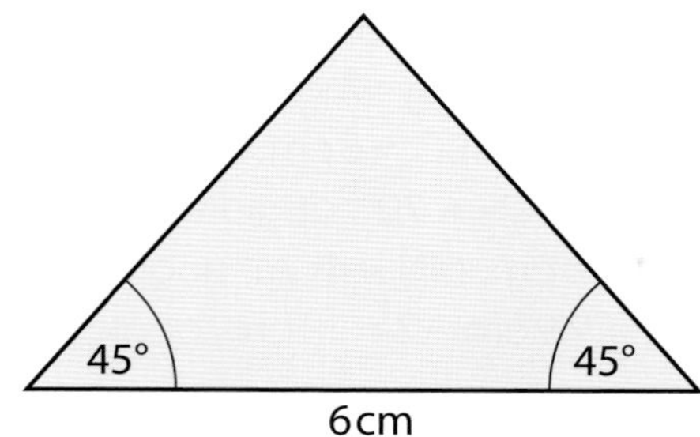

Skills practice B

1 Kim's dad is designing a slide for the garden.
This is his sketch of the slide.
a Use a scale of 1 cm to represent 1 m.
Make an accurate scale drawing of the side of the slide.
b How long must Kim's dad make the ladder?

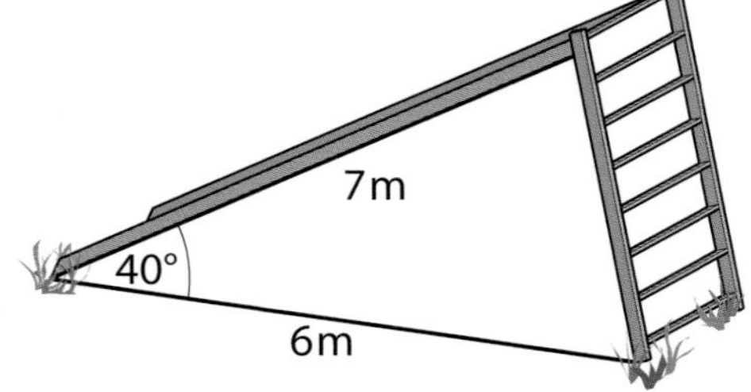

2 Barton is 6 km due North of Avonford.
Carville is 8 km due East of Avonford.
a Use a scale of 1 cm to 1 km to make a scale drawing.
b How far is it from Barton to Carville?

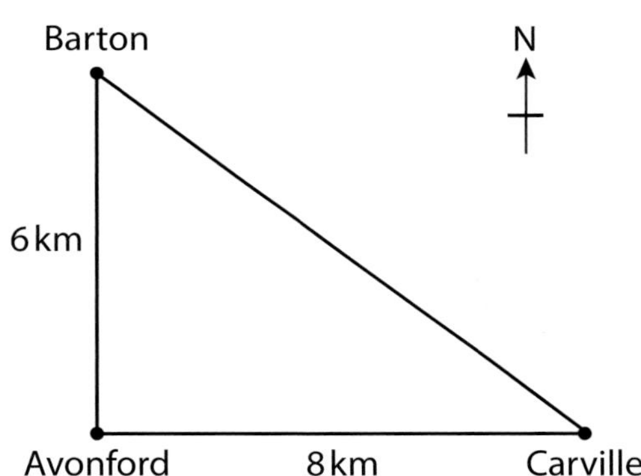

3 a Ten people are told to draw this triangle.
Do they all draw it exactly the same?

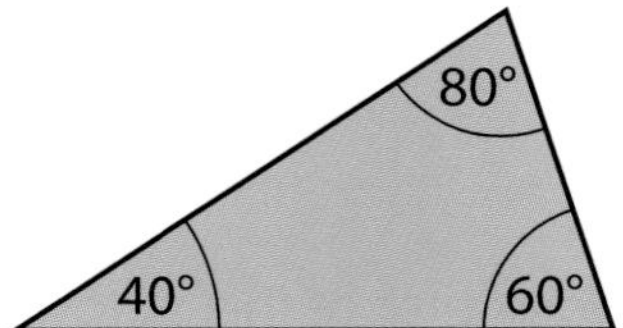

b Is the answer the same for this triangle?

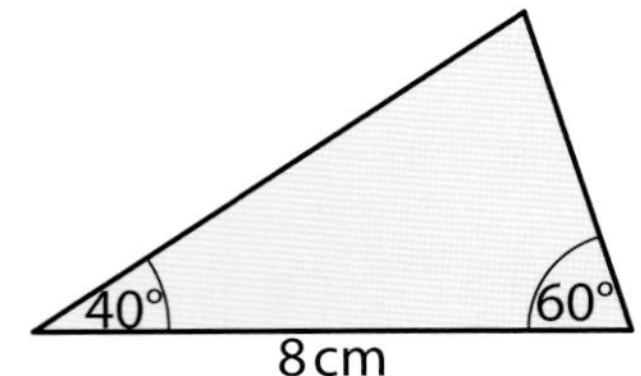

4 Construct these triangles.
In each case, say what is special about the triangle.

a AB = 8 cm, BC = 8 cm, angle ABC = 60°
b Angle PQR = 70°, angle QRP = 40°, QR = 8 cm
c LM = 8 cm, MN = 4 cm, angle LMN = 60°
d XY = 9.9 cm, YZ = 14 cm, angle XYZ = 45°

5 **a** Draw this triangle.
Measure the length of AB.

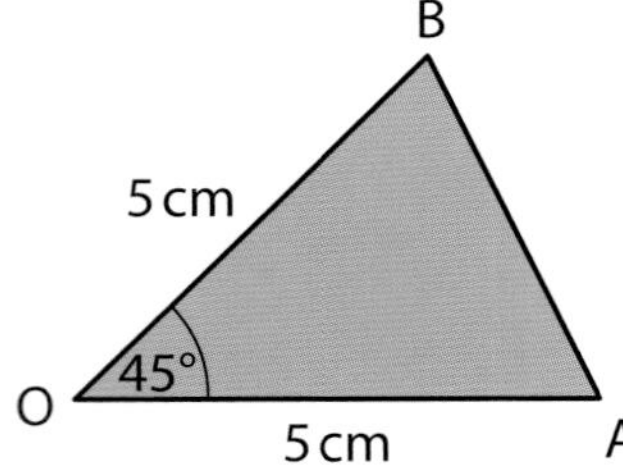

b Now add seven more triangles to make a regular octagon.

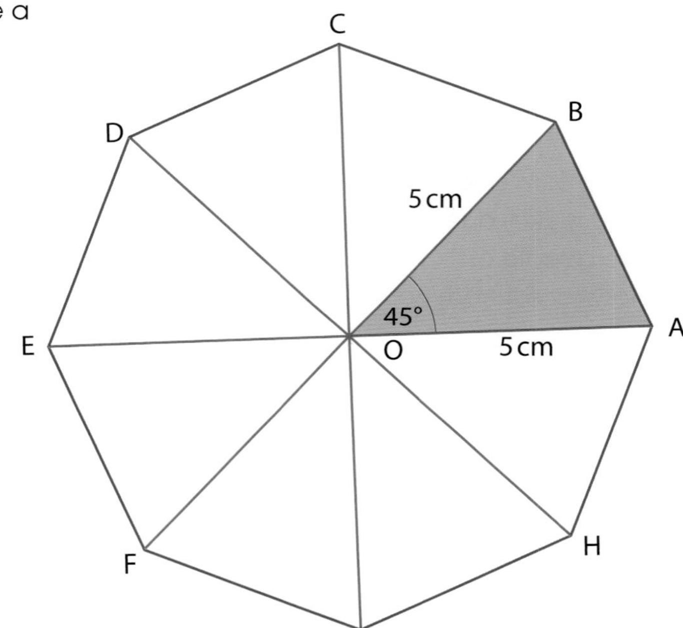

Measure the angles
i COF
ii CDE
iii DOF.

c What is the perimeter of the octagon?

Wider skills practice

1 **a** Draw these two triangles accurately.

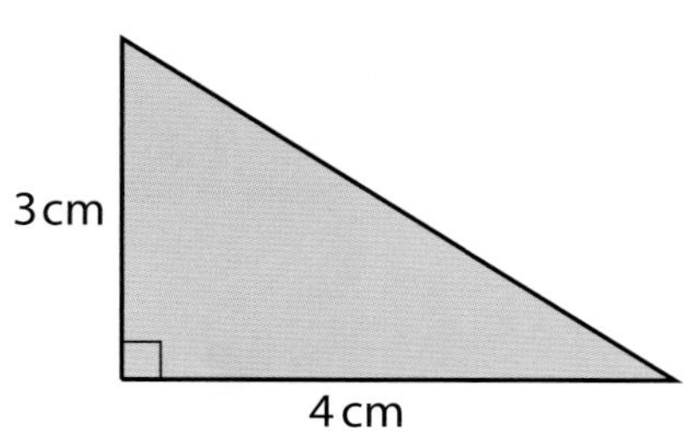

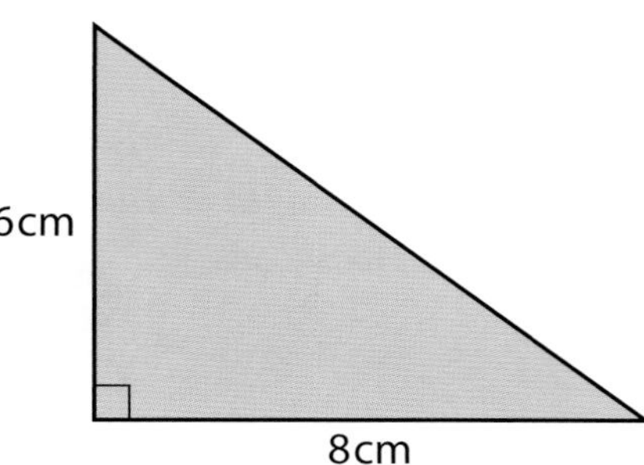

b Measure the other two angles.
What do you notice about the angles?

c Measure the other side.
What do you notice about the sides?

Applying skills

Problem solving

1 Wes is making a hanging basket support for his Design and Technology project.
He has made this sketch.
He is going to use a ready-formed hook.

a Make an accurate scale drawing of the frame.

b Use the diagram to find the length of metal strip that Wes needs.

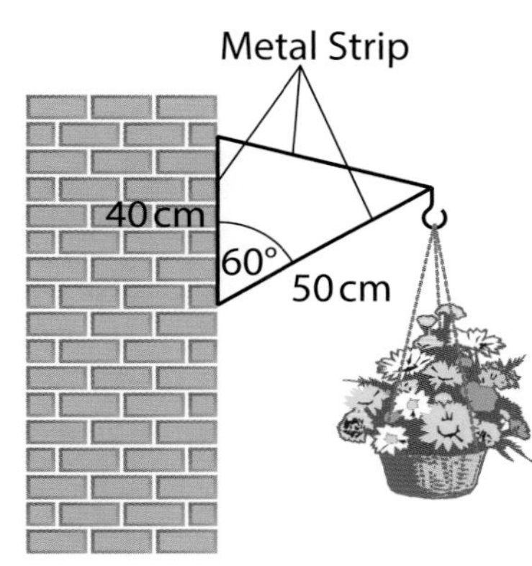

Reviewing skills

1 **a** Use a ruler and protractor to make an accurate drawing of ABDC.

b Measure

i the angle ABC

ii the length BC

iii the length CD.

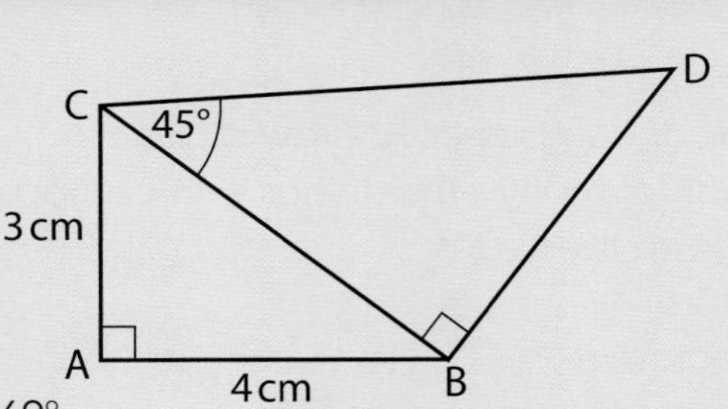

2 **a** Draw a triangle with angles of 60°, 60° and 60°.

b Measure its sides.

c What do you notice?
What sort of triangle have you drawn?

Unit 3 • Constructions with a pair of compasses • Band f

Building skills

Example outside the Maths classroom

House design

Toolbox

A circle has a very useful property.

All the points on the circumference are an equal distance from the centre.

That distance is the radius of the circle.

Part of a circle is called an **arc**.

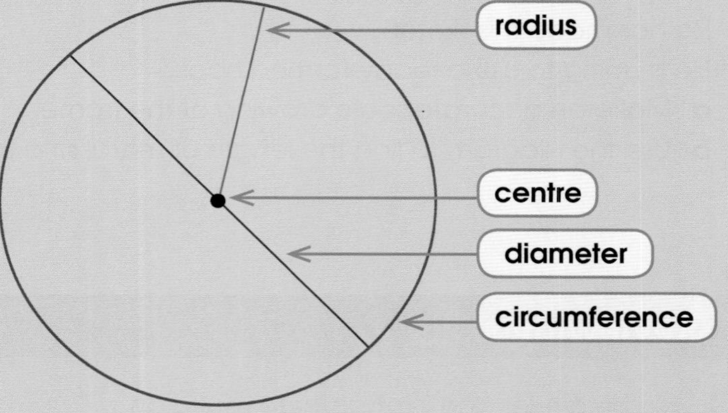

You use a pair of compasses to draw a circle.

You can construct many other things with compasses:

- triangles given their sides
- line bisectors
- angle bisectors
- perpendicular from a point **to** a line
- perpendicular from a point **on** a line

Example – Bisecting a line

Bisect the line AB.

Solution

- Open your compasses.
  Put the point on A.
  Draw arcs above and below the line.

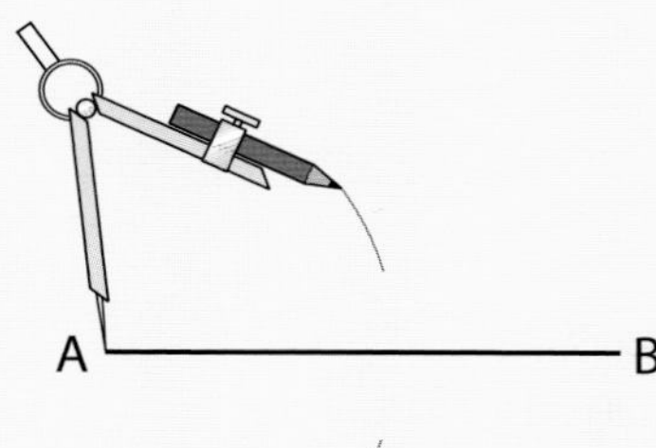

- Do not adjust the compasses.
  Put the point on B.
  Draw two more arcs to cut the first two.

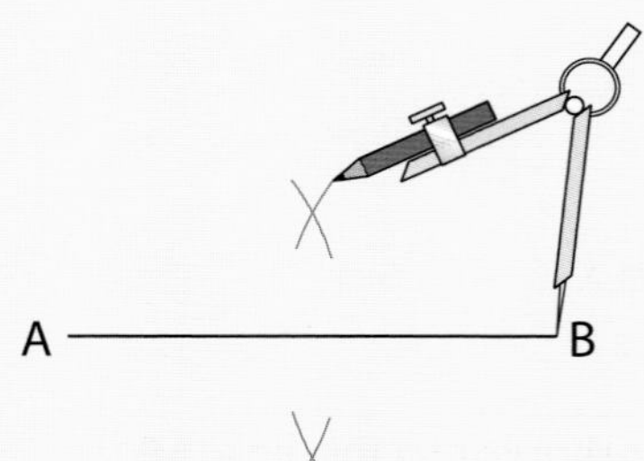

- The arcs meet at X and Y.
  Draw the line XY.

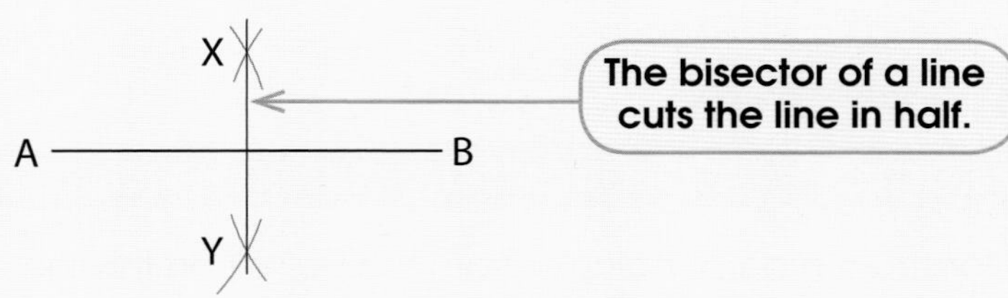

Example – Constructing a perpendicular from a point to a line

Construct the perpendicular from the point R to the line.

Solution

- Draw a line. Mark R away from the line.

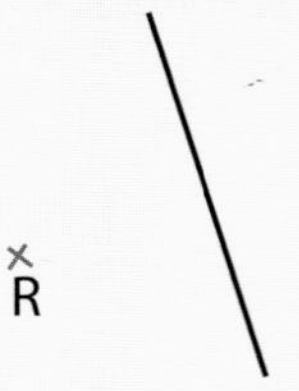

- Put the point of the compasses on R.
  Draw an arc.
  It intersects the line at A and B.

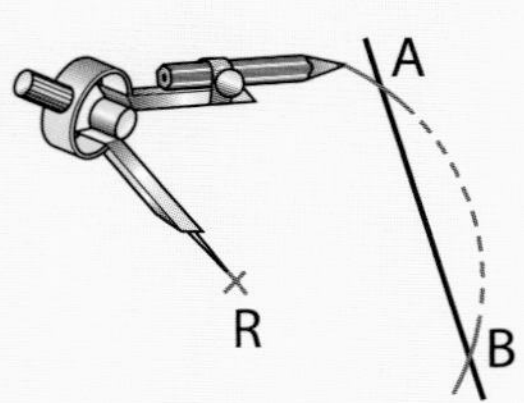

- With the point on A, draw an arc on the opposite side of the line from R.

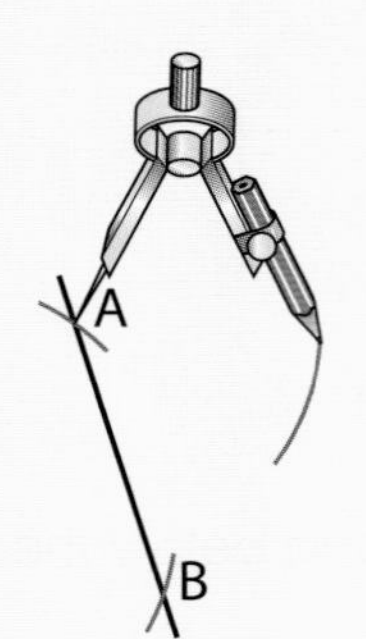

- Do not adjust the compasses.
  With the point on B, draw another arc.
  The arcs intersect at X.

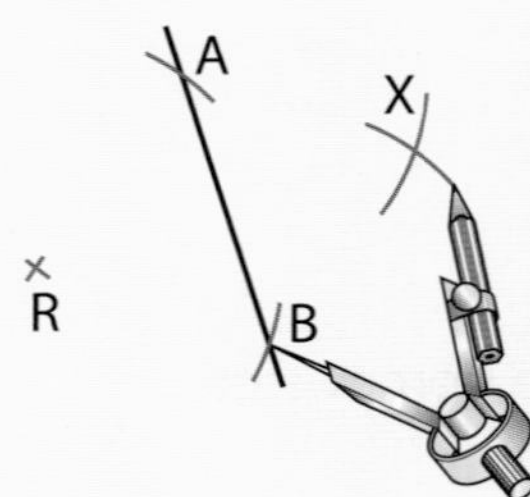

- Draw a line through R and X.

The line RX is perpendicular to the line AB.

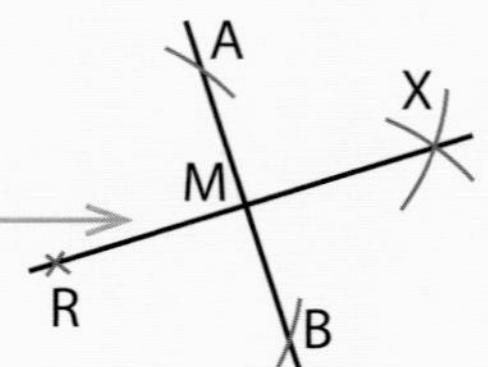

Example – Constructing a perpendicular from a point on a line

Construct a perpendicular from the point T on this line.

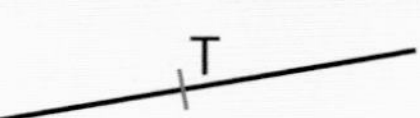

Solution

- Mark A and B.
  Each point is the same distance from T.

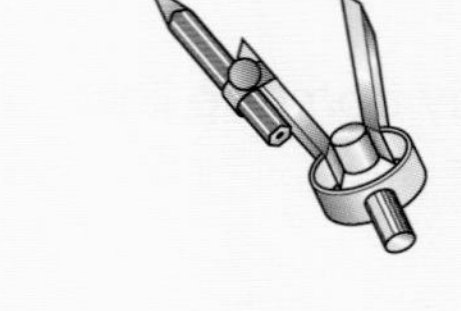

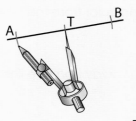

- Open the compasses wider.
  With the point on A draw an arc.

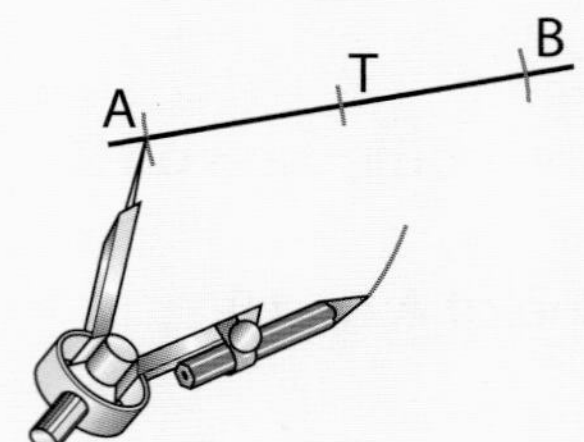

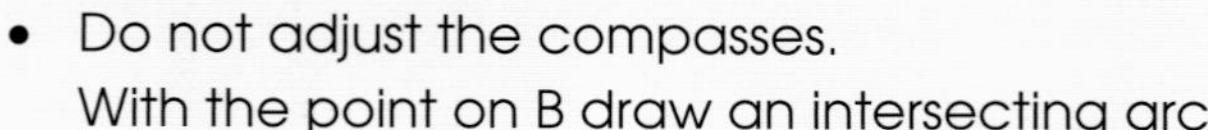

- Do not adjust the compasses.
  With the point on B draw an intersecting arc.

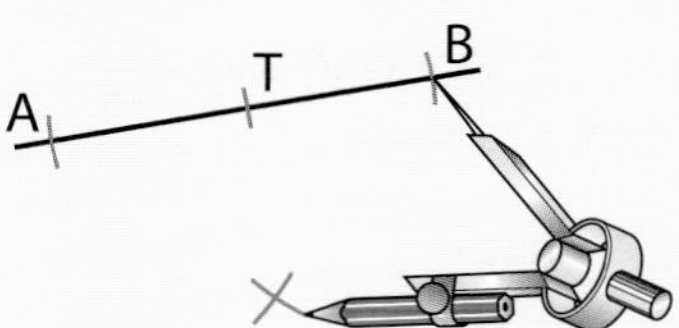

- Draw a line through T and the intersection X.

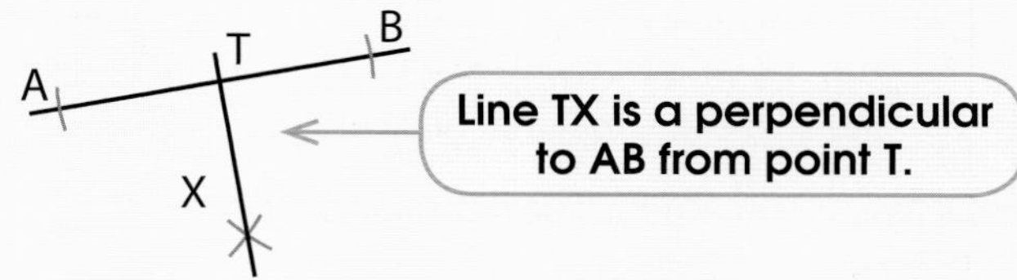

Example – Bisecting an angle

Bisect this angle.

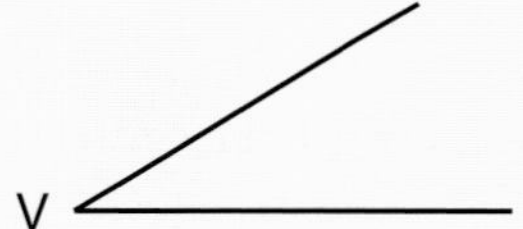

Solution

- Put the point of the compasses on V.
  Draw an arc cutting both lines.
- Put the point of the compasses on A.
  Draw an arc.
  Do not adjust the compasses.
  Put the point of the compasses at B.
  Draw an arc.

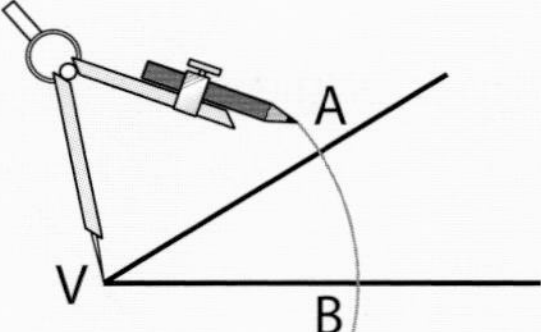

These two arcs meet at C.

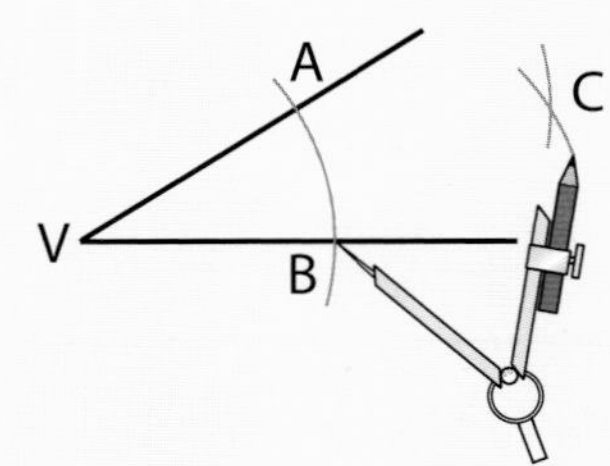

- Join V to C.

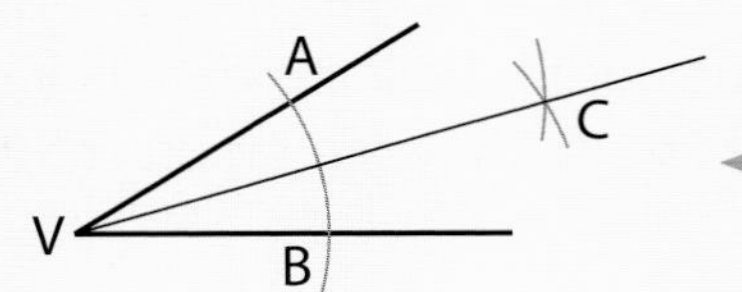

Example – Constructing a triangle

Make an accurate drawing of this triangle.

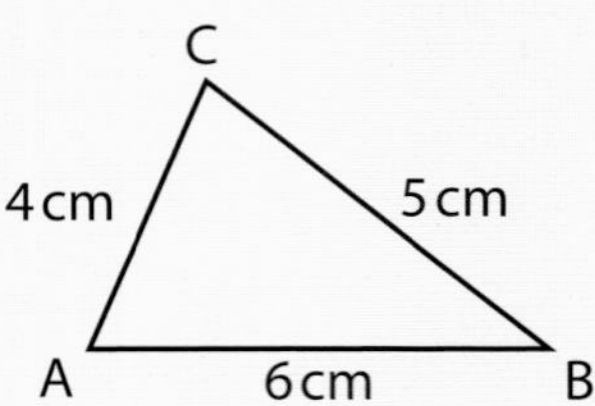

Solution

- Draw a line 6 cm long.

A 6 cm B

- Draw an arc of radius 4 cm and centre A.
  The point C is on this arc.

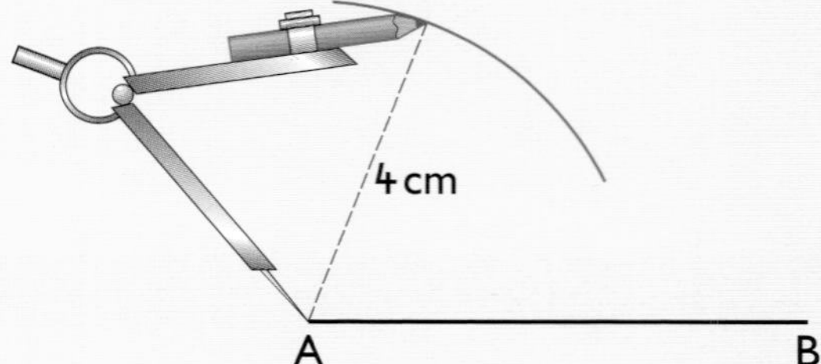

- Now draw an arc of radius 5 cm and centre B.
  C is on this arc too.

A B

- The two arcs meet at C.
  Join AC and BC to form the triangle.

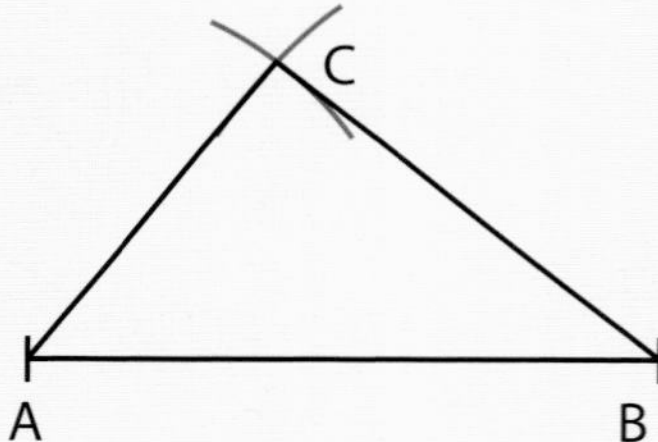

Remember:

- ✦ The bisector of a line crosses the mid-point of the line at 90°.
- ✦ The angle bisector divides the angle into two equal halves.
- ✦ Use the pair of compasses to mark the exact lengths of lines.
- ✦ Do not erase the arc marks. They are an essential part of the constructions.

Skills practice A

In this exercise you will need a ruler and a pair of compasses.

1 Draw these circles on the same diagram.
Use the same point as the centre of them all.
The circles form a pattern.
You can easily see if one of them is wrong.

- **a** Radius 8 cm
- **b** Radius 7 cm
- **c** Radius 6 cm
- **d** Radius 5 cm
- **e** Diameter 8 cm
- **f** Diameter 6 cm

2 Draw lines of these lengths.
Then bisect each line.

- **a** 8 cm
- **b** 10 cm
- **c** 12 cm

3 Draw a triangle with sides 6.5 cm, 6 cm and 5 cm.
Measure its angles.

4 Make an accurate drawing of this triangle.
Construct the bisector of each angle.
What do you notice about where these bisectors intersect?

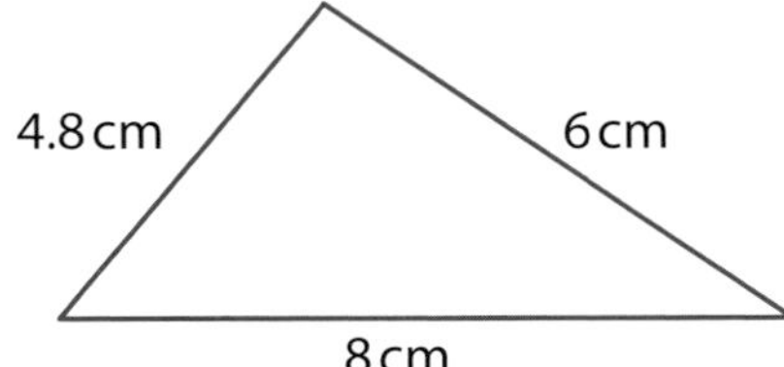

5
- **a** Draw a triangle with sides 10 cm, 10 cm and 10 cm.
- **b** Bisect its angles.
- **c** What angles have you now made?
- **d** Now make angles of 15° and 7.5° on your diagram.

6
- **a** Construct triangles with sides of these lengths.
  Describe each triangle.
  - **i** 6 cm, 6 cm and 6 cm
  - **ii** 6 cm, 8 cm and 8 cm
- **b** What happens if the sides are 6 cm, 5 cm and 12 cm?

7 Look at this regular hexagon.
The hexagon has been drawn inside a circle.

- **a** Draw a circle with a radius of 4 cm.
  Use a ruler and compasses to construct the hexagon inside it.
- **b** Bisect each of the angles at the centre of the hexagon.
  Use the bisectors to make a 12-sided regular polygon.

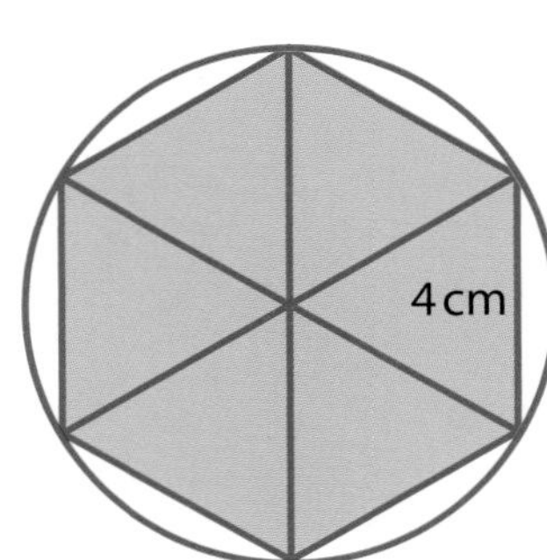

8 **a** Draw a line AB

Construct the perpendicular bisector of AB.

b Join A to X, X to B, B to Y and Y to A.

c What is the special name given to the quadrilateral AXBY?

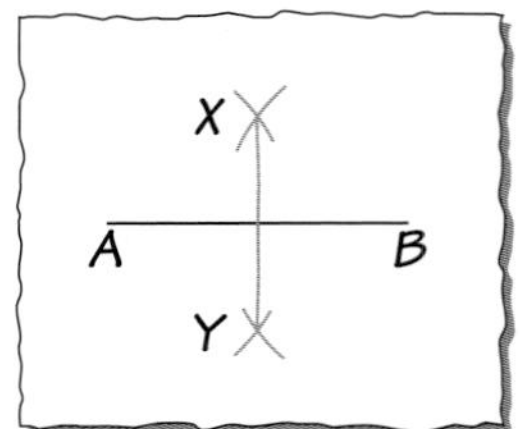

9 **a** Make an accurate drawing of this triangle.

b Construct the perpendicular bisector of AC.

c What do you notice?

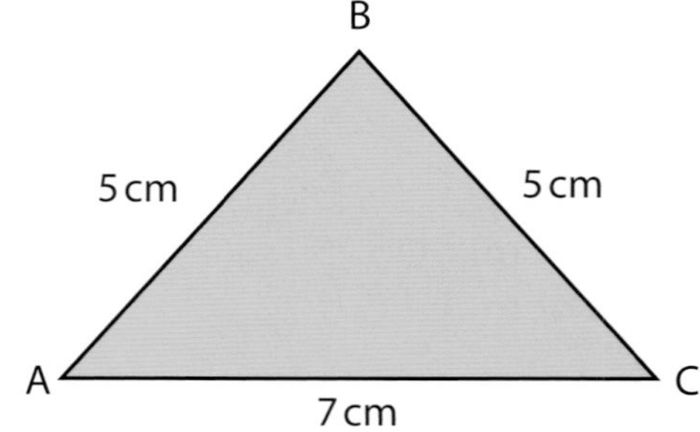

Skills practice B

In this exercise you will need a ruler and a pair of compasses.

1 Look at this diagram.

Martyn is leaving the cave in a rowing boat.

He is keeping as far from the rocks as he can.

a Copy the diagram.

b Bisect the angle to show his path to the sea.

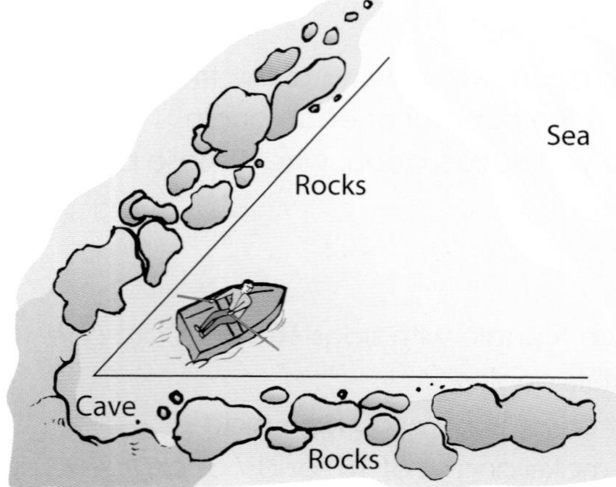

2 **a** Draw a triangle PQR where PQ = 10 cm, PR = 10 cm and QR = 12 cm.

b Construct a line from P at right angles to QR.

It crosses at S.

c Measure the lengths QS and PS.

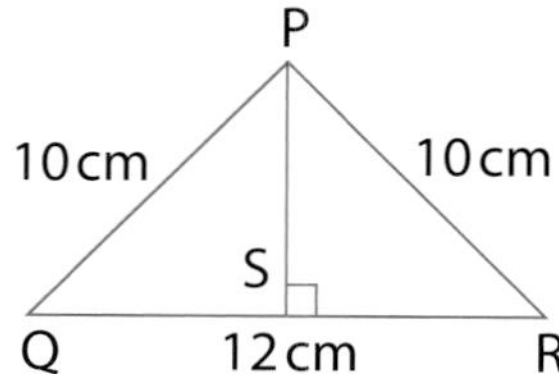

3 **a** Open your compasses to about 5 cm and draw a circle.

b Without adjusting the compasses, use them to make marks around the circumference.

Draw the first mark on the circumference as a starting point.

To make the second mark put the point of the compasses on the first mark, draw a mark, and so on.

c Join the marks with straight lines as shown to draw a six-pointed star.

4 Draw a regular hexagon in a circle.

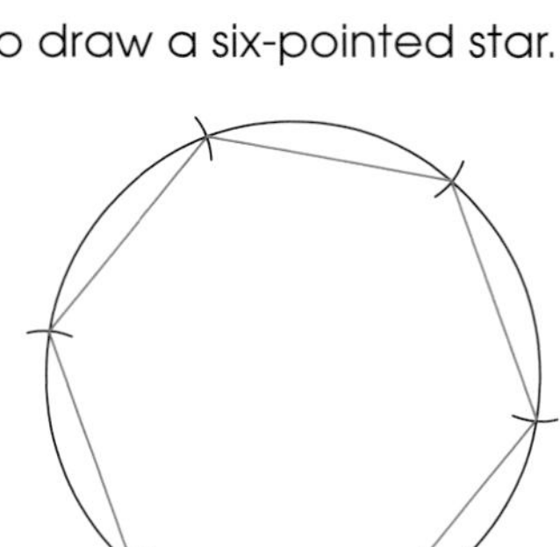

5 Draw this flower.

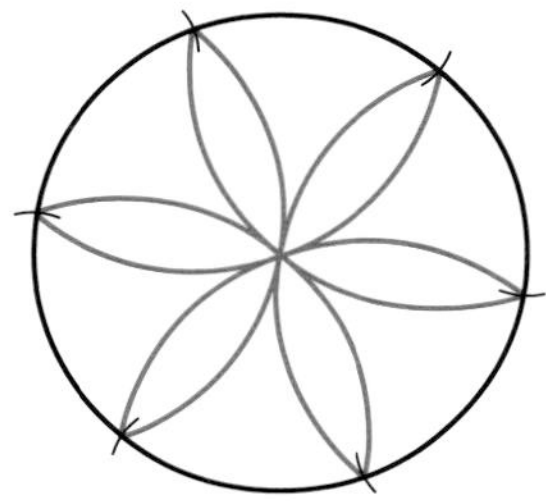

6 **a** Draw a triangle.

Label it ABC.

Construct the perpendicular bisectors of the three sides.

They should meet at a point.

Label it M.

b Draw a circle with centre M, going through the point A.

What do you notice?

This circle is called the **circumcircle** of the triangle.

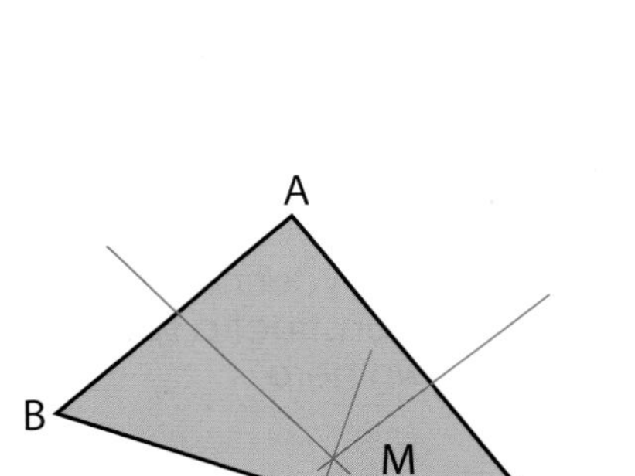

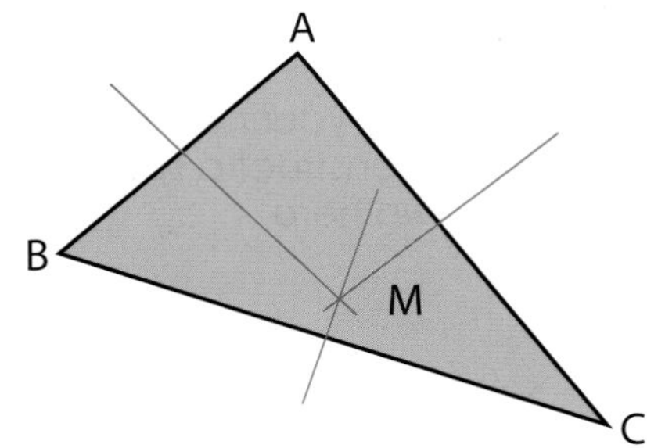

7 Rosie stands on the edge of the river opposite a tall tree at R.

She measures the angle of elevation of the top of the tree, T, as 50°.

Alan is 25 m behind Rosie at A.

He measures the angle of elevation of the top of the tree, T, as 23°.

a Make an accurate scale drawing of triangle ART.

b Extend the line AR.

Construct the perpendicular from T to this line.

c Find the height of the tree.

d What is the width of the river?

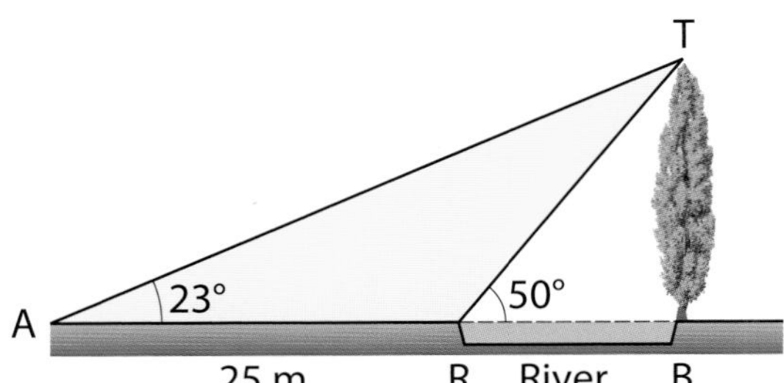

Reasoning

8 Construct this diagram accurately, using a ruler and a pair of compasses only.

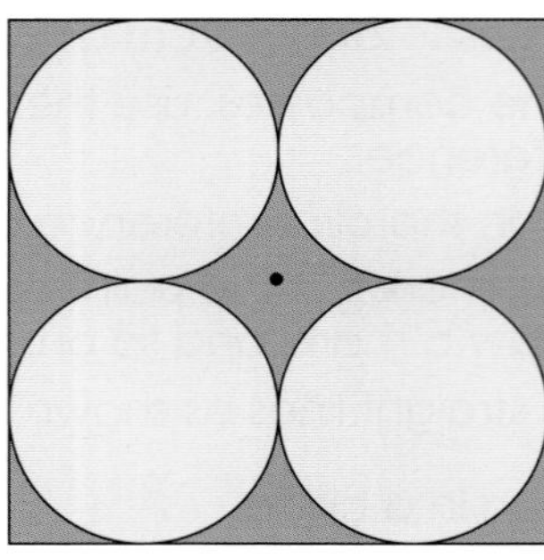

Reasoning

9 Use ruler and compasses to construct each of these angles.

a 120°

b 45°

c 75°

d 315°

e 345°

Wider skills practice

Reasoning

1 **a** Construct a triangle with sides of lengths 2 cm, 4 cm and 2.5 cm.

b Construct a triangle with sides of lengths 4 cm, 8 cm and 5 cm.

c Measure the angles in each triangle.
What do you notice? Why is this?

2 Draw a horizontal line 12 cm long.
Now, using only your ruler and a pair of compasses, construct a spider's web like the one shown here.

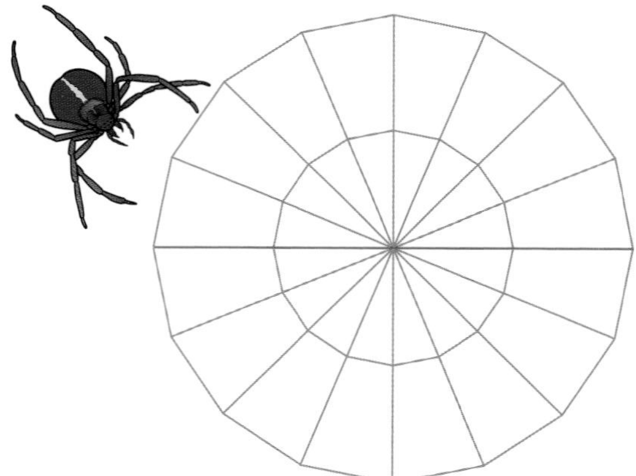

Applying skills

1 **a** Draw a circle of radius 6 cm with centre O.

b Draw a diameter AB and bisect it.
Label the points where the bisector meets the circle C and D (see diagram).

c What name is given to the quadrilateral ACBD?

d Bisect angle AOC.
Label the point where the bisector meets the circle E.
Extend the bisector to meet the circle again at F.

e Bisect angle COB.
The bisector meets the circle at G.
Extend it to meet the circle again at H.

f What shape is the polygon AECGBFDH?

Problem solving

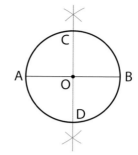

2 Draw this design for yourself. You will need to measure some of the lengths with a ruler. As well as the ruler, use only a pair of compasses.

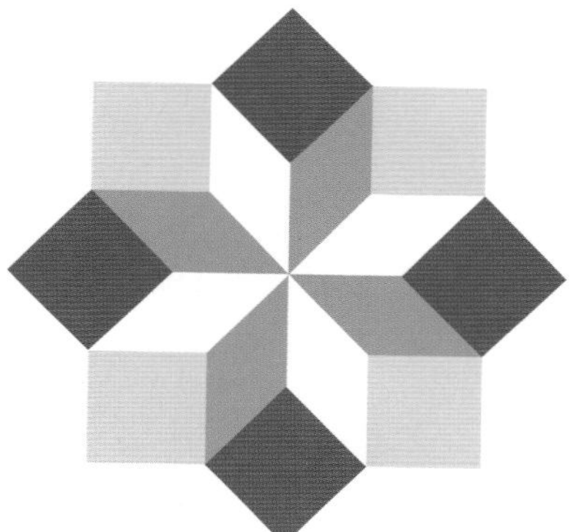

Reviewing skills

1 **a** Make an accurate drawing of this triangle.

b What type of triangle is it?

c Bisect angle A.
The angle bisector meets BC at D.
Mark point D on your triangle.

d Measure BD and DC.

e What do you notice?

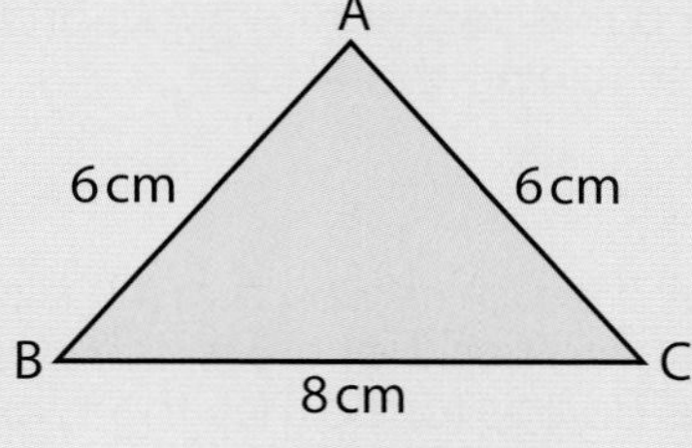

2 Look at this plan of a house.
A garden wall is built from M.
It is perpendicular to the house.
Copy the diagram and construct the line of the wall.

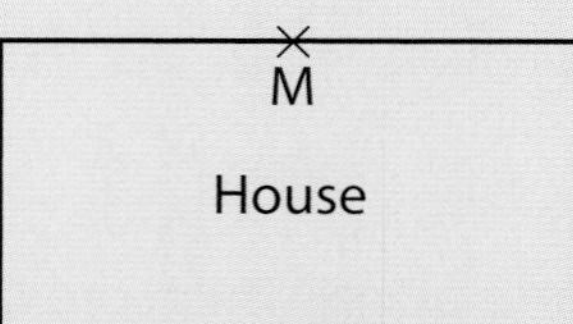

Building skills

Example outside the Maths classroom

Cosmology

Toolbox

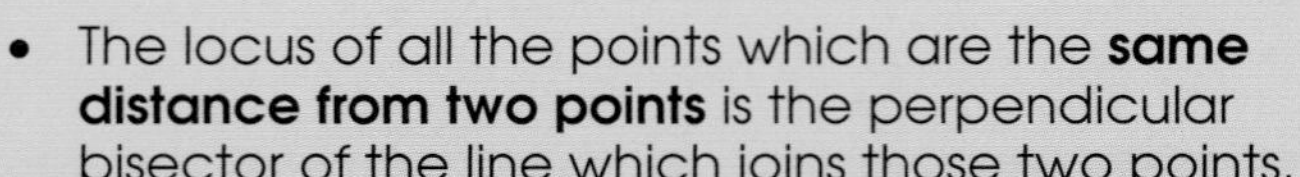

A **locus** is the path of all the points which obey a rule.

In two dimensions:

- The locus of all the points which are the **same distance from one point** is a circle.
- The locus of all the points which are the **same distance from two points** is the perpendicular bisector of the line which joins those two points.
- The locus of all the points which are the **same distance from a line segment** is two parallel lines with semicircles joining the ends.
- The locus of all the points which are the **same distance from two lines that cross** is the angle bisector of the angles formed by the two lines.
- The locus of all the points which are **the same distance from two lines that do not cross** is a line parallel to them, and midway between them.

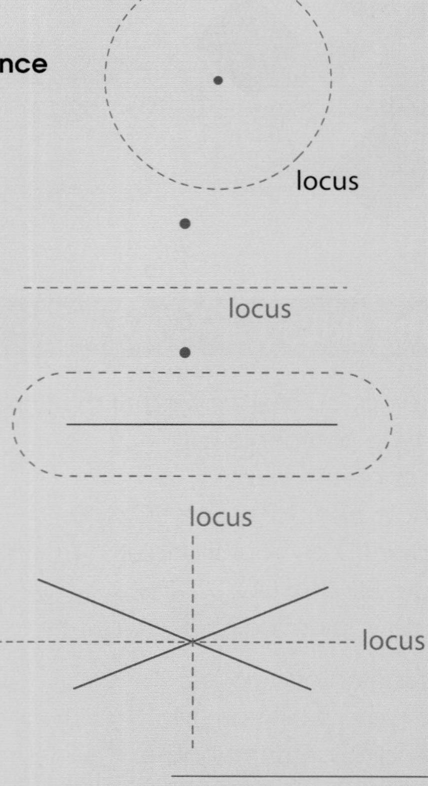

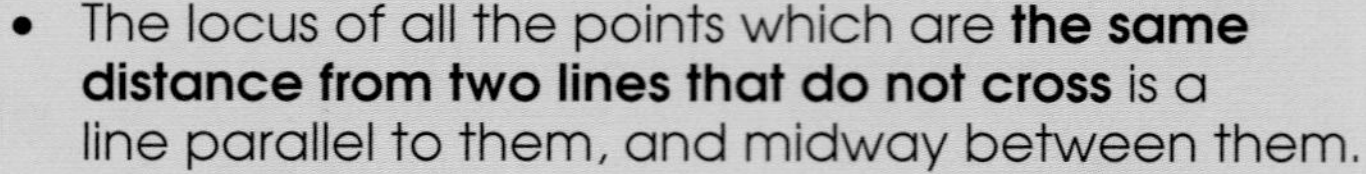

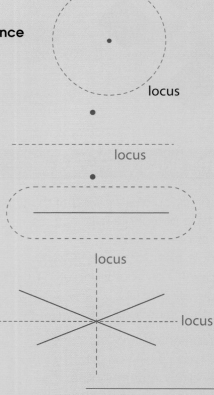

Example – Loci equidistant from two points

A and B are the positions of two radio beacons.

An aeroplane flies so that it is always the same distance from each beacon.

Draw two crosses to represent the beacons.

Construct a line to represent the aeroplane's path.

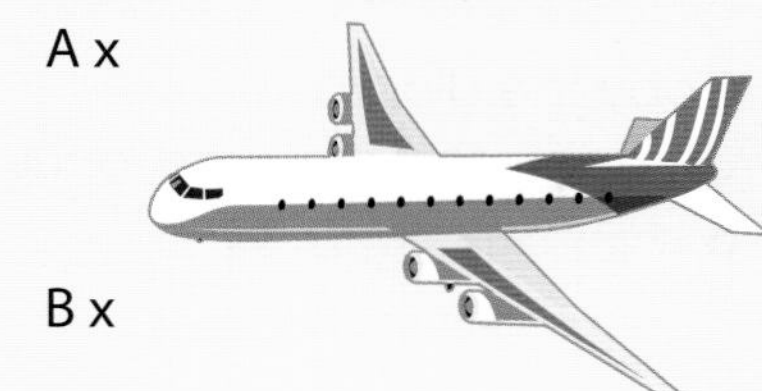

Solution

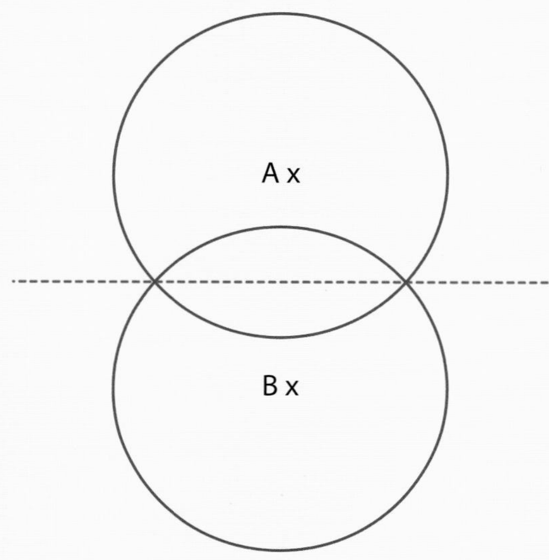

Remember:

- A locus is the path of points subject to a particular rule.

Skills practice A

1 Look at this diagram.

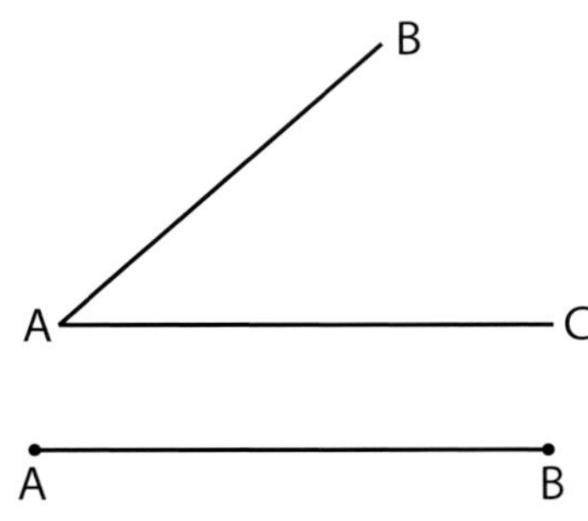

a Draw a sketch showing the locus of the points which are the same distance from AB and AC.

b Make an accurate drawing of this locus.

c What name is given to this locus?

2 Look at this diagram.

a Draw a sketch showing the locus of the points which are the same distance from A and B.

b Make an accurate drawing of this locus.

c What name is given to this locus?

3 Draw a line of length 6 cm.
Draw the locus of all the points that are 3 cm from this line.

4 Copy the diagrams.
On each one draw the locus of the points which are the same distance from the two points marked.

a

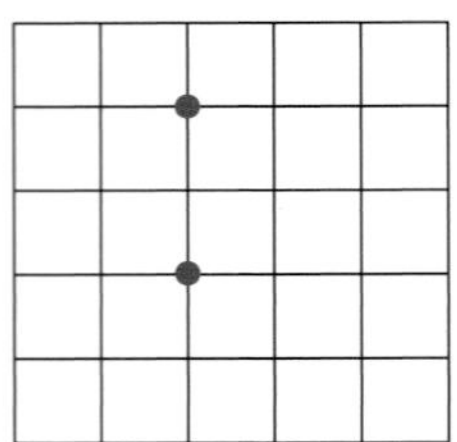

b

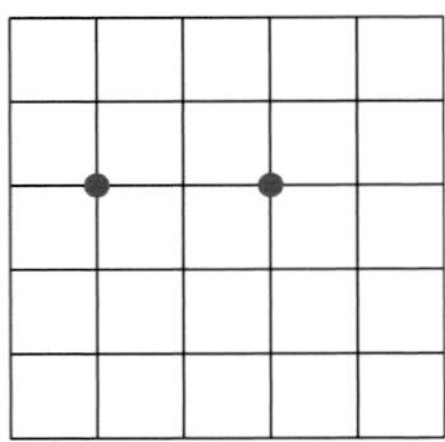

c

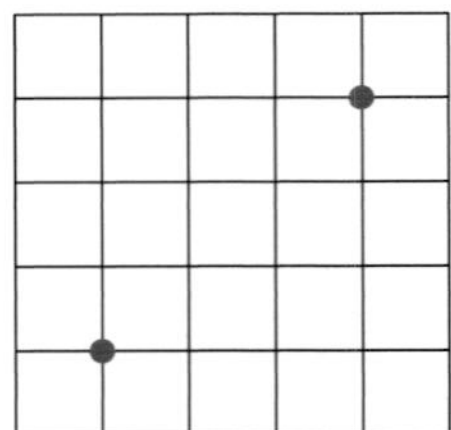

5 Draw two parallel lines that are 6 cm apart.
Draw the locus of all the points that are
- **a** the same distance from both lines
- **b** twice as far from one line as from the other.

Skills practice B

1 Describe the locus of
- **a** a ball bouncing along a level playground
- **b** the end of the minute hand on a clock
- **c** the light on a bicycle travelling along a level road
- **d** the end of a car's windscreen wiper.

Reasoning

2 Draw two lines, AB and CD, both 10 cm long and crossing one another at 45°.
- **a** Draw the locus of all the points which are the same distance from both lines and explain why you have drawn two lines for the locus.
- **b** What is the angle between the two locus lines?

3 **a** Construct an equilateral triangle with sides of length 10 cm.
- **b** Shade all points that are less than 1 cm from its perimeter (inside and outside).

4 **a** Make a sketch of this island.
- **b** A and B are the positions of radio transmitters.
Add these to your sketch.
- **c** Each transmitter can transmit signals a distance equal to AB.
Shade the regions on the island that will not receive signals.

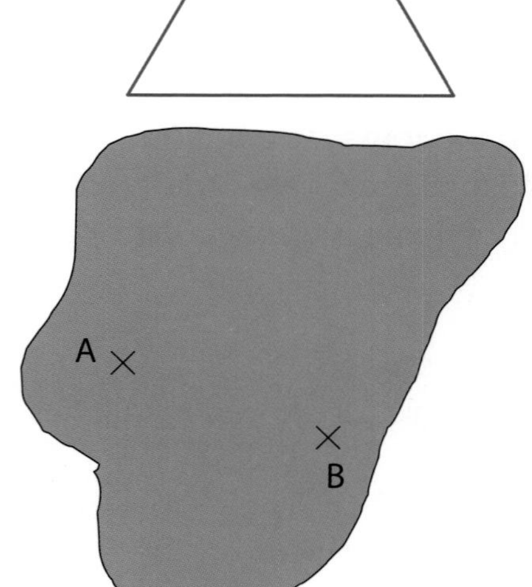

Reasoning

5 Look at the diagram.

a Lines AB and CD are both 10 cm long.
They cross one another at 60°, 3 cm from B and 2 cm from D.

b Draw an accurate version of this diagram. On your diagram, draw the locus of all the points which are the same distance from A and B.
Draw also the locus of all the points which are the same distance from C and D.

c What is the angle between the two locus lines?

d Use your locus lines to find the centre of a circle that passes through A, B, C and D.

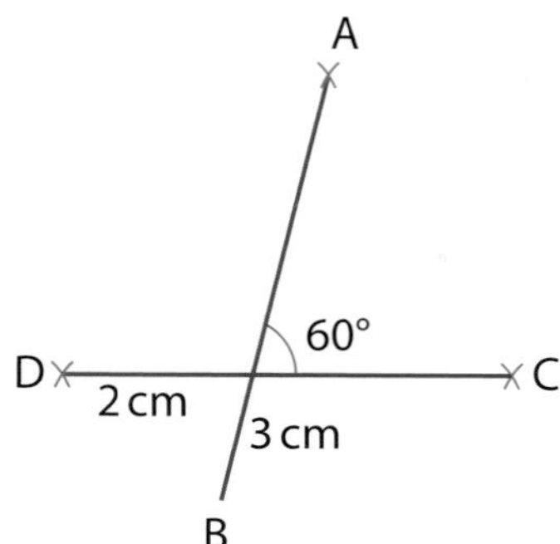

6 A goat is tied by a piece of rope 3.5 m long to a small tree in the centre of a garden lawn.
The lawn is a square, 8 m by 8 m.

a Use a scale of 1 cm to 50 cm to make a scale drawing of the lawn showing the area of grass that the goat can eat.

Another goat is tied by a 2 m rope to a post in one corner of the lawn.

b Show on your scale drawing the area of grass that this goat can eat.

c Can the two goats meet?

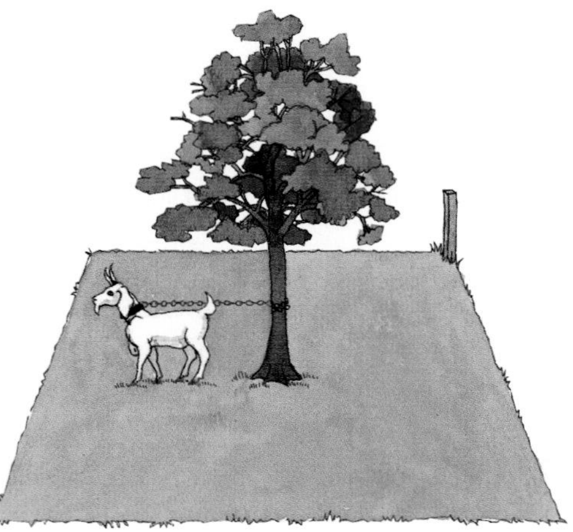

Wider skills practice

Reasoning

1 Draw a rectangle 3 cm wide and 6 cm long.
Draw the locus of points that are the same distance from AD and from DC.
Explain why this locus is not the same as the diagonal DB.

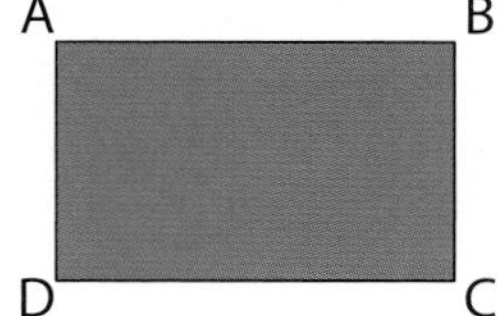

2 **a** Draw a line 7 cm long.

b Shade the region of points that are less than or equal to 3 cm from your line.

c Find the area and perimeter of the shaded region.

3 **a** Draw x and y axes from −5 to 5.

b Plot the lines $y = 3$ and $x = -1$.

c Draw the locus of the points that are the same distance from both lines.

Reasoning

4 Here is a plan of a garden.
The garden is 12 m long and 6 m wide.
The shed is 2 m wide.
The patio is the same size as the shed.

a Draw the plan to scale using a scale of 1 : 200.

b A tree is to be planted at least 4 m from the shed and at least 2 m from the patio.
Shade all the possible positions where the tree could be planted.

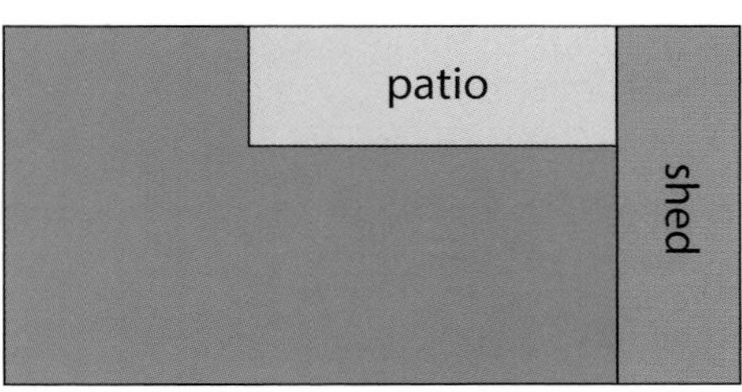

Applying skills

1

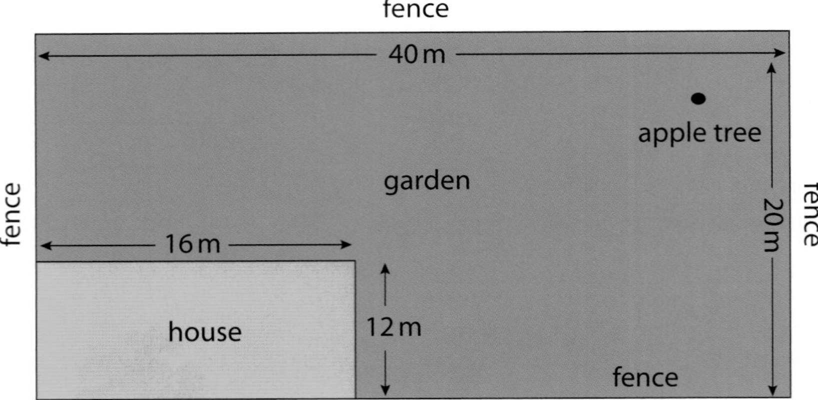

The diagram shows the plan of a house and a garden.

A tree is to be planted which must be at least 10 metres from the house and 5 metres from the fences.

It also has to be at least 6 metres from the apple tree, located in the corner of the garden, 4 metres from each of the two fences.

Make a scale drawing of the garden and shade the area in which the tree can be planted.

Reviewing skills

Make two accurate constructions of this triangle.

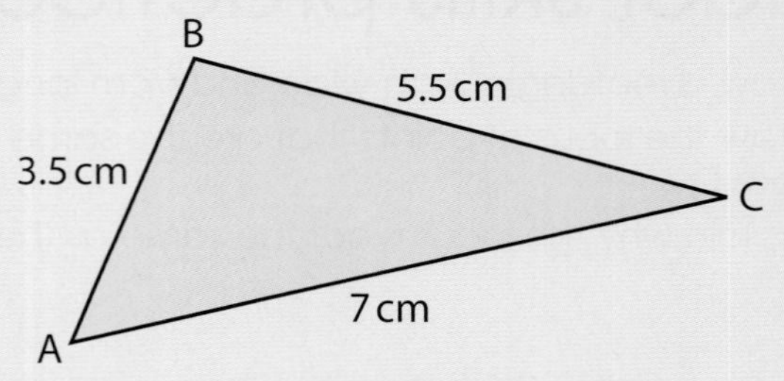

a On the first triangle, draw the three loci that show points the same distance from each of these pairs of points.

- **i** A and B
- **ii** B and C
- **iii** C and A

How do these three loci help you to find a point which is the same distance from A, B and C?

b On the second triangle draw the three loci that show the points the same distance from each of these pairs of lines.

- **i** AB and AC
- **ii** BA and BC
- **iii** CA and CB

How do these three loci help you to find a point which is the same distance from all three sides?

Strand 5 • Transformations

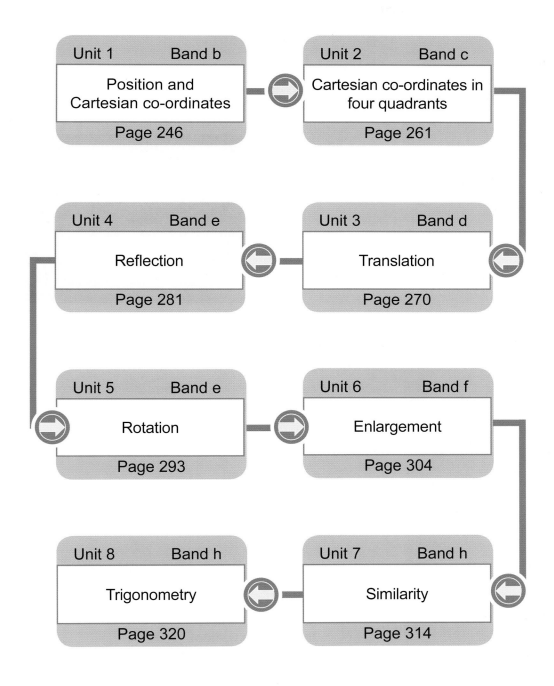

Unit 1	Band b	Position and Cartesian co-ordinates	Page 246
Unit 2	Band c	Cartesian co-ordinates in four quadrants	Page 261
Unit 3	Band d	Translation	Page 270
Unit 4	Band e	Reflection	Page 281
Unit 5	Band e	Rotation	Page 293
Unit 6	Band f	Enlargement	Page 304
Unit 7	Band h	Similarity	Page 314
Unit 8	Band h	Trigonometry	Page 320

Unit 1 • Position and Cartesian co-ordinates • Band b

Example outside the Maths classroom

Seat bookings

Toolbox

To describe a position in two dimensions you need two pieces of information.
These are often the distances, along and up.

Up

Along

Sometimes they are a distance and a direction.

Distance

Direction

You may want to give the position of a point or place.

Cartesian co-ordinates are used for the position of points.

For example, the point P has co-ordinates (3, 2).

Remember the x co-ordinate comes before the y co-ordinate.

The y axis is vertical.

The origin is the point (0, 0).

The x axis is horizontal.

Example – Using distance and direction to find a position

This is a map of a maze.
Each square is one pace.
Class 7G are visiting the maze.
They use the map to find their way around.

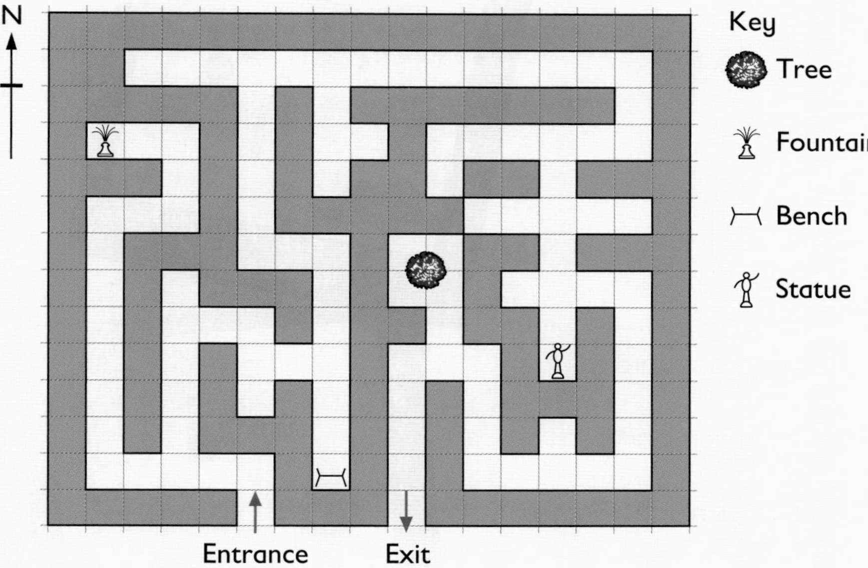

a Where do these instructions lead?

From the statue,
2 paces North,
2 paces East,
5 paces South,
4 paces West,
3 paces North,
2 paces West,
4 paces South.

b Write instructions to walk from the entrance to the fountain.
Use the compass directions and the number of paces.

Solution

a To the exit

b From the entrance, 2 paces North, 4 paces West, 7 paces North, 2 paces East, 2 paces North, 2 paces West.

Example – Finding a position using a grid

Hassan and Melinda are playing a game of Starships.

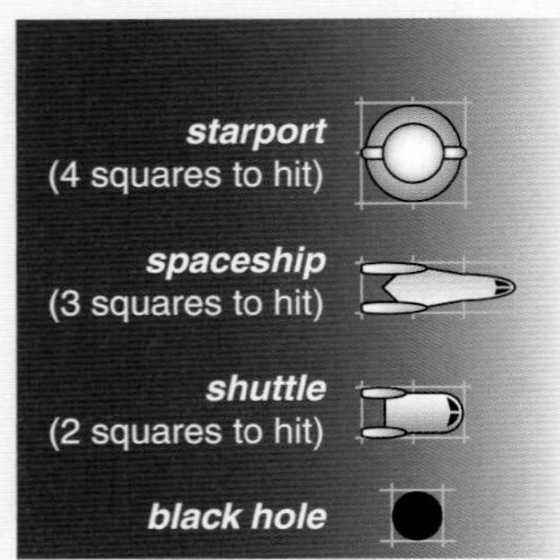

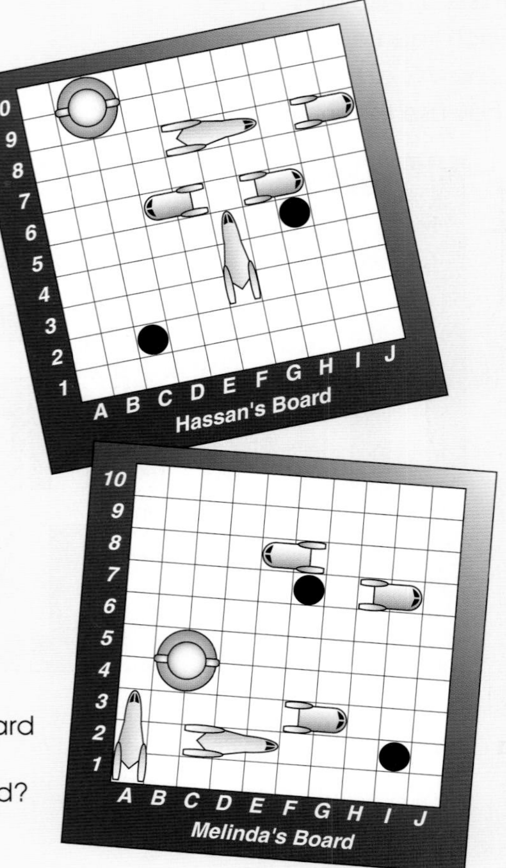

Hassan and Melinda take it in turns to name a square to eliminate on the other's board.

Hassan goes first and chooses square C5. He hits part of a starport on Melinda's board.

a What other squares should Hassan hit to eliminate the starport completely?

If a player hits the two black holes on their opponent's board they lose the game.

b Which two squares contain black holes on Melinda's board?

c What is contained in square D6 on Hassan's board?

Solution

a B4, B5 and C4 ← **The starport occupies four squares in total.**

b I2 and F7 ← **The black circles are black holes.**

c A shuttle

Example – Using co-ordinates to describe the position of a point

a What are the co-ordinates of points A to F?

b Which of the points (4, 1) or (1, 4) is inside the shape?

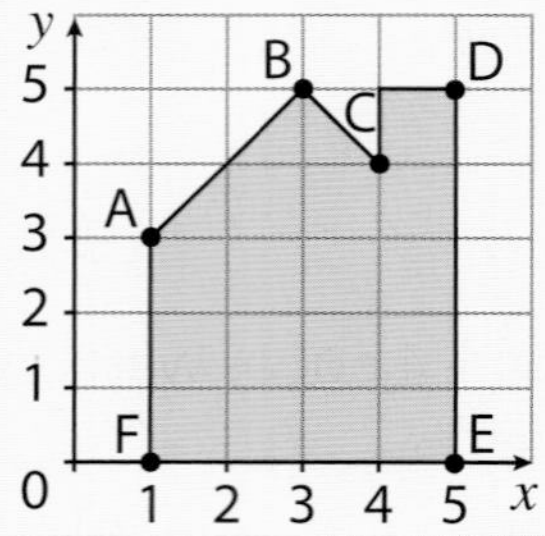

Solution

a The point A has co-ordinates (1, 3).

This is the x co-ordinate.

This is the y co-ordinate.

B(3, 5), C(4, 4), D(5, 5), E(5, 0), F(1, 0)

b (4, 1)

Example – Plotting co-ordinates on axes

a Draw a pair of axes from 0 to 5.
Plot the points A(2, 5), B(5, 5), C(5, 2) and D(2, 2) and join them in order.

b What shape have you drawn?

Solution

a

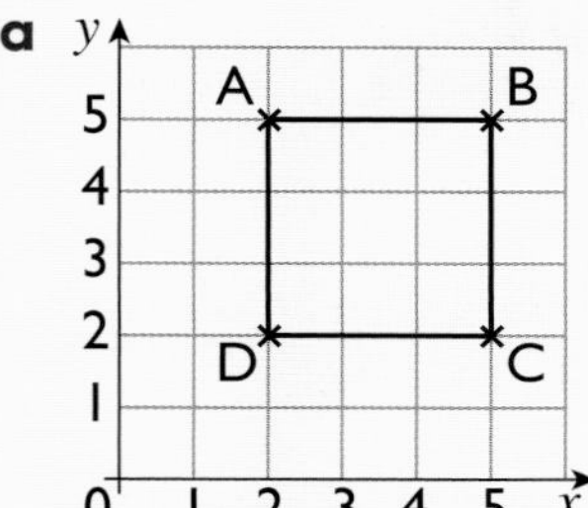

b A square

Remember:

✦ When you write the co-ordinates of a point, the x co-ordinate comes first.

Skills practice A

1 Class 7G has a new teacher.
She draws a seating plan.
Sophie sits in position E3.
Pete sits in column C.

a Which row does Pete sit in?

b Which position does Pete sit in?

Samir sits in position F6.

c Who does Samir sit next to?

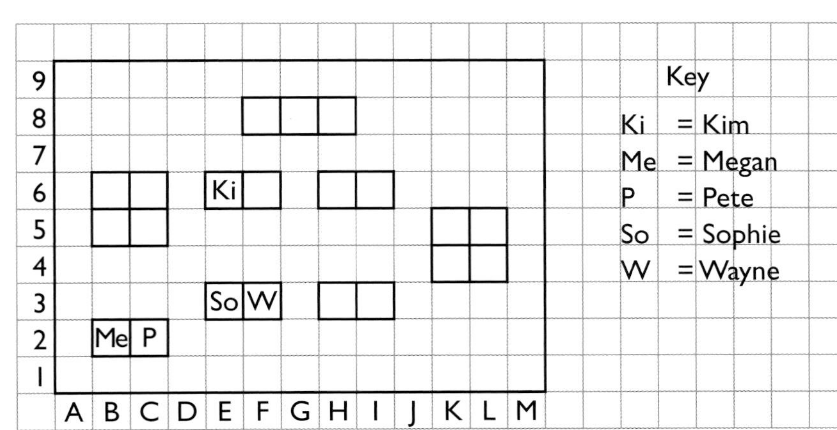

The other children sit in these seats:

John	L4	Alan	I3	Harry	C5	Tim	B5	Karl	K4
Mercy	B6	Humza	H6	Lucy	H8	Andy	F8	Christina	K5
Jo	G8	Jack	I6	Ali	L5	Meena	C6	Michelle	H3

d Copy and complete the teacher's seating plan using this information.

e On squared paper, draw a seating plan for your classroom.
Make a list of all the people in the class, and give their positions.

2 The diagram shows part of a seating plan in an aeroplane.

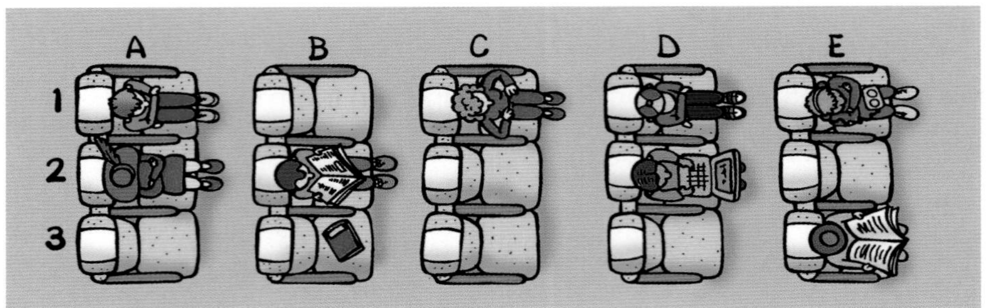

a What colour hair has the person in seat B2?
b What is the man in seat D2 doing?
c Which seats are empty?
d In which seats are people wearing hats sitting?

3 Here is a map of Skull Island.
Blackbeard and his pirate crew are using the map to try to find the buried treasure.

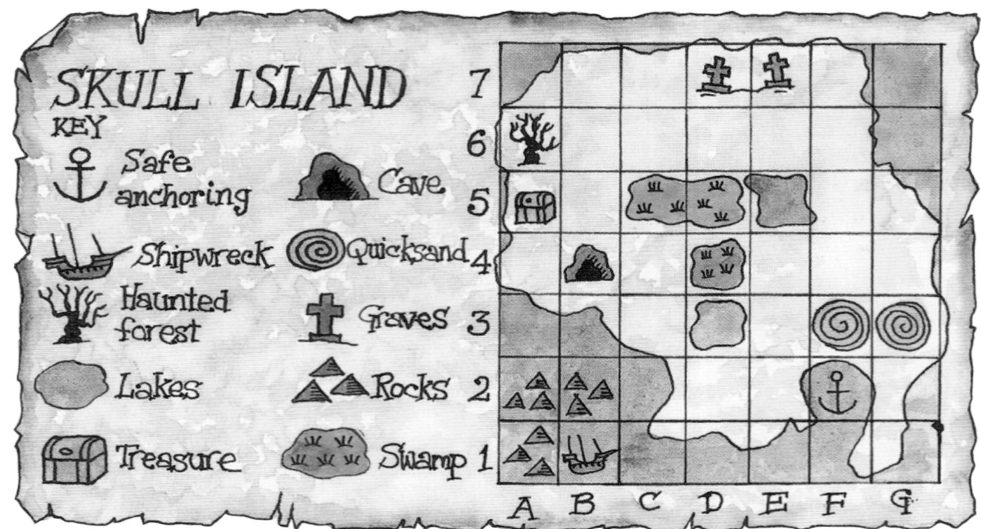

a The pirates want to land on the island.
In which square should they leave their ship?
b In which squares are the dangerous rocks?
c What is in square B4?
d Where on the island is there quicksand?
e Where would you be if you were at these squares?
i E7 **ii** D5 **iii** A6
f In which square is the treasure?
g Write down which squares the pirates should pass through to reach the treasure safely.

4 Sally is rearranging her bedroom.
She draws a plan of her room.
Each large square is 1 metre × 1 metre.

a What is in square B2?

b What is in square C3?

c Which squares show the position of her bed?

d Where is Sally's desk?

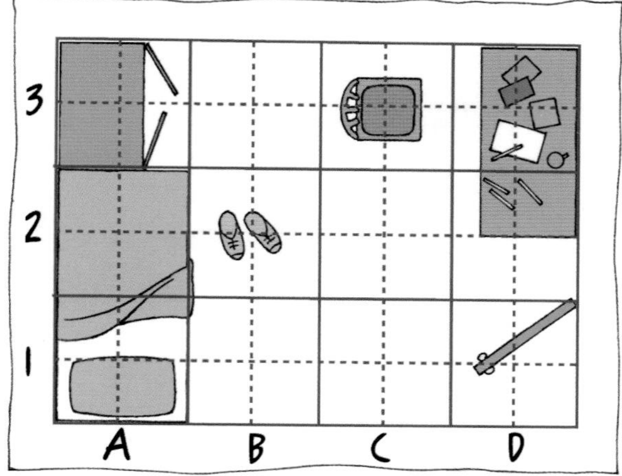

5 a Use the letter square to find out what these messages say.
The first word has been done for you.

8	a	c	r	j	b	w	k	g
7	j	z	v	h	b	f	y	h
6	o	n	a	g	s	u	a	o
5	q	g	b	j	p	l	x	e
4	l	y	f	k	d	f	v	c
3	s	i	e	t	i	c	z	p
2	w	q	n	e	k	x	d	m
1	d	i	l	m	r	t	u	h
	A	B	C	D	E	F	G	H

i H2, C3, C3, F1 / D1, H5 / A8, D3 / F1, D7, C3 / E6, H1, A6, E5, E6.
M e e t

ii D1, G6, D3, H7, A3 / B3, E6 / F3, H6, A6, F5!

iii F3, A6, D1, C3 / D3, A6 / H2, G7 / E5, A8, E1, D3, B4.

iv D1, G7 / E4, A6, H8 / B1, E6 / B8, C6, C1, F5, H5, E4 / H2, G6, F2.

b Use the letter square to write your own message.
Give the message to a friend to solve.

6 Write down the co-ordinates of each of the points plotted below.

a

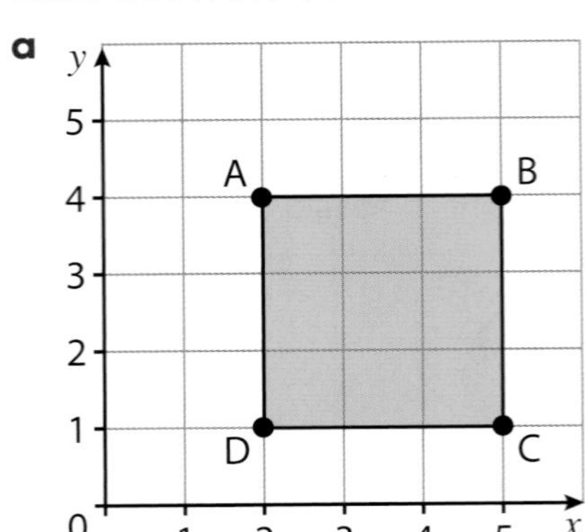

b

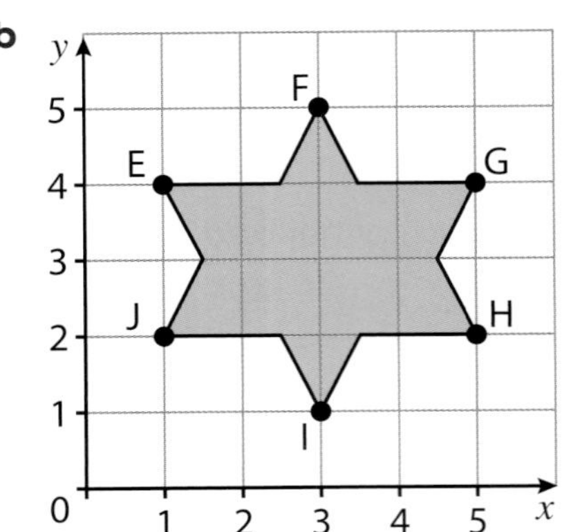

c

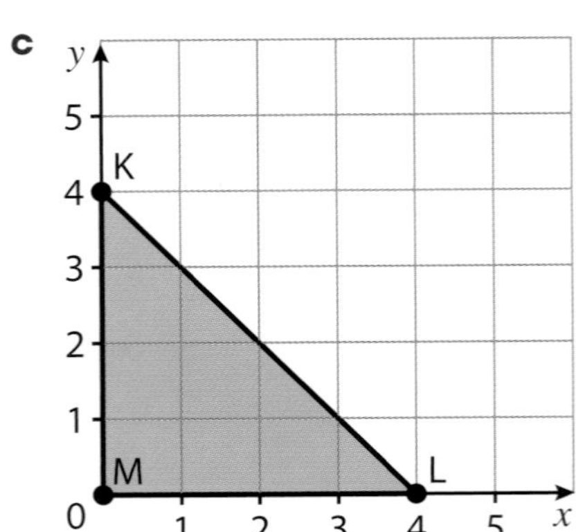

d

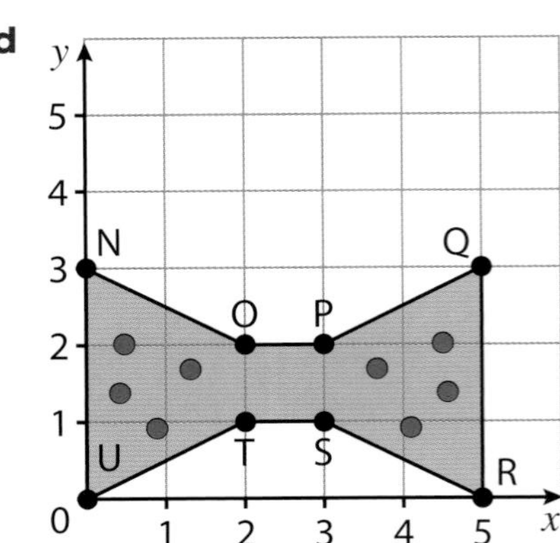

7 Draw a pair of axes from 0 to 10.

a Plot these points and join them in order.

(3, 4) → (2, 2) → (0, 0) → (1, 0) → (3, 2) → (2, 0) → (3, 0) → (4, 2) →

(4, 3) → (6, 3) → (7, 4) → (7, 2) → (6, 0) → (8, 0) → (8, 1) → (9, 0) →

(10, 1) → (10, 2) → (9, 5) → (7, 6) → (5, 6) → (4, 7) → (4, 8) → (3, 7) →

(2, 8) → (2, 7) → (1, 6) → (0, 6) → (1, 5) → (2, 5) → (3, 4) → (3, 2) →

b Plot these points and join them up as you go.

(7, 0) → (8, 2) → (8, 1) → (8, 3) → (9, 4)

c Now put dots at (2, 6) and (0, 6). What have you drawn?

8 Draw a pair of axes from 0 to 5.

a Plot these points and join them in order.

$(2, 4\frac{1}{2})$, $(4\frac{1}{2}, 4\frac{1}{2})$, $(2\frac{1}{2}, 1\frac{1}{2})$, $(0, 1\frac{1}{2})$ and back to $(2, 4\frac{1}{2})$

b Describe the shape you have drawn.

Skills practice B

Reasoning

1 This is a map of Mr Mac's Farm

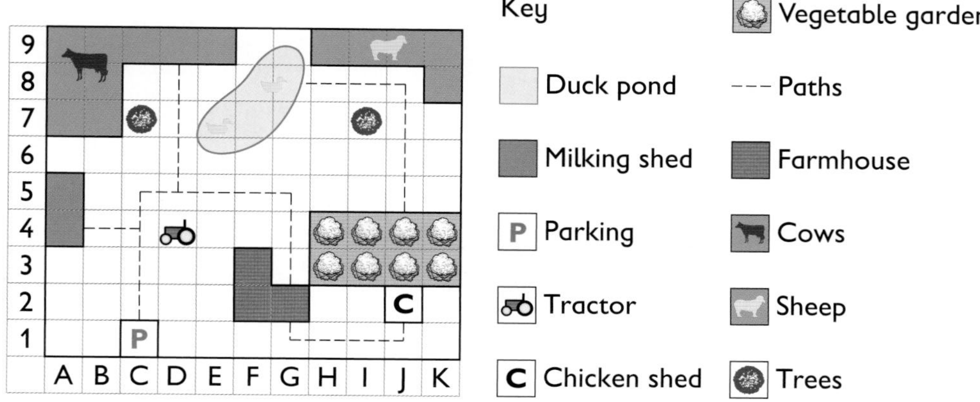

Class 7G are visiting the farm.
They use the map to find their way around.
The teacher asks these questions.

1 *Where do we park the bus?*
2 *In which squares are the sheep?*
3 *What is in square H4?*
4 *Where is the chicken shed?*
5 *What will you see in these squares: F7, G2, C7 and D4?*

a These are Alan's answers.
Some of Alan's answers are wrong.
i How can you tell?
ii What are the correct answers?

1 G7
2 H9, I9, J9, K9, K8
3 Chickens
4 J2
5 F7 tractor, G2 tree, C7 ducks, D4 farmhouse

b Mrs Mac makes a special cream tea for the class.
The class walk from the milking shed to the farmhouse for tea.
Which squares do they walk through?
(Keep to the paths!)

2 This map is part of the London A–Z.

a King's Cross station is given in the index as 2 J 67.
Explain how the referencing system works.

b In which square is St Pancras Hospital?

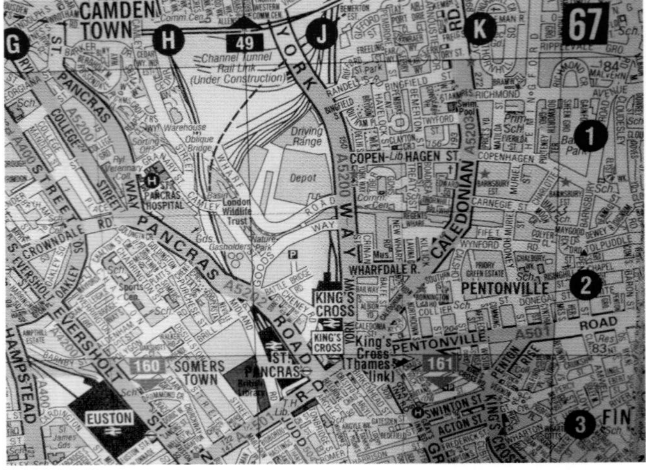

Reasoning

3 This is a spreadsheet from a computer screen.

File Edit View Insert Format Tools Data Window Help — 9:53 am — Microsoft Excel

F4 = =SUM(B4:E4)

Ice cream sales in one week.xls

	A	B	C	D	E	F	G
1	**Day**	**vanilla**	**chocolate**	**raspberry**	**strawberry**	**totals**	
2							
3							
4	**Monday**	18	23	10	12	63	
5	**Tuesday**	20	20	8	15		
6	**Wednesday**	13	5	2	4		
7	**Thursday**	25	25	18	17		
8	**Friday**	40	32	20	36		
9							
10	**totals**	116	105				
11							
12							

The spreadsheet shows how many ice creams a shop sold last week.
Each rectangle is called a cell.

a What number is in cell D5?

b Which cell shows the number of raspberry ice creams sold on Thursday?

c Why does 'F4' appear above the grid?

Look above the grid.
You can see the instruction =SUM(B4:E4).
This means 'add all the numbers from cells B4 to E4 together'.

d How many ice creams were sold altogether on Monday?

e Cell E10 is for the total number of strawberry ice creams.
What instruction needs to be in cell E10?

4 Tim draws a map of his town.
The red lines show the roads in his town.

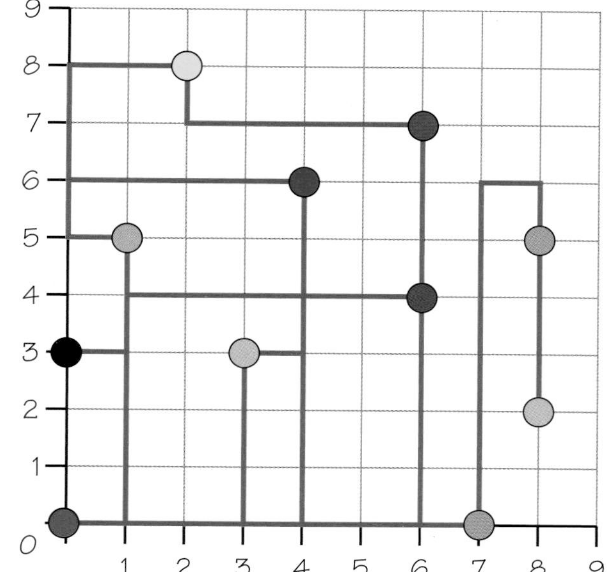

a What is at these co-ordinates?

i (3, 3) **ii** (6, 4) **iii** (4, 6)

b Write down the co-ordinates of these places.

i Park **ii** Shop **iii** Woods

c Tim starts at his house and walks to (1, 0) and turns left.
He then walks to (1, 4) and turns right.
What is at the end of this road?

d How does Michelle get from her house to the cinema?

e Tim and Michelle have an argument.

i Who is wrong?

ii Why does their mistake matter?

5 Write down the co-ordinates of the points which make this bow-tie.

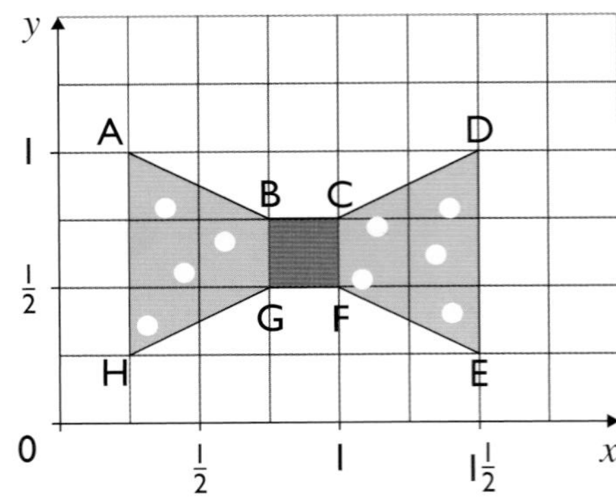

6 Here is a map showing seven towns and the roads between them.

The distances along the axes are in kilometres.

Elmlow is at the point (10, 30).

Melton is at (25, 0).

a What are the co-ordinates of

  i Whiston

  ii Midchurch?

b How far is it from

  i Thornhill to Senbury

  ii Midchurch to Cookfield?

c A new town is to be built at (45, 35).

  i How far will the new town be from Thornhill?

  ii How far will the new town be from Cookfield?

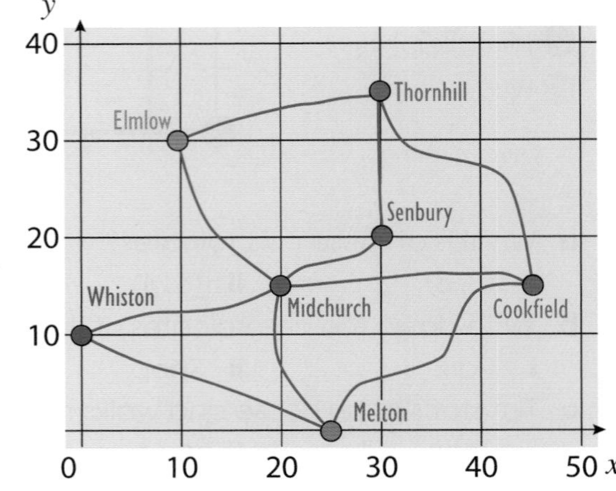

7 Sophie and Alan are playing a game of Find the Gold.

They each place four gold bars on their grid.

The winner is the first person to find the hidden gold.

Here is Sophie's grid.

a Alan guesses the point (2, 4).

  Does Sophie answer 'Yes' or 'No'?

b Next turn, Alan guesses the point (4, 3).

  What does Sophie answer?

c Alan tries a point next to (4, 3).

  List the four possible points.

  For each one say 'Yes' or 'No'.

d Write down the co-ordinates of all the points with gold.

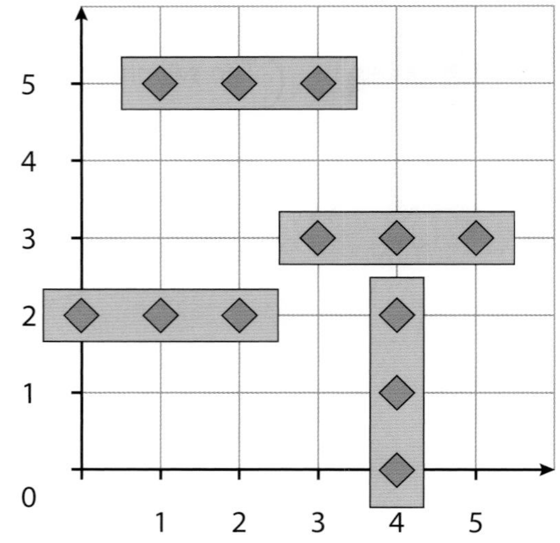

Wider skills practice

Reasoning

1 This grid is used for a game of noughts and crosses.
Ali and Jo had these turns.

3			
2			
1			
	A	B	C

Ali (X)	A3, A1, C1
Jo (O)	B2, A2

Jo wins the game.
Where did she put her last O?

2 Draw axes, using the values of 0 to 14 for x and 0 to 6 for y.

a Plot these points and join them in order.
(0, 3), (4, 6), (9, 6), (13, 3), (9, 0), (4, 0) and back to (0, 3)

b Describe the shape you have drawn.

c Are all the sides the same length?

d Are all the angles the same size?

Applying skills

Problem solving

1 Here is a spreadsheet.
Each rectangle is called a cell.

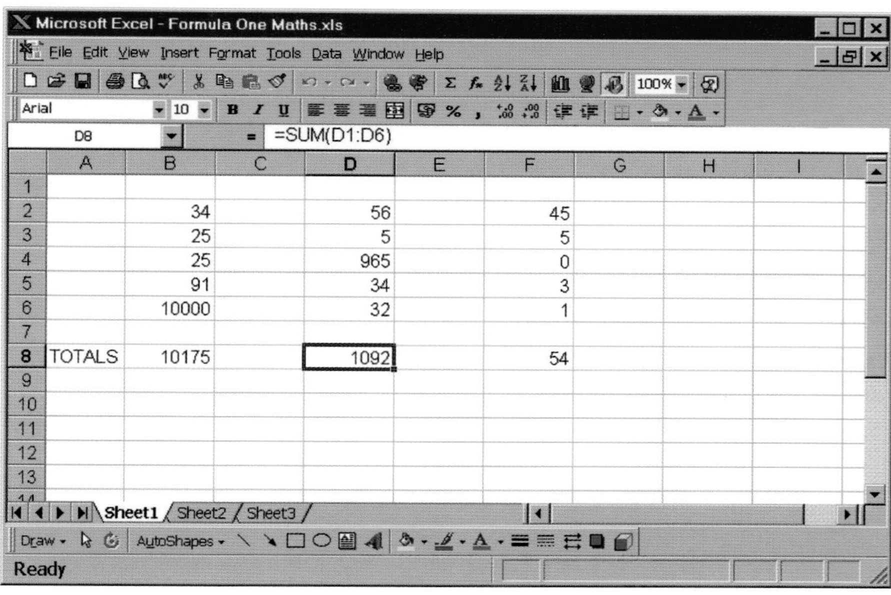

	A	B	C	D	E	F	G	H	I
1									
2		34		56		45			
3		25		5		5			
4		25		965		0			
5		91		34		3			
6		10000		32		1			
7									
8	TOTALS	10175		1092		54			
9									
10									
11									
12									
13									

a What number is in cell B3?

b What number is in cell D5?

c In which cell is the word TOTALS?

d Why does the reference D8 appear above the grid?

e Above the grid you can see =SUM(D1:D6).
Can you explain this?

f What is 34 + 25 + 25 + 91 + 10000?

g Write down the references of the cells which contain even numbers.

Problem solving

2 Many road atlases use grid references to describe positions of places.
This is a grid of squares over a map of part of Great Britain.
The large squares are 100 km across.
Each large square is divided into small squares.

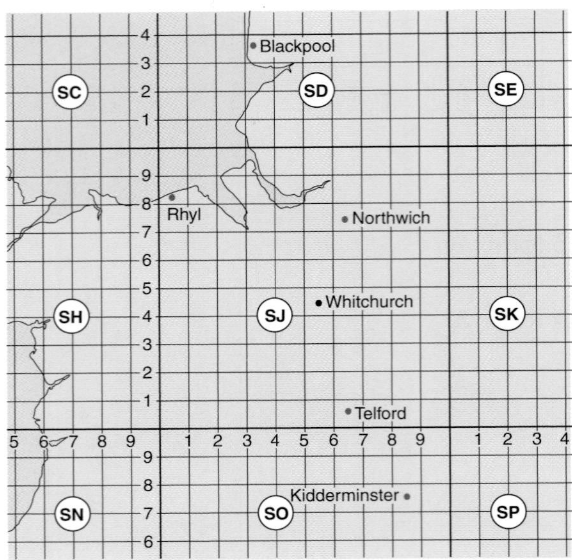

The position of Whitchurch is described as SJ54.

a Write down the position of each of these places.
  i Northwich
  ii Telford
  iii Rhyl
  iv Blackpool
  v Kidderminster

b Explain how the system works.

c How wide is each small square?

d How many small squares are there in each large square?

e How accurate is a grid reference such as SJ54?

A two-figure grid reference gives the approximate position of anywhere in Britain.

f Can you give an approximate position for your school?
Which large grid square is it in?

Reviewing skills

1 Here is the seating plan of a cinema.

SCREEN

a How many seats are there in row C?

b How many seats are there between the two aisles in row D?

c Jenny has seat E8.
How many rows are there in front of her?

d Matt has seat F5.
How many seats are there between him and the nearer aisle?

e Alison has seat A1 and Charlie has seat A6.
How many seats are there between them in that row?

f Write down the seats where you can sit with no-one next to you.

g Find the pairs of seats where two people can sit by themselves.

2 Look at this map.

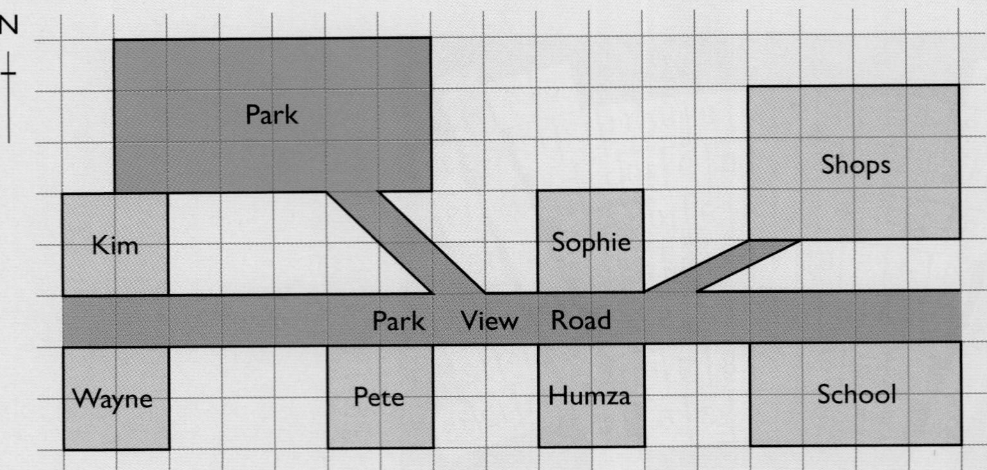

a Who lives North of Wayne?
b Who lives South of Sophie?
c What is East of Humza's house?
d Kim's best friend lives in the house to the West of Pete.
Who is Kim's best friend?
e Humza walks to the park.
In which direction does he go?
f Sophie is the same age as the person who lives South-West of her.
Who is the same age as Sophie?

3 Write down the co-ordinates of each of the points plotted below.

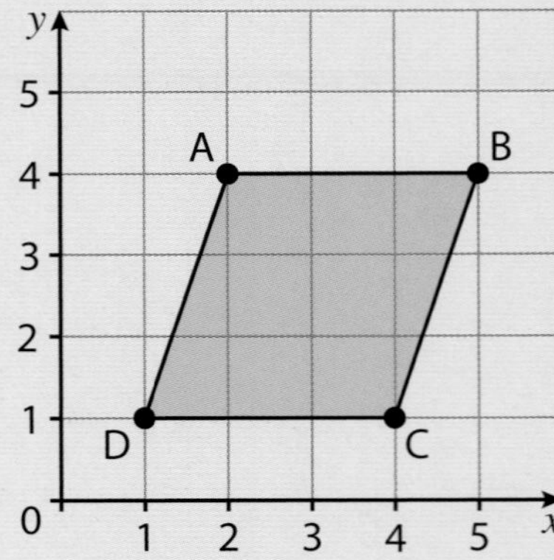

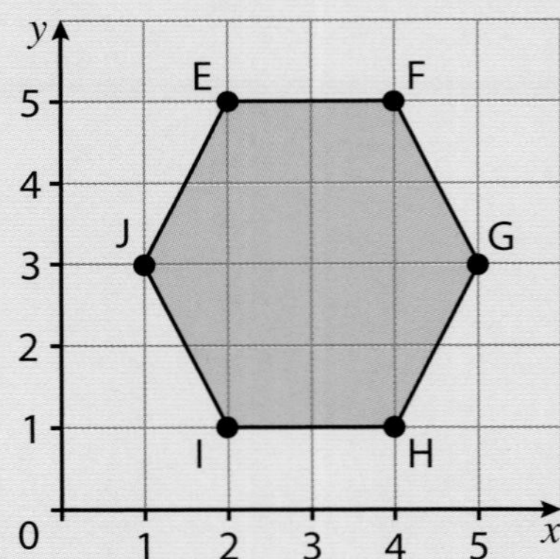

4 Draw a pair of axes from 0 to 6.
a Plot the points (1, 2), (4, 2) and (4, 5).
b Mark a fourth point to make a square.
What are the co-ordinates of this point?

Unit 2 • Cartesian co-ordinates in four quadrants • Band c

Building skills

Example outside the Maths classroom

Global positioning satellites

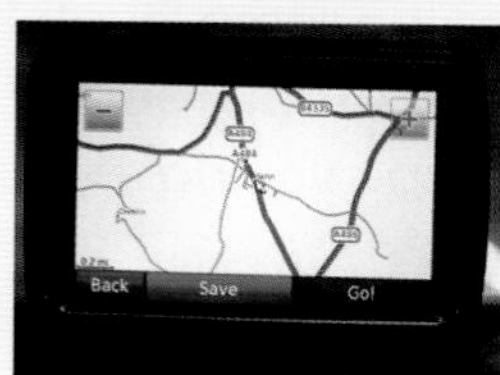

Toolbox

Cartesian co-ordinates can be extended to a larger grid.
The x axis is extended to the left and the y axis is extended downwards.
This gives four quadrants.
Q has co-ordinates (–2, –3).
R has co-ordinates (–4, 2).
S has co-ordinates (4, –2).

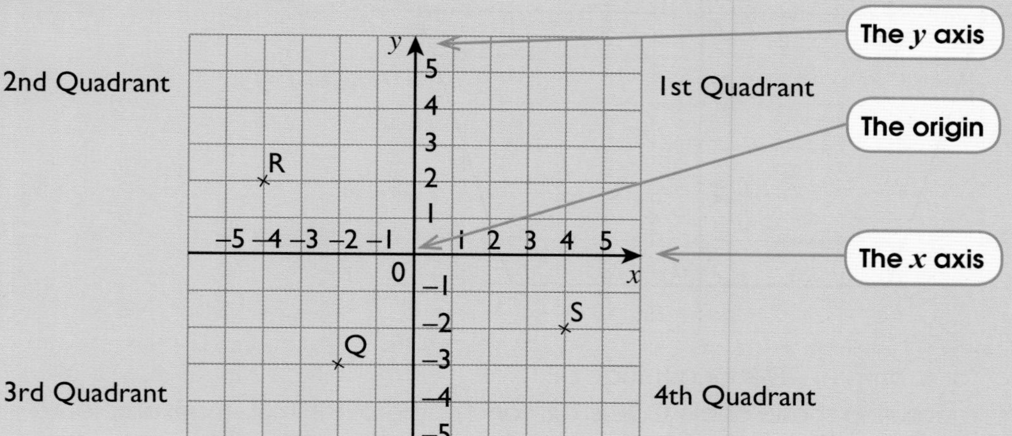

Example – Reading co-ordinates in all four quadrants

Look at this star.
What are the co-ordinates of the points A, C and E?

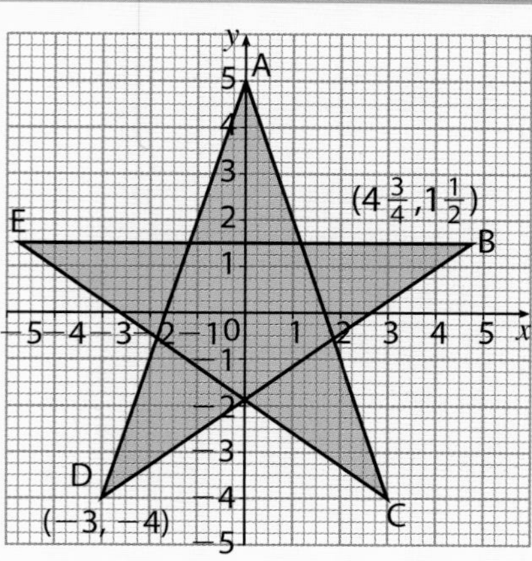

Solution

A is (0, 5). ← **A is on the y axis.**

C is (3, –4).

E is $(-4\frac{3}{4}, 1\frac{1}{2})$. ← **E has the same y co-ordinate as B.**

Example – Plotting co-ordinates in all four quadrants

Draw axes using values of –6 to 6 for both x and y.

a Plot these points and join them in order:
A(4, 6), B(6, 0), C(4, –6), D(–4, –6), E(–6, 0), F(–4, 6) then back to A.

b Are all the sides the same length?

c Are all the angles exactly equal?

d Describe the shape you have drawn.

Solution

a

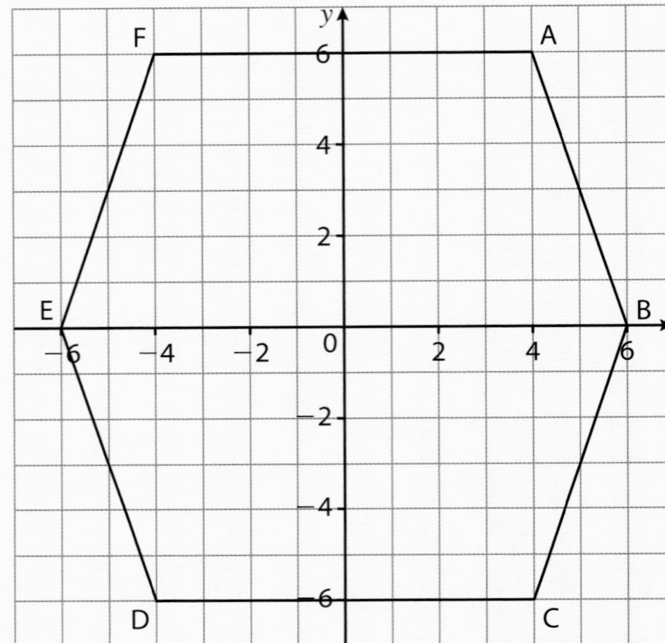

b No, for example, CD is longer than DE.

c No, measuring shows angle CDE is 108° and angle DEF is 143°, for example.

d The shape has six sides so it is a hexagon.
As the lengths of the sides and the angles are different, it is an irregular hexagon.

Remember:

- The x axis is positive to the right of the origin.
- The y axis is positive above the origin.

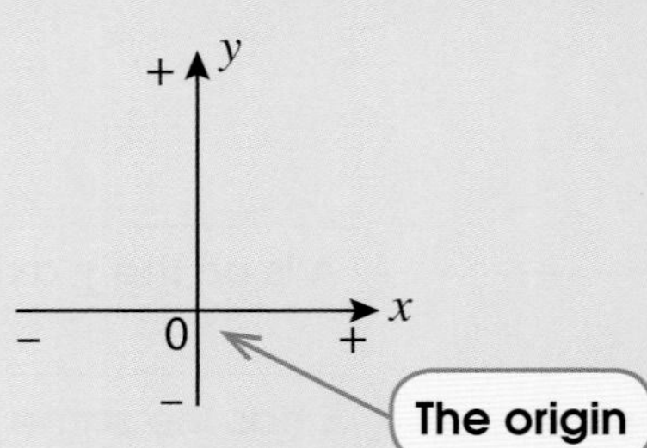

Skills practice A

1 **a** Write down the co-ordinates of points A, B, C and D.
b Which point is in the third quadrant?

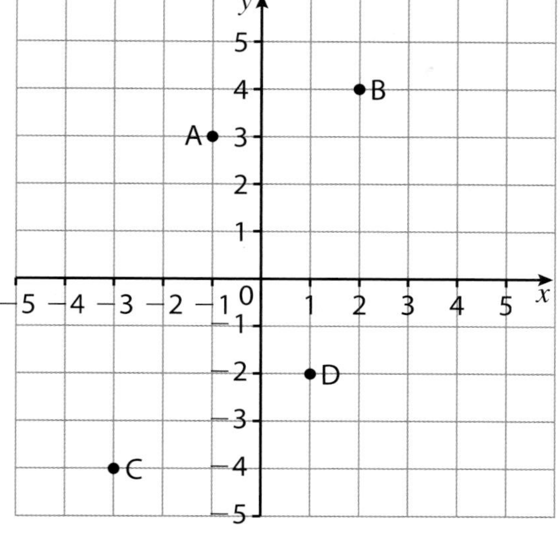

2 Write down the co-ordinates of the vertices of these shapes.

a

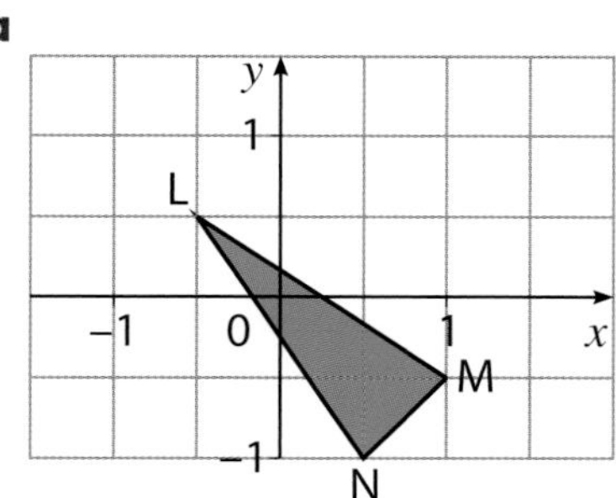

b

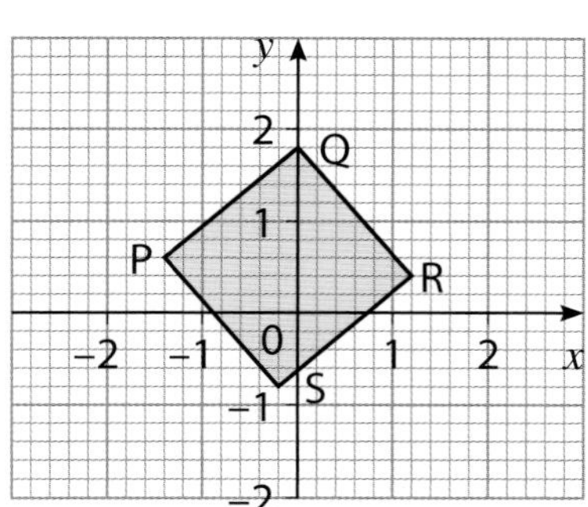

3 Write down the co-ordinates of the points A, B, C and D.

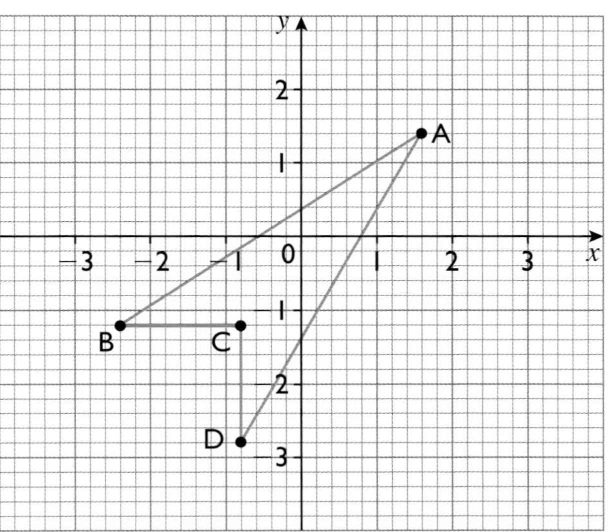

4 Write down the co-ordinates of the vertices of these shapes.

a

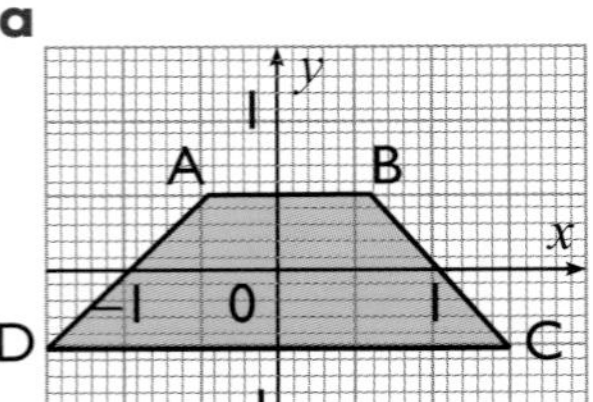

b

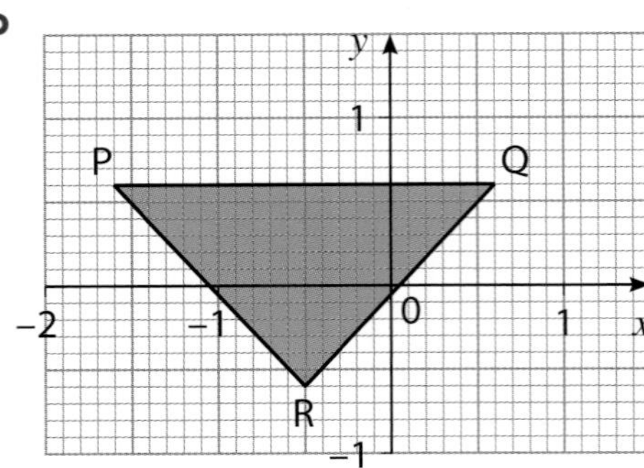

Reasoning

5 On this grid, points A, B and C are three points of a rectangle.

a What are the co-ordinates of points A, B and C?

b What are the co-ordinates of the fourth point of the rectangle?

c How long is the rectangle?

d How wide is the rectangle?

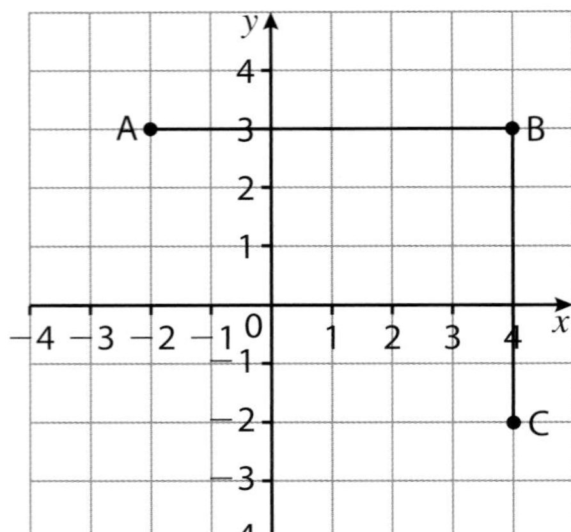

Reasoning

6 Draw axes, using values of –1 to 6 for x and –4 to 4 for y.

a Plot these points and join them in order.

(–1, –1), (2, 3), (6, 0)

b Mark a fourth point to make a square.

What are the co-ordinates of this point?

7 Write down the co-ordinates of the points A, B, C and D shown on the grid below.

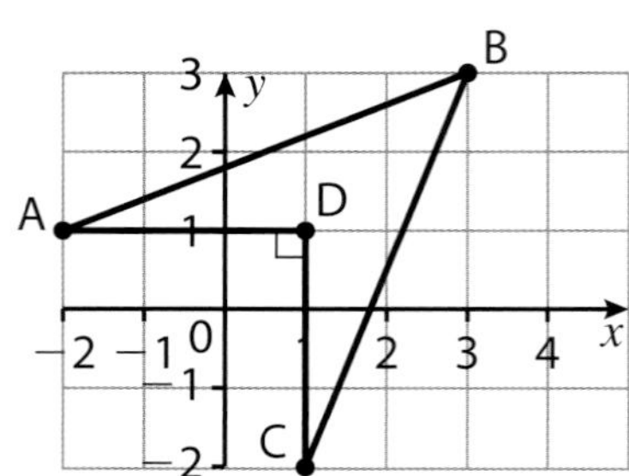

8 Write down the co-ordinates of the vertices A, B, C, D, E, F, G, H, I, J and K.

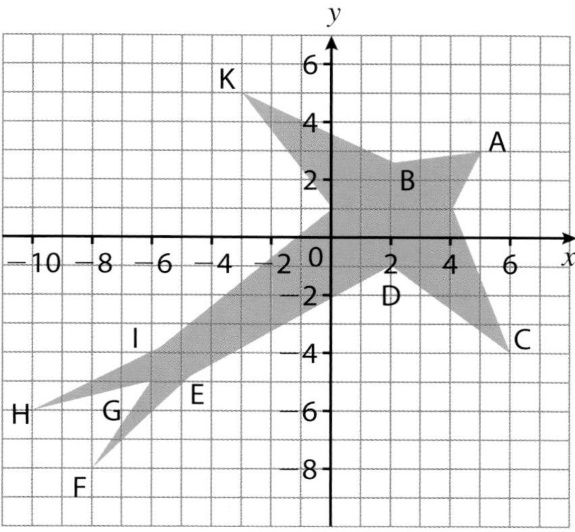

9 **a** Write down the co-ordinates of the vertices U, V, W, X, Y and Z.

b This is a picture of a flat fish. Put in the two eyes.

What are the co-ordinates of the eyes?

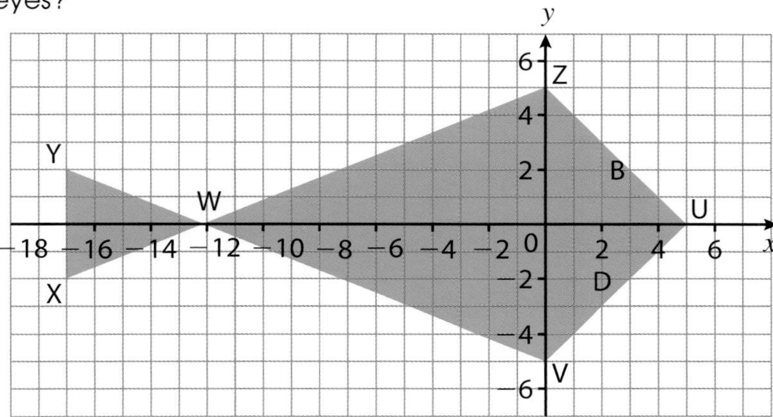

10 Look at this grid.

a What are the co-ordinates of point D?

Copy the grid and plot the points A, B, C, D and E.

Join them in order.
The shape is half of a four-pointed star.

b Mark three new points on the grid and join them to complete the star.

c Write down the co-ordinates of the three new points.

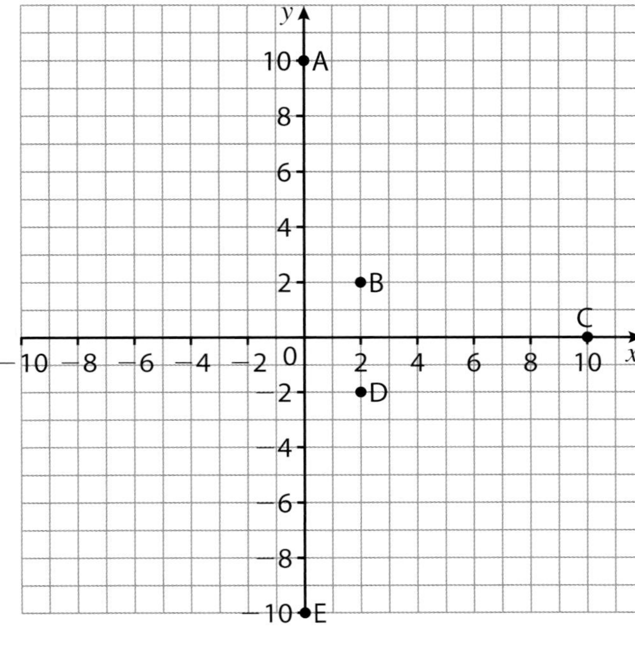

Skills practice B

1 **a** Draw an x axis from −10 to 10 and a y axis from −6 to 4.

b Plot these points and join them in order.

(−10, 1), (6, 1), (7, 3), (9, 3), (10, 1), (10, −1), (9, −3), (7, −3), (6, −1), (−5, −1), (−5, −5), (−6, −5), (−6, −2), (−7, −2), (−7, −3), (−8, −3), (−8, −2), (−9, −2), (−9, −4), (−10, −4), (−10, 1)

c What have you drawn?

Reasoning

2 Draw axes with values for both x and y from −5 to 5.

Plot these points and join them in order.

A(−3, 1), B(0, 5), C(4, 2)

These are three vertices of the square ABCD.

a What are the co-ordinates of D?

b What are the co-ordinates of the mid-point of each side of the square?

3 **a** What is the name of this shape?

b Write down the co-ordinates of points A, B, C, D, E, F, G and H.

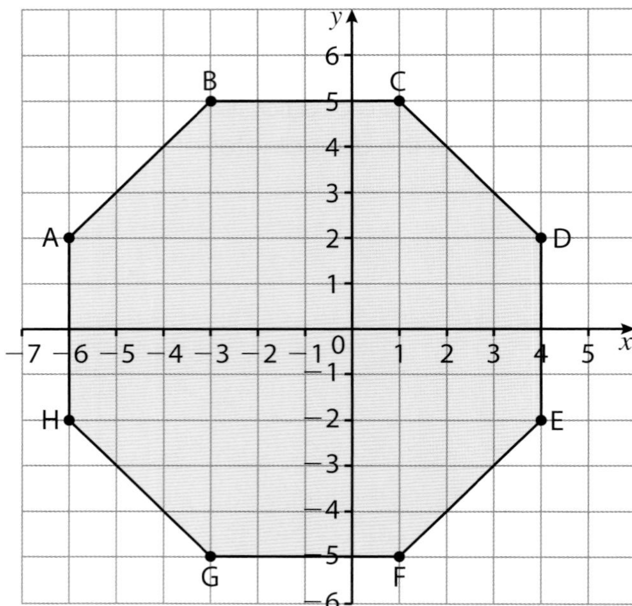

Reasoning

4 Draw axes with values for both x and y from −6 to 6 using the same scale.

Plot these points and join them in order.

A(−6, 0), B(−2, 4), C(2, 4), D(6, 0), E(2, −4)

A sixth point, F, is needed to make a symmetrical hexagon.

FE is of equal length to BC and parallel to BC.

a What are the co-ordinates of F?

Add F to your graph and complete the hexagon.

b How many lines of reflection symmetry has the hexagon?

c Are all the sides the same length?

Are all the angles equal?

Reasoning

5 Draw axes with values for both x and y from –5 to 5.
Plot these points and join them in order.
X(–3, 4), Y(0, 3), Z(5, 3)
These are three vertices of the parallelogram WXYZ.
a What are the co-ordinates of W?
b What are the co-ordinates of the mid-points of
i XZ
ii YW?

6 **a** Copy the four-pointed black star.
b Write down the co-ordinates of the points L, M and N.
c Mark L, M and N on your diagram.
d Join LM and LN and extend the lines to reach the four-pointed star.
e Mark nine more points and draw lines to complete an eight-pointed star.

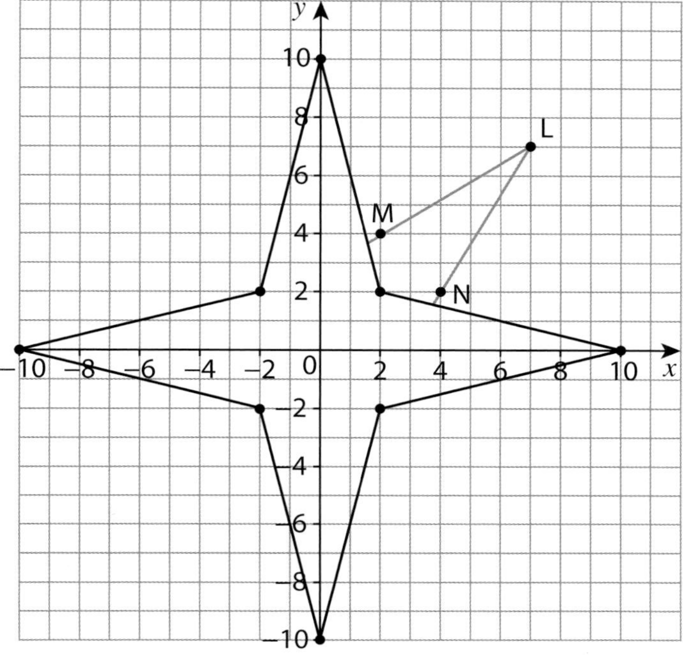

Wider skills practice

Reasoning

1 Three points of a rectangle are (–2, 1), (5, 1) and (–2, –2).
What are the co-ordinates of the fourth point of the rectangle?

Reasoning

2 **a** Copy the diagram and mark
i three points with the same x co-ordinate as A
ii three points with the same y co-ordinate as B.
b What do you notice?

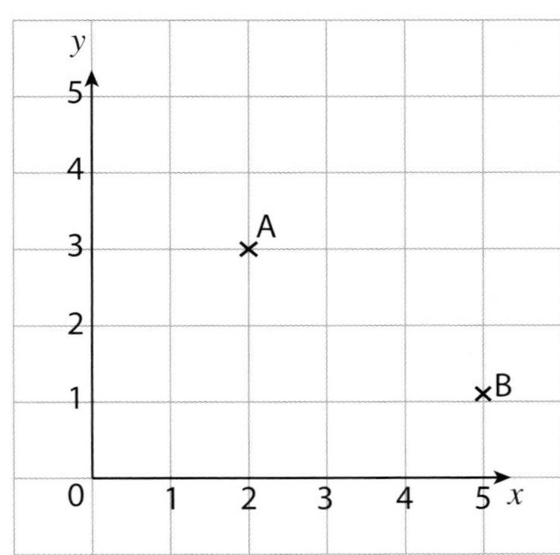

Applying skills

1 The four stars of this constellation are called the Square of Pegasus.

a Write down their co-ordinates.
Is it really a square?

b Copy the graph with the four stars on it.
Draw both the x and y axes from –4 to 4.

Nearby, at $(-3\frac{1}{2}, 2)$, is the Andromeda galaxy.

c Mark it on your graph.
You can see the galaxy on a clear night.
Keep your graph to help you find it.

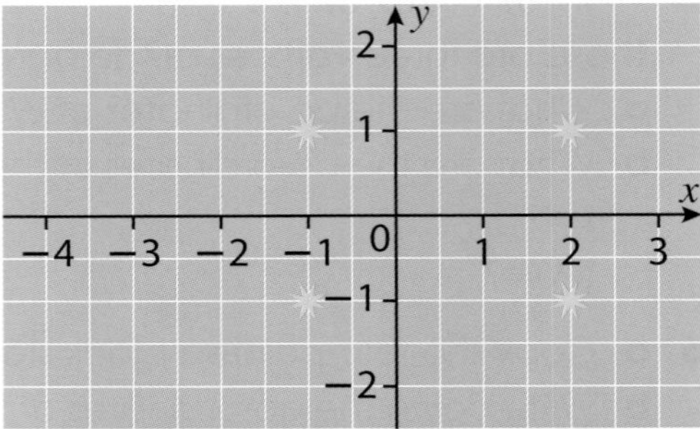

Andromeda galaxy

Reviewing skills

1 The point (1, 3) is in the first quadrant.
In which quadrants are these points?

a (–3, –6)
b (5, –10)
c (–3, 8)
d (100, 100)

2nd quadrant	1st quadrant
3rd quadrant	4th quadrant

2 Write down the co-ordinates of the points A, B, C, D, E, F, G, H, J and K.

3 **a** Draw x and y axes, each from –5 to 5 using the same scale.

b Plot the points (–1, 4), (–2, 2), (1, –2) and (1, 3).

c Join the points in order, and join the last point to the first.

d What shape have you drawn?

Unit 3 • Translation • Band d

Building skills

Example outside the Maths classroom

Chess moves

Toolbox

A **transformation** moves an object according to a rule.
One transformation is a **translation**.
In the diagram object PQR is translated to the image P′Q′R′.
The object slides without turning.
The translation is described as 3 units right and 2 units up.

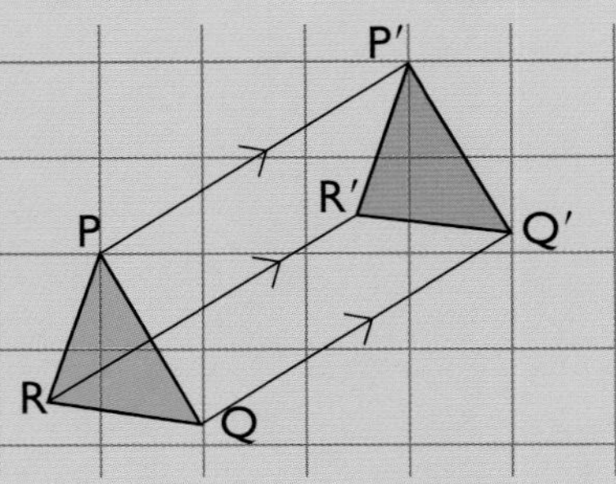

It can also be written as $\begin{pmatrix}3\\2\end{pmatrix}$

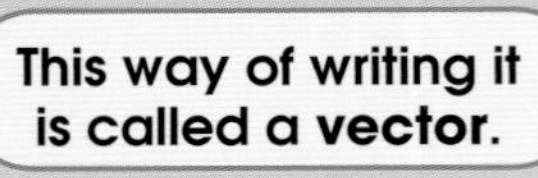

This way of writing it is called a vector.

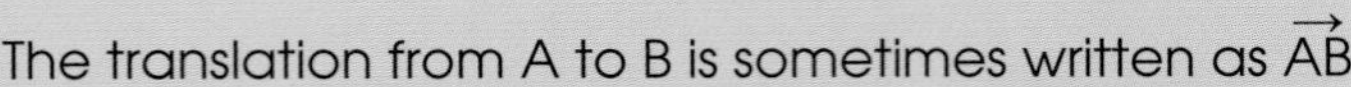

The translation from A to B is sometimes written as $\overrightarrow{AB}$.
The image and the original object are **congruent**.
This means they are the same shape and the same size.

Example – Describing a translation

Describe the translation that maps

a A to C
b B to C
c C to A.

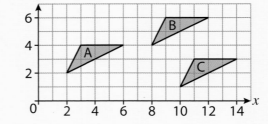

Solution

a The translation 8 units right and 1 down maps A to C, so $\begin{pmatrix}8\\-1\end{pmatrix}$.

b The translation 2 units right and 3 down maps B to C, so $\begin{pmatrix}2\\-3\end{pmatrix}$.

c The translation 8 units left and 1 up maps C to A, so $\begin{pmatrix}-8\\1\end{pmatrix}$.

Example – Translating an object

On a copy of this grid

a Translate A 5 units left and 2 up.
Label this triangle B.

b Translate B 4 units right and 2 up.
Label this triangle C.

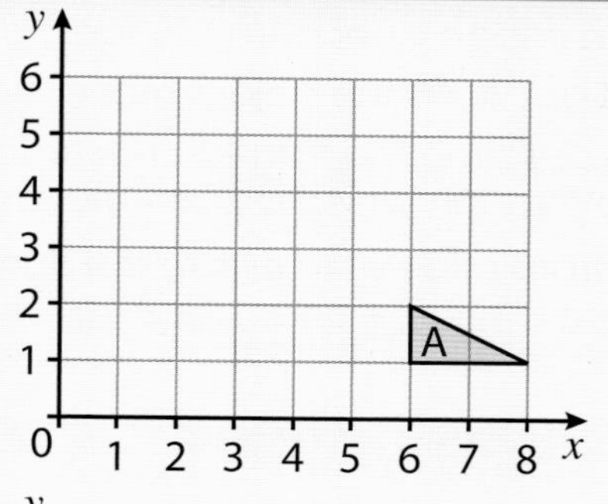

Solution

a Translating A 5 units left results in the blue triangle.
Translating the blue triangle 2 units up results in image B.

b Translating B 4 units right results in the green triangle.
Translating the green triangle 2 units up results in image C.

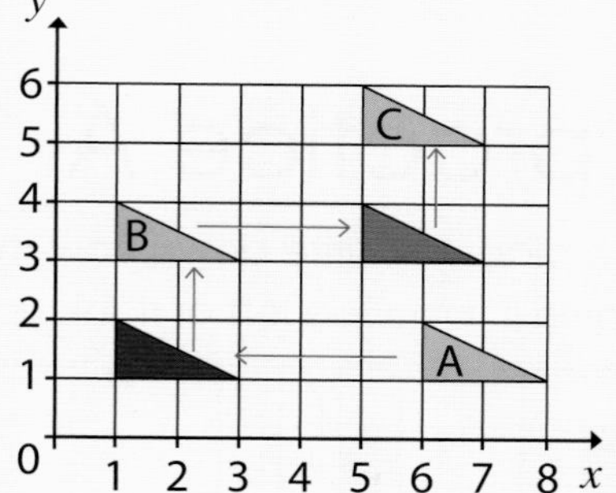

Example – Translating an object using a vector

a Draw and label an x axis from –5 to 7 and a y axis from 0 to 7.
Plot and label these points.
(1, 0), (3, 1), (4, 3), (5, 1)
Join the points in order to form a quadrilateral.
Label it A.

b Translate shape A by $\begin{pmatrix} 2 \\ 4 \end{pmatrix}$.
Label the image B.

c Now transform shape B by $\begin{pmatrix} -7 \\ 0 \end{pmatrix}$.
Label the image C.

d Describe the single transformation that maps A to C.

Solution

a–c

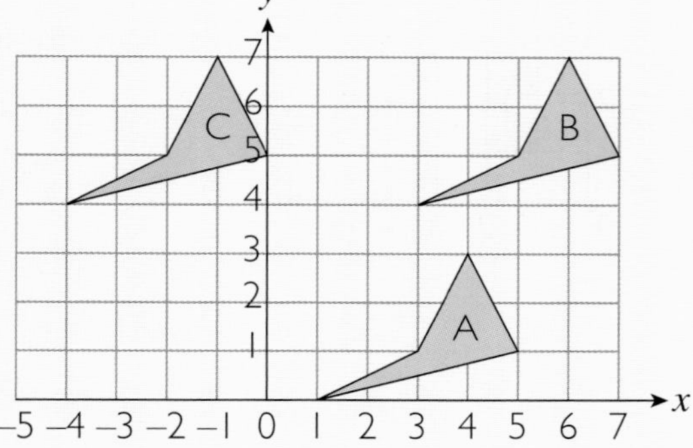

d The transformation A to C is a translation $\begin{pmatrix} -5 \\ 4 \end{pmatrix}$.

Remember:

- The image after a translation is congruent to the original object.
- All points on the object move according to the same instructions.
- To describe the translation, start by counting or measuring the units along and up through which a point moves.
- A translation of 4 units to the right and 5 up can be written as $\begin{pmatrix} 4 \\ 5 \end{pmatrix}$. In this notation, right is +, left – and up is +, down –.

Skills practice A

1 For each diagram write down how many units the blue shape has moved. Write down if it moved left or right, up or down.

a

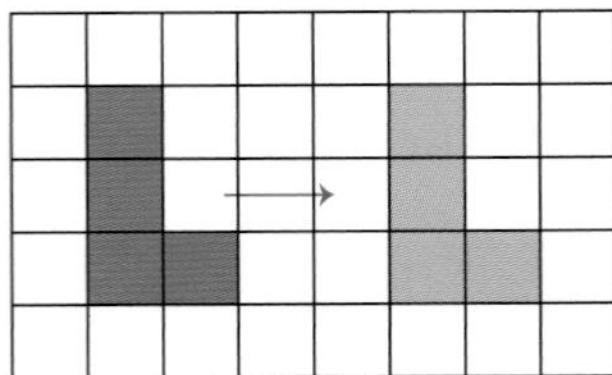

b

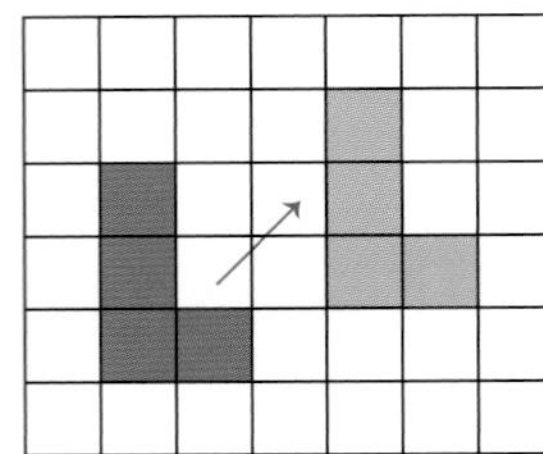

c

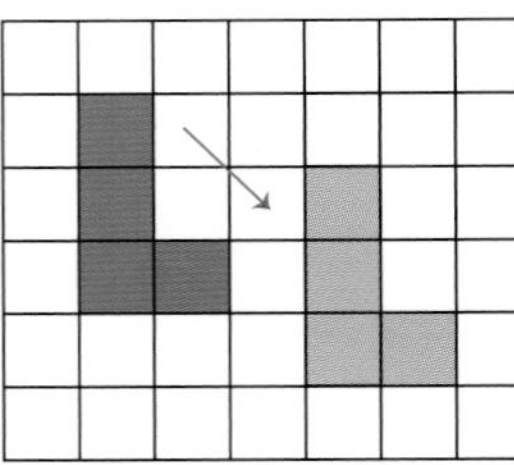

d

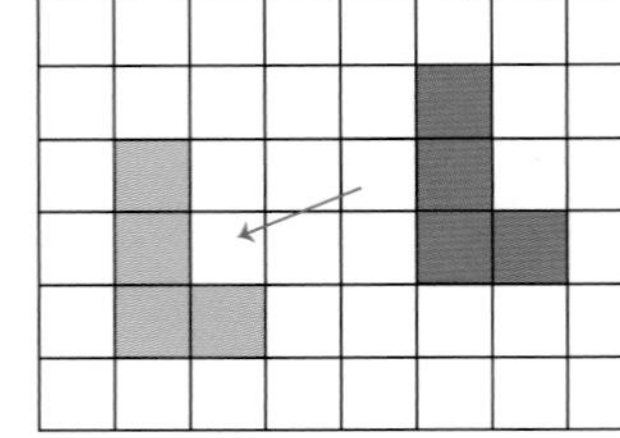

e

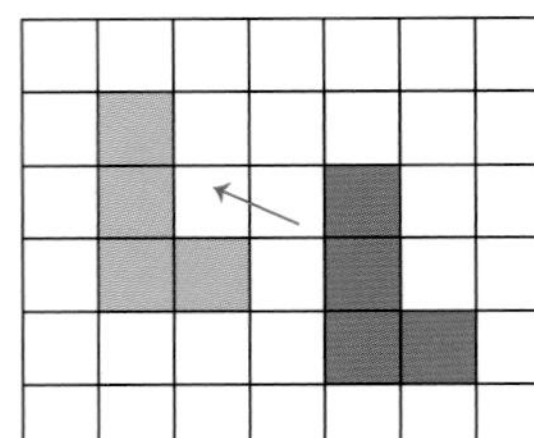

f

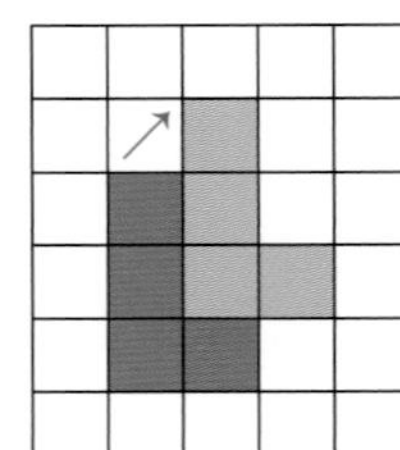

2 Copy the diagram, extending the grid as necessary. Draw the shape after

a a translation of 4 to the right and 1 up

b a translation of 2 to the left

c a translation of 3 down

d a translation of 6 down and 6 to the left.

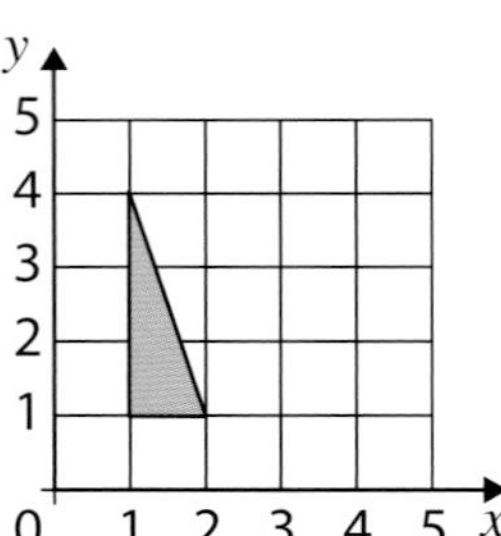

3 Copy this diagram, extending the grid as necessary.
Draw shape A after each of the transformations given in the table.
Label each of the images.

Translation	Left/right	Up/down
A → B	3 right	1 up
A → C	8 right	3 down
A → D	2 left	5 down
C → E	4 left	4 down
C → F	5 left	1 up
F → G	8 right	4 down
B → H	7 left	2 down

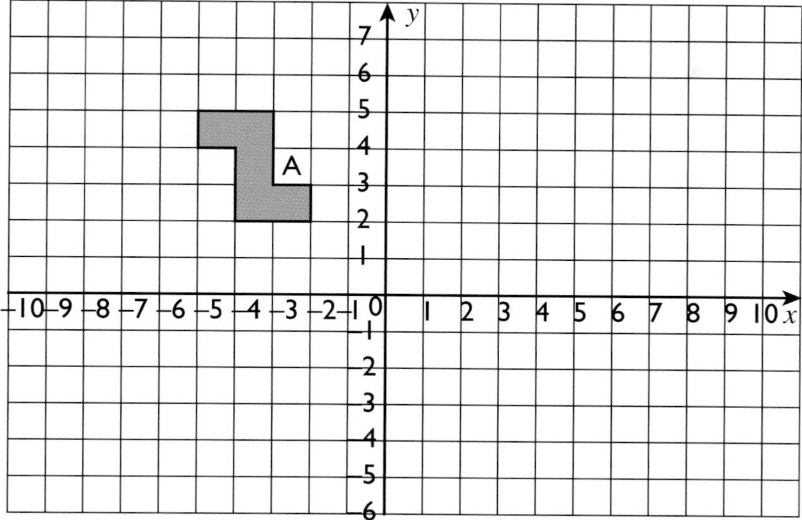

Reasoning

4 **a** Describe these transformations.
In each case say how many left or right and how many up or down.

i C to B
ii A to B
iii D to A
iv C to D
v B to D
vi D to C
vii B to C

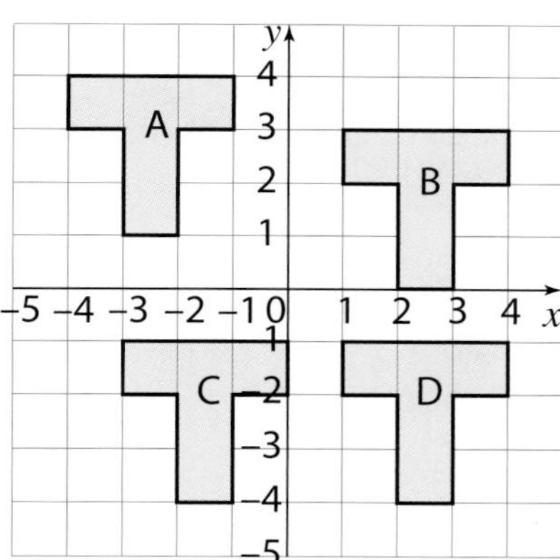

b How could you work out the answers for parts **vi** and **vii** using your answers for parts **i** to **v**?

5 Look at these shapes.

a Copy this table and complete it to describe each translation.

Translation	Translation vector
A → B	$\begin{pmatrix} 8 \\ 1 \end{pmatrix}$
A → C	
A → H	
C → B	
C → G	
C → D	
E → D	
E → F	
G → D	
G → A	
G → H	

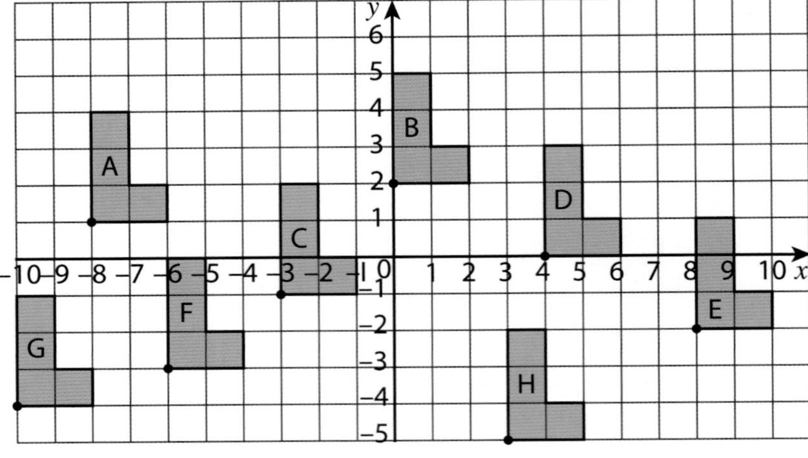

b What are the co-ordinates of the point marked at the bottom left corner of each shape?

6 **a** Draw and label x and y axes from –9 to 9.

Plot these following points and join them to make a triangle.

(–8, 2), (–6, 7), (–3, 2)

Label this triangle W.

b Translate triangle W 11 to the right and 2 up.

Label the new triangle X.

c What translation will map X to W?

7 **a** Draw and label x and y axes from –10 to 10.

Plot the points A(3, 10), B(5, 10), C(6, 7), D(6, 5), E(3, 5) and F(2, 7).

Join them in order to form the hexagon ABCDEF.

After a translation, A is mapped to A′(–3, 4).

b Describe this translation.

c Plot and draw image A′B′C′D′E′F′.

A′B′C′D′E′F′ is translated by $\begin{pmatrix} 2 \\ -7 \end{pmatrix}$.

The image is A″B″C″D″E″F″.

d Draw A″B″C″D″E″F″ and write down its co-ordinates.

e Describe the single transformation that maps ABCDEF to A″B″C″D″E″F″.

8 Draw x and y axes from –5 to 5.

Plot these co-ordinates and join them to make a rectangle.

(1, 2), (1, 5), (3, 5), (3, 2)

Label this rectangle P.

a Translate rectangle P by $\begin{pmatrix} -5 \\ -3 \end{pmatrix}$.

Label the image Q.

b Translate rectangle P by $\begin{pmatrix} -1 \\ -5 \end{pmatrix}$.

Label the image R.

c Translate rectangle R by $\begin{pmatrix} 2 \\ 4 \end{pmatrix}$.

Label the image S.

d Describe the translation which maps P to S.

e Translate rectangle Q by $\begin{pmatrix} 0 \\ 3 \end{pmatrix}$.

Label the image T.

f Describe the translation which maps T to P.

Skills practice B

1 Look at this wallpaper pattern of repeated cars.

Car A is mapped to C by a translation of 4 right and 1 down.

a Describe the translations which map

i A to B **ii** A to D

iii C to D **iv** E to D

v C to B **vi** B to E

vii F to E **viii** F to A.

b Two of these translations are the same translation as D to C.

Which are they?

Reasoning

2 Hannah is playing a computer game.

She is shooting down aeroplanes with a gun (G).

She moves the gun 3 right and 2 up to shoot down A.

From A she moves the gun to B, C, D and E.

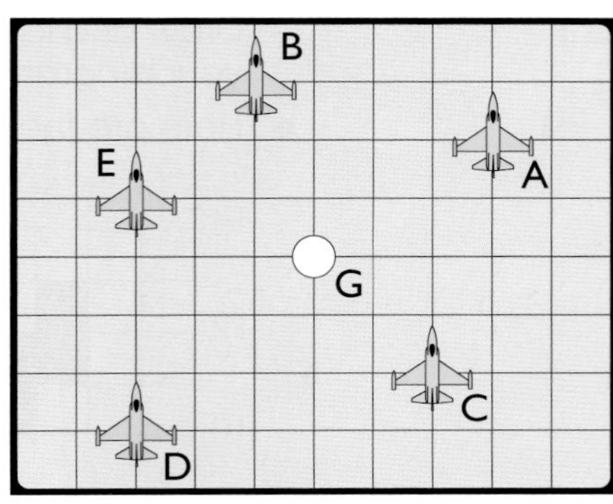

a Describe the moves from

i A to B **ii** B to C **iii** C to D **iv** D to E.

b What order could Hannah shoot them in to minimise the distance her gun travels?

3 John is designing a computer game.

Shapes appear at the top of the screen.

The aim of the game is to get as many horizontal lines as possible completely filled by the shapes.

You can use any translation to move them to a final position at the bottom of the screen. Shapes must not overlap.

John is testing his game.

This picture shows where the first shape appears (at the top of the screen) and where John places it.

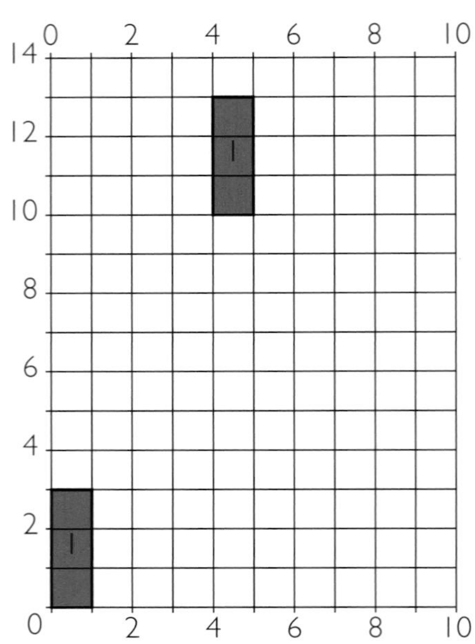

a Describe the translation of shape 1.

This is where the next two shapes appear.

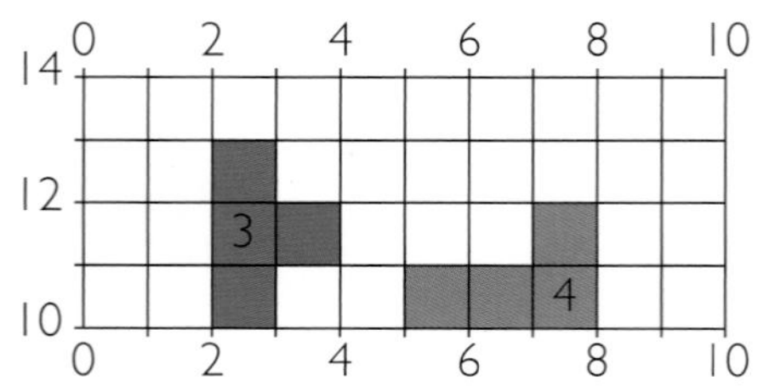

b This picture shows where the second shape appears (at the top of the screen) and where John places it.

Describe the translation of shape 2.

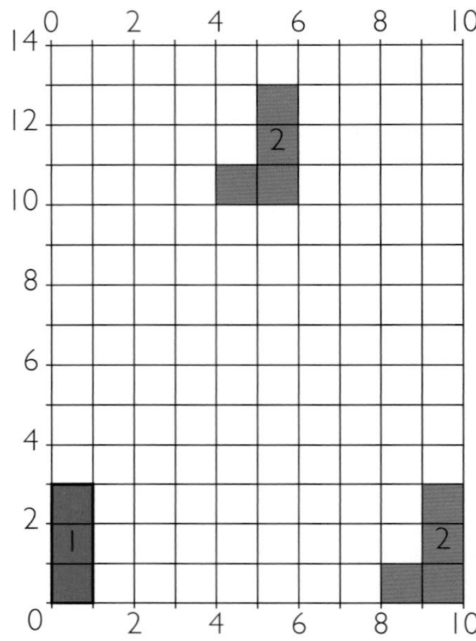

c John translates shape 3 by 5 to the right and 10 down.

Copy the diagram in part **b**.

Draw and label the final position of shape 3.

d He translates shape 4 by 4 to the left and 10 down.

Draw and label the final position of shape 4 on your diagram.

e Here are the next three shapes.

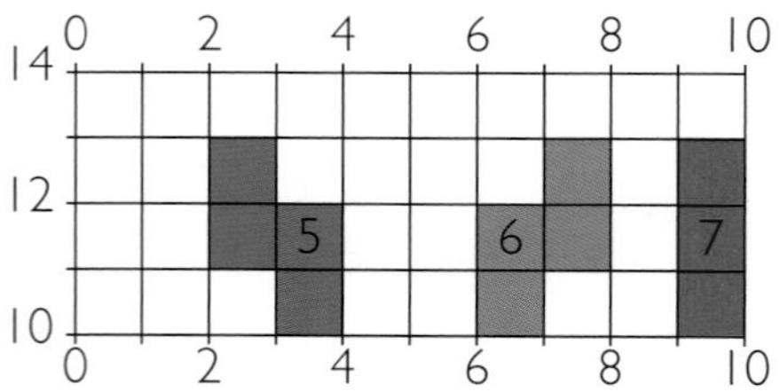

Decide where to place each one.

i Draw the final position of each shape on your diagram.

ii Describe the translation of each shape.

Reasoning

4 In this puzzle the S has been translated by $\begin{pmatrix} 4 \\ 2 \end{pmatrix}$.

a Draw the remaining tiles after these translations.

Translate M by $\begin{pmatrix} 1 \\ 2 \end{pmatrix}$

Translate I by $\begin{pmatrix} -1 \\ 2 \end{pmatrix}$

Translate L by $\begin{pmatrix} -3 \\ 2 \end{pmatrix}$

Translate E by $\begin{pmatrix} -1 \\ 2 \end{pmatrix}$

What word have you made?

b Write down the translations which move the letters in this word to make SLIME on the bottom row.

c Apply these translations to the letters in SLIME.

i Translate S by $\begin{pmatrix} 4 \\ 6 \end{pmatrix}$ **ii** Translate L by $\begin{pmatrix} 1 \\ 6 \end{pmatrix}$ **iii** Translate I by $\begin{pmatrix} -1 \\ 6 \end{pmatrix}$

iv Translate M by $\begin{pmatrix} -3 \\ 6 \end{pmatrix}$ **v** Translate E by $\begin{pmatrix} -1 \\ 6 \end{pmatrix}$

What word have you made this time?

5 **a** Copy the diagram.

Shape A is translated by the vector $\begin{pmatrix} -2 \\ 4 \end{pmatrix}$.

b Draw the image of shape A. Label it B.

Shape B is translated by $\begin{pmatrix} -3 \\ -2 \end{pmatrix}$.

c Draw the image of shape B. Label it C.

d What single translation maps shape C to shape A?

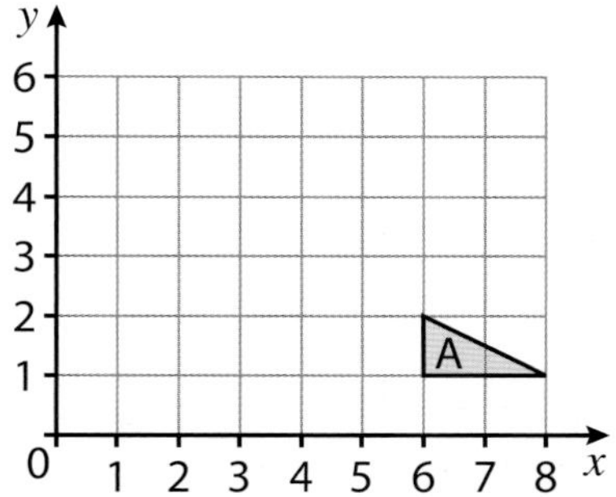

Reasoning

6 **a** Describe these transformations.

i C → A

ii A → B

iii C → B

b Compare your answer **iii** in part **a** to those for **i** and **ii** together. What do you notice?

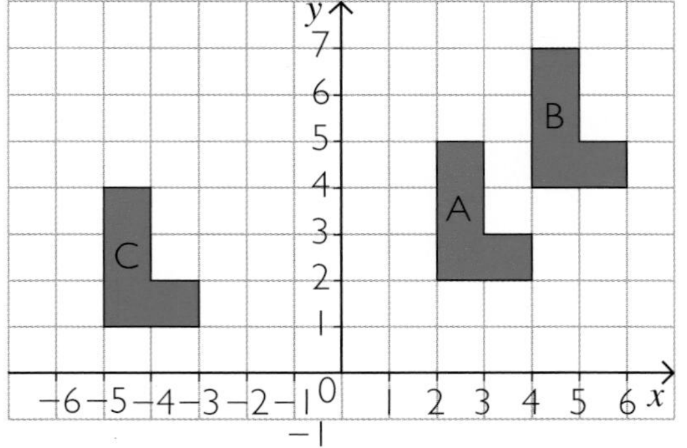

7 In this computer game, the rescue helicopter P picks up survivors in the sea.

a Move the helicopter to survivor A.
What is the translation vector $\overrightarrow{PA}$?

b Pick up all the other survivors in order.
Describe each journey as a translation vector.

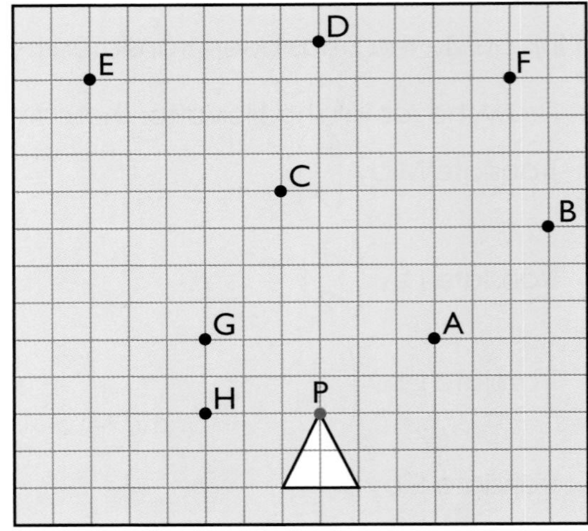

Wider skills

Reasoning

1 **a** Write down a translation vector where the object does not move.

b Write down a sequence of three translations where the shape returns to its original position.

2 Look at this grid.

+	–	X	÷	/	:	space
)	T	U	V	W	X	Y
(	S	F	G	H	I	Z
?	R	E	start	A	J	1
.	Q	D	C	B	K	2
,	P	O	N	M	L	3
0	9	8	7	6	5	4

a Use these vectors on the grid to form the word SLIMMER.

$\begin{pmatrix}-2\\1\end{pmatrix}, \begin{pmatrix}4\\-3\end{pmatrix}, \begin{pmatrix}0\\3\end{pmatrix}, \begin{pmatrix}-1\\-3\end{pmatrix}, \begin{pmatrix}0\\0\end{pmatrix}, \begin{pmatrix}-2\\2\end{pmatrix}, \begin{pmatrix}-1\\0\end{pmatrix}$

b Decode this message.

$\begin{pmatrix}0\\2\end{pmatrix}\begin{pmatrix}-1\\-2\end{pmatrix}\begin{pmatrix}1\\-1\end{pmatrix}\begin{pmatrix}-2\\3\end{pmatrix}\begin{pmatrix}1\\-4\end{pmatrix}\begin{pmatrix}-1\\2\end{pmatrix}\begin{pmatrix}5\\3\end{pmatrix}\begin{pmatrix}-3\\-4\end{pmatrix}\begin{pmatrix}-1\\-1\end{pmatrix}\begin{pmatrix}0\\1\end{pmatrix}\begin{pmatrix}0\\1\end{pmatrix}$

$\begin{pmatrix}-1\\1\end{pmatrix}\begin{pmatrix}5\\2\end{pmatrix}\begin{pmatrix}-2\\-3\end{pmatrix}\begin{pmatrix}-3\\0\end{pmatrix}\begin{pmatrix}1\\0\end{pmatrix}\begin{pmatrix}4\\3\end{pmatrix}\begin{pmatrix}-4\\-3\end{pmatrix}\begin{pmatrix}2\\0\end{pmatrix}\begin{pmatrix}-3\\1\end{pmatrix}\begin{pmatrix}5\\1\end{pmatrix}\begin{pmatrix}-6\\-3\end{pmatrix}$

Applying skills

Problem solving

1 Here is a race track for a game. You move around this racetrack as follows.

i Start at one of the positions on the start line (in blue). The example shown starts at position 3.

ii Choose a starting vector and move according to that vector. The example shows a move with translation vector $\begin{pmatrix} 0 \\ 3 \end{pmatrix}$

iii Choose your next vector by changing each component of your previous move by 1 in each direction or left alone.

iv After a move of $\begin{pmatrix} 0 \\ 3 \end{pmatrix}$ any of the following moves are possible.

$\begin{pmatrix} 0 \\ 3 \end{pmatrix}, \begin{pmatrix} -1 \\ 3 \end{pmatrix}, \begin{pmatrix} 1 \\ 3 \end{pmatrix}, \begin{pmatrix} 0 \\ 2 \end{pmatrix}, \begin{pmatrix} -1 \\ 2 \end{pmatrix}, \begin{pmatrix} 1 \\ 2 \end{pmatrix}, \begin{pmatrix} 0 \\ 4 \end{pmatrix}, \begin{pmatrix} -1 \\ 4 \end{pmatrix}, \begin{pmatrix} 1 \\ 4 \end{pmatrix}$

The example has chosen $\begin{pmatrix} 1 \\ 4 \end{pmatrix}$

a List all the possible translation vectors for the next move.

b Choose your own starting position and find the minimum number of moves needed to get all round the track.

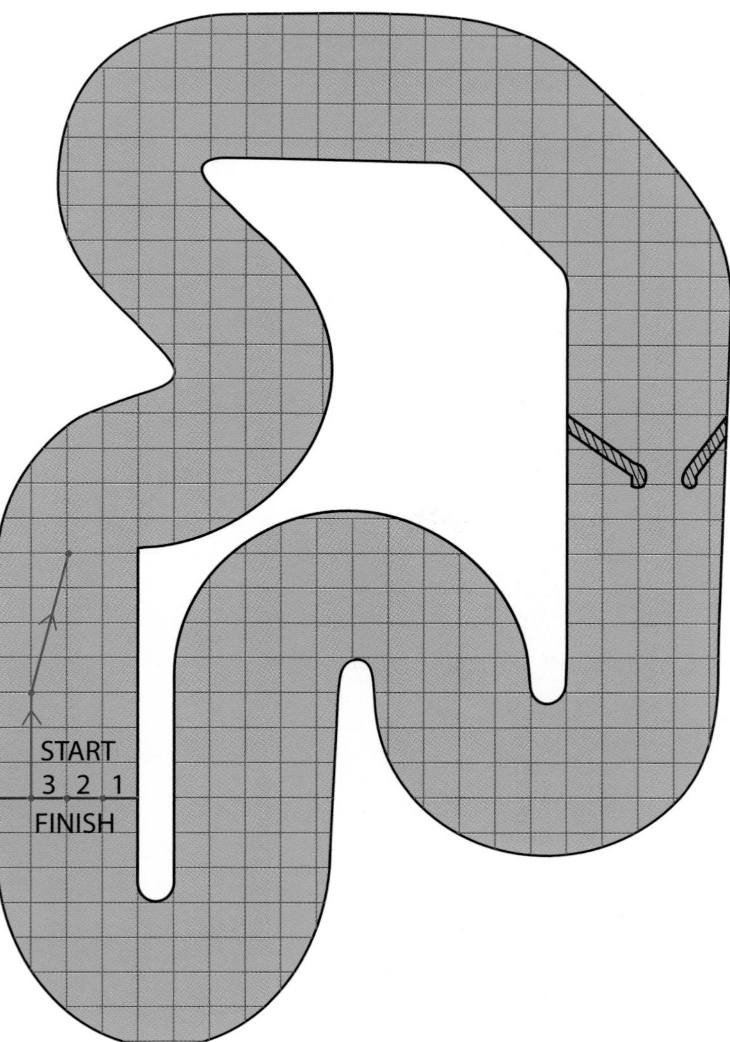

Reviewing skills

1 a Describe each of these translations.
   i A to D
   ii B to A
   iii E to C
   iv C to E
   v F to E
   vi D to B

  b Why is A to C impossible?

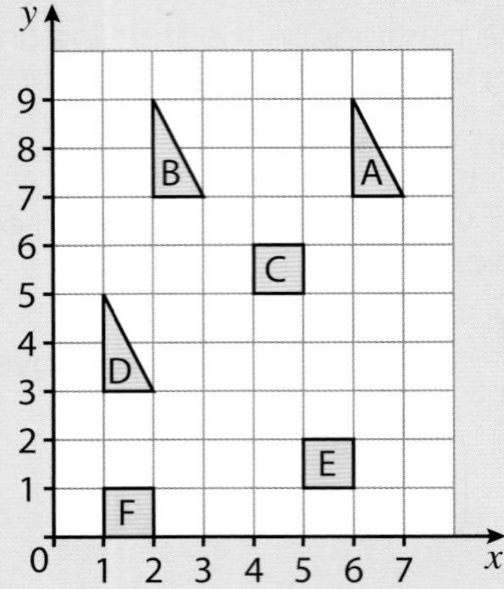

2 Draw x and y axes from –3 to 8.
On the same grid draw these images.

a Shape A after translation by $\begin{pmatrix} 1 \\ 4 \end{pmatrix}$.

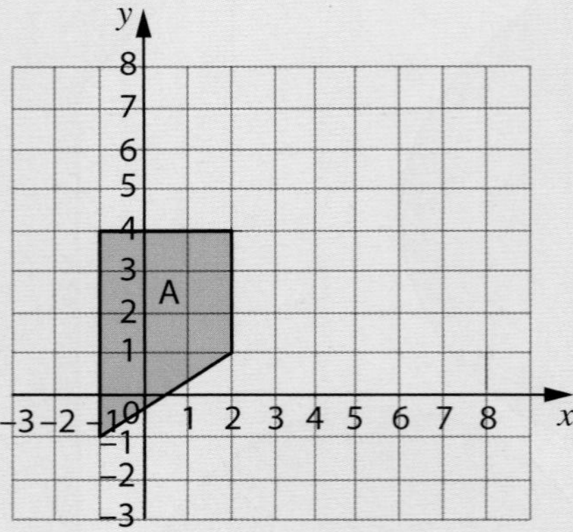

b Shape B after translation by $\begin{pmatrix} 0 \\ 3 \end{pmatrix}$.

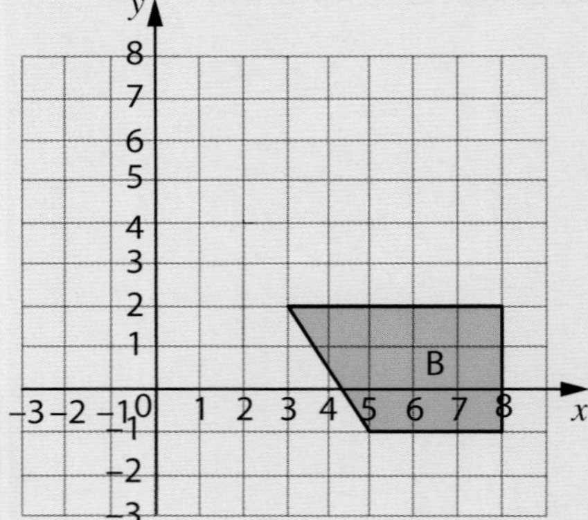

c Shape C after translation by $\begin{pmatrix} -6 \\ 0 \end{pmatrix}$.

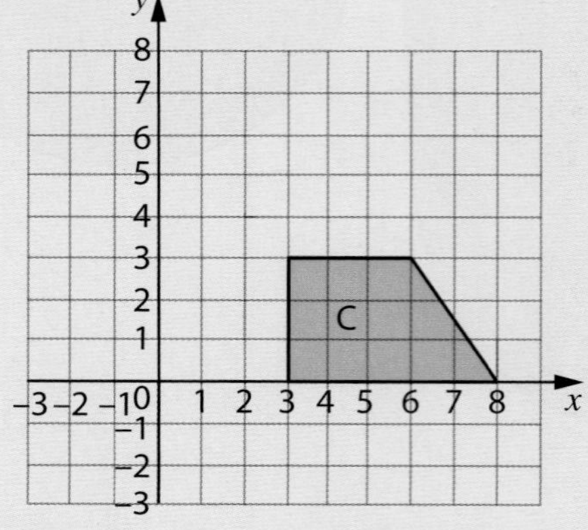

d Shape D after translation by $\begin{pmatrix} -3 \\ -6 \end{pmatrix}$.

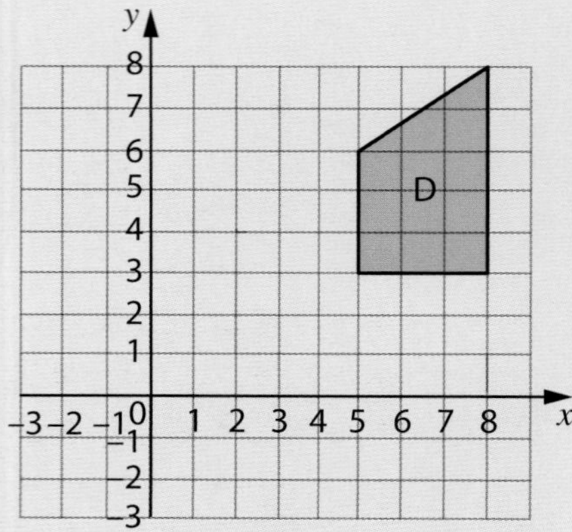

Building skills

Example outside the Maths classroom

Kaleidoscopes

Toolbox

Reflection is a transformation.

When an **object** is reflected in a line, its **image** is formed on the other side of the line. Each point is at the same distance from the line.

The line is called the **mirror line** or the **line of reflection**.

The image and the original object are **congruent**.

They are the same shape and the same size.

The line of reflection does not have to be horizontal or vertical, it can be in any direction.

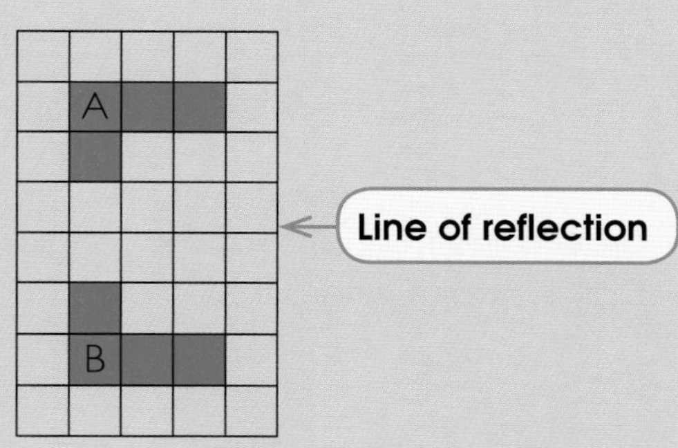

Common lines used for reflection are shown in the diagram below.

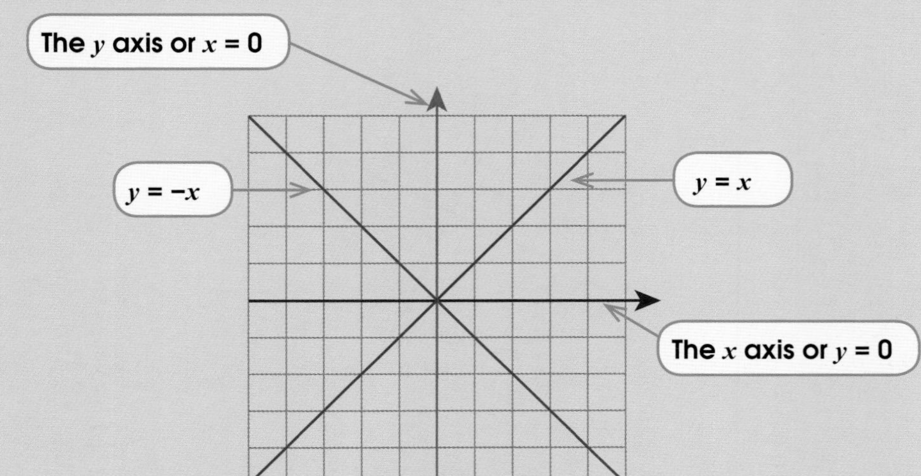

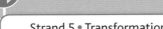

Other commonly used lines of symmetry are

- vertical lines parallel to the y axis e.g. $x = 3$, $x = -2$
- horizontal lines parallel to the x axis e.g. $y = -3$, $y = 4$.

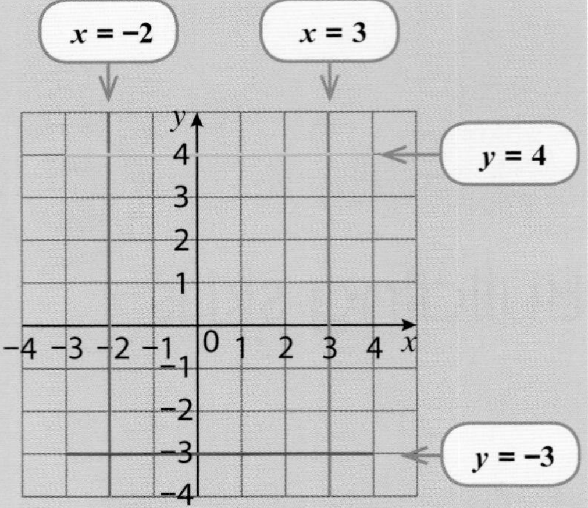

It is often obvious where the image should be, but sometimes you may need to do a drawing. To reflect a point P in a line l, take these steps.

- Draw a line through P at right angles to l. Call the point where they cross X.

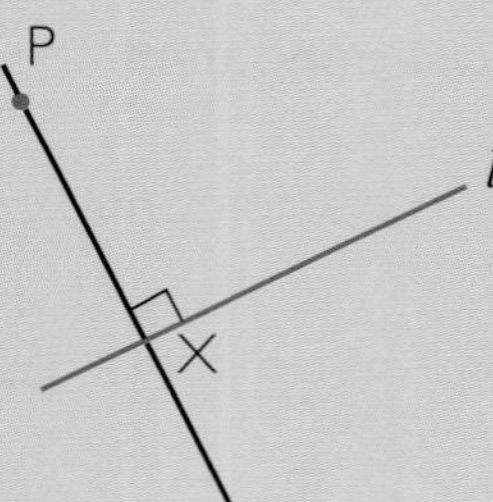

- Find P′ where P′X = PX. P′ is the image of P.

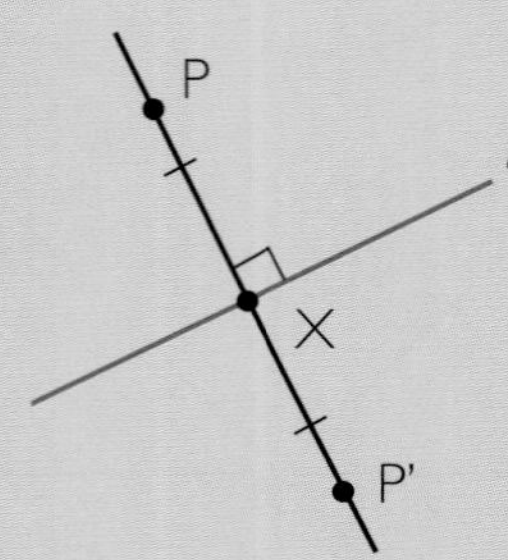

Example – Reflecting a shape on a co-ordinate grid

a Copy this diagram.
Draw the image of the triangle when it is reflected in the y axis.

b Are the two triangles congruent?
How do you know?

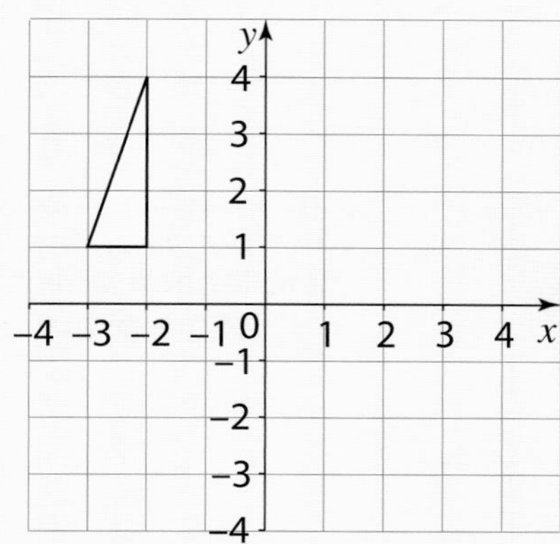

Solution

a

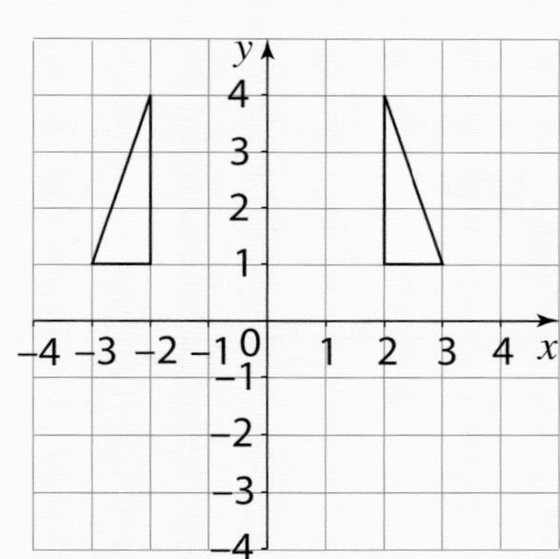

b When an object is reflected, all the lines and angles on the object and its image are the same size.
So they are congruent. ← **Congruent shapes are the same size and shape.**

Example – Reflecting a shape in a mirror line

Copy the diagram.
Draw the image of the shape after it is reflected.

Solution

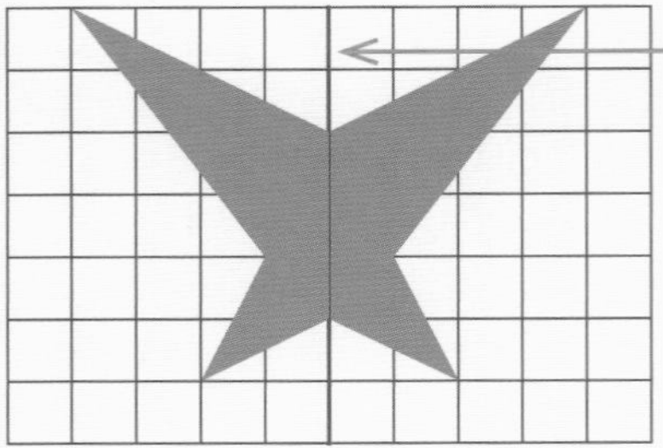

The points on the line of reflection stay in the same place.

Example – Finding lines of reflection

For each of these diagrams show the line of reflection and give its equation.

a

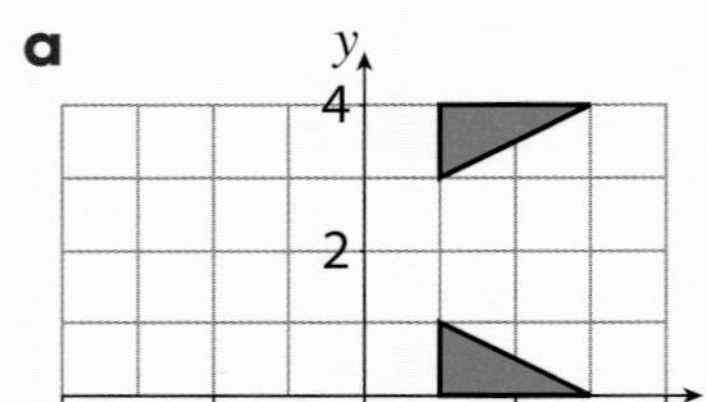

b

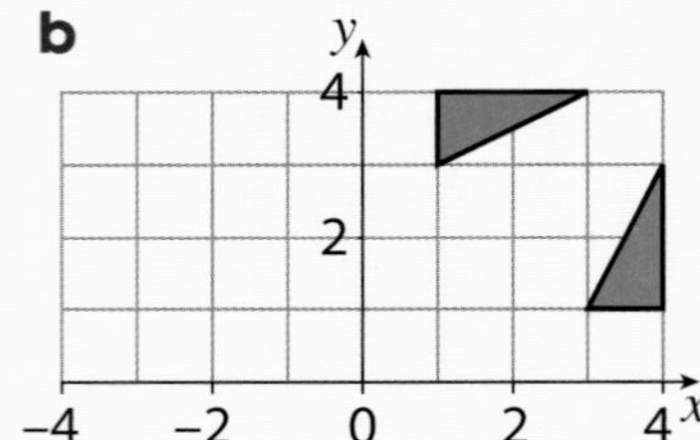

c

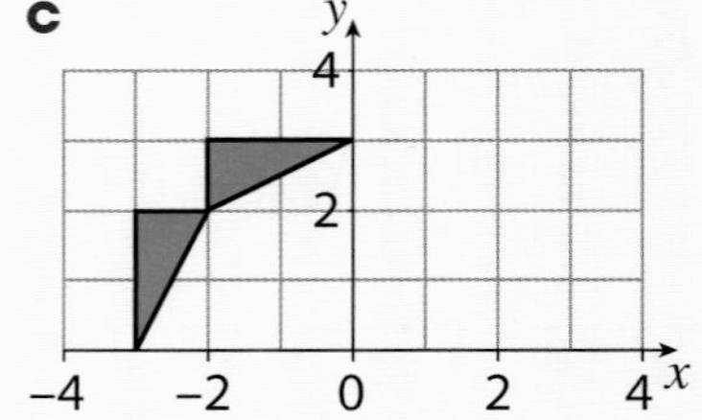

Solution

a

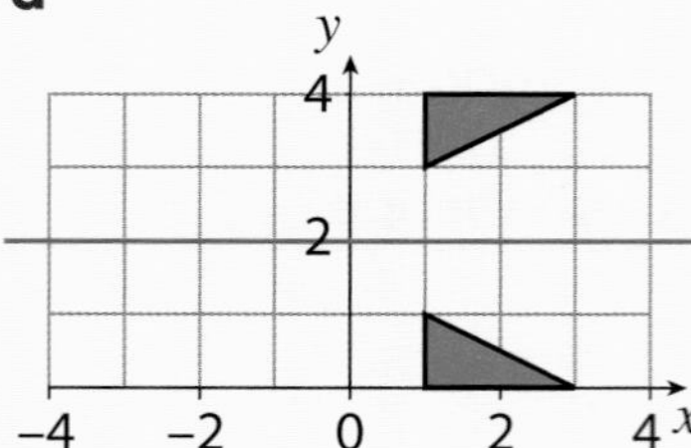

The line of reflection is $y = 2$.

b

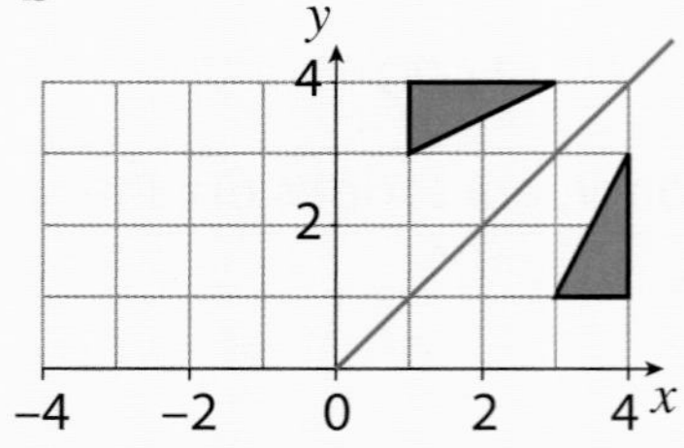

The line of reflection is $y = x$.

c

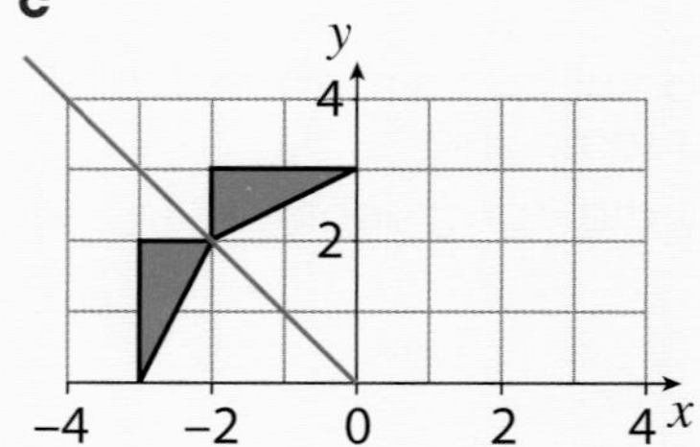

The line of reflection is $y = -x$.

Remember:

- The line joining a point and its image is at right angles to the mirror line.
- A point and its image are the same distance from the mirror line.
- To fully describe a reflection you must state the mirror line.

Skills practice A

1 Copy these shapes and reflect them in the mirror line.

a

Mirror line

b

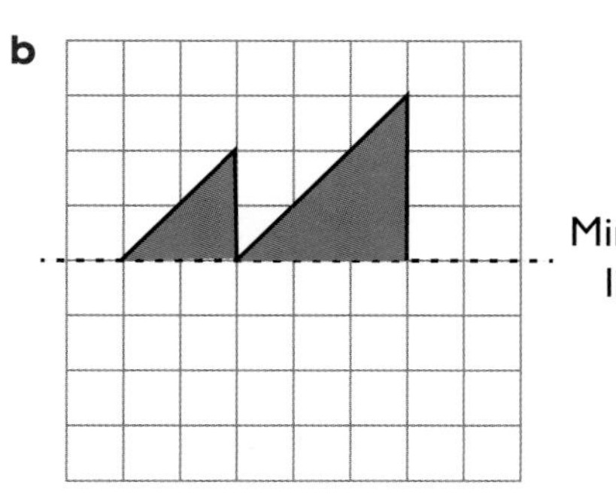

c

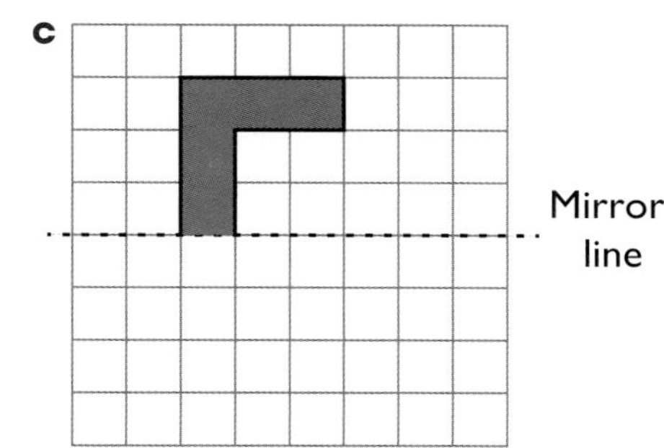

d

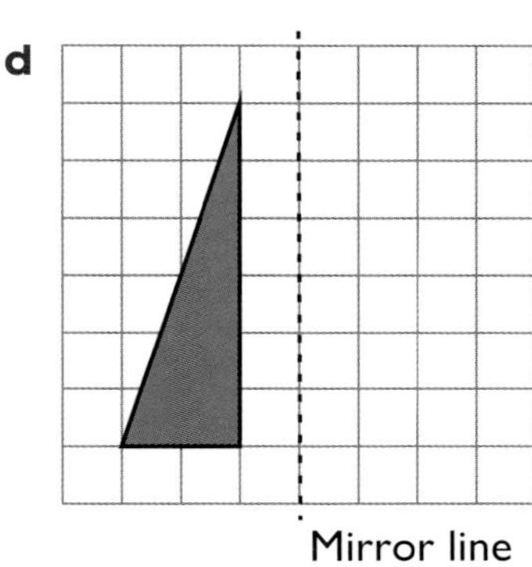

e

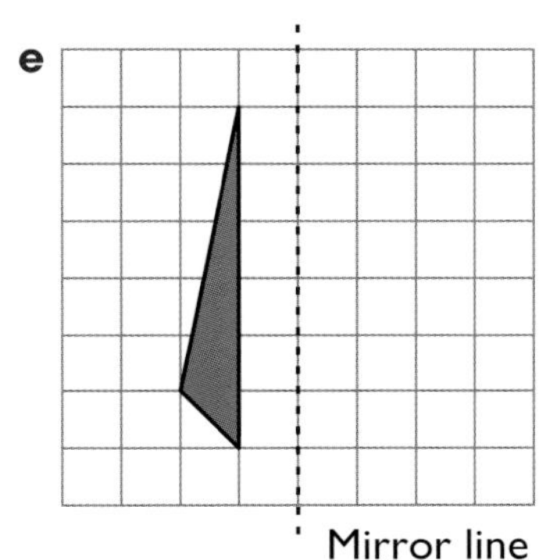

f

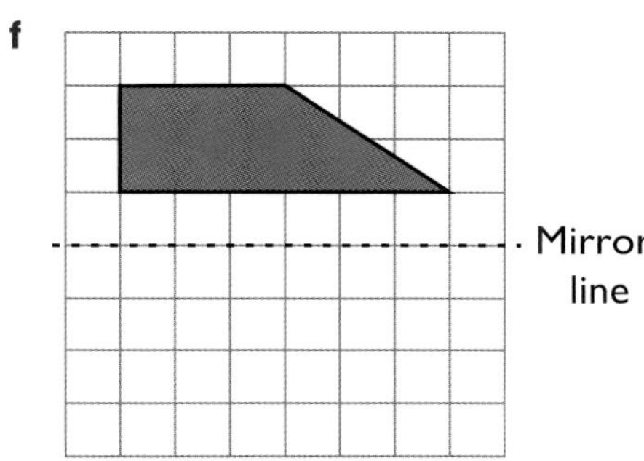

2 Copy these diagrams. Reflect each shape in the line shown and so complete a symmetrical shape.

a

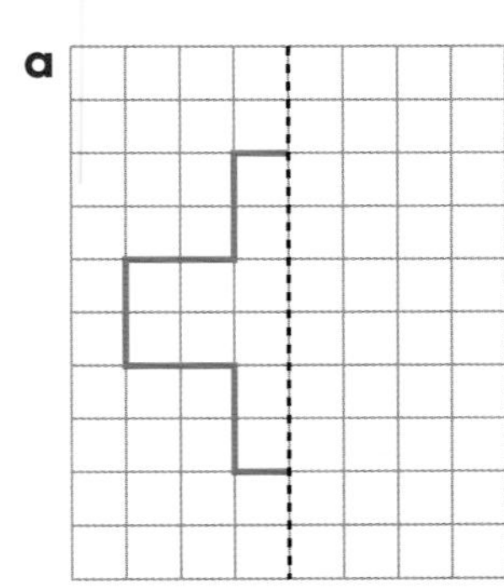

b

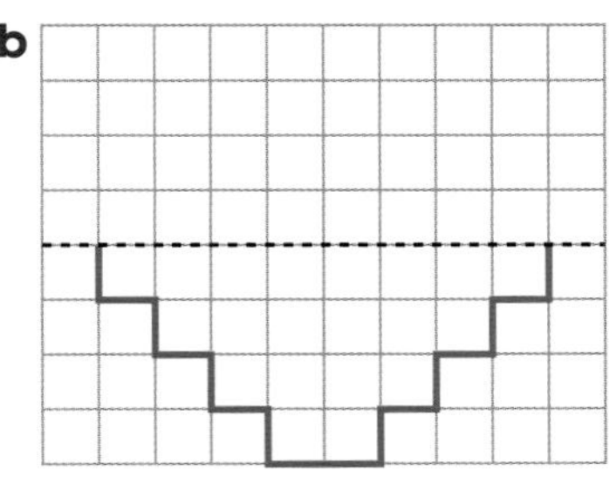

3 **a** Copy this diagram.

b Reflect the shape in the y axis.

c Now reflect both shapes in the x axis.

d Describe the pattern you have made as clearly as you can.

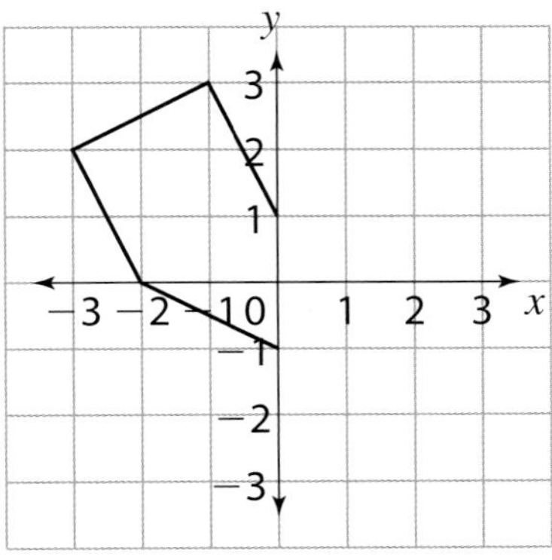

4 **a** Describe fully the transformation from shape A to shape B.

b Describe fully the transformation from shape A to shape D.

c Describe fully the transformation from shape B to shape C.

d In which quadrant is shape C?

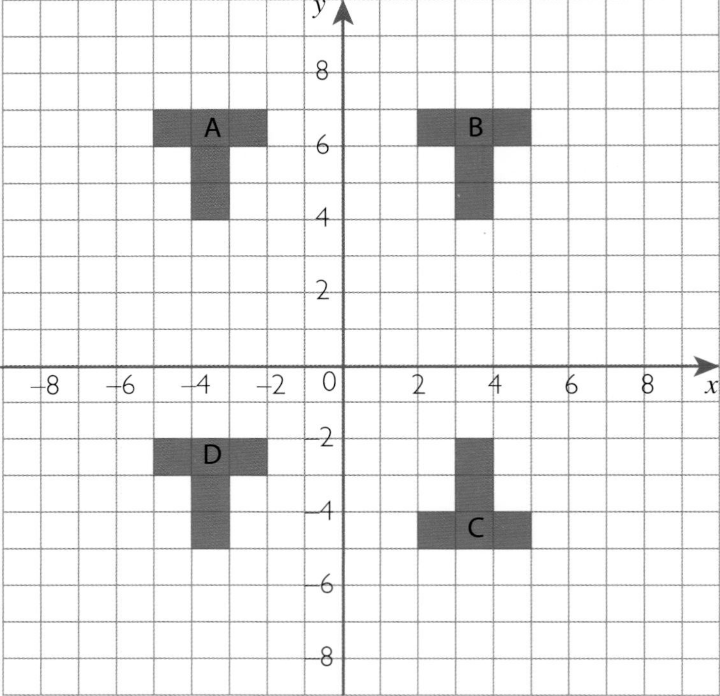

5 Copy this diagram.

a Reflect shape P in the y axis.
Label the image Q.

b Reflect Q in the x axis.
Label the image R.

c Reflect R in the y axis.
Label the new image S.

d Reflect S in the x axis.
What do you notice?

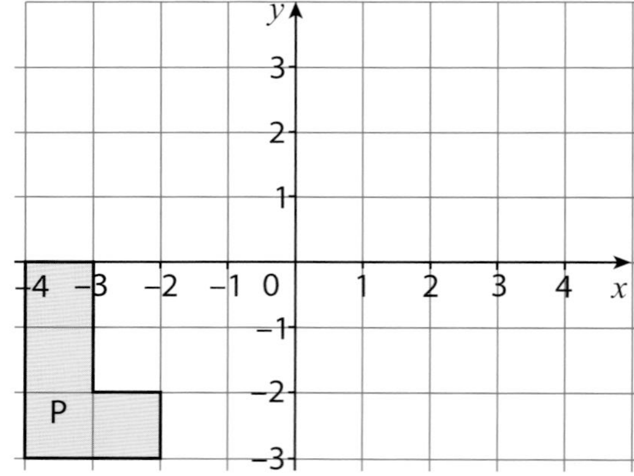

Reasoning

6 Elaine is designing a patchwork pattern.

a Copy the diagram.
Reflect the pattern in the y axis.

b Now reflect the total pattern in the x axis.

c How can you extend the pattern further?

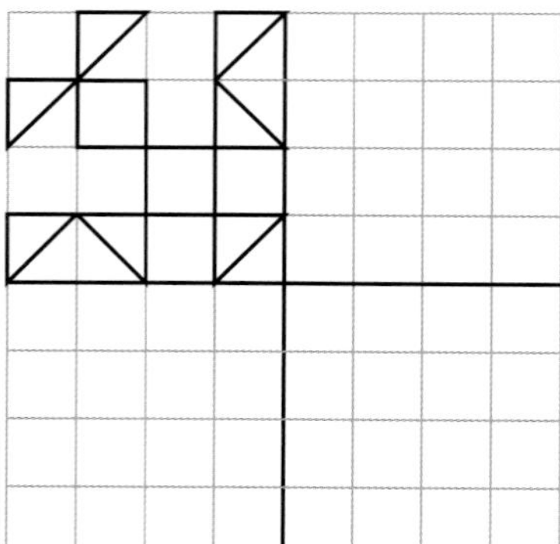

7 **a** Draw x and y axes from –8 to 8.
Plot these points and join them to form a triangle.
(–7, –2), (–2, –3), (–5, –6)
Label the triangle A.

b Reflect A in the x axis and label the image B.

c Reflect B in the y axis and label the image C.

d Reflect C in the x axis and label the new image D.

e Reflect D in the y axis.
What do you notice?

8 Reflect this shape in both lines of symmetry.
Then reflect one of the images in the other line of symmetry to complete the picture.

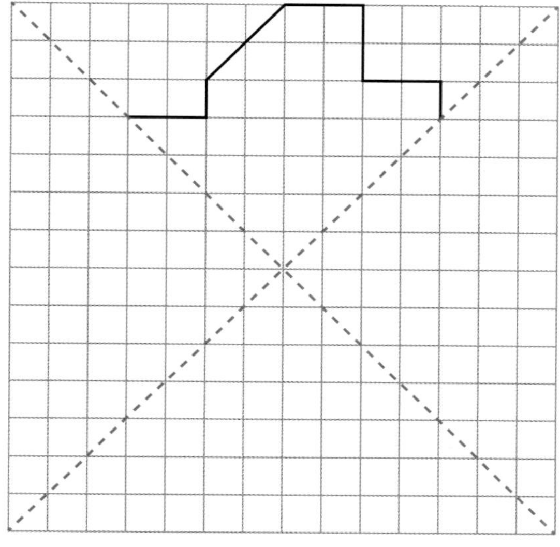

Skills practice B

Reasoning

1 **a** Copy this diagram.

b Reflect triangle V in the y axis and label its image W.

c Reflect W in the y axis.
What do you find?

d Is it always true that two reflections bring you back to your starting place?

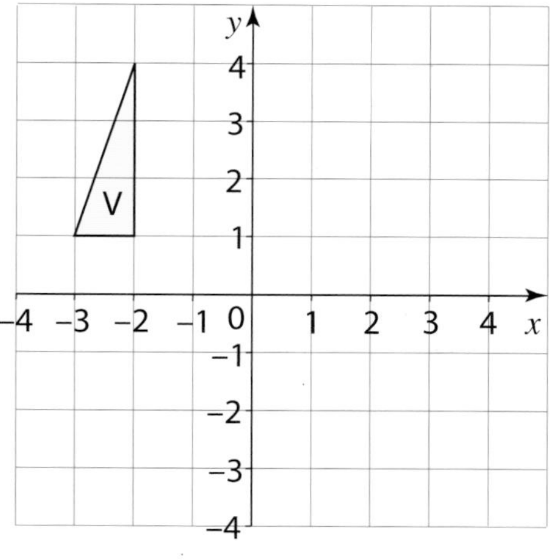

2 a Copy this diagram.

b Reflect the shape in the y axis.
What can you say about the image?

c Reflect the new image in the x axis.
Look carefully at the result.
What can you say about the image?

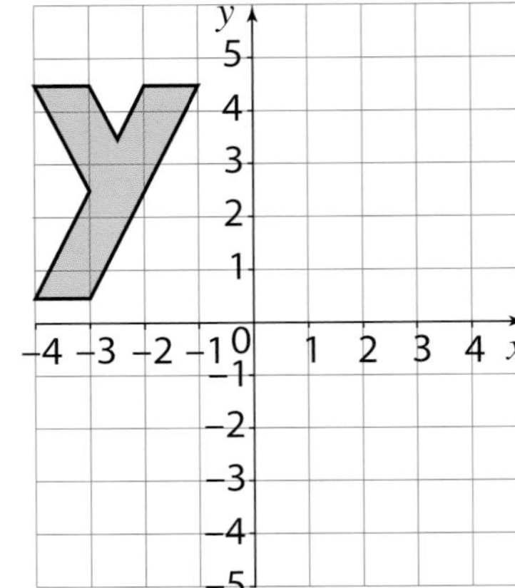

3 a Copy this diagram.

b Reflect the shape in the y axis.
What can you say about the image?

c Reflect the new image in the x axis.
Look carefully at the result.
Is it 'back to front'?

d Why is the letter A always the 'right way round' but not the letter Y?

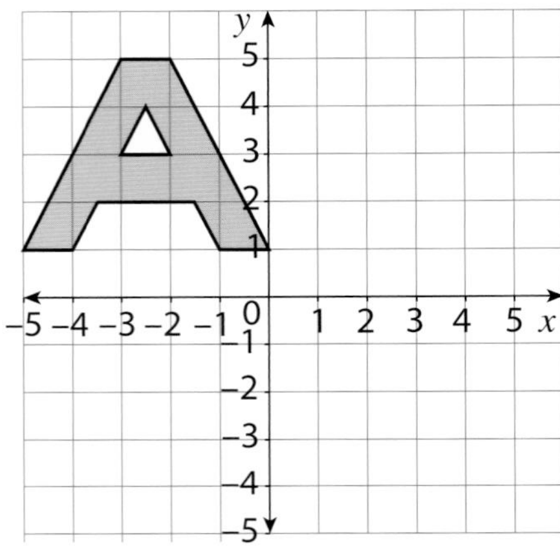

4 Look at this diagram.

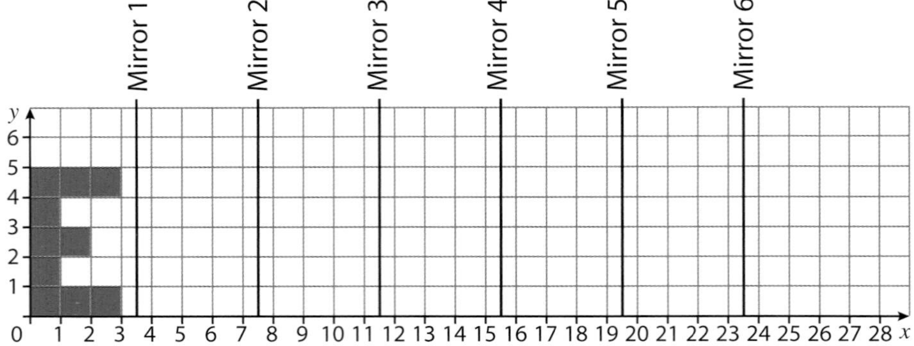

a The E shape has 12 vertices.
Write down the co-ordinates of the vertices.

b The equation of mirror line 1 is $x = 3\frac{1}{2}$.
What are the equations of the other five mirror lines?

c Copy the diagram.
Reflect the E shape in mirror 1.

d Reflect the image from part **c** in mirror 2.

e Reflect the image from part **d** in mirror 3.

f Continue the process with mirrors 4, 5 and 6.

g Describe what is happening.

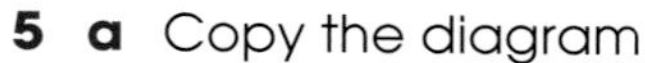

5 **a** Copy the diagram.

b Reflect the shape in the y axis.

c Now reflect both shapes in the x axis.

d Describe the shape you have drawn.

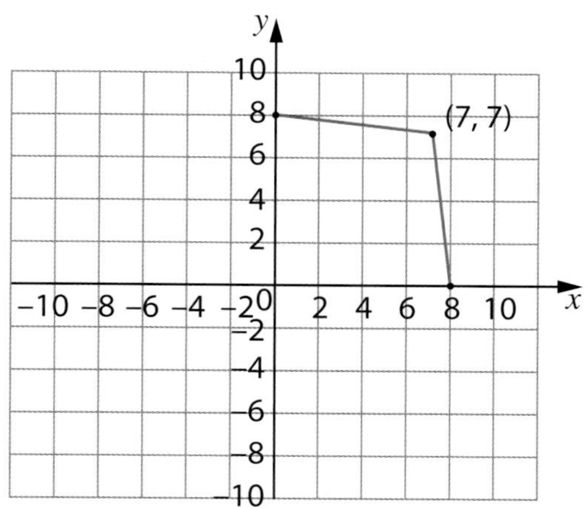

6 Katie is making a pattern by reflecting triangles in mirror lines.
Triangle A_0 is reflected in M_1 to form A_1, triangle A_1 is reflected in M_2 to form A_2, and so on.

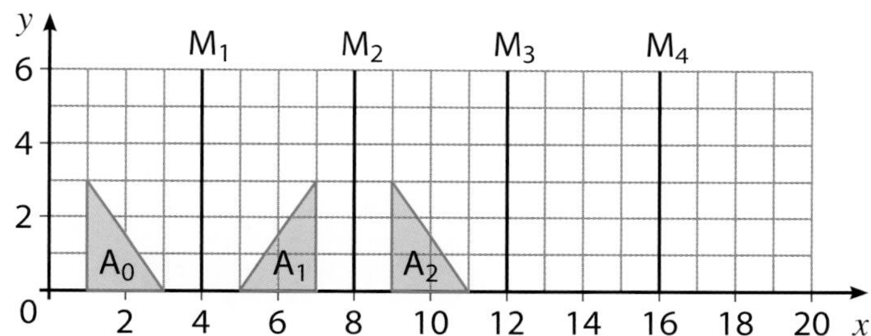

a Copy the diagram and draw triangles A_3 and A_4.

b Describe the transformation which maps A_0 to A_3.

c Write down the equations of these mirror lines

i M_1 **ii** M_2.

d Reflect triangle A_0 in $y = 3$ to form triangle B.

e Describe combinations of two transformations which map

i B to A_1 **ii** B to A_3.

7 Each of the diagrams shows a shape and its image after reflection.
Give the equation of each of the mirror lines.

a

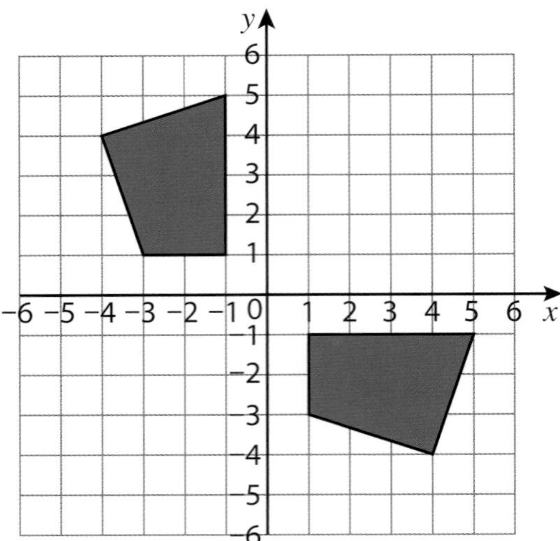

b

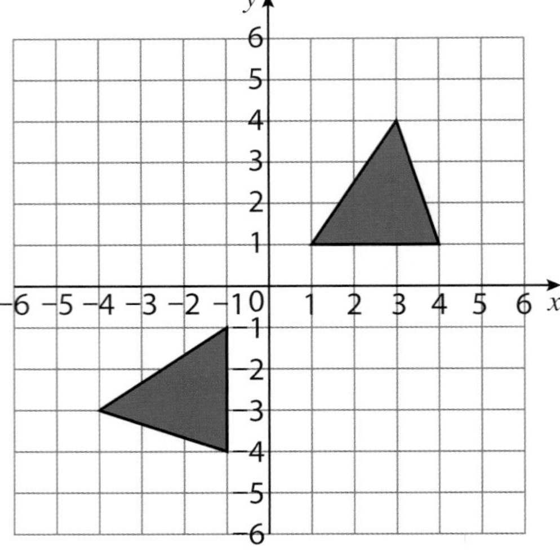

8 **a** Copy this word on to squared paper in pencil.

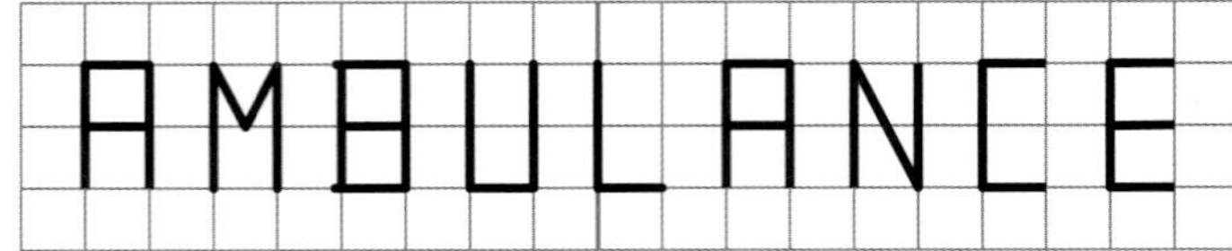

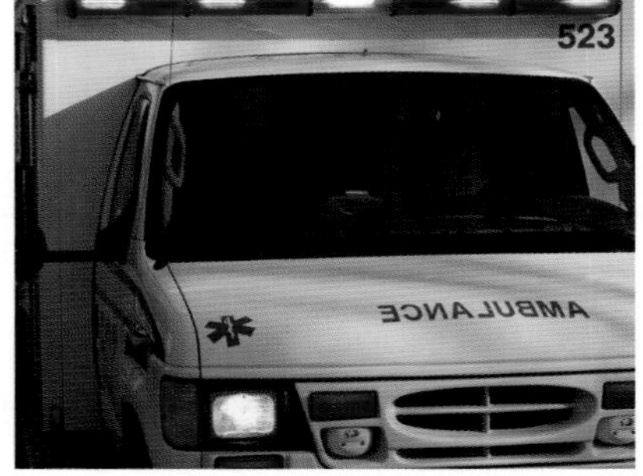

b Now draw its reflection in the red line in ink.

c Why do emergency vehicles have writing like this on the front?

Wider skills practice

1 James is designing a quilt cover.
He wants pictures of angels on it.
The diagram shows part of his design.

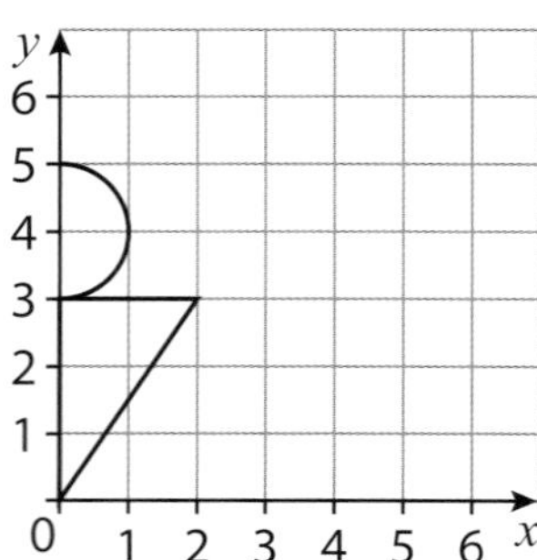

a Draw and label axes from −10 to 10.
Copy the diagram on to your grid.

b Reflect the triangle in

i the x axis

ii the y axis

iii the y axis and then the x axis.

c Reflect the semicircle in the y axis.
Now you have a complete angel.

d Translate the angel

i 7 right and 2 down

ii 7 left and 2 up.

2 **a** Copy the diagram and reflect the triangle in the y axis.

b What sort of triangle have you drawn?

c What happens if angle a is 60°?

d What happens if angle a is 45°?

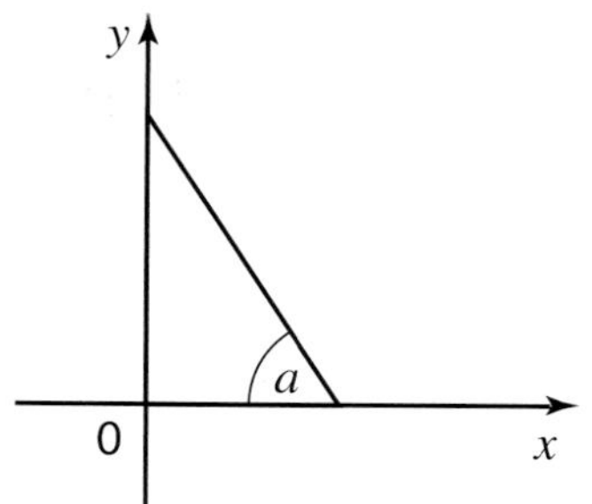

Reasoning

3 **a** Copy the diagram.

b Triangle B is the image of A when it is reflected in the x axis.

Draw triangle B.

c Triangle C is the image of B when it is translated by $\begin{pmatrix}3\\3\end{pmatrix}$.

Draw triangle C.

d Triangle D is the image of A when it is translated by $\begin{pmatrix}6\\0\end{pmatrix}$.

Draw triangle D.

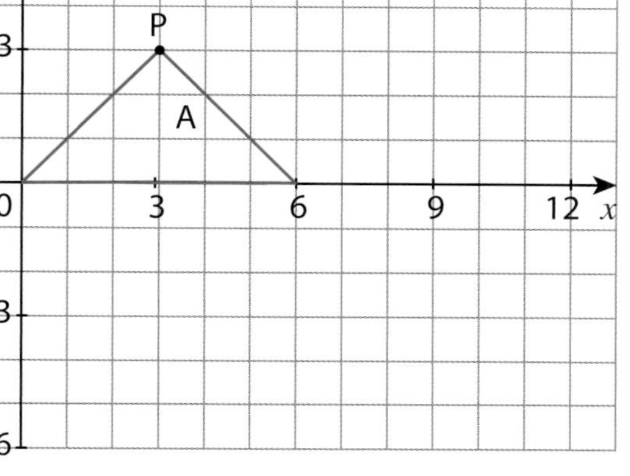

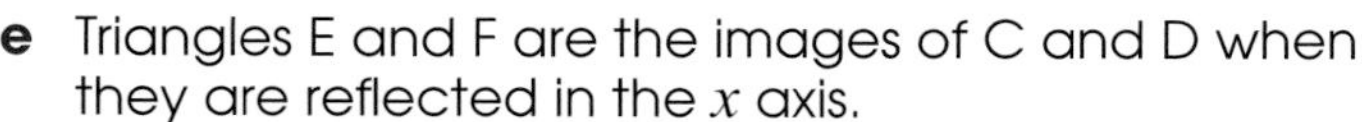

e Triangles E and F are the images of C and D when they are reflected in the x axis.

Draw triangles E and F.

f Describe the shape you have drawn.

Measure the sides and the angles.

Explain how you know it is not regular.

g The vertex P of triangle A is (3, 3).

Imagine this point is moved to a new position.

As a result the final shape is regular.

Use measurement and drawing to find the new co-ordinates of P.

Applying skills

Problem solving

1 **a** Draw and label axes from –10 to 10.

b Draw a shape on the grid which is not symmetrical.

c Reflect the shape in the x axis.

d Transform the image using a translation.

e Is it possible to move the final image back to the original shape using a single transformation?
If it is, can you describe the transformation which would do this?

2 Draw an L shape on the right side of mirror line m_1.
Reflect it in m_1.
Draw the image.
Reflect the image in m_2.
Reflect the new image in m_3.
Investigate what happens when other shapes are reflected successively in the parallel mirror lines m_1, m_2 and m_3.

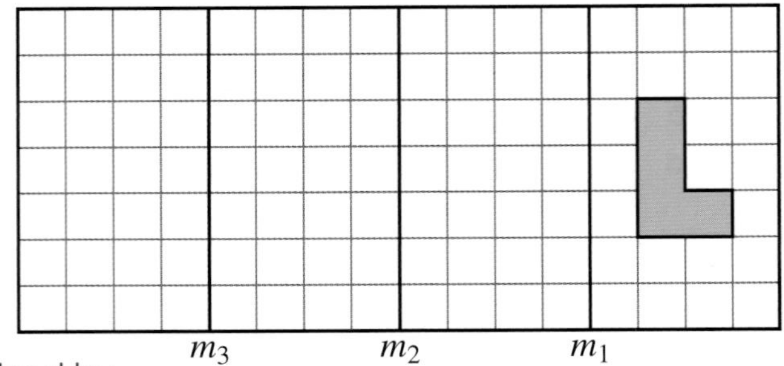

Can the resulting combined transformation be described by

a a translation followed by a reflection

b a reflection followed by a translation

c a single reflection?

Reviewing skills

1 **a** Draw and label axes from −8 to 8.

b Plot the following points and join them to form a triangle.

(1, 2), (3, 7), (6, 1)

Label the triangle A.

c Reflect triangle A in the x axis and label the image B.

d Write down the co-ordinates of the vertices of triangle B.

e Translate triangle B by $\begin{pmatrix} -5 \\ -1 \end{pmatrix}$.
Label the image C.

f Reflect triangle C in the x axis.
Label the image D.

g Describe the transformation that maps D on to A.

2 Sam is designing a window.
This is what he has in mind.
He draws shape A and then uses it to make the rest of the pattern using reflections.

a In what line does he reflect A to get B?

b In what line does he reflect B to get C?

c In what line does he reflect C to get D?

d What single line of reflection could he now use to complete his design?

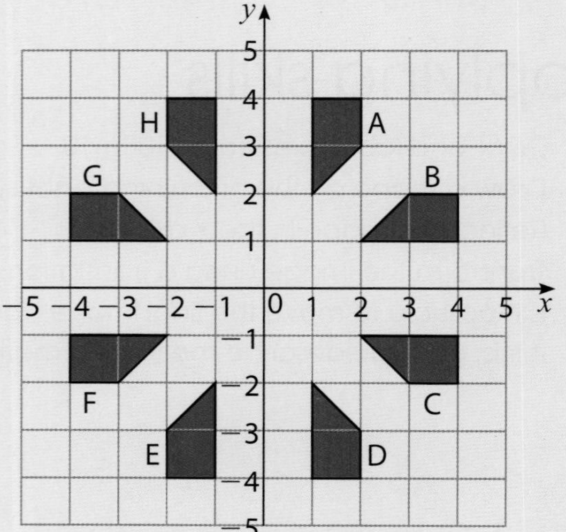

Unit 5 • Rotation • Band e

Building skills

Example outside the Maths classroom

Art

Toolbox

Another transformation is a **rotation**.

Rotation is a movement made by **turning** an object.

The flag here has been rotated four times, each time by a quarter turn.

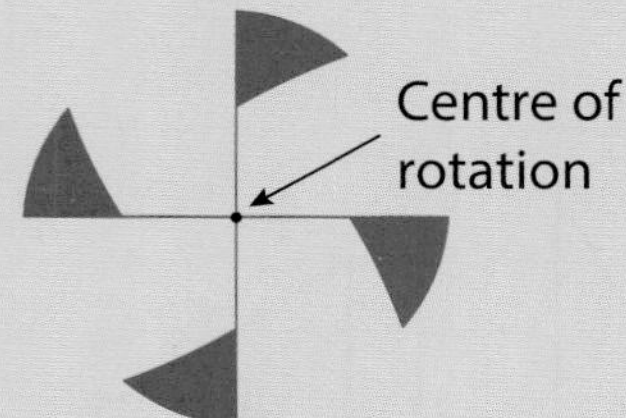

A full turn is 360°, so

- a quarter turn is 90° (a right angle)
- a half turn is 180°
- a three-quarters turn is 270°.

The flag has been rotated by holding the end of the stick still. This is the **centre of rotation**.

In this diagram the red triangle is the image of the green triangle when it is rotated through 90° clockwise about the origin.

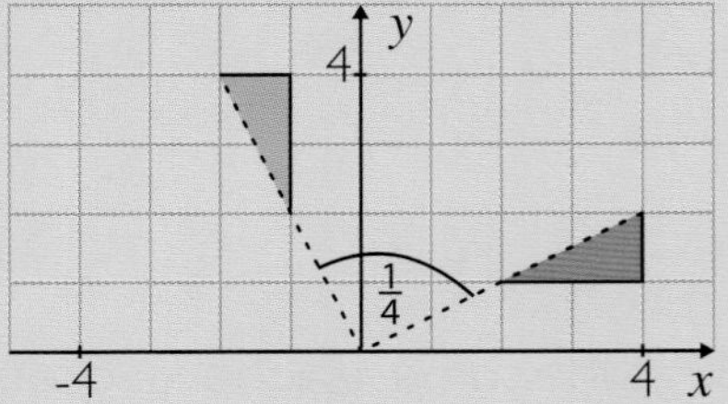

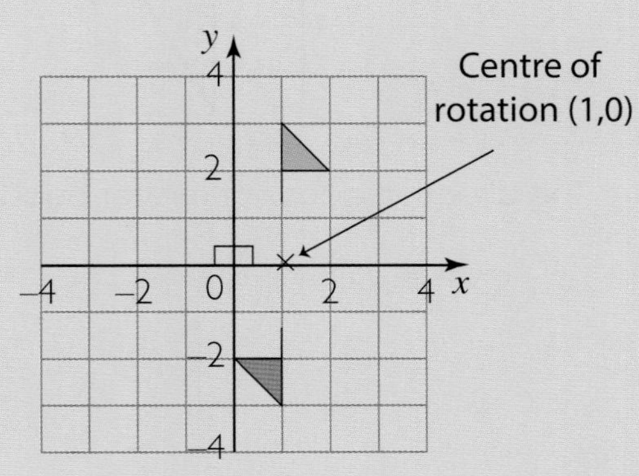

In this diagram the red flag is the image of the green one when it is rotated through 180° about the point (1, 0).

A rotation of 180° can be **clockwise** or **anticlockwise**.

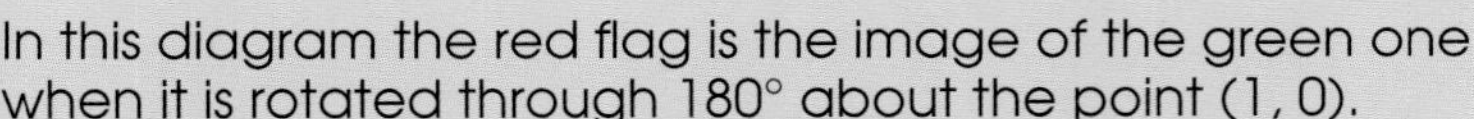

Objects and their images as a result of a rotation are always **congruent**.

Example – Rotating a shape about the origin

a Rotate shape A by 90° in a clockwise direction, centre (0, 0). Label the resulting image B.

b Rotate shape A by 180°, centre the origin.
Label the resulting image C.
Why does the direction not matter for this rotation?

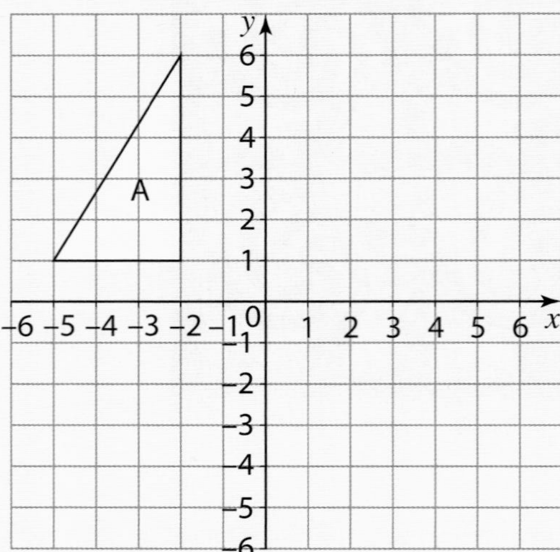

Solution

a,b

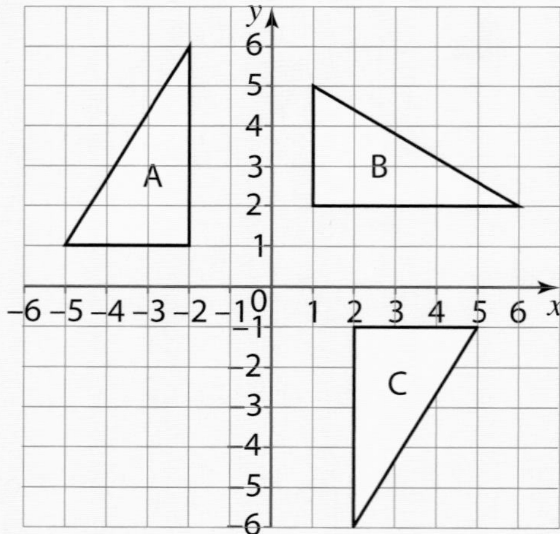

The direction does not matter because a turn of 180° is the same whichever direction you turn.

Example – Describing a rotation

This clock has three hands.
What angles do the hands turn through in

a 1 minute
b 30 seconds
c 15 seconds?

Solution

		Second hand	Minute hand	Hour hand
a	**1 minute**	360°	360° ÷ 60 = 6°	6° ÷ 60 = 0.1°
b	**30 seconds**	180°	3°	0.05°
c	**15 seconds**	90°	1.5°	0.025°

Remember there are 60 seconds in 1 minute and 60 minutes in 1 hour.

Example – Rotating a point

The point A is (2, 6).

It is rotated through 90° clockwise about the point (4, 2).

Find the co-ordinates of its image A′.

Solution

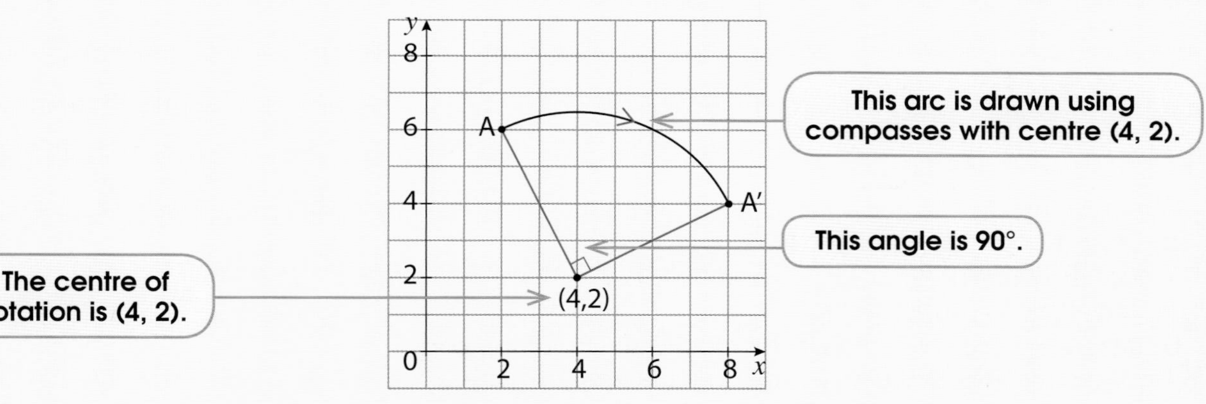
y
8
6
4
2
0
2
4
6
8
x
A
A′
(4,2)
This arc is drawn using compasses with centre (4, 2).
This angle is 90°.
The centre of rotation is (4, 2).

Its image A′ is (8, 4).

Remember:

- You must always give the centre of rotation.
- You must say whether a rotation is clockwise or anticlockwise.
- Use tracing paper to help you with rotations.

Skills practice A

1 Match these arrows with the descriptions below.

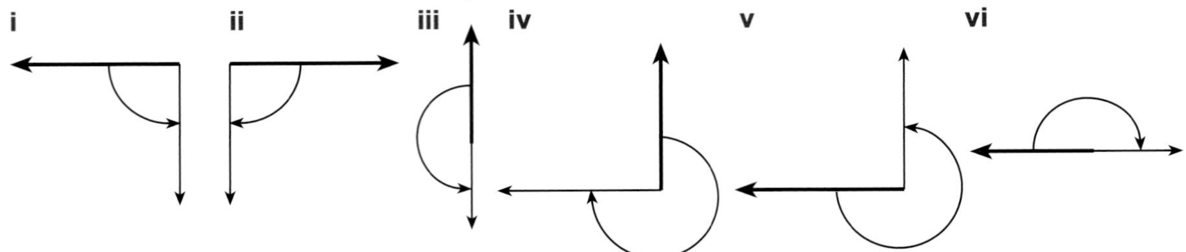

a A quarter turn clockwise
b A three-quarter turn anticlockwise
c A half turn anticlockwise
d A quarter turn anticlockwise
e A half turn clockwise
f A three-quarter turn clockwise.

2 Copy this diagram.

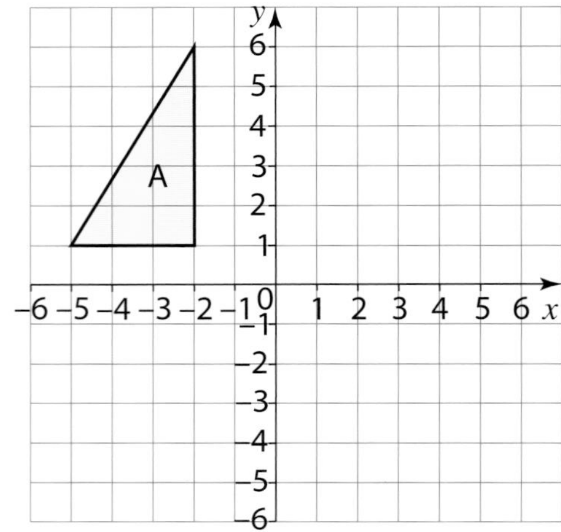

a The triangle A is rotated through a quarter turn in a clockwise direction.
Draw its image and label it B.
b The triangle A is rotated through a quarter turn in an anticlockwise direction.
Draw its image and label it C.
c Describe the rotation that maps triangle B on to triangle C.

3 **a** Draw x and y axes from –6 to 6.
Plot the points A(2, 1), B(2, 5) and C(4, 1).
Join them to form a triangle.
b Rotate the triangle through 90° clockwise about the origin.
Label the image A′B′C′ and write down its co-ordinates.
c Rotate A′B′C′ by a further 90° clockwise about the origin.
Label the image A″B″C″ and write down its co-ordinates.

4 **a** The vertices of triangle A are (0, 0), (1, 0) and (1, 4).
Draw the triangle on a grid.
b Draw the image of A when it is rotated through 90° in a clockwise direction, centre the origin.
c Draw the image of A when it is rotated through 180°, centre the origin.
d Draw the image of A when it is rotated through 270° in a clockwise direction, centre the origin.
e What happens when you rotate A through 360°, centre the origin?

Reasoning

5 **a** Draw x and y axes from –5 to 5.

Plot these co-ordinates and join them to make a trapezium.

(1, 2), (1, 5), (3, 5), (5, 2)

Label it A.

b Rotate trapezium A by a quarter turn clockwise, centre the origin.

Label the image B.

What are B's co-ordinates?

c Rotate trapezium A by a half turn, centre the origin.

Label the image C.

What are C's co-ordinates?

d What single transformation would map B on to C?

6 Describe fully these transformations.

State whether each of the transformations is a rotation, translation or reflection.

In each case give full details of the transformations.

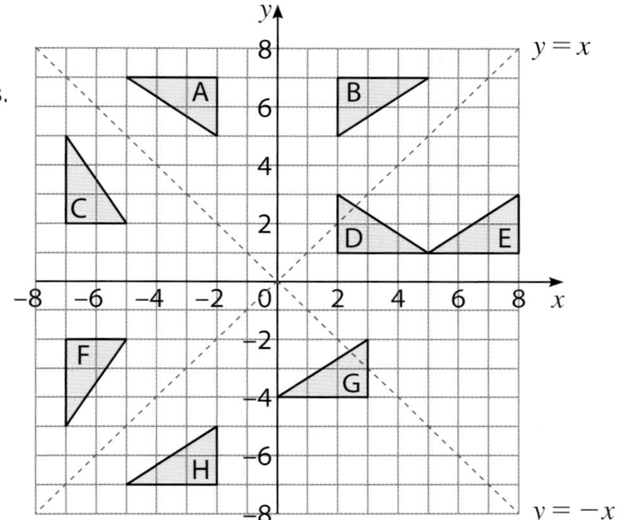

a A → B
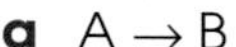

b A → H

c G → E

d C → B

e G → H

f A → F

g F → H

h B → H

i B → D

j D → E

Reasoning

7 **a** Describe each of these rotations.

i A to B

ii A to C

iii B to A.

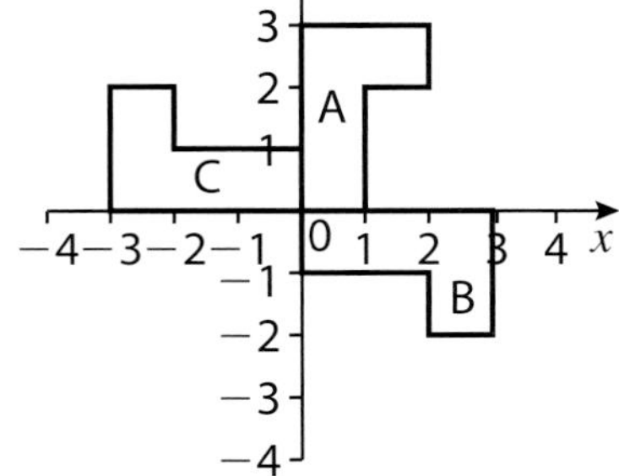

b Draw and label x and y axes from –5 to 5.

Copy shape A on to your diagram.

Now rotate A through 180° about the origin.

Label the image D.

c **i** Is shape D congruent to shape A?

ii Explain your answer.

8 Draw this arrow after each of these transformations.

Make it four squares long each time.

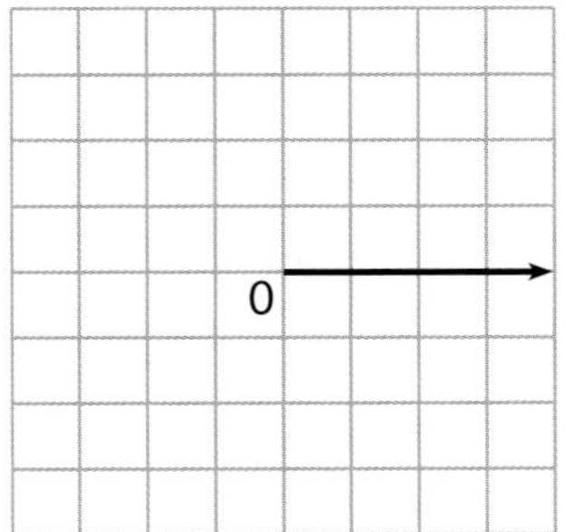

a A rotation 90° clockwise, centre the origin.

b A rotation 270° anticlockwise, centre the origin.

c A rotation 180° clockwise, centre the origin.

d A rotation 180° anticlockwise, centre the origin.

Skills practice B

Reasoning

1 Look at this diagram.

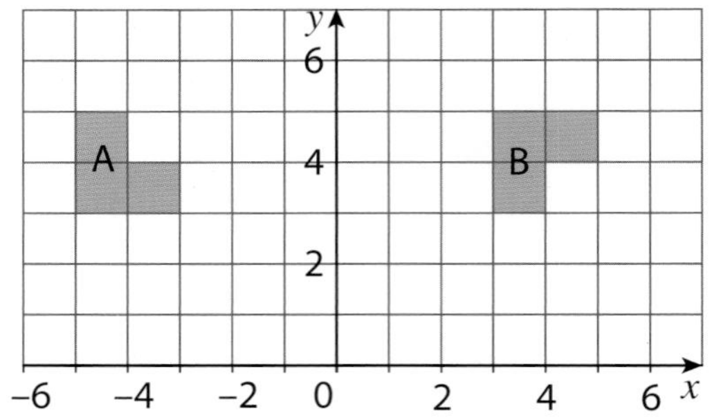

a Describe the transformation from shape A to shape B.

b What are the co-ordinates of shape B?

c Emily says that shape B is congruent to shape A. Gemma says that Emily is wrong because the shape is upside down. Say who is right and explain why.

2 Draw a pair of axes from –6 to 6.

a Plot these points and join them to form a letter L.
(0, 0), (–5, 0), (–5, 6), (–3, 6), (–3, 2), (0, 2)
Label this object A.

b Rotate A by a right angle clockwise, centre the origin. Label the image B.

c Rotate B by a right angle clockwise, centre the origin. Label the image C.

d What single transformation maps A to C?

3 For this question you will need a protractor and a pair of compasses.
The diagram shows a plan of a children's roundabout.
O and T show the starting positions of Owain and Tom.

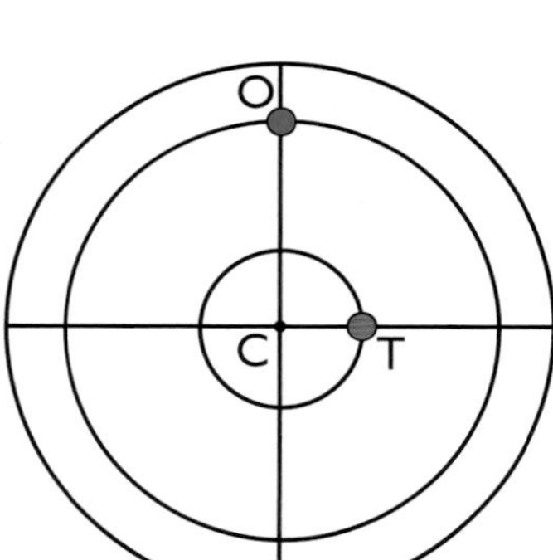

a Make a copy of the diagram.
On your diagram show the positions of Owain and Tom after each of these transformations from their starting positions.

i A clockwise rotation of 90° about C. Label them O_1 and T_1.

ii A clockwise rotation of 120° about C. Label them O_2 and T_2.

iii A clockwise rotation of 290° about C. Label them O_3 and T_3.

b Describe *two* transformations for each of these mappings.

i O_1 to O_2

ii T_1 to T_3

iii O_2 to O

iv O_3 to O_1

4 You can plot the movement of the stars on a co-ordinate grid.

In the diagram you can see how the Plough rotates in six hours.

The centre of rotation is the origin.

Each star rotates a quarter turn (90°) in an anticlockwise direction.

a Copy and complete this table to show how the co-ordinates of the Plough have rotated.

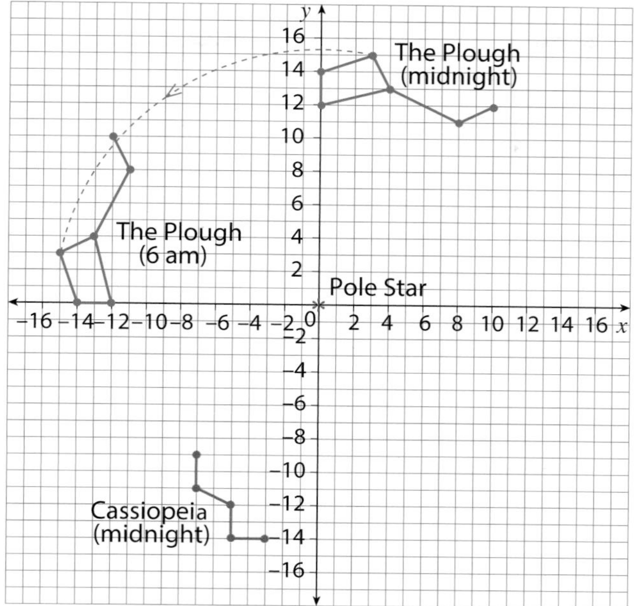

The Plough	
midnight →	**6 a.m.**
(10, 12)	(−12, 10)
(8, 11)	
(4, 13)	
(3, 15)	(−15,3)
(0, 12)	
(0, 14)	(−14, 0)

b Make a similar table for Cassiopeia and draw its position at 6 a.m.

5 Draw x and y axes from −5 to 5.

Plot these points and join them to form an L shape.

(1, 1), (3, 1), (3, 2), (2, 2), (2, 4), (1, 4)

Label it A.

a Rotate shape A 90° clockwise about the origin. Label the image B.

b Rotate shape B 90° clockwise about the point (1, −1). Label the image C.

c Rotate shape C 90° anticlockwise about the point (−1, −1). Label the image D.

d Rotate shape D 180° about the point (−1, 1). Label the image E.

e Describe fully the rotation which maps shape E back on to shape A.

Reasoning

6 Draw and label x and y axes from −8 to 8.

Plot and label these points and join them to form a quadrilateral ABCD.

A(2, 1), B(4, 0), C(7, 5), D(4, −2)

a Rotate ABCD clockwise through 90° about the origin. Label the image A′B′C′D′.

b Rotate ABCD through 180° about the origin. Label the image A″B″C″D″.

c Compare the object co-ordinates and corresponding image co-ordinates. Explain how the rotations about the origin change the co-ordinates.

7 **a** Draw and label axes from –8 to 8.

b Plot these points and join them to form a triangle.

(1, 2), (3, 7), (6, 1)

Label the triangle A.

c Reflect triangle A in the x axis and label the image B.

d Rotate triangle B by 90° clockwise about the origin. Label the image C.

e Rotate triangle C by 270° anticlockwise about the origin. Label this image D.

f What single transformation would map triangle B straight to triangle D?

g Reflect triangle D in the y axis.

Where does triangle D map to?

8 **a** Do rotations always have to take place about the origin?

b You need three pieces of information to specify a rotation. What are they?

c Look at this diagram.

Describe these rotations.

i A to B

ii A to C

iii B to D

iv C to E

v E to F

vi D to G

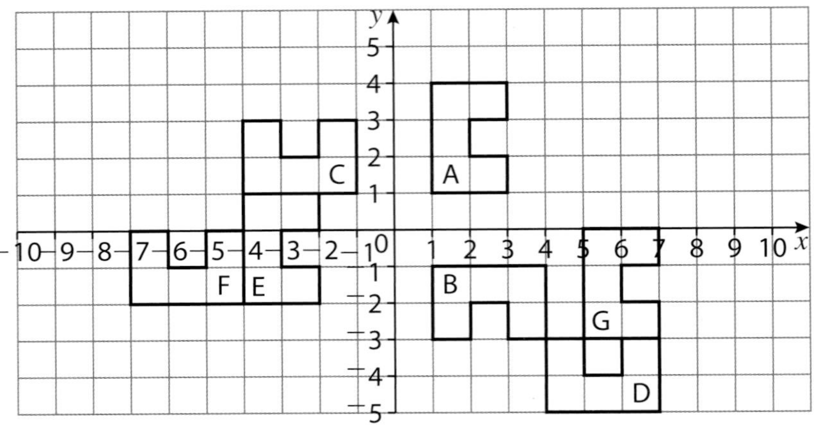

Wider skills practice

1 Copy this grid.

Fill in the grid with the answers to the clues.

When you have finished, the green squares will give the name of a regular shape

Clues

1 This shape has six sides.

2 A square has rotation symmetry of ______ 4.

3 A square is a four-sided ______.

4 Use a mirror for this.

5 This moves a shape to a new position.

6 You measure these with a protractor.

7 A turn.

8 To draw a shape accurately.

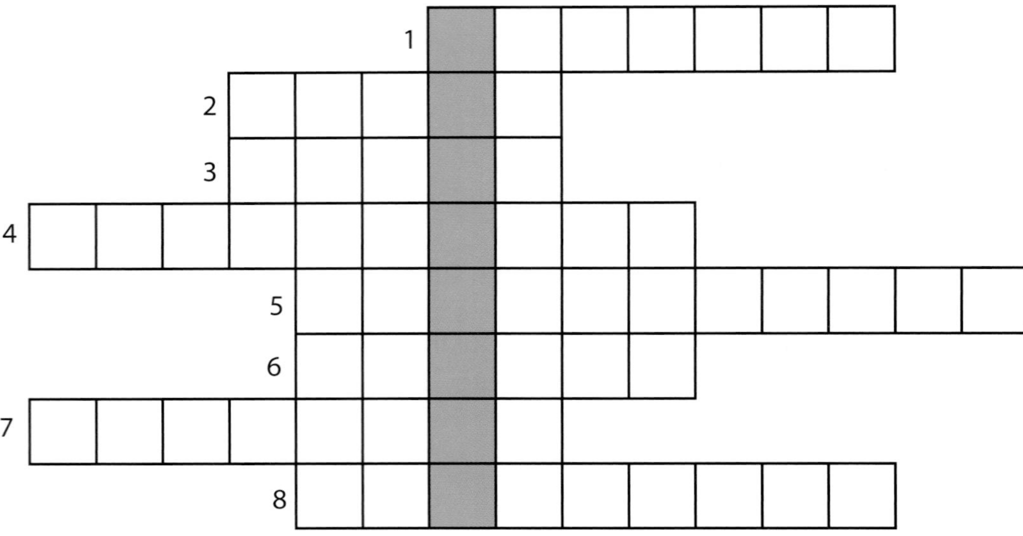

Reasoning

2 All the shapes on this grid are congruent.

a Describe *fully* each of these transformations.

i C to D

ii G to C

iii D to G

iv E to G

b Think of a different transformation which maps C to D besides your answer to part **i** above.

Why can both transformations work?

c Find two shapes where the transformation could be either a reflection or a rotation.

Explain your answer fully.

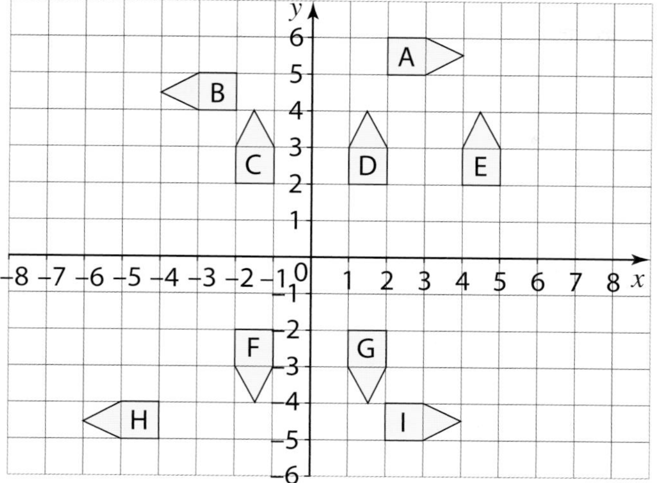

3 Describe fully these transformations.

a K to L
b A to B
c K to B
d I to B
e K to I
f C to M
g F to E
h F to I
i A to L
j H to G
k C to D
l J to I
m F to J

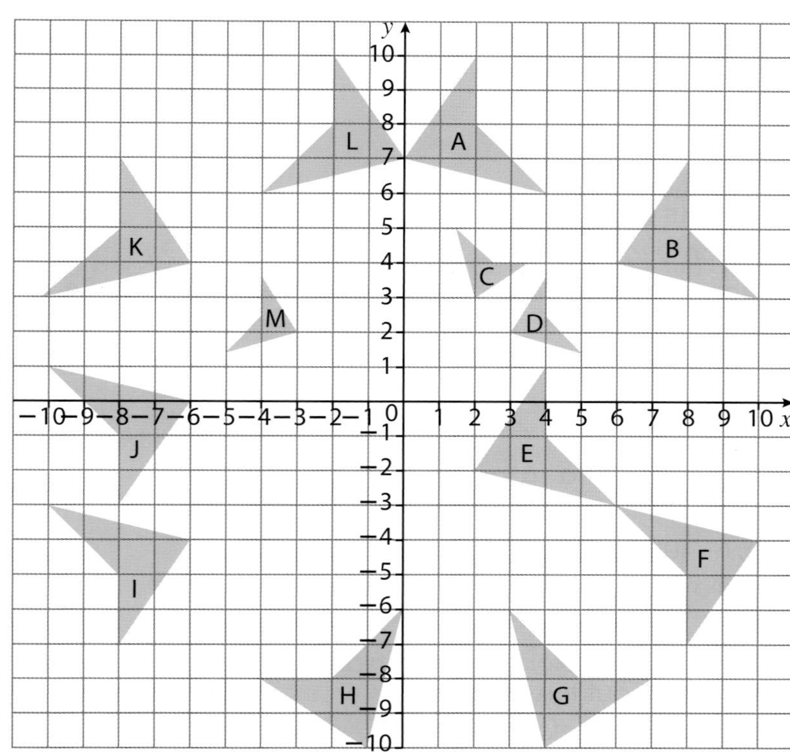

Applying skills

Problem solving

1 A tetromino is made by joining four squares edge to edge.
This is a tetromino.

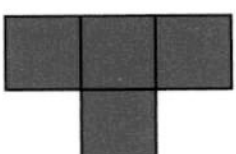

a Draw all of the different tetrominoes.

Four T-shaped tetrominoes make a square.

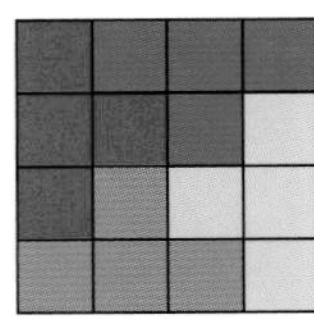

b Can you make squares from the other tetrominoes?

2 Look at the picture of a spanner turning a bolt.
The bolt head is a regular hexagon.
Point C is the centre of rotation.

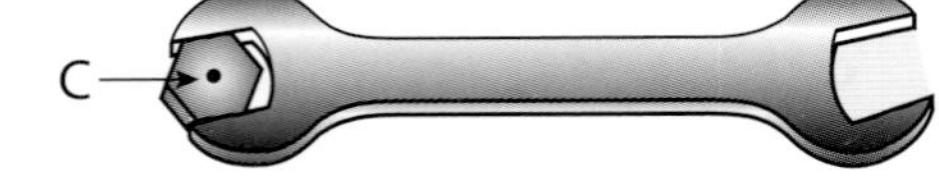

a The spanner is rotated 60° clockwise about C.
What can you say about the *appearance* of the bolt before and after the rotation?

b Write down three other clockwise rotations which do not change the appearance of the bolt.

c Write down two anticlockwise rotations which do not change the appearance of the bolt.

d Give a *full* description of the symmetry of the bolt.

Problem solving

3 A car drives round a race track.
A to B is a translation of 1 km due East.

a Describe the transformation C to D.

B to C is a rotation of 180° about Q.

b Describe the transformation D to A.

c Use a combination of transformations to describe the movement of the car from

i A to C

ii B to D.

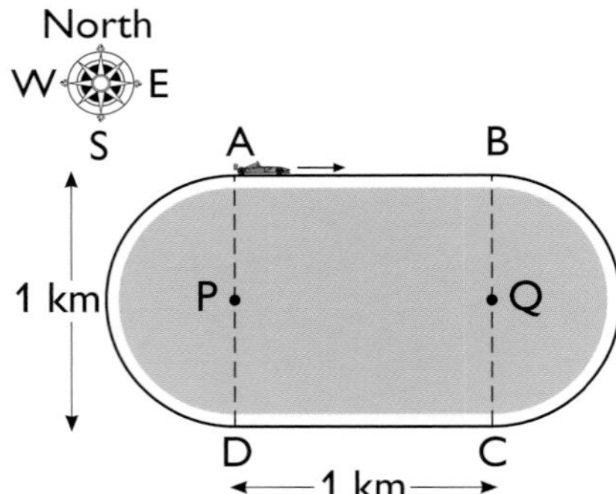

Problem solving

4 Draw and label a set of axes from –16 to 16 for both x and y axes.
Plot these co-ordinates and join them in order.

(–16, 4), (–13, 4), (–10, 3), (–7, 0), (–10, 0), (–13, –2)

You have drawn the constellation called *Camelopardalis* at 6 p.m.
Plot and label the Pole Star at the origin.

a Draw and label the position of *Camelopardalis* at midnight when it has rotated a quarter turn in an anticlockwise direction about the pole star.

b Draw and label the position of *Camelopardalis* after another six hours.
Describe fully the transformation which takes *Camelopardalis* from its 6 p.m. position to its 6 a.m. position.
What do you notice about the co-ordinates of *Camelopardalis* at 6 p.m. and 6 a.m.?

Reviewing skills

1 Copy this flag.
Rotate the flag about point A.

a a quarter turn anticlockwise

b a half turn clockwise

c a three-quarter turn clockwise.

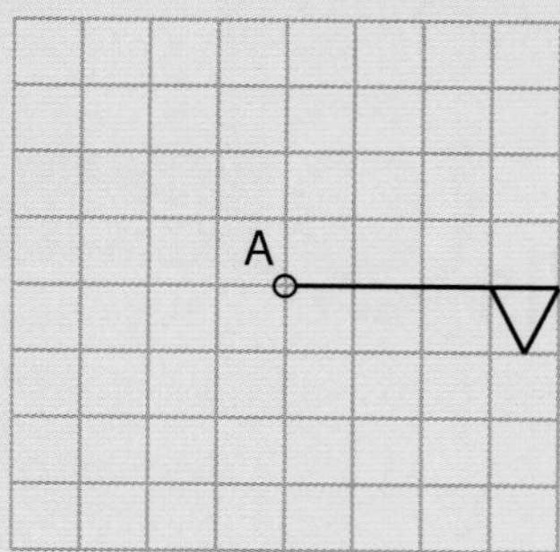

2 The triangle A in the diagram is rotated through 90° anticlockwise with centre of rotation (4, 5).

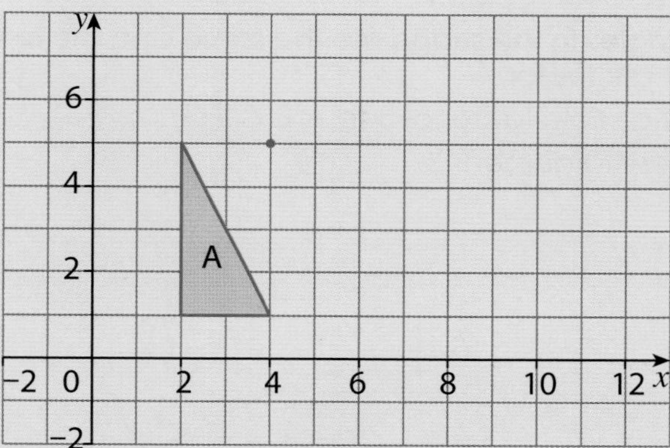

Its image is B.

The triangle B is rotated through 90° clockwise with centre of rotation (7, 2).

Its image is C.

Find the single transformation that maps A onto C.

Building skills

Example outside the Maths classroom

TV screens

Toolbox

Enlargement is a transformation that changes the size of an object.

One shape is an enlargement of another if **all the angles in the shape are the same** and the **lengths of the sides** have all been increased by the **same scale factor**.

In this diagram the pentagon ABCDE is enlarged by a scale factor of 3 to A′B′C′D′E′.

The position of the image depends on the centre of enlargement X.

Because the scale factor of enlargement is 3

- side A′B′ is three times as long as AB
- side E′D′ is three times as long as ED, etc.
- the distance XB′ is three times the distance XB
- the distance XD′ is three times the distance XD, etc.

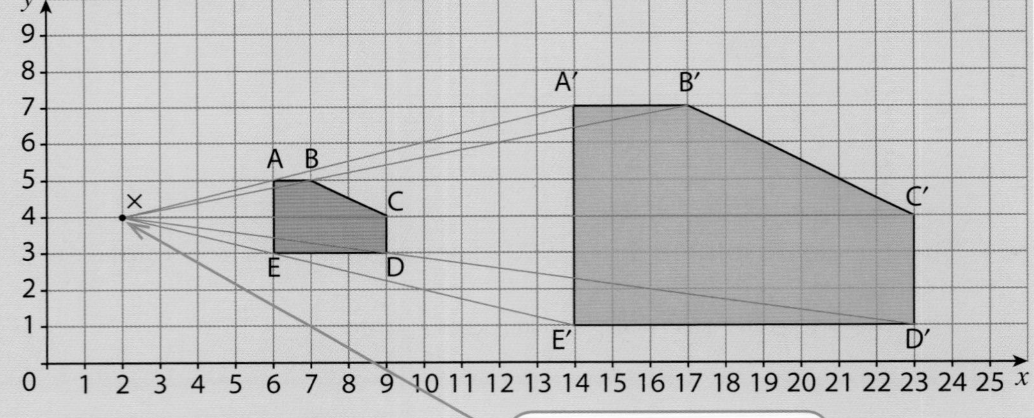

The term enlargement is also used in situations where the shape is made smaller.

In these cases the scale factor is a fraction, like $\frac{1}{2}$ or $\frac{1}{3}$.

In the diagram the scale factor for the enlargement of A′B′C′D′E′ to ABCDE is $\frac{1}{3}$.

Example – Finding the scale factor of an enlargement

The shape on the right shows an enlargement from centre P.

a What is the scale factor of the enlargement?

b What happens if you change the centre of enlargement?

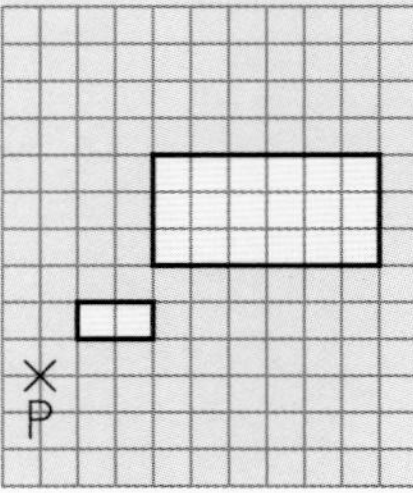

Solution

a The scale factor is 3 because all the sides of the second shape are three times as long as those of the first shape.

b The image would be in a different place.

Example – Enlarging a shape

Plot the points A(2, 4), B(4, 4), C(4, 6) and D(2, 6).

Join the points to form a square.

Enlarge ABCD using (0, 5) as the centre of enlargement and a scale factor of 2. Label the image A′B′C′D′.

a What are the co-ordinates of the points A′, B′, C′ and D′?

b What shape is A′B′C′D′?

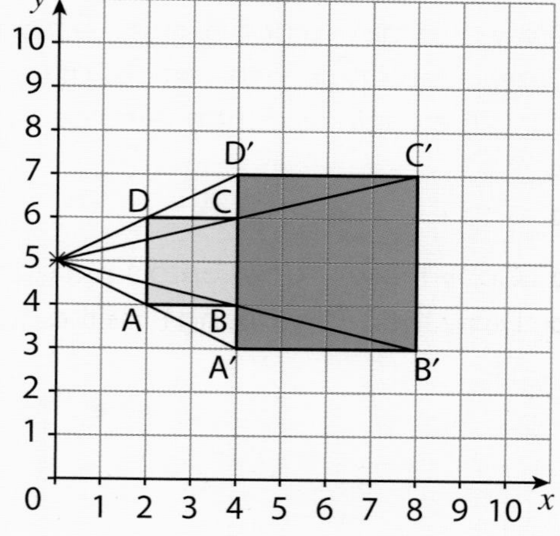

Solution

a A′(4, 3), B′(8, 3), C′(8, 7), D′(4, 7)

b A square

Example – Describing an enlargement

Emma draws the quadrilateral ABCD and then she transforms it as shown to A′B′C′D′.

a Describe fully the transformation from ABCD to A′B′C′D′.

b What can you say about the distance of the object and the image from the centre of enlargement?

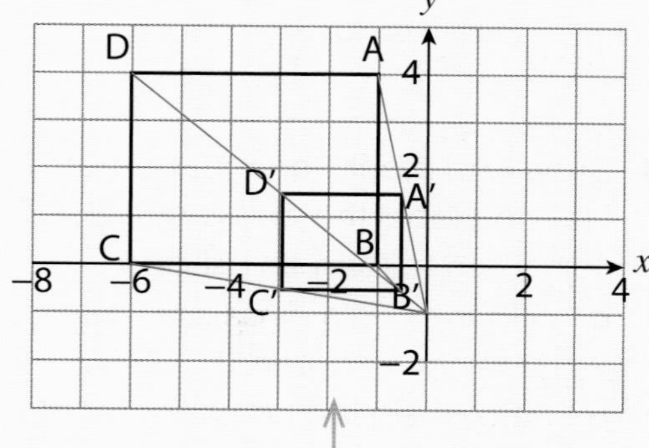

Solution

a The enlargement scale factor is $\frac{1}{2}$.

The centre of enlargement is (0, –1).

b The image is half the distance that the object is from the centre.

The lines joining the image and the object all meet at the centre of enlargement.

Remember:

- To draw an enlargement you need to know the scale factor.
- To determine where the image is drawn you need to know the centre of enlargement.
- When a shape is enlarged all the angles remain the same.
- If an enlargement has a fractional scale factor it is a reduction in size.

Skills practice A

Reasoning

1 A shape is enlarged. The scale factor is 2.
What happens to
a the lengths
b the angles?

2 Draw and label a pair of axes, with x from –7 to 10 and y from –4 to 10.
Draw a rectangle with vertices at (–2, 3), (3, 3), (3, –1), (–2, –1).
Using the origin as centre, enlarge the rectangle with a scale factor of
a 2
b 3.

3 **a** Copy these shapes onto squared paper.
b Draw the image, using P as the centre of the enlargement.

i
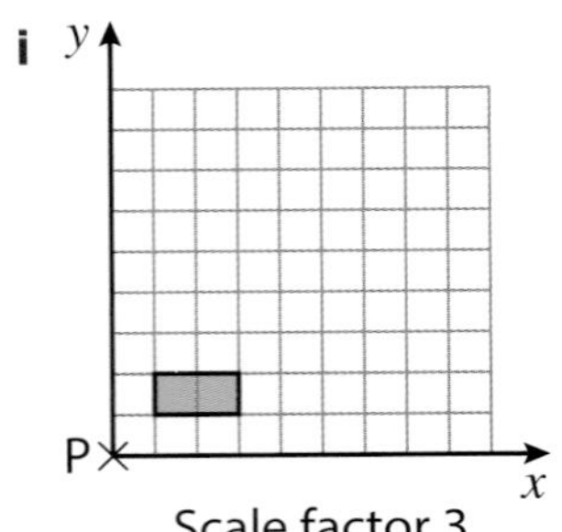

Scale factor 3

ii
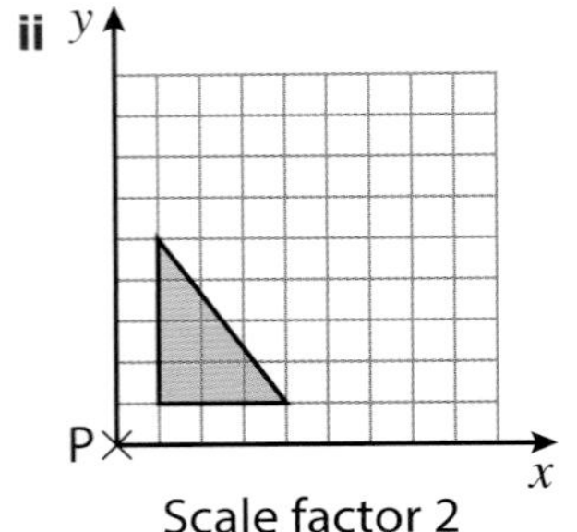

Scale factor 2

iii
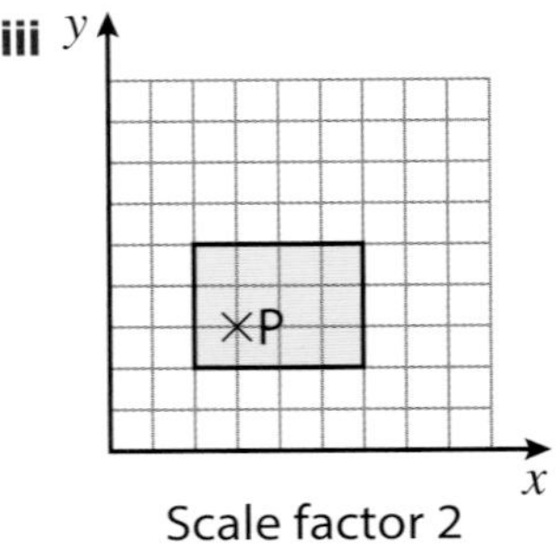

Scale factor 2

4 Draw and label x and y axes from 0 to 15.
Plot and label the points A(3, 2), B(3, 7), C(4, 7), D(4, 3), E(7, 3) and F(7, 2).
Join them to form an L shape.
a With the origin as centre, enlarge object ABCDEF by a scale factor of 2.
Label the image A′B′C′D′E′F′.
b Write down the co-ordinates of A′, B′, C′, D′, E′ and F′.
Compare them with the object co-ordinates.
What do you notice?

5 Copy each of these shapes onto a grid and enlarge it by a scale factor of 3.

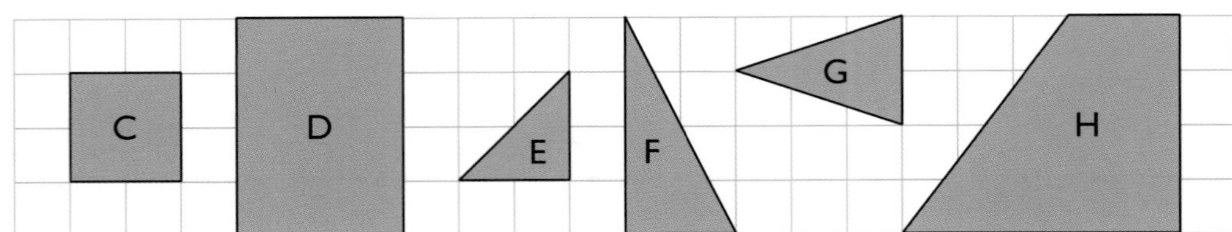

Reasoning

6 Draw and label x and y axes from 0 to 12.

a Plot these points and join them to form triangle A.
(3, 7), (5, 7), (5, 10)

b Plot these points and join them to form triangle B.
(5, 1), (11, 1), (11, 10)

c Explain why B is an enlargement of A.

d Write down the scale factor of enlargement.

e Find the co-ordinates of the centre of enlargement.

7 Copy this diagram.

a Enlarge shape P with scale factor 2 and centre of enlargement (0, 2). Label the image Q.

b Enlarge Q with scale factor $\frac{1}{2}$ and centre of enlargement (8, 2). Label the image R.

c Describe the transformation that maps P to R.

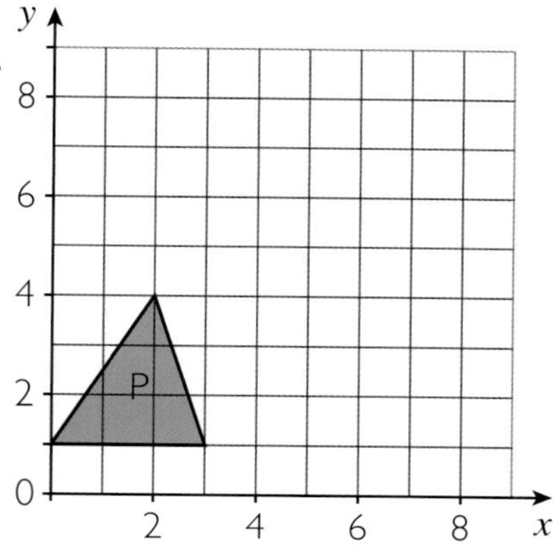

8 Copy each of these diagrams.
Enlarge each shape by a scale factor of $\frac{1}{2}$ using the origin as the centre of enlargement.

a

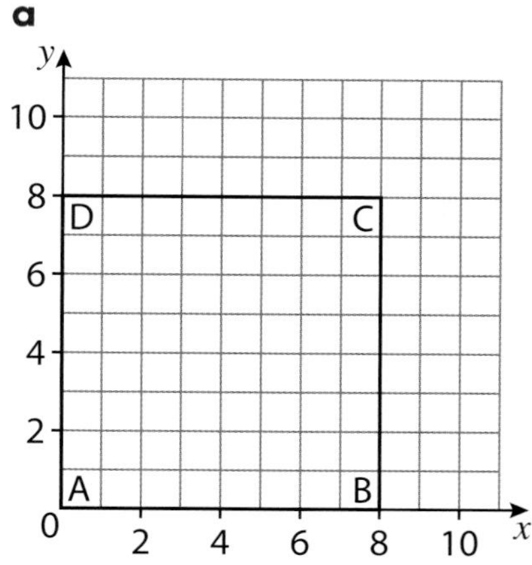

b

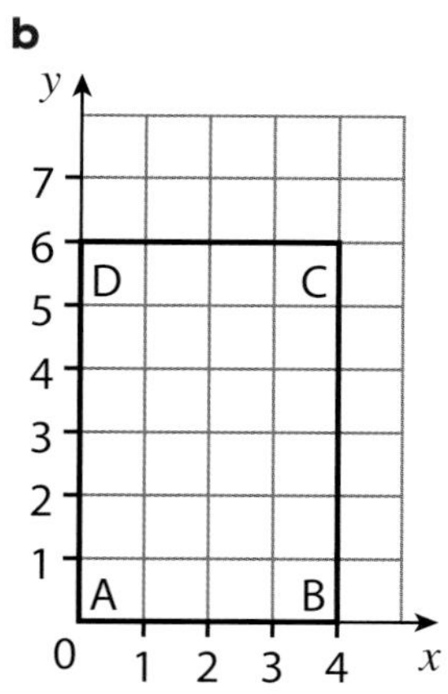

c

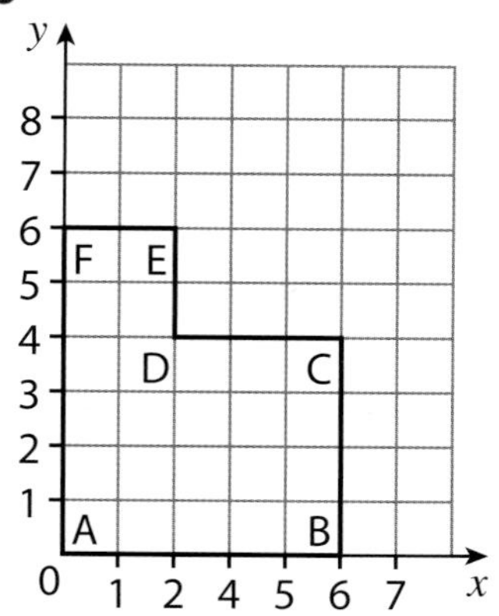

Skills practice B

1 Write down the scale factor of enlargement of each of these signs.

2 The co-ordinates of a polygon are P(2, 5), Q(3, 9), R(8, 7) and S(9, 2).
After an enlargement with the origin as centre, P is mapped to P′(8, 20).

a What is the scale factor?

b Write down the co-ordinates of Q′, R′ and S′.

With the origin as centre, PQRS is enlarged to form image P″Q″R″S″.

c Write down the co-ordinates of P″, Q″, R″ and S″ for each of these scale factors.

i 5

ii 10

iii 20

iv 100

Reasoning

3 Draw and label x and y axes from 0 to 15.

a Plot these points and join them to form quadrilateral Q.
(6, 8), (6, 11), (8, 12), (9, 11)

b Plot these points and join them to form the quadrilateral Q′.
(9, 3), (9, 9), (13, 11), (15, 9)

c Explain why Q′ is an enlargement of Q.

d Write down the co-ordinates of the centre of enlargement.

e Work out the scale factor of enlargement.

Reasoning

4 Look at these two model cars.
The angles marked in yellow are used in part **c**.

a Copy and complete this table.

Measurement	Red car	Blue car
length of car	3 cm	
height of car		
diameter of wheel		
length of aerial		

b Compare the measurements of the two cars.

c Copy and complete this table.

Measurement of angle between	Red car	Blue car
aerial and roof	30°	
bonnet and front windscreen		
rear window and boot		
bottom and rear wheel arch		

d Compare the angles in the two cars.

5 Look at this Christmas card.

The main picture (picture 1) contains a reduced version of itself (picture 2), which contains a reduced version (picture 3) and so on.

a Measure the height of the trees in pictures 1, 2, 3 and 4.

b What is the scale factor of the enlargement from picture 1 to picture 2?

What about picture 2 to picture 3?

And picture 3 to picture 4?

c Measure the heights of the snowmen in picture 1.

Use your answers to part **b** to calculate the heights of the snowmen in pictures 2, 3 and 4.

6 A rectangular park is 2.5 km by 2 km.

On a map the longer side of the park measures 5 cm.

a What is the scale of the map?

b How long is the shorter side on the map?

c Find the area of the park on the map.

d Find the real area of the park.

e Is the scale for the areas the same as that for the lengths?

7 **a** Draw and label an x axis from 0 to 10 and a y axis from 0 to 8.

Plot these points and join them to form triangle T.

(6, 8), (6, 4), (10, 6)

b Measure the angles of triangle T.

c Enlarge triangle T by scale factor $\frac{1}{2}$ using (1, 8) as the centre of enlargement.

d Measure the angles of the image.
What do you notice?

8 Draw an x axis from 0 to 16 and a y axis from 0 to 8.

a Draw the trapezium whose vertices are P(2, 8), Q(8, 8), R(5, 2) and S(2, 2).

b Using (14, 5) as the centre of enlargement

i enlarge PQRS by scale factor $\frac{1}{3}$ and label the image P′Q′R′S′

ii enlarge PQRS by scale factor $\frac{2}{3}$ and label the image P″Q″R″S″.

c P′Q′R′S′ is an enlargement of P″Q″R″S″.

i What are the co-ordinates of the centre of enlargement?

ii What is the scale factor?

Reasoning

9 Triangle A is enlarged to form triangle B.

a What is the scale factor of enlargement?

b Write down the co-ordinates of the centre of enlargement.

c Are triangles A and B congruent?

Explain your answer.

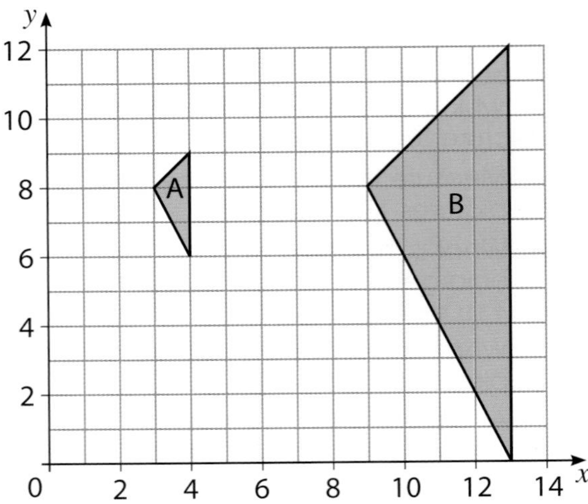

Wider skills practice

1 Amy's picture is $1\frac{1}{2}$ cm high and 3 cm wide.

Which of these pictures are enlargements of Amy's picture?

a

b

c

Reasoning

2 The enlargement reading on a photocopier is 100% when the copy is to be the same size as the original.

When the reading is 120% then each length is increased by 20%.

a What enlargement reading do you use if you want

i each length decreased by 10%

ii a 5 cm line increased to 7 cm

iii a 6 cm line reduced to 4.5 cm?

b On the original a rectangle is 20 cm × 12 cm. The enlargement is set to 125%.

i What is the area of the rectangle on the original?

ii What is the area of the enlarged rectangle?

iii What is the percentage increase of the area in the enlargement?

Applying skills

1 Here is a drawing of a coffee table (it is an elevation) and a photograph of a coffee table.

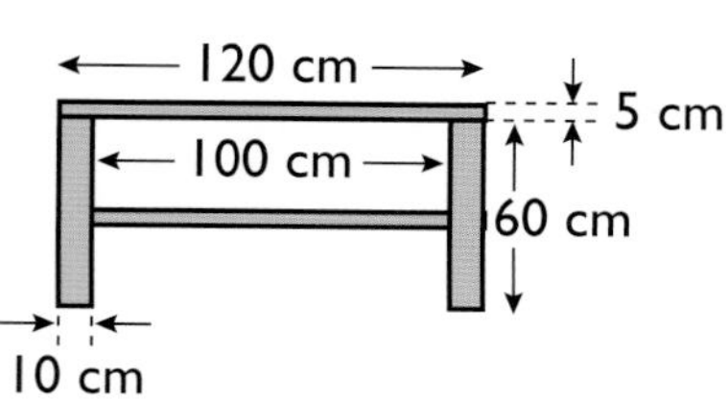

Caitlin made the drawing in Design and Technology.
Does the photograph match Caitlin's drawing?
Explain your answer.

2 The diagram shows a film being projected.

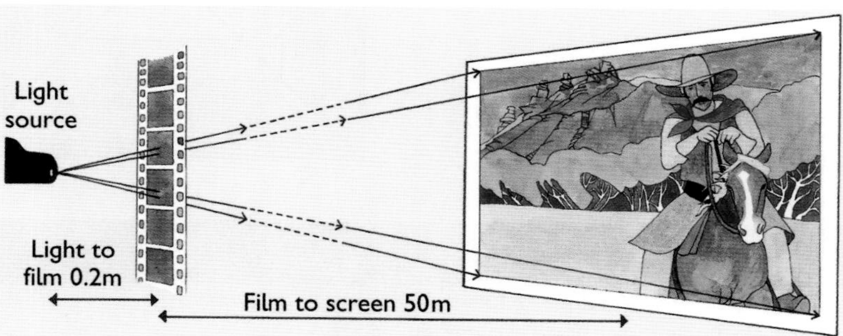

An object on the film is enlarged to form its image on the screen.
The light source is the centre of enlargement.

a Look at the distances on the diagram.
Explain why the scale factor of the enlargement is 251.

b The width of the picture on the film is 16 mm.
What is the width of the picture on the screen?
Give your answer in metres.

c On the screen, the cowboy's hat is 1 m wide.
How wide is the hat on the film?
Give your answer to the nearest millimetre.

3 **a** Look at this picture of a tunnel.
A friend has not seen the picture.
Describe the picture to your friend.

b Measure the width and height of each rectangle in the tunnel picture.
Start with the smallest and end with the biggest.
Write your answers in a copy of this table.

Width	Height
1 cm	2 cm

What is the scale factor of the enlargement from the largest to the smallest?
Where is the centre of enlargement?

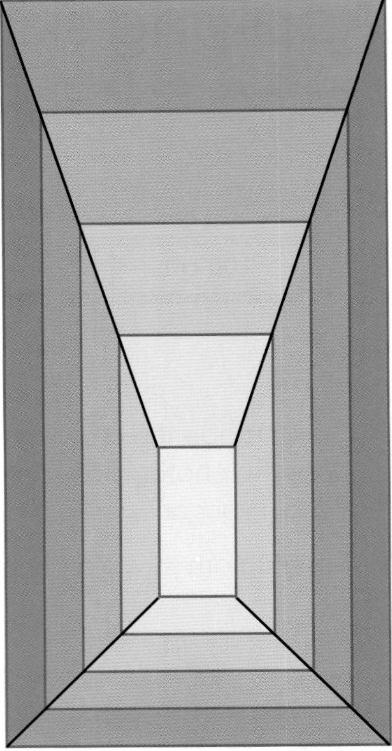

Reviewing skills

1 In the diagram there are five different shapes and their enlargements.
a Match the shapes to their enlargements.
b Name the shapes.
c Work out the scale factor of enlargement for each shape.

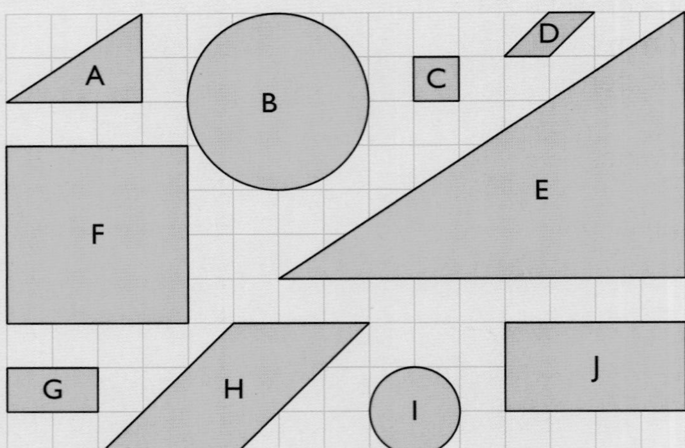

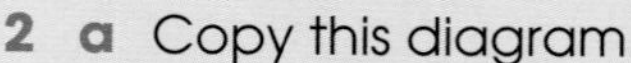

2 **a** Copy this diagram.

b Draw an enlargement of XYZ, scale factor 2, centre C.

Label the image X′Y′Z′.

c Draw an enlargement of XYZ, scale factor 3, centre C.

Label the image X″Y″Z″.

d What transformation maps X′Y′Z′ to X″Y″Z″?

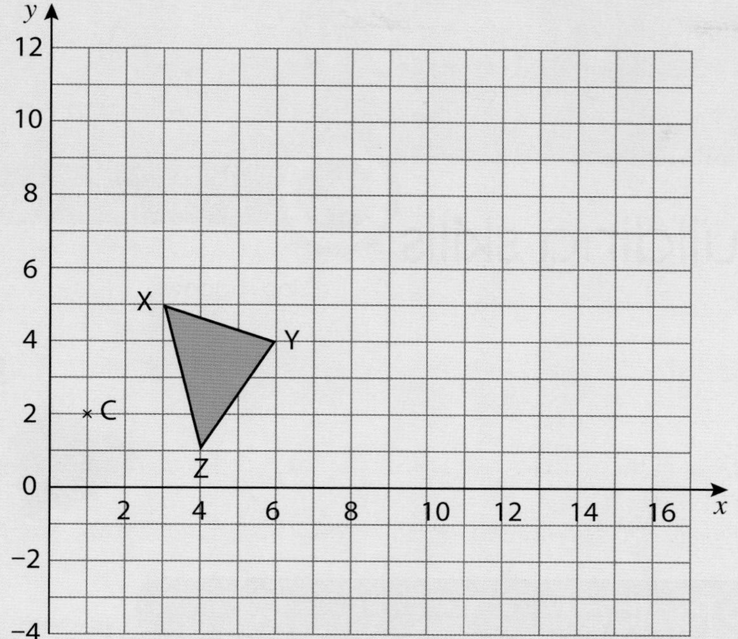

3 P is the centre of enlargement for each of these pairs of shapes.
Find the scale factor of the enlargement from the larger shape to the smaller in each case.

a

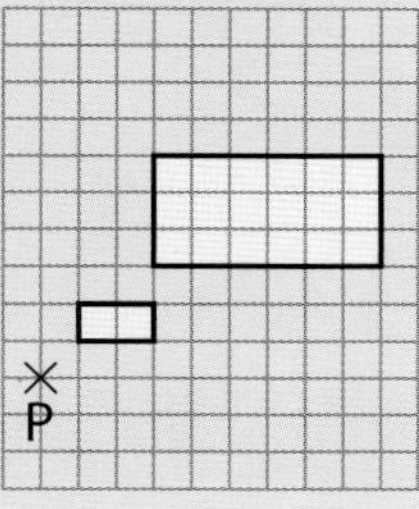

b

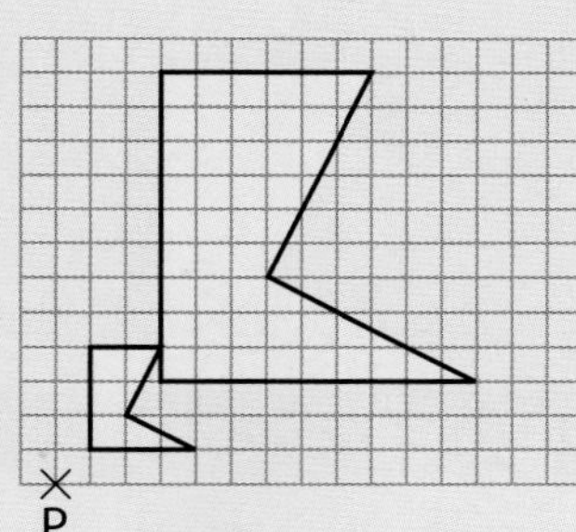

c

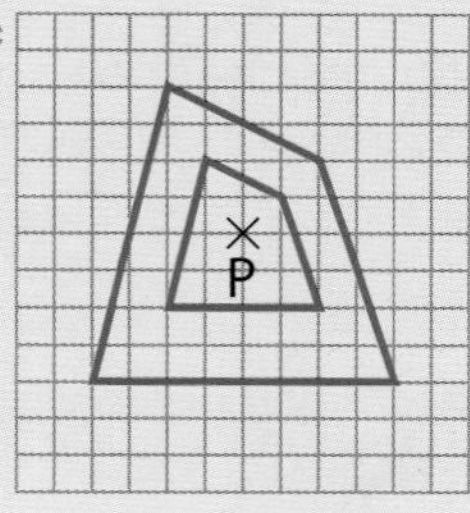

4 Klaudia draws shape A.
She transforms it to the image B.

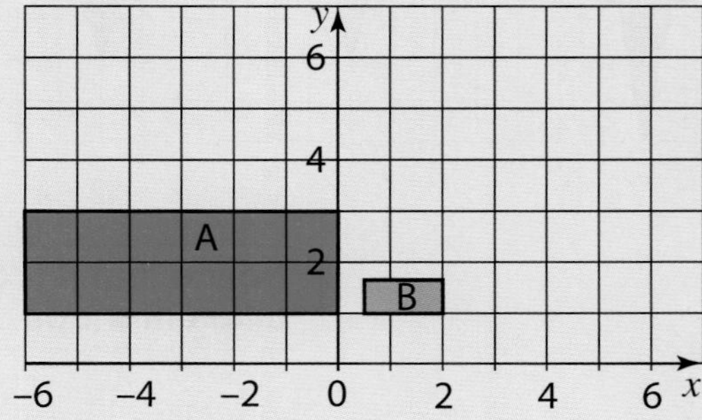

Describe fully the transformation which maps shape A to shape B.

Unit 7 • Similarity • Band h

Building skills

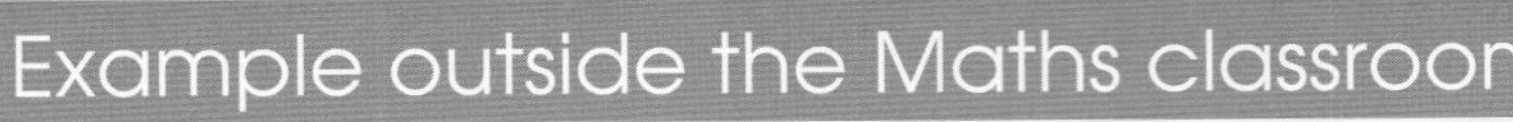

Example outside the Maths classroom

Using shadows

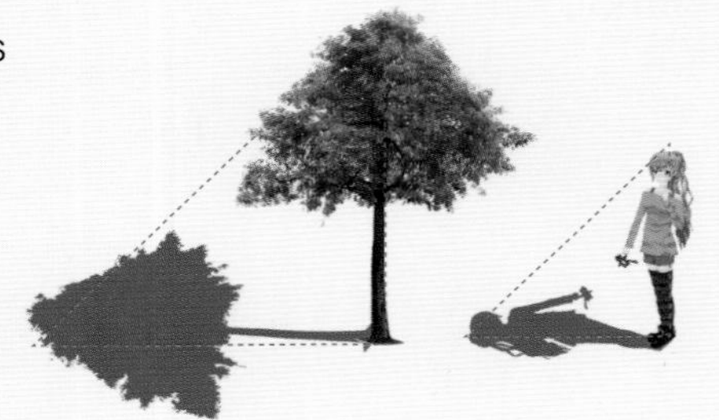

Toolbox

When you enlarge a figure with a scale factor of 3, all the lengths are three times longer but the corresponding angles remain the same.

They are the same shape but different sizes.

The figures are **similar**.

The angles in the two figures are the same and the lengths of the sides are all in the same ratio.

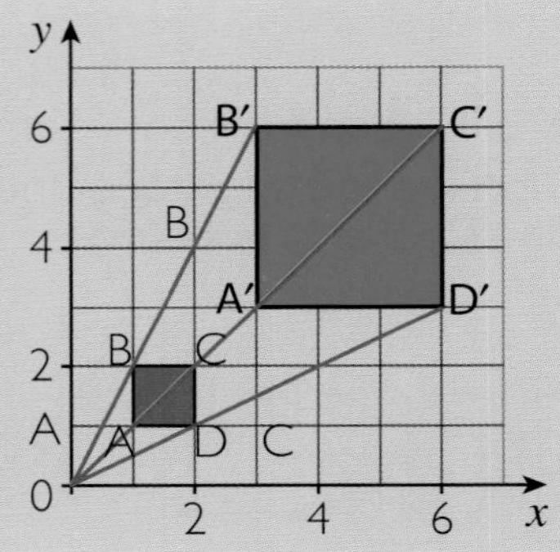

Example – Similar and congruent figures

Sarah has drawn the stars on this grid.

a Which stars are congruent to A?

b Which stars are similar to A?

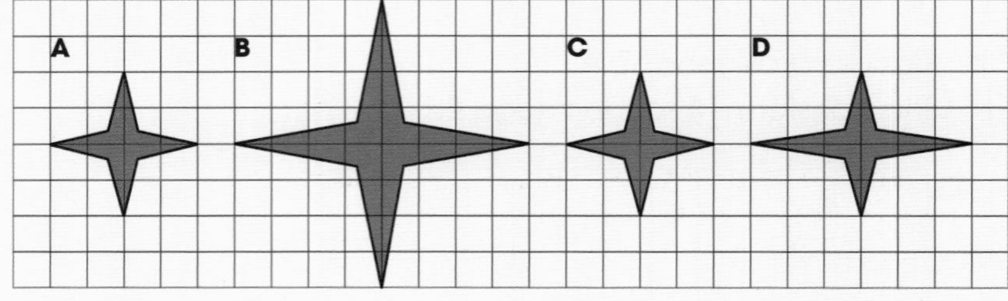

Notice that D is a different shape.

Solution

a Congruent figures are the same shape and size.

Only A and C are congruent.

b Similar figures are the same shape but they may be different sizes.

A, B and C are all similar.

Notice that congruent figures are also similar but similar figures need not be congruent.

Example – Using ratios of similar figures

These triangles are similar.
What is the value of x?

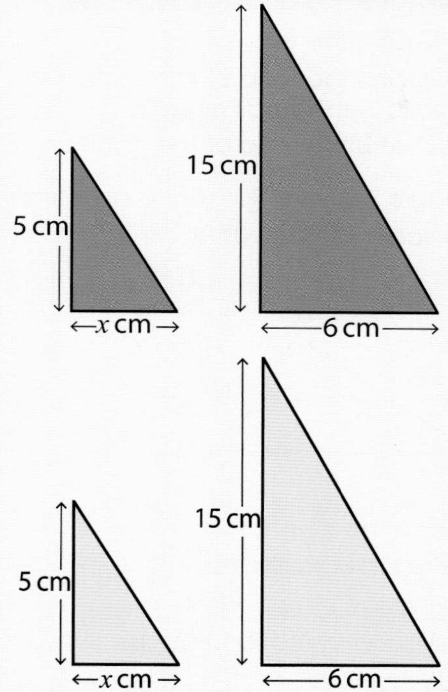

Solution

Method 1

The ratio of the vertical sides, small : large is 5 : 15 = 1 : 3.

So the sides of the large triangle are three times those of the small triangle.

For the bases

$3x = 6$

$x = 2$ ← **The base of the small triangle is 2 cm long.**

Method 2

In the large triangle the ratio of the sides,

base : height is 6 : 15 = 1 : 2.5.

So the height of each triangle is 2.5 times the base.

For the small triangle

$2.5x = 5$

$x = 2$ ← **The base of the small triangle is 2 cm long.**

Remember:

- Similar figures have the same shape but may be different sizes.
- The ratios of the sides of similar figures are the same.
- When a figure is enlarged, the image is similar to the original figure.

Skills practice A

1 a Which rectangles are similar to rectangle A?

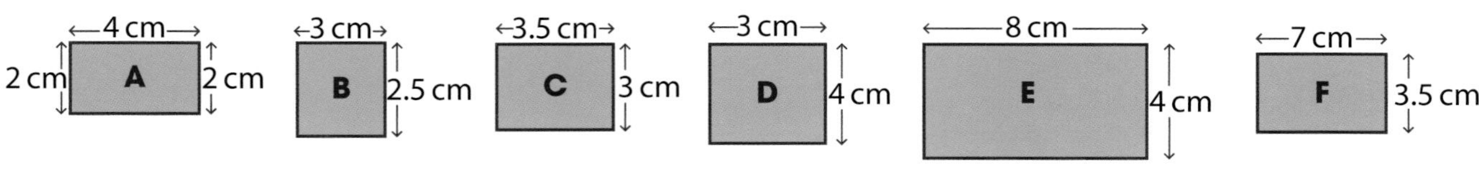

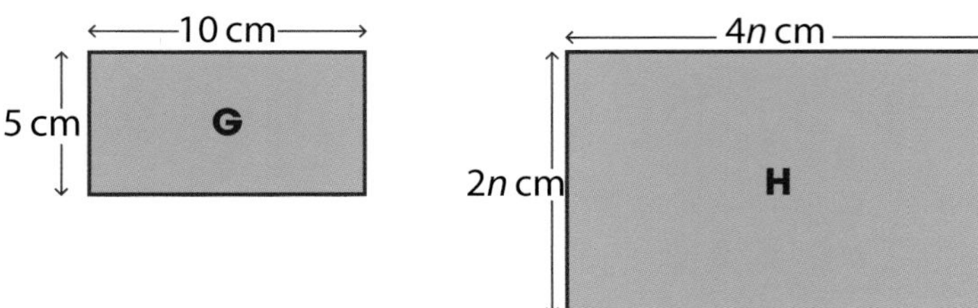

b What is the scale factor for an enlargement mapping A on to each of those rectangles?

2 These two triangles are similar.

a What is the value of x?

b What is the value of y?

c What is the ratio of the sides in the large triangle to the sides in the small triangle?

d What is the ratio of the angles of the two triangles?

12 cm
x cm
9 cm
y cm
10 cm
6 cm

3 Triangle ABC is enlarged to give triangle PQR.
The scale factor is 3.

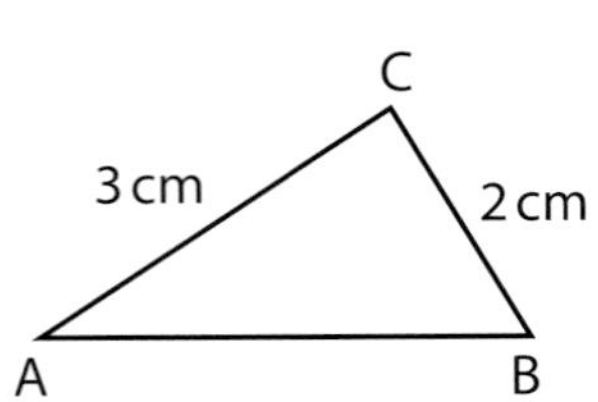

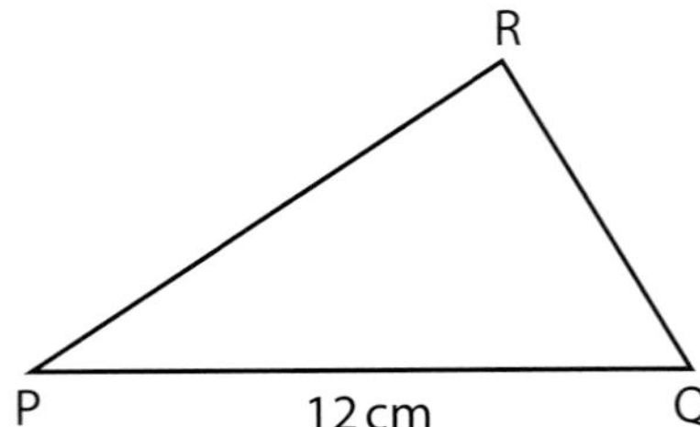

a Find the length PR.

b Find the length RQ.

c Find the length AB.

d What is the ratio of the perimeter of ABC to the perimeter of PQR?

4 The triangles in each of these pairs are similar.
Find the sides marked x.

a

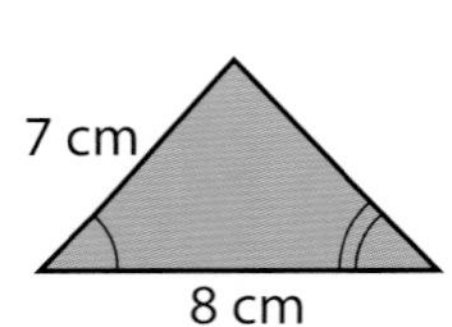

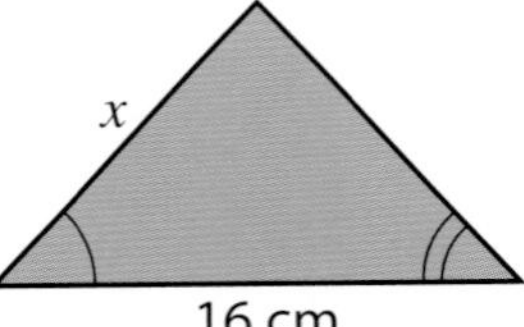

b

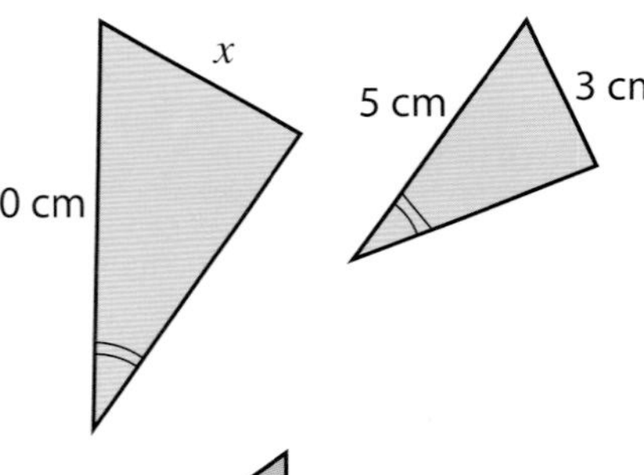

c

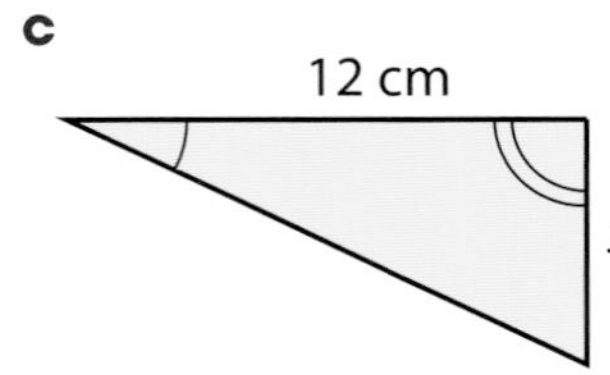

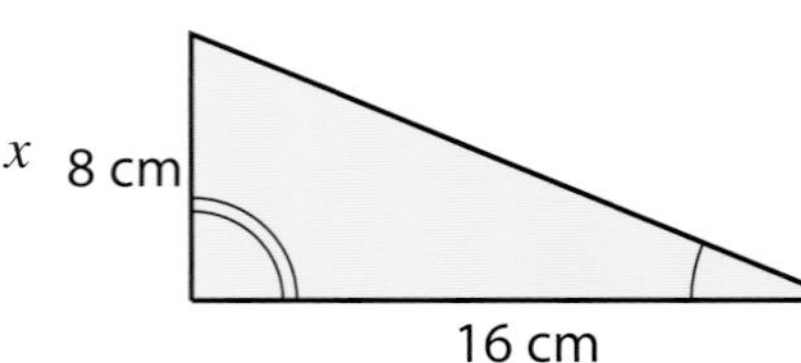

d

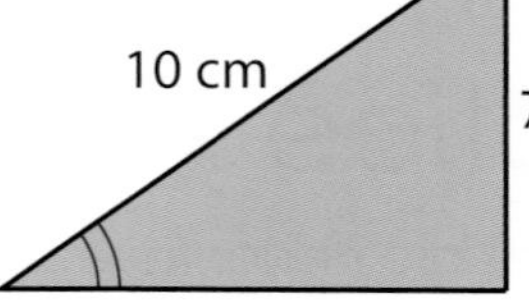

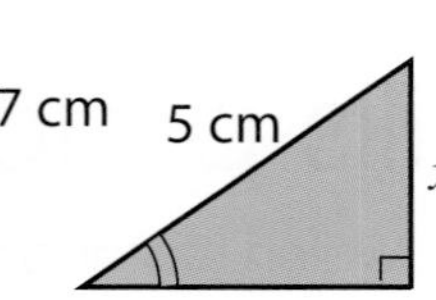

Skills practice B

1 A photograph is 10 cm by 15 cm.

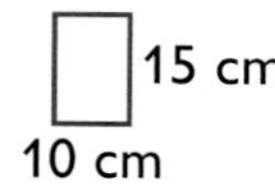

a The diagrams show some possible enlargements. Find the missing measurements.

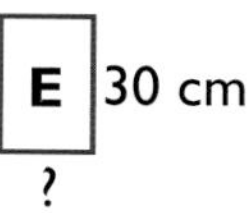

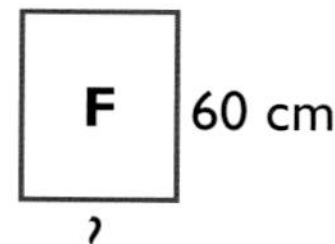

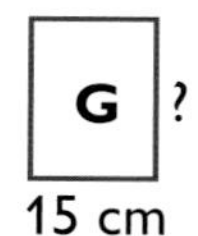

b What is the ratio, the width of E : the width of F?

c Calculate the areas of the photographs E, F and G.

Reasoning

2 **a** Show that these two triangles are similar.

b Find the values of p and q.

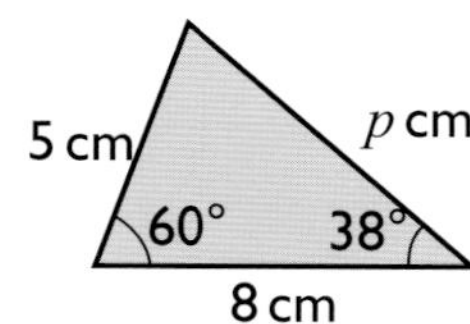

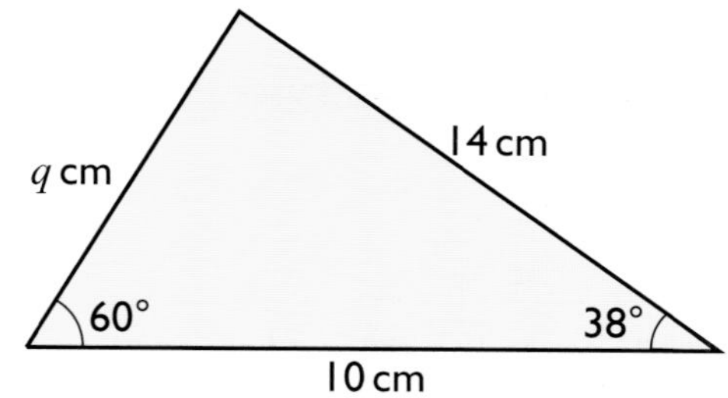

3 **a** Draw these triangles accurately.

i

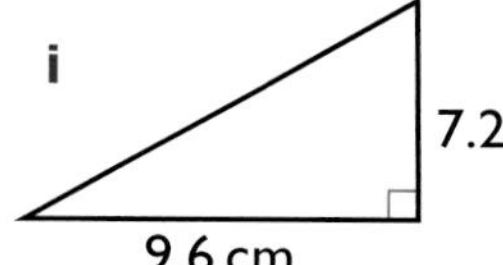

ii

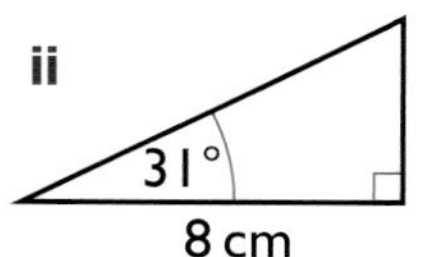

iii

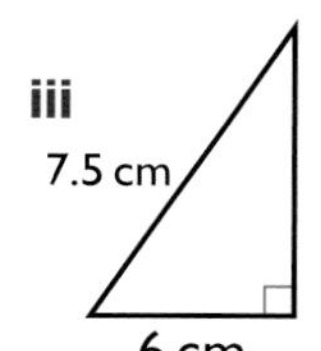

iv

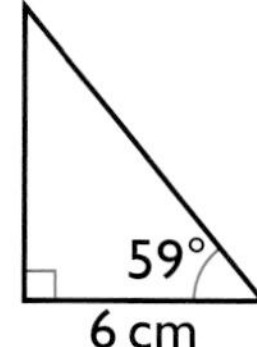

v

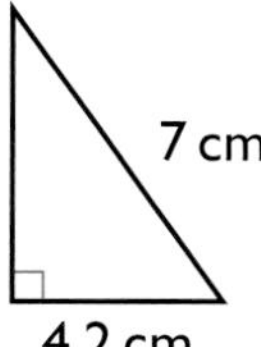

vi

3.6 cm

6 cm

b Measure the lengths of the unknown sides to the nearest 0.1 cm.

c Sort the triangles into two groups of similar triangles.
Are any of these triangles congruent?

Reasoning

4 Say whether each of these statements is true or false.
For those that are false, draw a diagram showing a counterexample.

a All squares are similar.
b If the angles of two triangles are the same, the triangles are similar.
c If the angles of two rectangles are the same, the rectangles are similar.
d All congruent figures are similar.
e All similar figures are congruent.
f All equilateral triangles are congruent.
g All right-angled triangles are similar.
h A figure is transformed by one of translation, reflection, rotation or enlargement.
The figure and its image are congruent.

Reasoning

5 You need two sheets of A4 paper, a ruler and a pair of scissors.
Copy this table and complete it with your results.

Size of paper	Longer side (mm)	Shorter side (mm)	Longer ÷ shorter
A4			
A5			
A6			

Measure the sides of a sheet of A4 paper.
Work out the ratio of longer side ÷ shorter side.
Cut the second sheet of A4 paper in half as shown in the diagram.
Each half is now an A5 sheet.
Measure the sides and work out the ratio longer side ÷ shorter side for an A5 sheet.
Cut one of the A5 sheets in half (see diagram) to get two A6 sheets.
Measure the sides and work out the ratio longer side ÷ shorter side for an A6 sheet.
What do you notice about the three ratios that you have worked out?

Wider skills practice

1 These rectangles are similar.

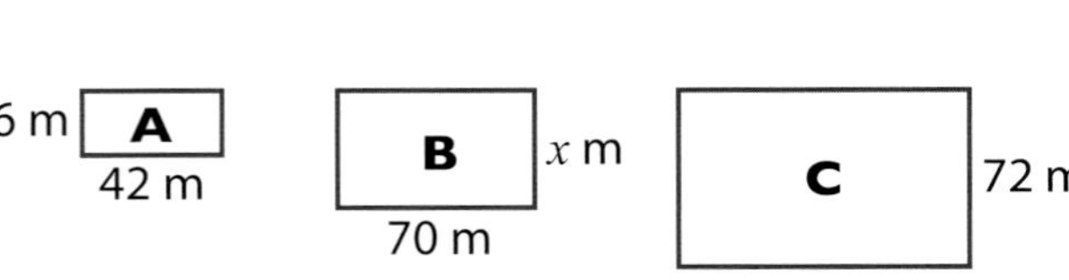

a Find the lengths marked x and y.
The rectangles represent gardens.
b **i** Find their perimeters.
ii Find the ratio of their perimeters.
c **i** Find the areas of the gardens.
iii Find the ratio of the areas.
d Fauzia does scale drawings with a scale of 1 : 2000.
What are the measurements of her scale drawings?

Problem solving

2 Two cuboids are similar.
The ratio of their lengths is $1 : n$.
a What is the ratio of their surface areas?
b What is the ratio of their volumes?
c Do these rules apply to any similar shapes?

Applying skills

Problem solving

1 Tom is a film star. He sends out souvenir models of his Oscar.

The souvenirs are packed in boxes.

There are two sizes, large and small.

The boxes are similar.

On the front of each box is a photograph of Tom.

Photograph C is an enlargement of photograph D.

Photograph C

Photograph D

a What is the ratio, width : height, for photograph C?

b What is the height of photograph D?

The small box is 10 cm by 8 cm, and 12 cm high.

The ratio of the heights of the two boxes is 2 : 1.

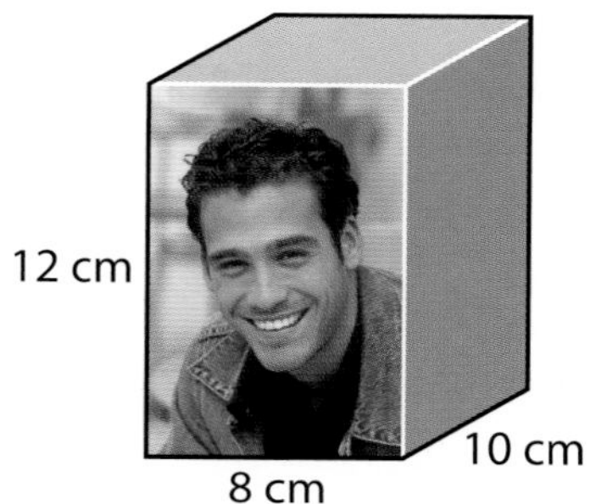

c What are the dimensions of the larger box?

d What is the surface area of

i the larger box

ii the smaller box?

e **i** What is the ratio of the surface areas (large : small) in its simplest form?

ii What is the connection between this and the 2 : 1 ratio of the heights?

f Calculate the volume of each box and write down the ratio of their volumes in its simplest form.

g Find a connection between this ratio and the ratio of the heights.

Reviewing skills

1 These pairs of rectangles are similar.

a

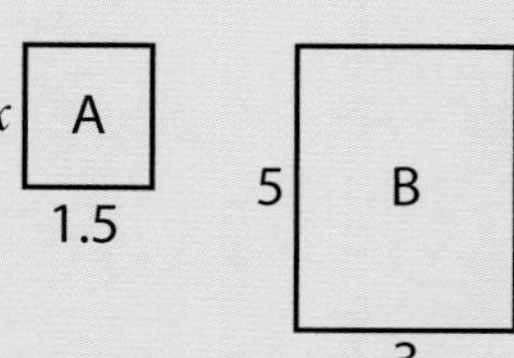

b

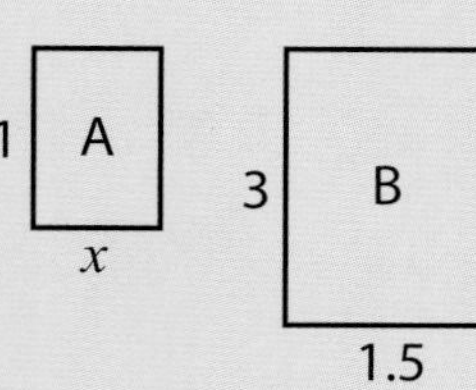

c

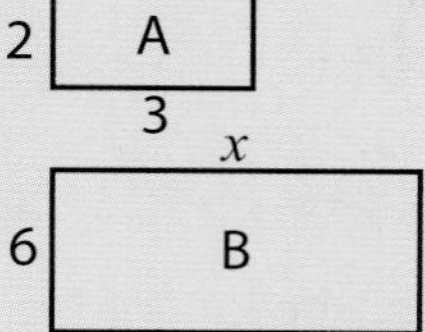

d

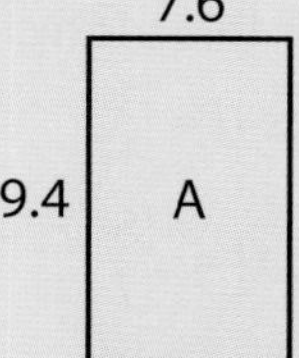

e

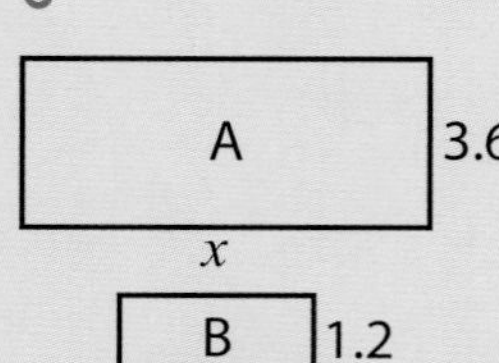

f

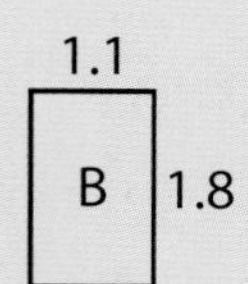

For each pair

i find the length marked x

ii work out the ratio, length of rectangle A : length of rectangle B

iii work out the ratio, width of rectangle A : width of rectangle B.

Building skills

Example outside the Maths classroom

Size of the Earth

Toolbox

Each side in a **right-angled triangle** has a name.

- The longest side is called the **hypotenuse (H)**.
- The side opposite the marked angle is called the **opposite (O)**.
- The remaining side, next to the marked angle, is called the **adjacent (A)**.
- The ratio of the lengths of the sides are given special names.

$$\text{cosine (cos) } \theta = \frac{\text{adjacent}}{\text{hypotenuse}}$$

$$\text{tangent (tan) } \theta = \frac{\text{opposite}}{\text{adjacent}}$$

$$\text{sine (sin) } \theta = \frac{\text{opposite}}{\text{hypotenuse}}$$

These ratios are constant for similar right-angled triangles.

You can use this information to find angles and lengths in triangles.

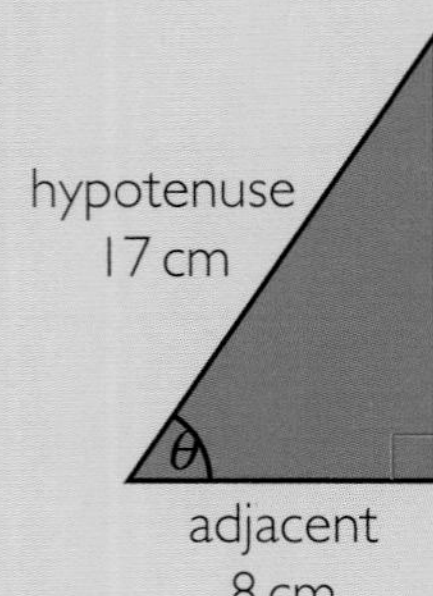

For this triangle

$\cos\theta = \frac{8}{17}$

$\sin\theta = \frac{15}{17}$

$\tan\theta = \frac{15}{8}$

You will find keys for sin, cos and tan on your calculator.

Example – Using trigonometry to find a length

John is a window cleaner. His ladder is 6 m long.

The angle between the ladder and ground is 70°.

Find the height that his ladder reaches up the wall.

Solution

$$\sin\theta = \frac{\text{opposite}}{\text{hypotenuse}}$$

$$\sin 70° = \frac{y}{6}$$

so $y = 6 \times \sin 70°$ ← **Multiply by 6**

$= 6 \times 0.94$ ← **sin 70° = 0.94 from a calculator.**

$= 5.64\text{ m}$

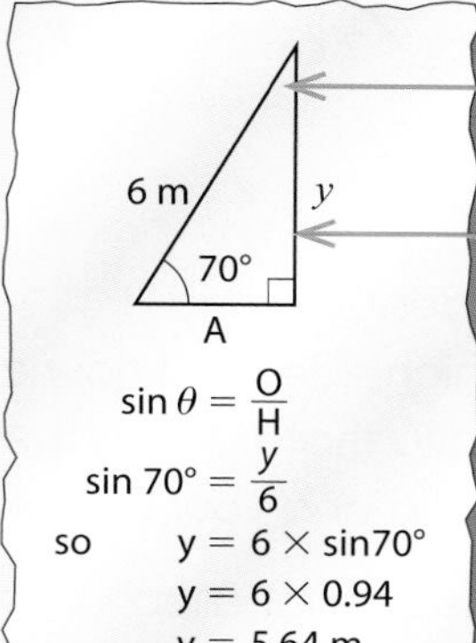

The ladder is the hypotenuse. It is 6 m long.

The height the ladder reaches up the wall is the opposite side to the angle of 70°.

The distance from the bottom of the ladder to the wall is adjacent side.

Example – Using trigonometry to find an angle

a Look at the diagram. Which ratio would you use to find θ?

b Find θ.

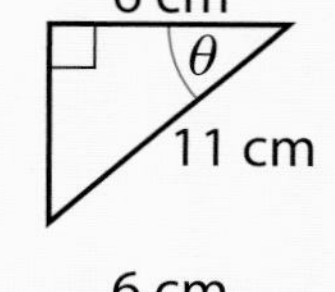

Solution

a Use $\cos\theta = \dfrac{\text{adjacent}}{\text{hypotenuse}}$

The adjacent side is 6 cm.

The hypotenuse is 11 cm.

6 cm
θ
11 cm

b $\cos\theta = \dfrac{6}{11}$

$\theta = \cos^{-1}\left(\dfrac{6}{11}\right)$

$\theta = 56.9°$

You can write $\theta = \cos^{-1}\left(\dfrac{6}{11}\right)$ or $\theta = \text{arccos}\left(\dfrac{6}{11}\right)$.

Find this using your calculator.

Remember:

- When using the trigonometric ratios to find lengths or angles, start by labelling the hypotenuse, the opposite side and the adjacent side.
- Make sure your calculator is set to degrees and not radians.

Skills practice A

1 Look at these right-angled triangles.

i

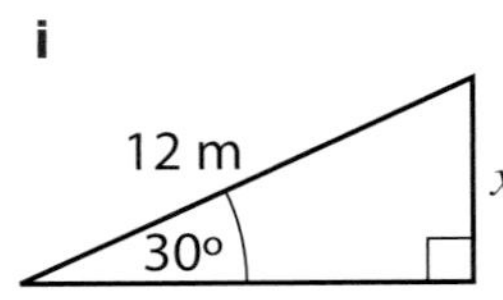

ii

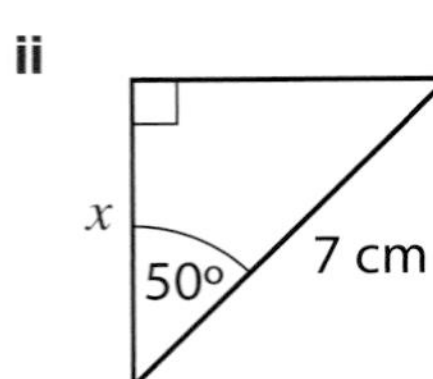

iii

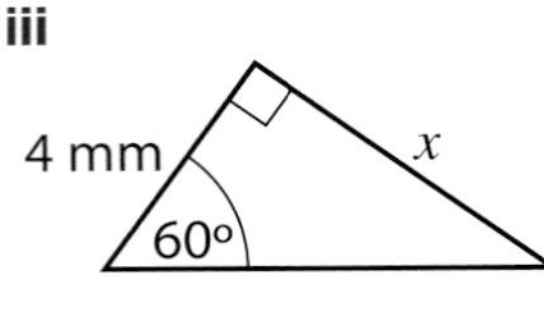

For each triangle

i make a sketch and label the sides O, A and H

ii write down (using O, A or H) which side you know and which side you want to find

iii write down the trigonometric ratio that involves these two sides

iv use the correct trigonometric ratio and your calculator to find the side length that you want.

2 For each of these triangles, find the side labelled x.

a

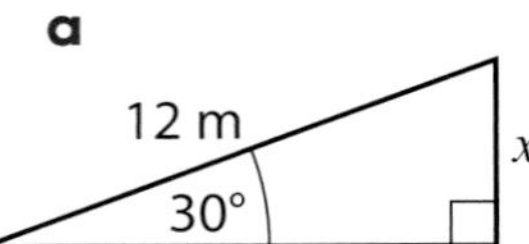

b

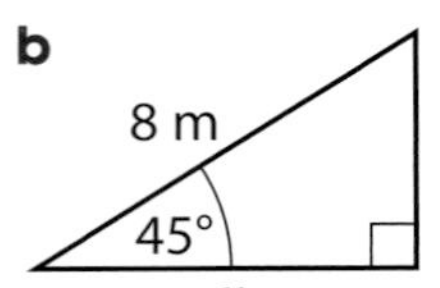

c

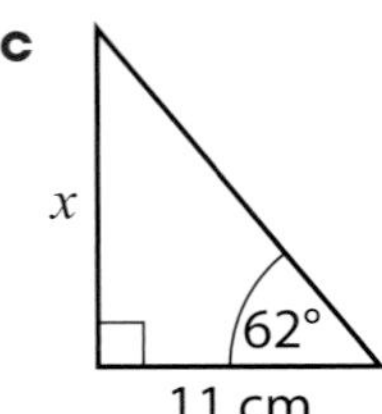

3 For each of these triangles, find the side labelled x.

a

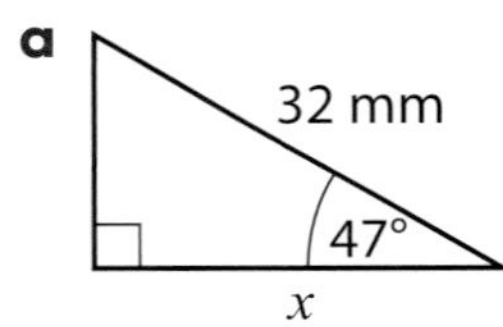

b

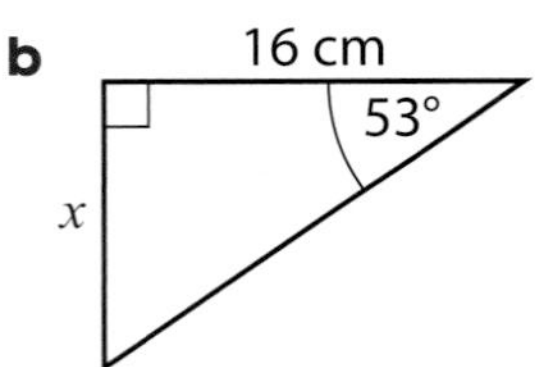

c

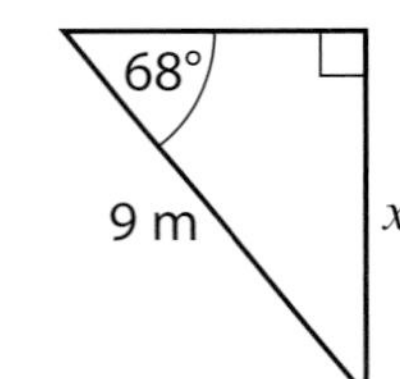

4 For each of these triangles, find the lengths labelled x and y.

a

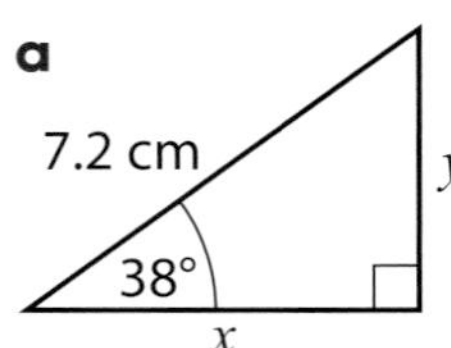

b

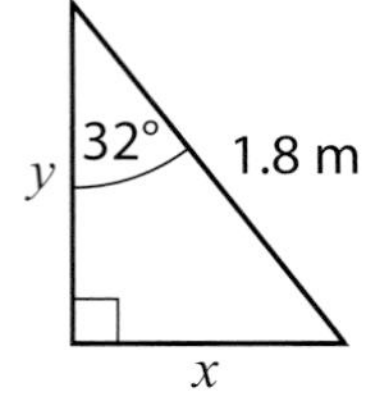

c

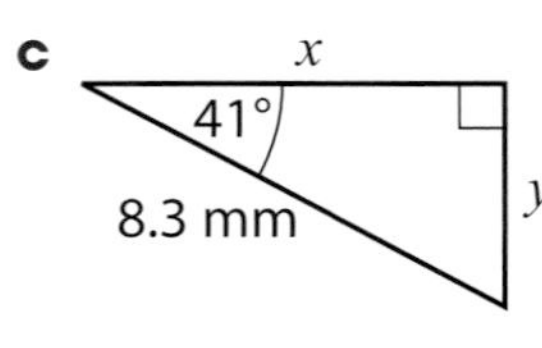

5 For each of these triangles, find the lengths labelled x and y.

a

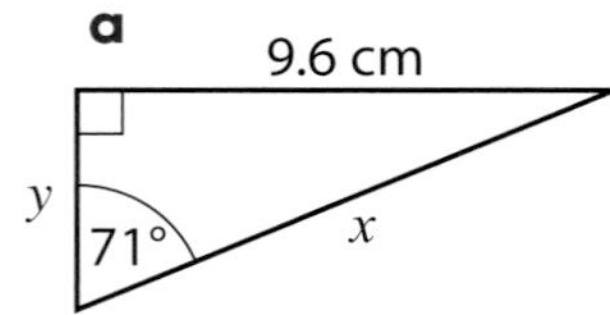

b

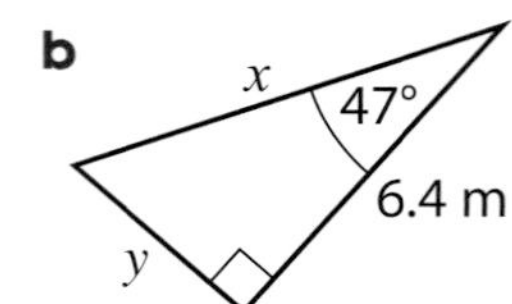

c

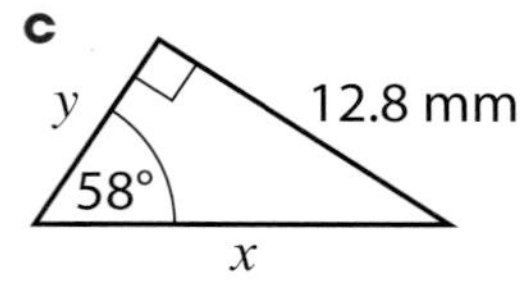

6 Use your calculator to find θ.

a $\sin\theta = 0.5$ **b** $\cos\theta = 0.707$ **c** $\tan\theta = 1.732$

d $\sin\theta = 0.866$ **e** $\cos\theta = 0$ **f** $\sin\theta = 1$

g $\cos\theta = 0.5$ **h** $\tan\theta = 0.577$ **i** $\tan\theta = 0.5$

7 Find angle θ in each of these triangles.

a

9 cm
15 cm
θ

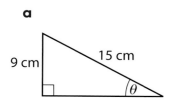

b

5 m
θ
2 m

c

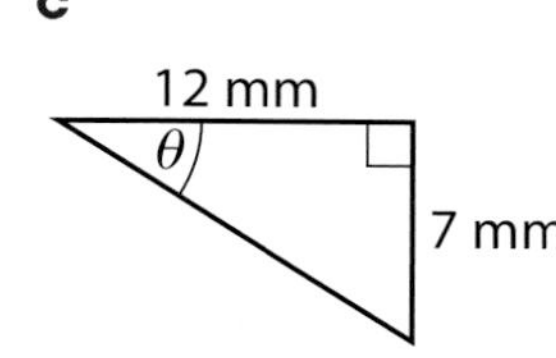

8 Find angle θ in each of these triangles.

a

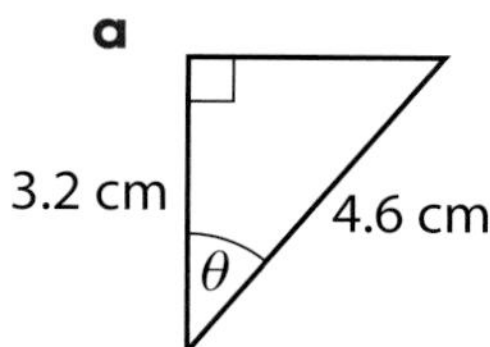

b

7.6 m
θ
11.8 m

c

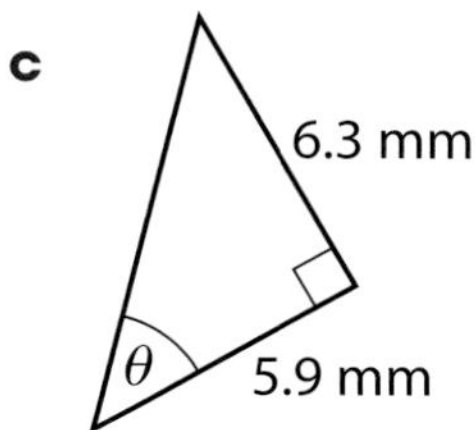

Skills practice B

1 Amy is flying a kite.
What is the height of the kite above Amy's hand?

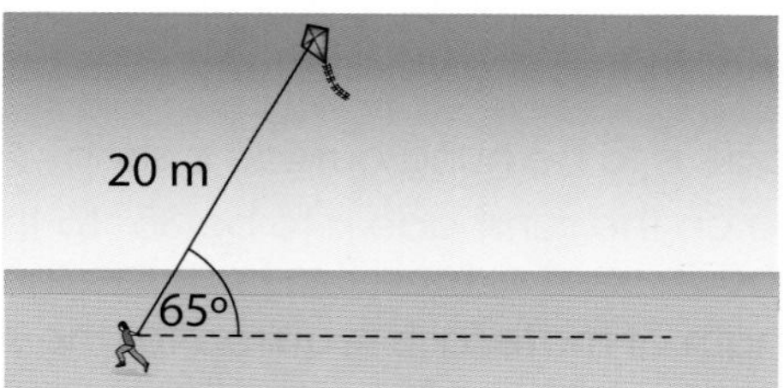

2 Kamil is standing 12 m away from a tree.
Find the height of the tree.

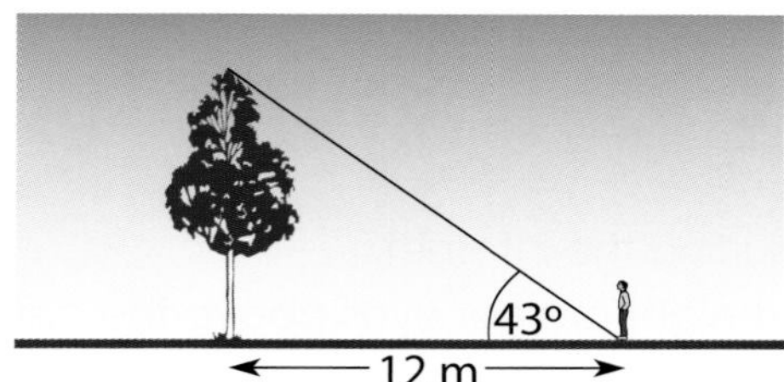

3 Steve is in a small boat.
It is 1500 m away from the bottom of a cliff.
The height of the cliff is 230 m.
Find the angle of elevation of the top of the cliff from the boat.

4 A ship is 8 km from port on a bearing of 036°.
How far North and how far East is the ship from the port?

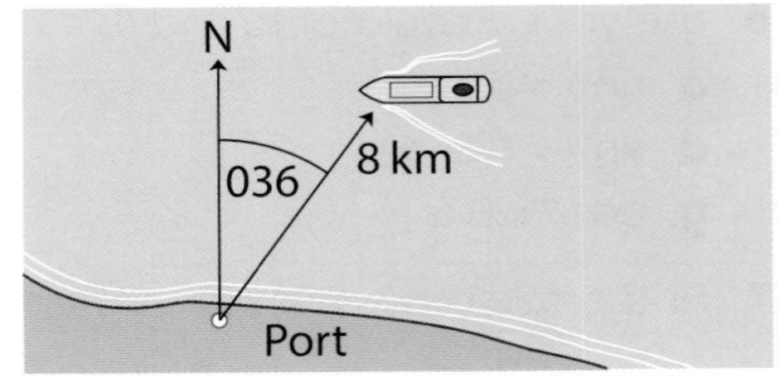

5 A man-made ski slope covers 30 m of land and is 42.4 m high at the top.
Assuming the slope is constant, what angle does the slope make with the ground?

Reasoning

6 Look at triangle ABC.
The angle B is 90°.
AB and AC are both 1 cm.

a Explain why
  i $AC = \sqrt{2}$ cm
  ii angle C = 45°.

b Use triangle ABC to show that $\sin 45° = \dfrac{1}{\sqrt{2}}$.

c Find the values of tan 45° and cos 45°.

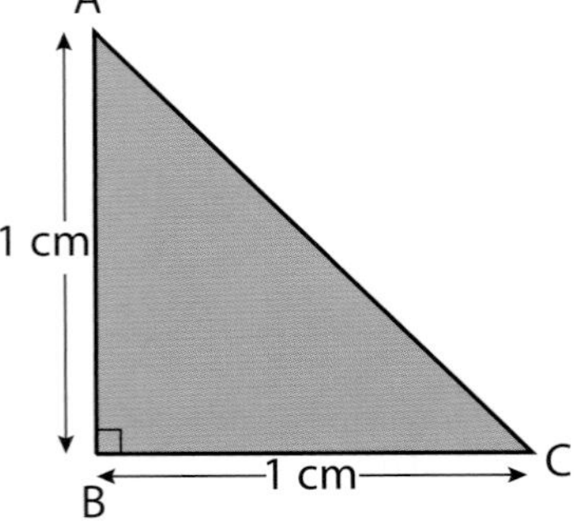

Reasoning

7 Say whether each of these statements is true or false.

a This triangle shows you that sin 60° can be $\dfrac{5}{5} = 1$.

b $\sin 90° = 1$

c $\tan 0° = 0$

d Whatever the value of θ, $\sin\theta$ cannot be greater than 1.

e Different triangles give different values for $\tan\theta$.

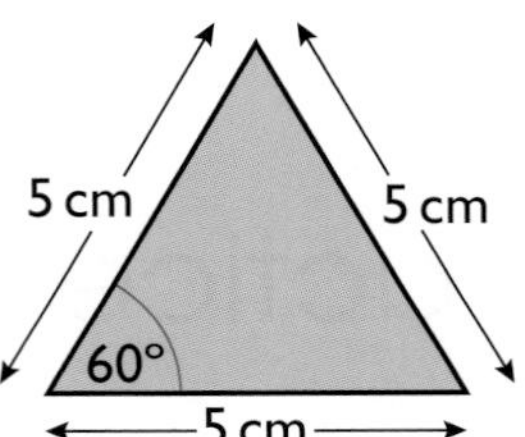

Problem solving

8 A sunshade is to be hung outside a shop.
The slope on the sunshade is to be 35° to the horizontal and the sunshade must extend 3 m from the shop front.
What length of material is required for the sunshade?

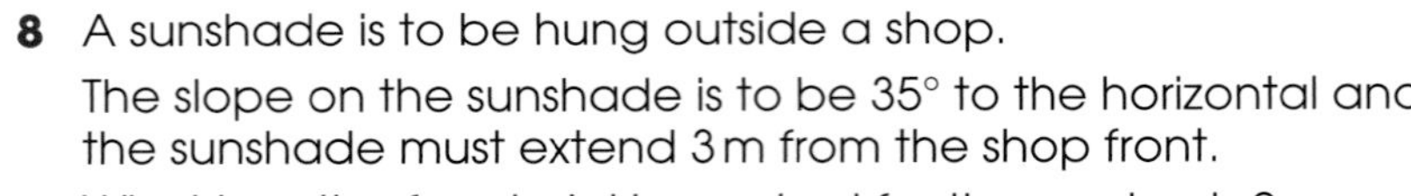

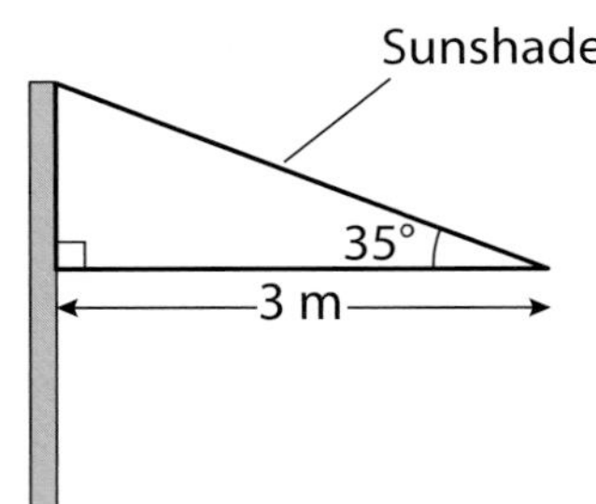

9 Hazel is surfing the internet.

a What height are her eyes above the centre of the screen?

b Find the horizontal distance from her eyes to the centre of the screen.

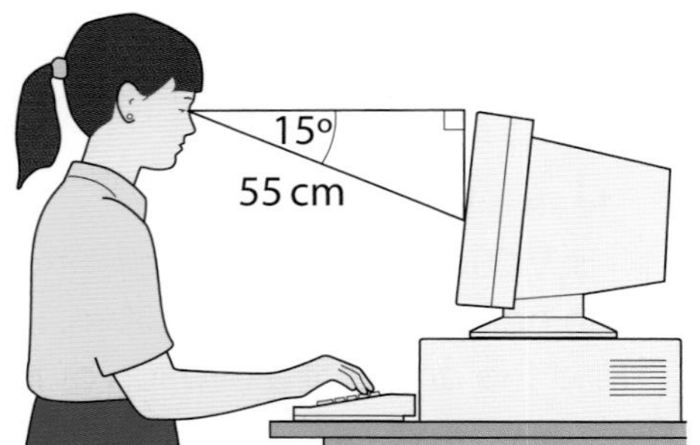

Wider skills practice

1 The diagram shows a form of spiral.
It continues with more triangles.

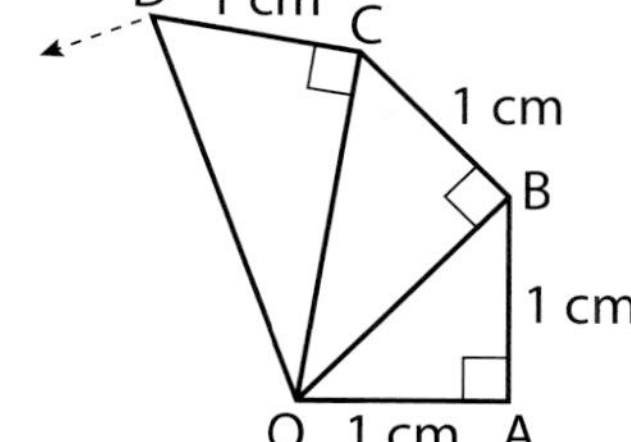

a Work out the lengths of
- **i** OB
- **ii** OC
- **iii** OD.

Use square root signs in your answers when you need to.

b How many triangles are needed to complete
- **i** 180° of the spiral
- **ii** 360° of the spiral?

2 Look at the triangle LMN.
Angle LNM is a right angle.

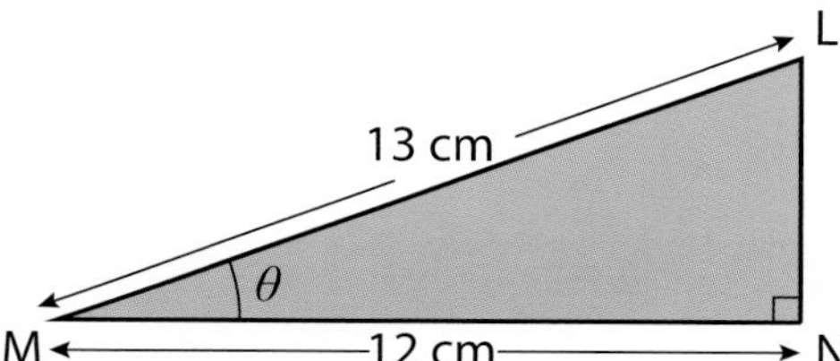

a Find the length of the side LN without using trigonometry.
b Write $\tan\theta$, $\sin\theta$ and $\cos\theta$ as fractions.
c Now use your calculator to find θ three times, using the three trigonometric ratios.
Check that you get the same answer each time.

3 **a** Draw x and y axes using the same scale for each.
b Using values of x from 0 to 4, draw the line $y = 2x + 3$.
c Find the angle that the line $y = 2x + 3$ makes with the x axis.

4 Look at this diagram.
It shows a circle with a chord AB of length 8 cm.
The radius of the circle is 5 cm.
Find θ.

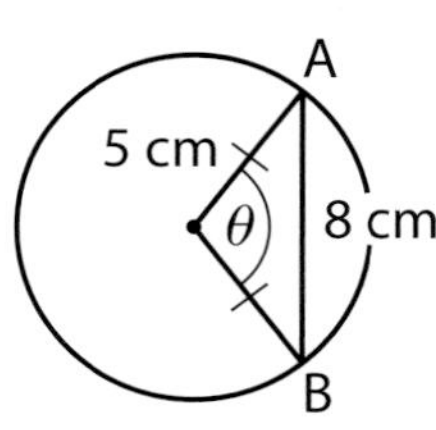

Reasoning

5 PQR is an equilateral triangle.
All its sides are 2 cm long.
M is the mid-point of QR and so QM = MR = 1 cm and angle PMQ = 90°.

a Explain why
- **i** angle Q = 60°
- **ii** PM = $\sqrt{3}$ cm.

b Use triangle PQM to show that $\sin 60° = \frac{\sqrt{3}}{2}$.
c Find cos 60° and tan 60°.
d Explain why angle QPM is 30°.
e Still using triangle PQM, find sin 30°, cos 30° and tan 30°.

P
2 cm
2 cm
Q
M
R

Applying skills

Problem solving

1 Danielle cycles from Abbotville to Bradbury along a straight road. Bradbury is 4 km North and 3 km East of Abbotville.

- a Find the bearing of her journey.
- b How far does she cycle?
- c What is the return bearing?

2 Look at this diagram.

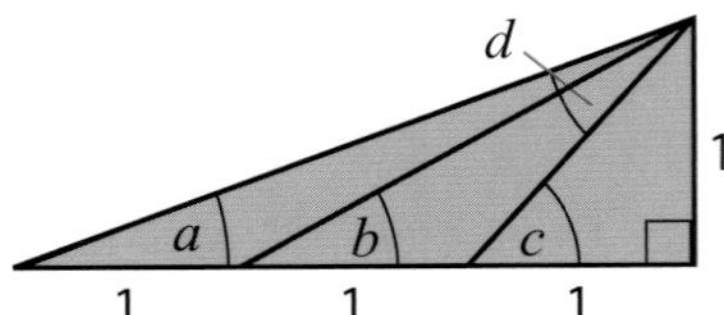

- a Use trigonometry to find the value of $a + b + c$.
- b Use your answer to part **a** to prove that $b = d$.

Reviewing skills

1 Find the missing lengths in each of these triangles.

a
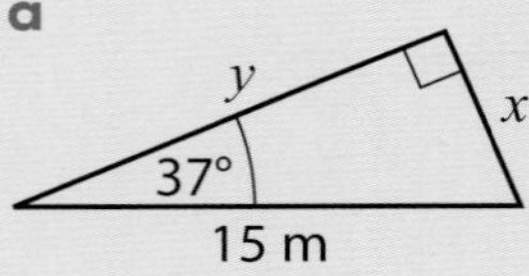

b
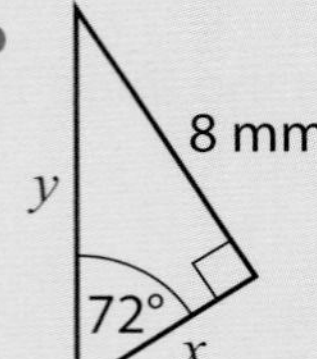

c
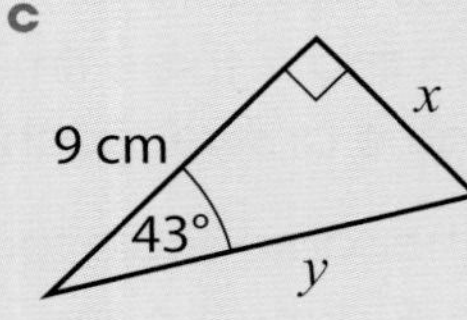

2 Use your calculator to find θ to one decimal place.

- a $\cos\theta = 0.296$
- b $\tan\theta = 0.545$
- c $\sin\theta = 0.741$
- d $\cos\theta = 0.118$

3 A rectangular field is 160 m long and 120 m wide. A straight path along one diagonal cuts across it.

- a Find the angle the path makes with the longer side.
- b Work out the length of the path.

Strand 6 • Three-dimensional shapes

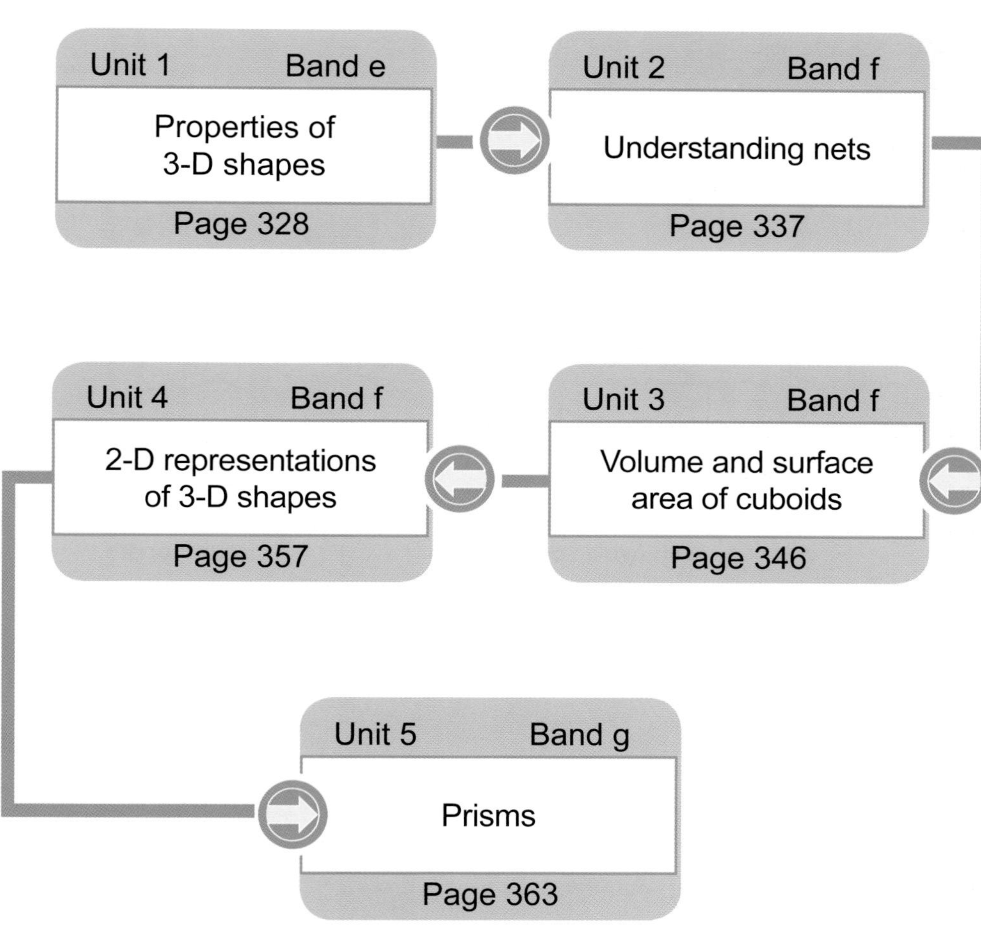

Building skills

Example outside the Maths classroom

Mineral structure

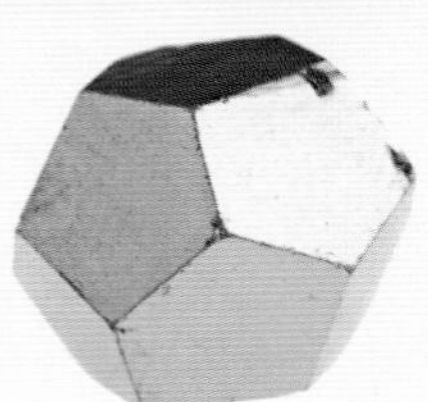

Toolbox

3-D shapes

A **prism** is a three-dimensional (3-D) shape with a constant cross-section.

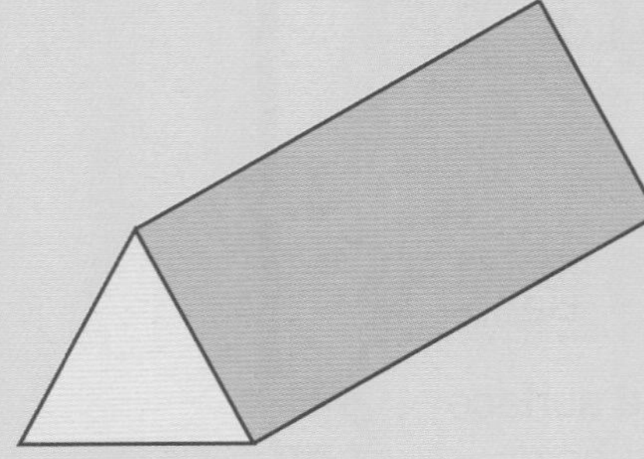

A triangular prism

This cuboid is a rectangular prism

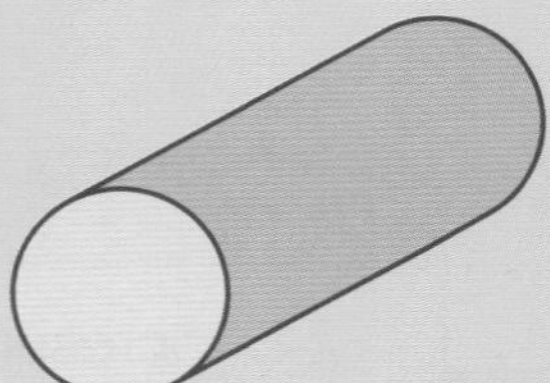

A cylinder is a circular prism

Not all 3-D shapes are prisms.

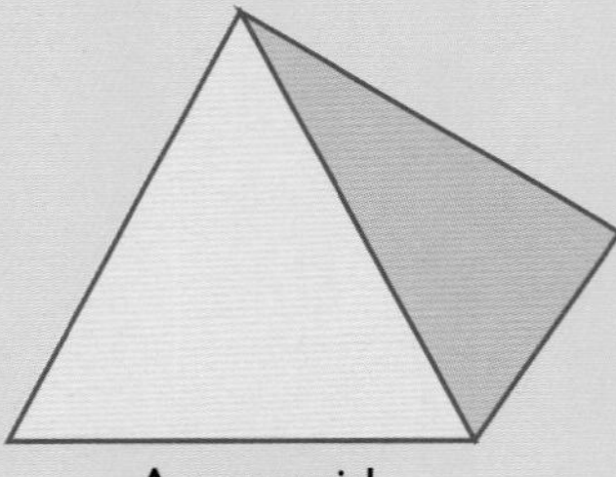

A pyramid

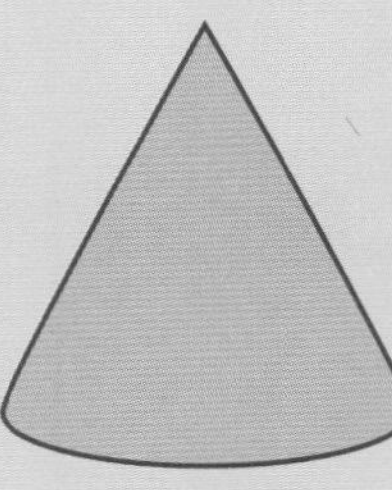

A cone

Euler's rule

$V + F = E + 2$

where

V is the number of vertices

F is the number of faces

E is the number of edges.

A **plane** is a completely flat surface, like the top of a table or a flat wall.

Just as in two-dimensional (2-D) shapes there may be a line of symmetry; some three-dimensional (3-D) shapes have a **plane of symmetry**.

The plane of symmetry slices the shape into two identical pieces.

One way to show a 3-D object on paper is an **isometric drawing**. This uses a triangular grid.

Example – Thinking in three dimensions

This object is a square-based pyramid.

a You cut the pyramid in the middle horizontally. What shape is the cut?

b You cut the pyramid in the middle vertically. What shape is the cut?

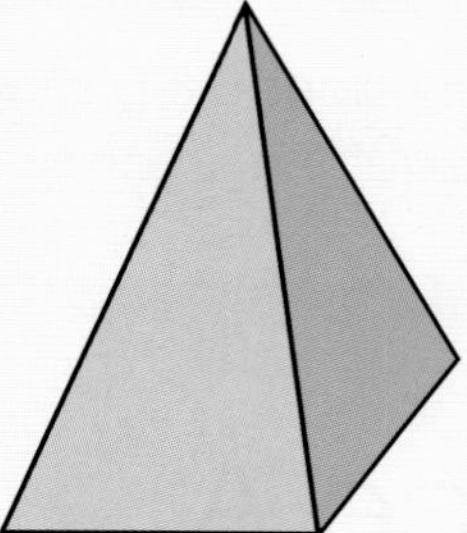

Solution

a The cut is a square.

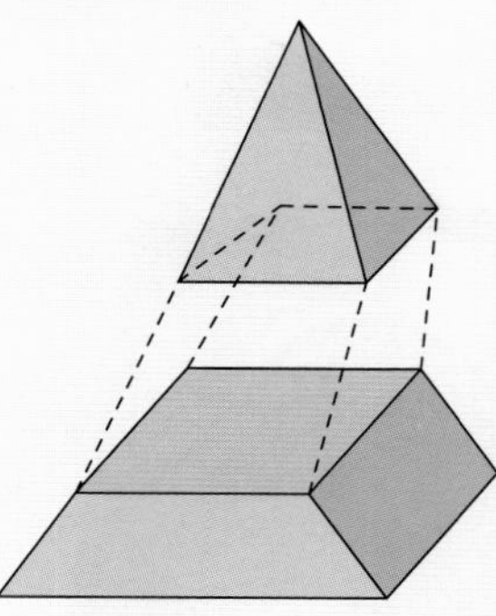

b The cut is an isosceles triangle.

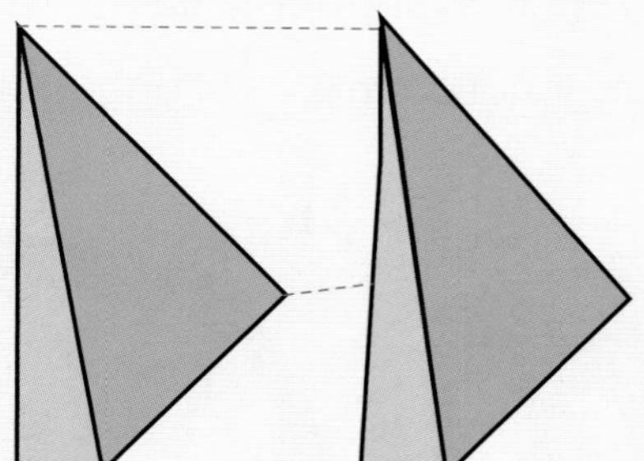

Example – Completing 3-D shapes

Add cubes to these shapes so that each mirror is a plane of symmetry.

Draw the new shapes on isometric paper.

a

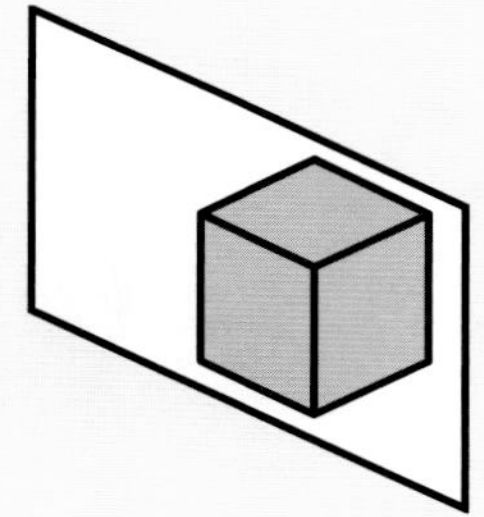

b

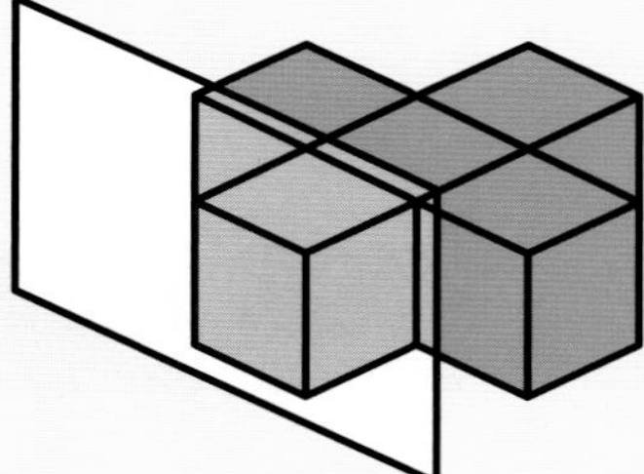

Solution

a

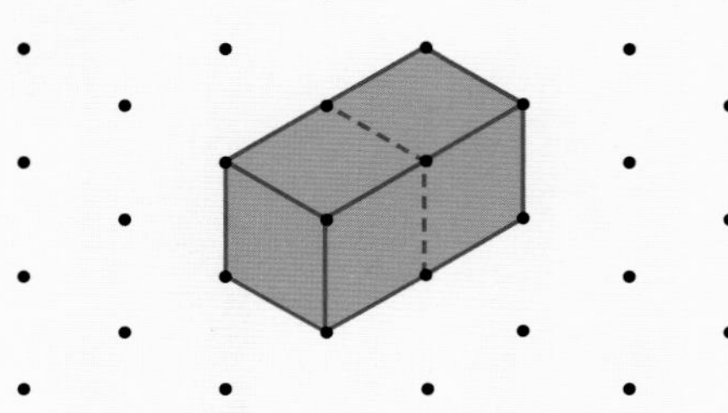

b

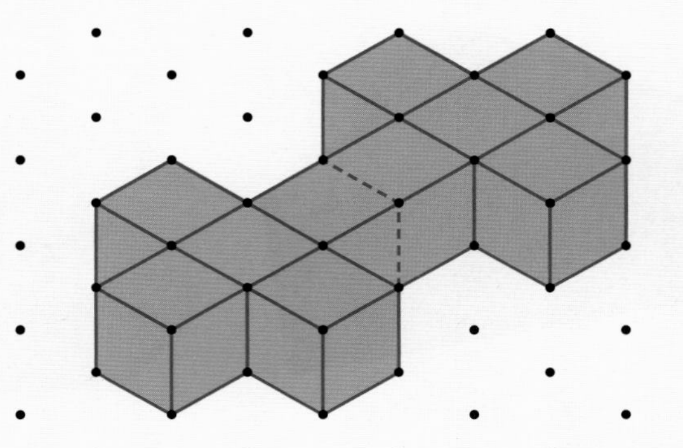

Remember:

- A plane of symmetry slices a 3-D shape into two identical pieces.
- There are no horizontal lines in an isometric drawing.

Skills practice A

1 Match these names to the shapes below.

cuboid | hexagonal prism | triangular prism | cube | octagonal prism

a

b

c

d

e

f

g

h

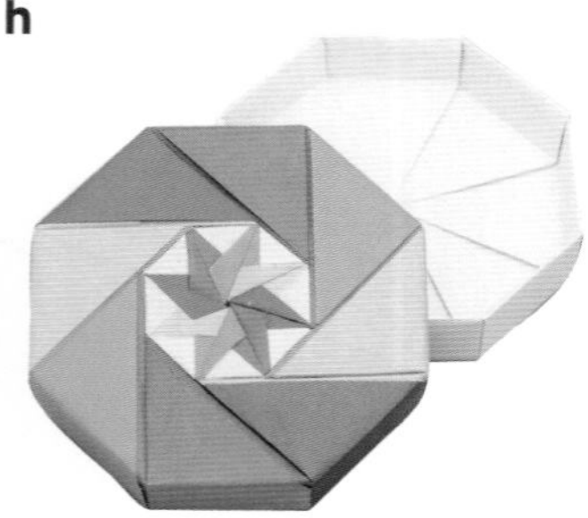

i

2 Each of these letters is a solid.

i Make a sketch of each solid and draw all of its planes of symmetry.

ii How many vertical and horizontal planes of symmetry does each solid have?

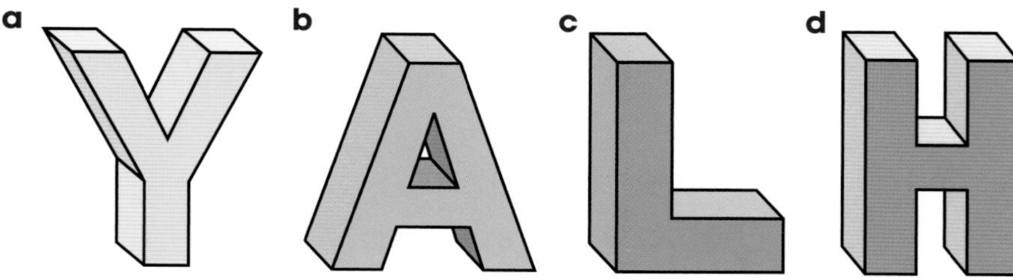

3 Look at this square-based pyramid.

a Using letters, identify the four vertical planes of symmetry (one is shown).

b Explain why this solid has no horizontal plane of symmetry.

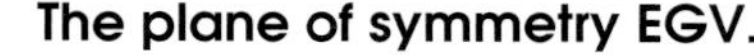

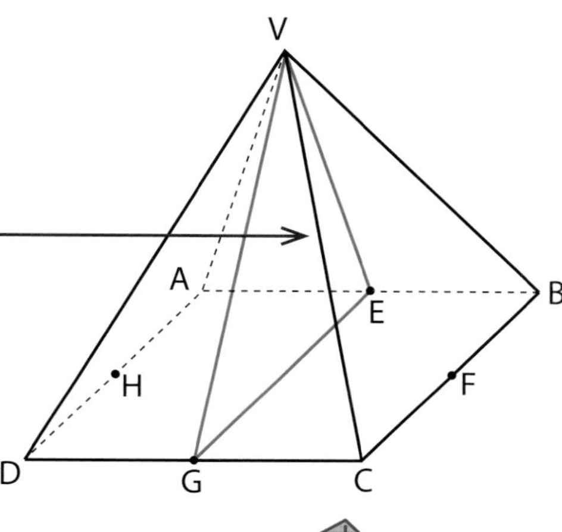

4 Look at this triangular prism.

a How many faces, edges and vertices does it have?

b Show that it obeys Euler's rule.

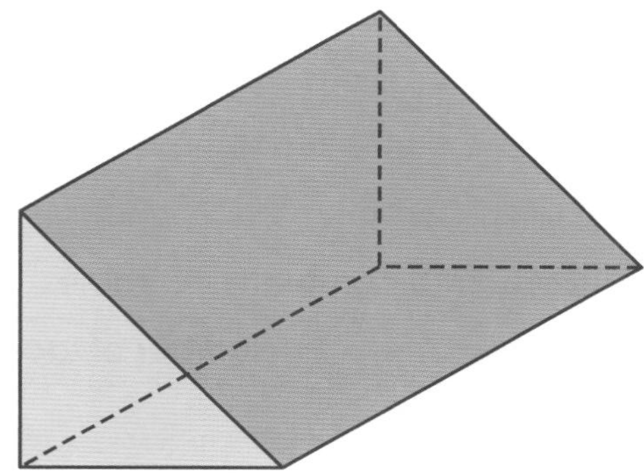

5 Use isometric paper to make 3-D representations of these shapes.

a

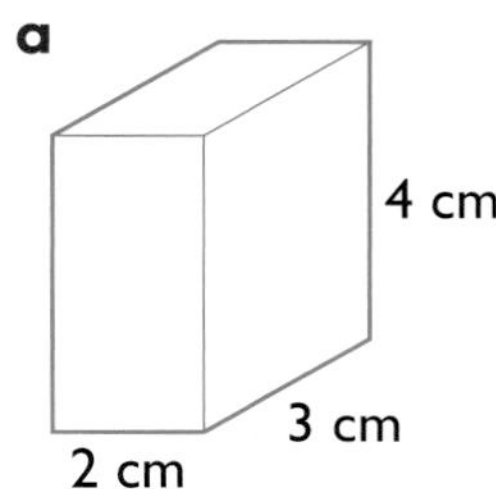

b

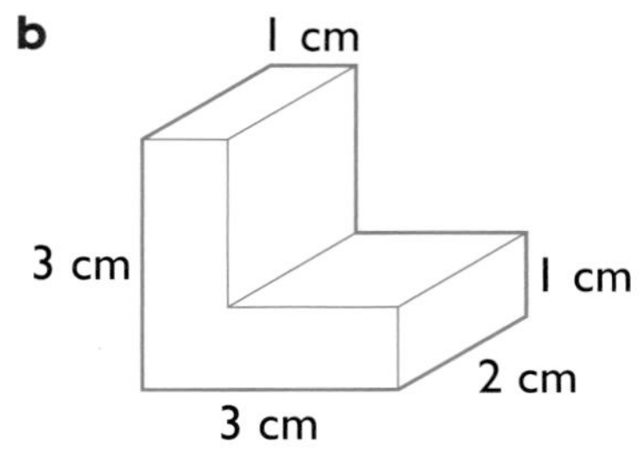

c

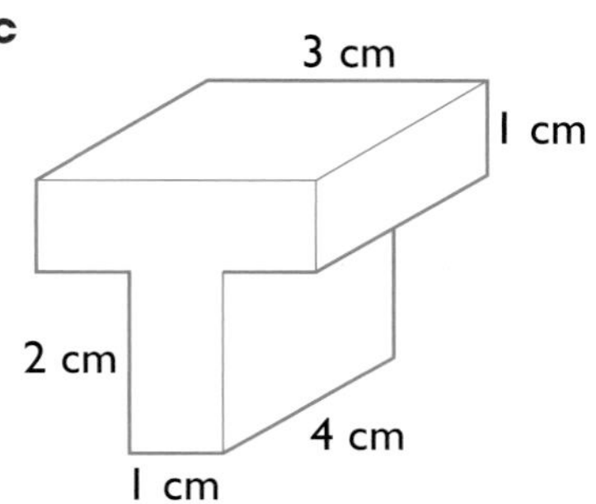

6 Michelle makes two cubes.
She glues them together, face to face.

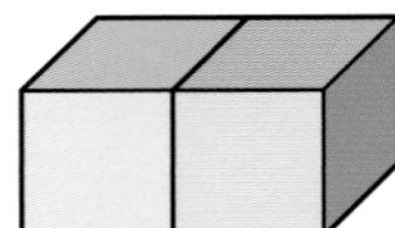

a What shape does she make?

b She makes two more cubes and adds them to the first two.

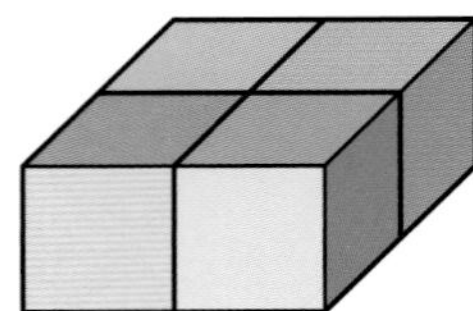

What shape does she make now?

c Michelle makes four more cubes.
She adds them to the first four to make a bigger cube.
Where does she place the new cubes?

7 Here are the five *platonic solids*.

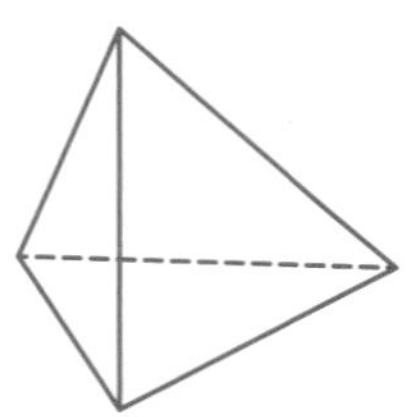
Tetrahedron

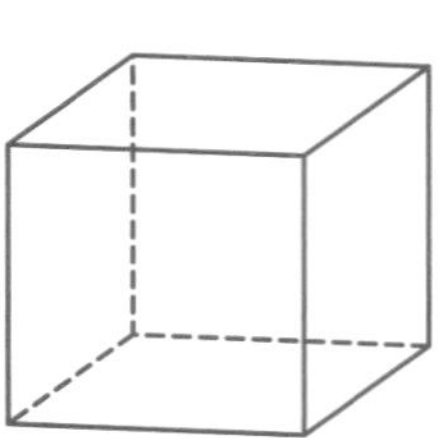
Cube

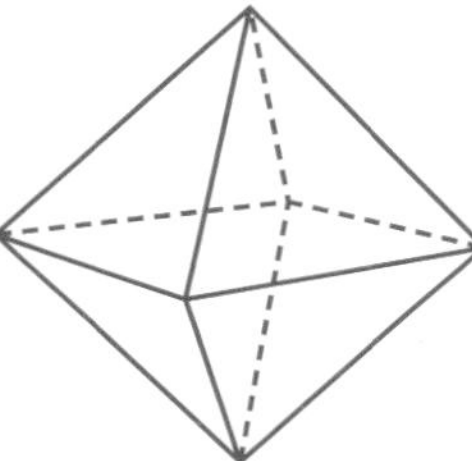
Octahedron

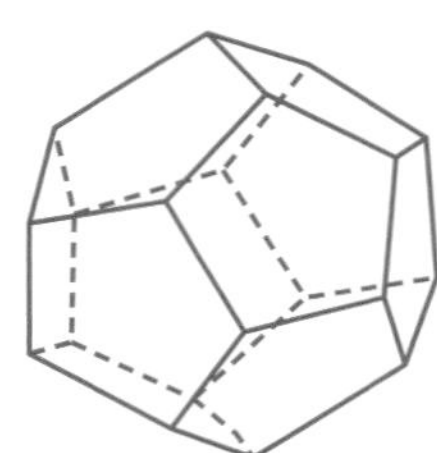
Dodecahedron

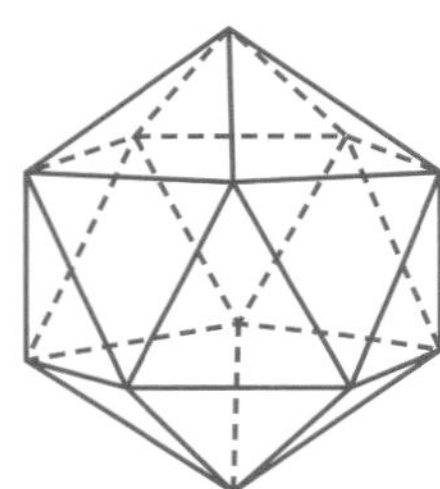
Icosahedron

Copy and complete this table.

Shape	Number of faces (*F*)	Shape of each face
tetrahedron		
cube		
octahedron		
dodecahedron		
icosahedron		

Plato was a Greek philosopher who was born in the fifth century BC.

8 Draw these cuboids on isometric paper.

a

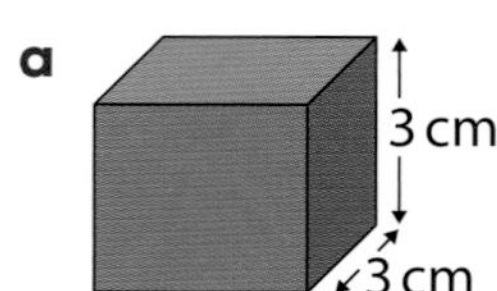

b

9 Draw these cuboids on isometric paper.

a

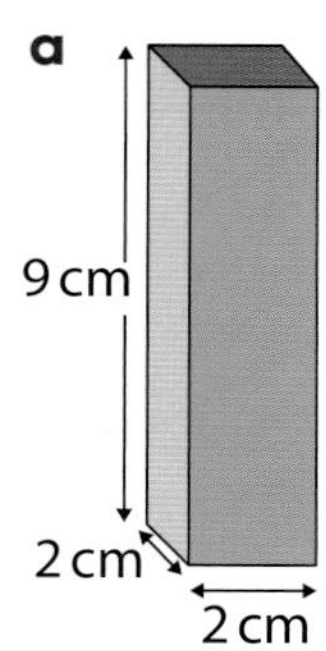

b

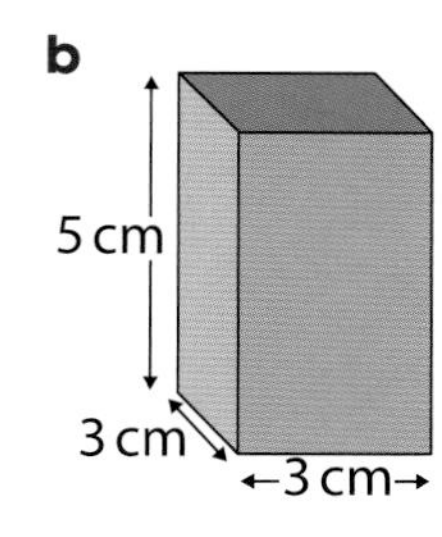

c

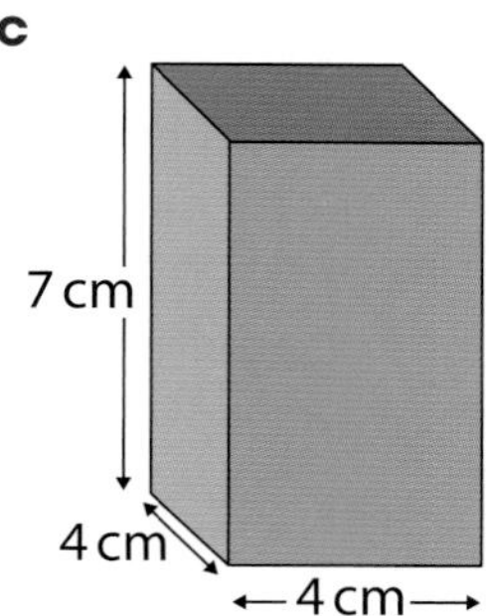

Skills practice B

1 **i** Complete each solid so that it is symmetrical in the mirror.

ii Draw the completed solid on isometric paper.

a

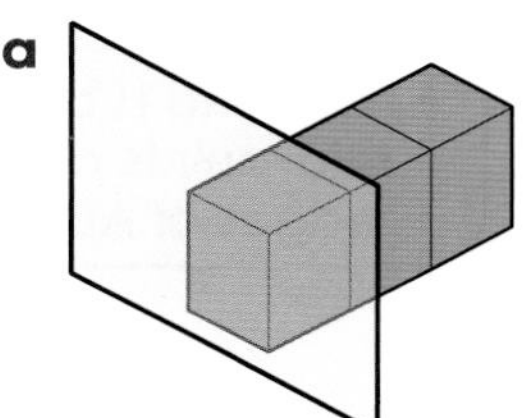

b

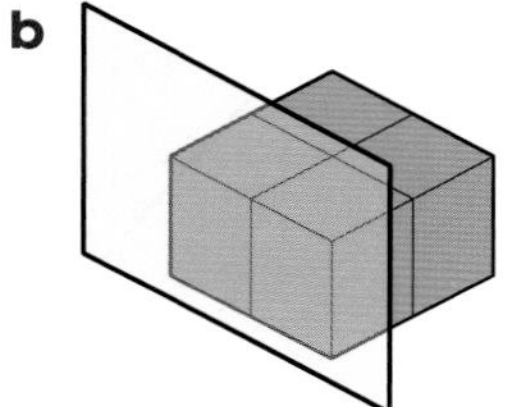

c

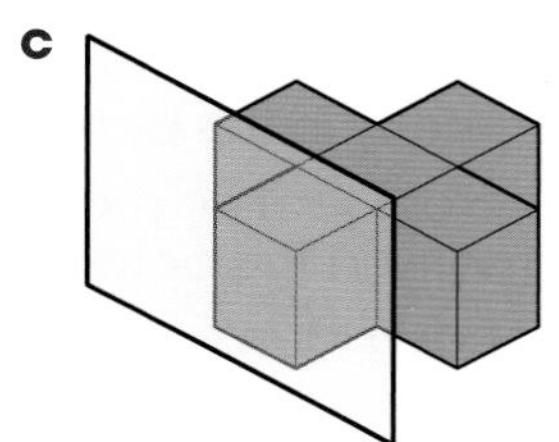

d

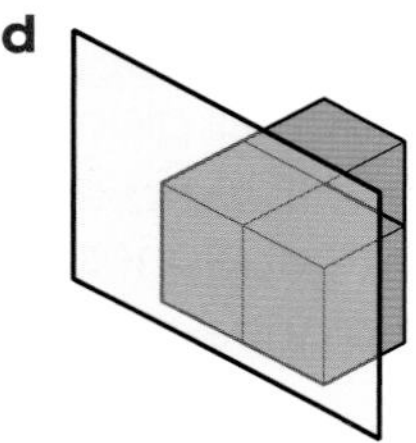

2 This shape is a cuboid with one corner removed.

a How many faces, edges and vertices does it have?

b Show that it obeys Euler's rule.

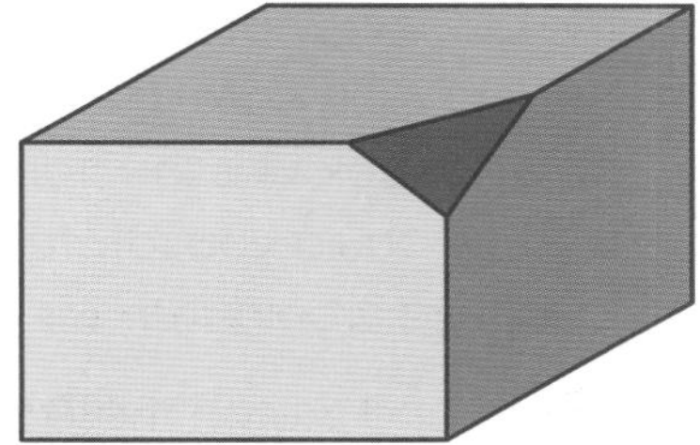

3 Use isometric paper to make 3-D representations of these shapes.

a

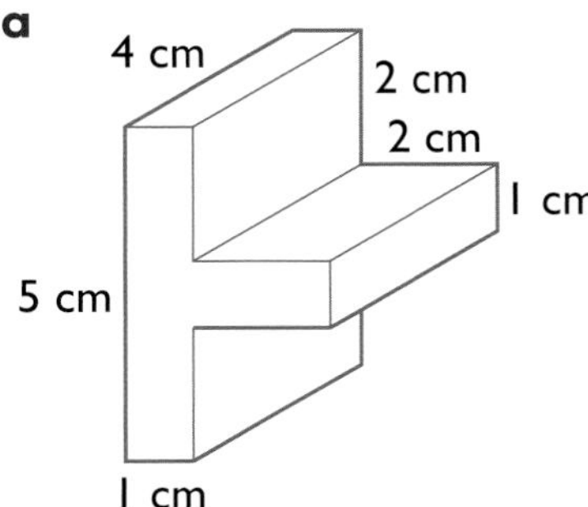

b

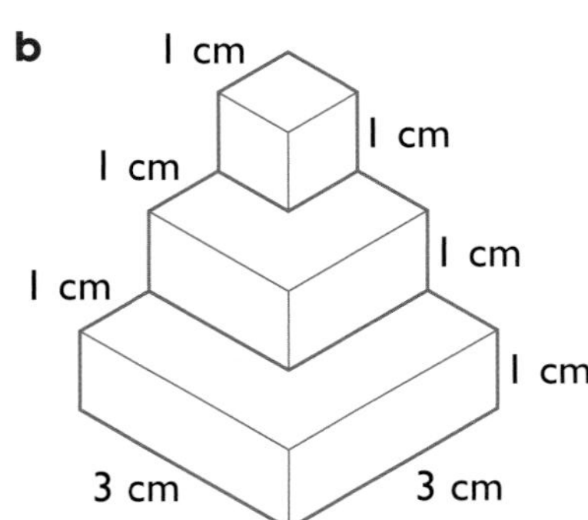

4 Different shaped triangles can be made by joining three vertices of a cube.

AEG is a right-angled triangle.

a How many different shaped triangles can be made by joining three vertices of the cube?

How many are there of each type?

How many are there altogether?

b What about for a cuboid?

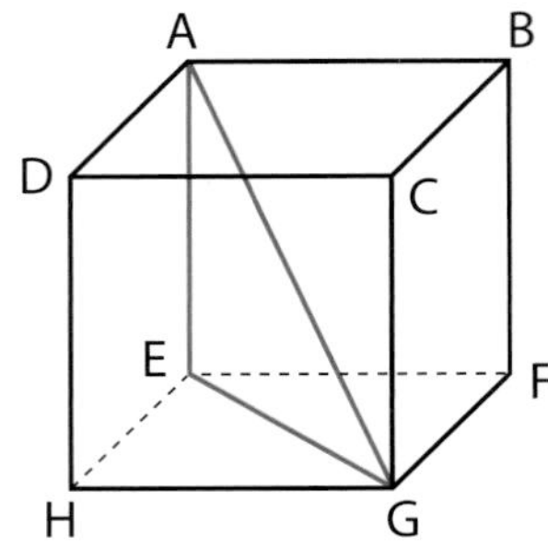

5 Look at this regular octahedron.

It is made by joining two identical square-based pyramids at their square bases.

a Using letters, identify all the vertical and horizontal planes of symmetry.

Using separate sketches, show each plane of symmetry.

b Describe the shape of the cross-section made by the horizontal plane of symmetry in this octahedron.

c Describe the shape of the cross-section made by the vertical planes of symmetry in this octahedron.

d Which planes of symmetry have identical shapes?

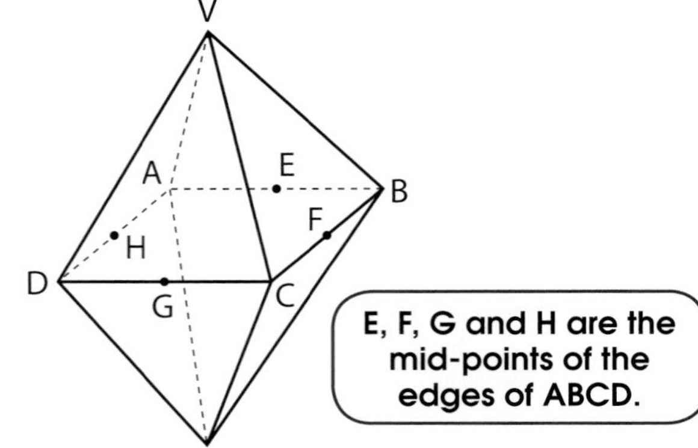

6 Look at this cube.

The line AB goes through the centre of the top and bottom of the cube.

The cube is rotated 90° clockwise about this line.

The line AB is called an axis of rotation.

a What colour is the front face of the cube now?

b How could you rotate the cube so that the top face is orange?

Where would the axis of rotation be?

7 Name and sketch these solids.

a A prism has ten faces and nine planes of symmetry.

Its cross-section is a regular polygon.

b A pyramid has seven faces and six planes of symmetry.

Its base is a regular polygon.

Wider skills practice

1 A solid shape is made from four identical cubes joined together.

Cubes are always joined as shown.

a Find the solid with the largest surface area.

b Find the solid with the smallest surface area.

c Draw the solids in **a** and **b** on isometric paper.

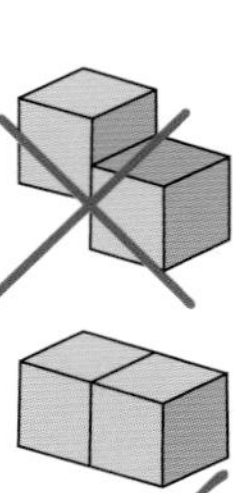

Applying skills

Problem solving

1 **a** Three cubes are stuck together face to face.
Draw all of the possible shapes on isometric paper.

b A fourth cube is added.
Draw all of the possible shapes on isometric paper and mark on the planes of symmetry.

c How many of the possible solids are

i prisms

ii cuboids

iii cubes?

Reviewing skills

1 Match these names to the shapes below.

cuboid | cylinder | hexagonal prism | irregular prism | equilateral triangular prism | isosceles triangular prism

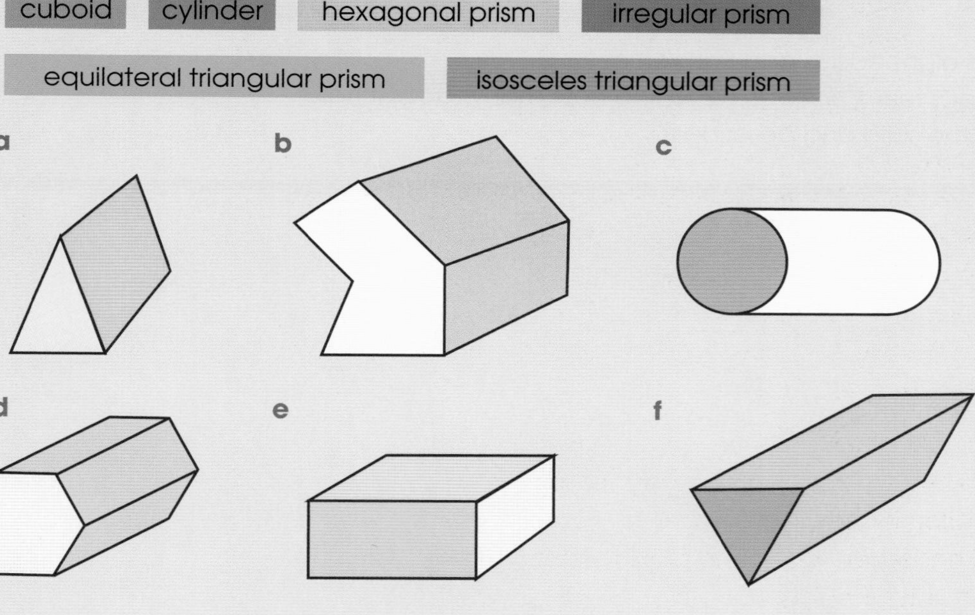

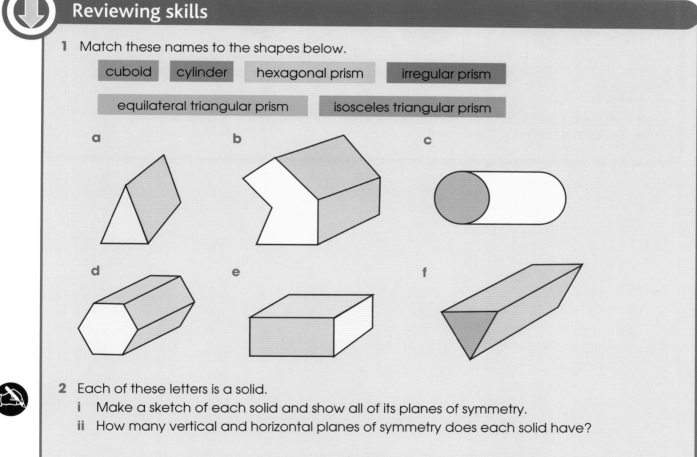

2 Each of these letters is a solid.

i Make a sketch of each solid and show all of its planes of symmetry.

ii How many vertical and horizontal planes of symmetry does each solid have?

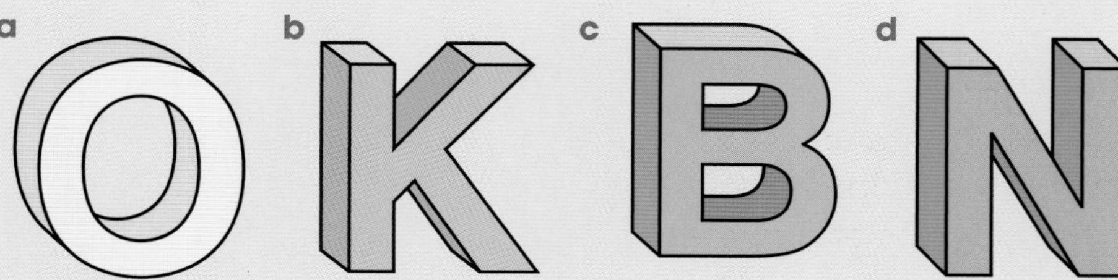

3 Two of these solids can be joined to make a cuboid.
Draw the cuboid on isometric paper and colour it to show how the two fit together.

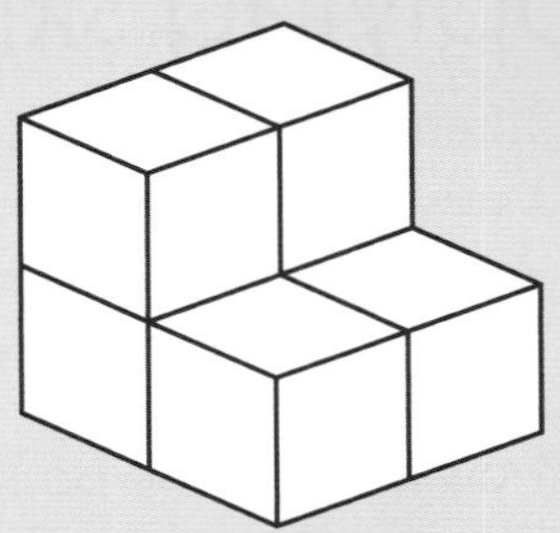

4 **i** Complete each solid so that it is symmetrical in the mirror.
ii Draw the completed solid on isometric paper.

a

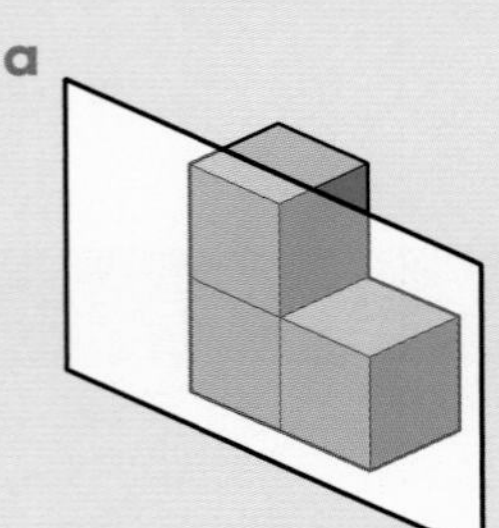

b

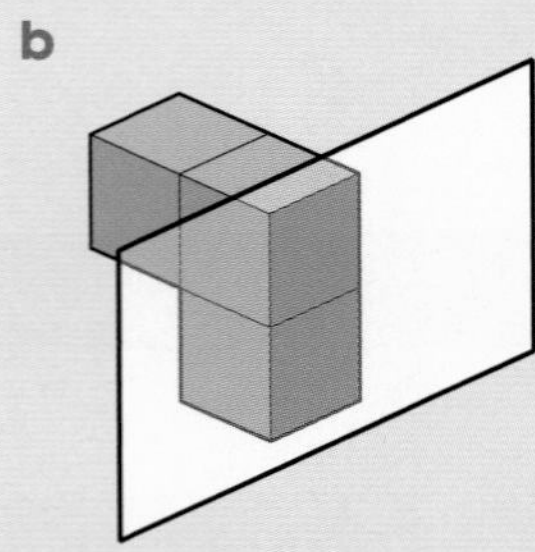

c

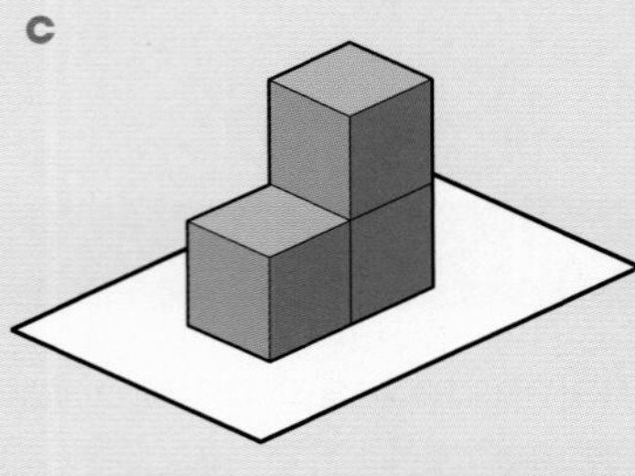

5 A solid has eight edges and five vertices.
a Use Euler's rule $V + F = E + 2$ to find how many faces it has.
b Sketch the solid and describe it.

Building skills

Example outside the Maths classroom

Sandbags

Toolbox

2-D nets are designed to be folded to make 3-D objects.

The same object may have different nets.

Whatever their arrangement, the sections of the 2-D design must match the faces of the 3-D object that is to be made.

Nets need flaps to enable the 3-D shape to become rigid.

Clever designers can design the flaps to make the box rigid without glue or staples.

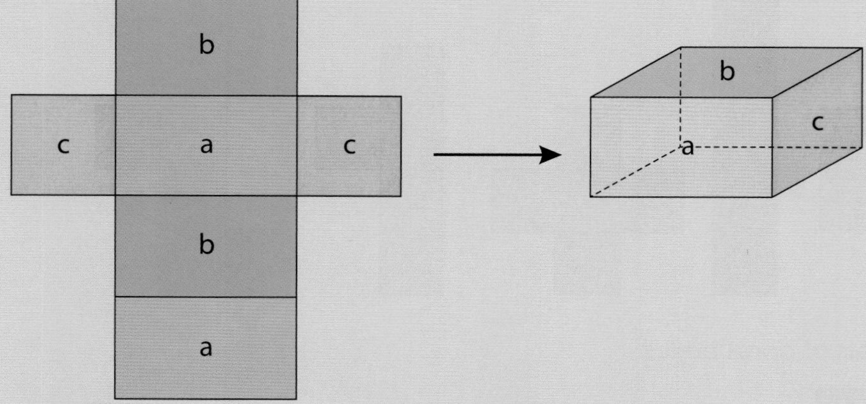

Nets can be used to find the surface area of shapes.

Example – Adding flaps

Is there a relationship between the number of edges of a net and the number of flaps needed to make the solid rigid?

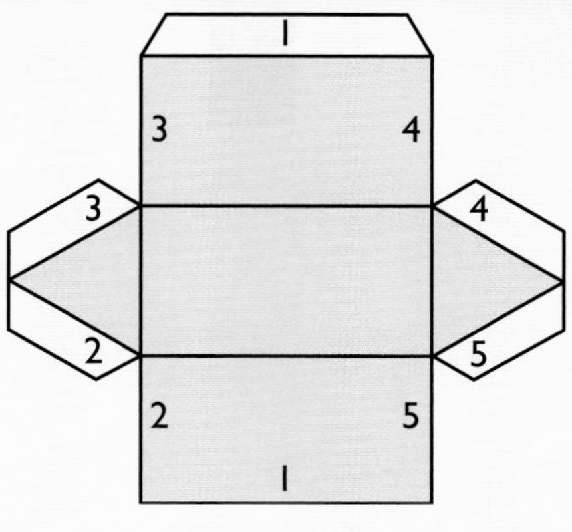

Solution

Yes; the minimum number of flaps needed is half the number of edges that need to be joined because each edge joins with another edge.

Example – Designing a net

A pentomino is made by connecting five congruent squares along their edges.

a Draw as many different pentominos as you can.

b Which pentominos can be a net of an open box?

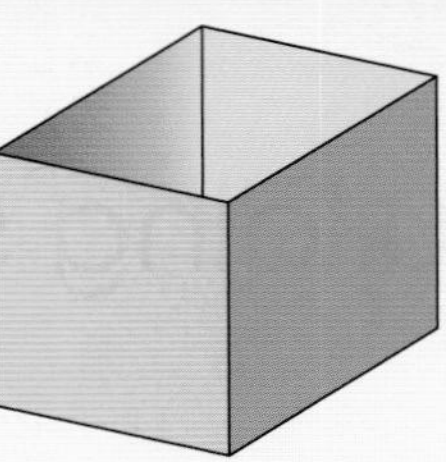

Solution

a There are twelve different pentominos.

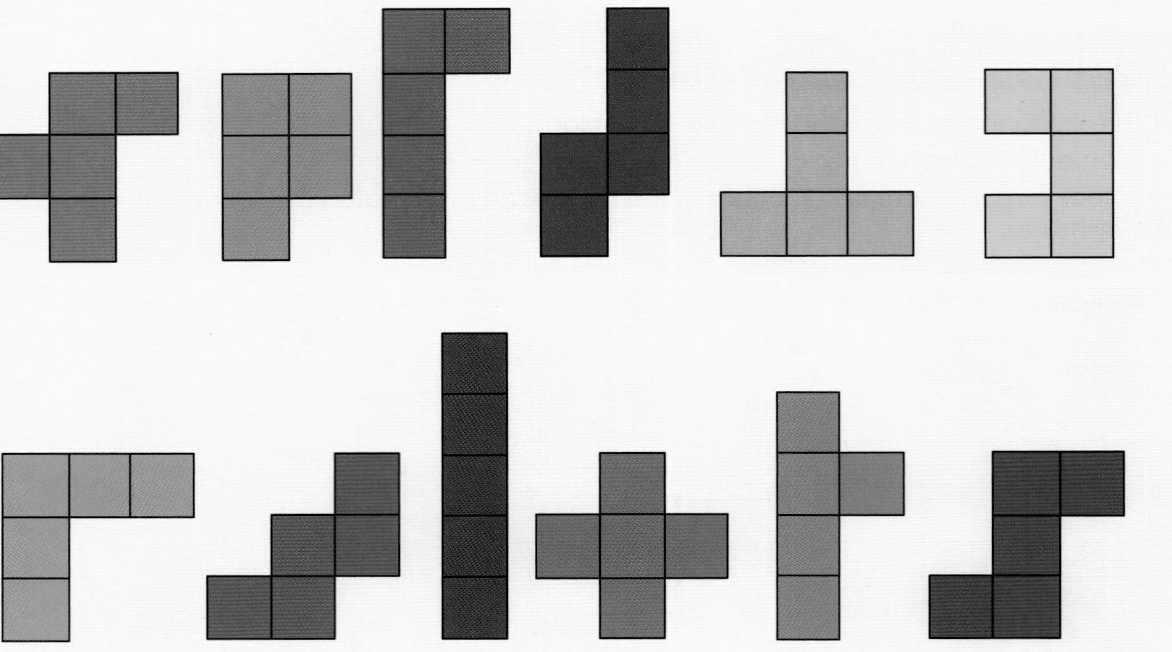

b These pentominos can be nets of open boxes.

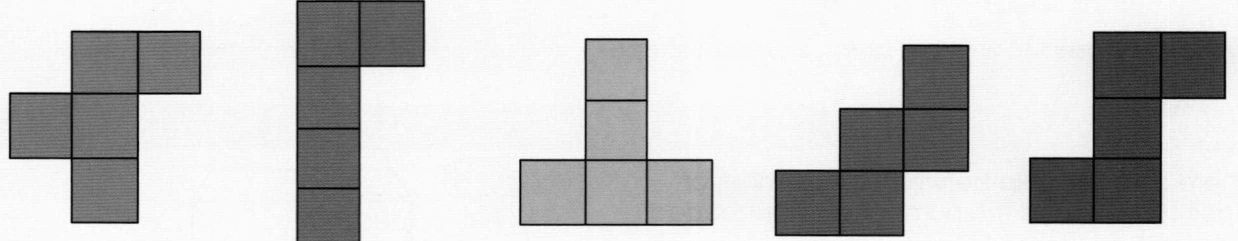

Example – Matching the dimensions of a net and its solid

These are the nets of two cuboids.

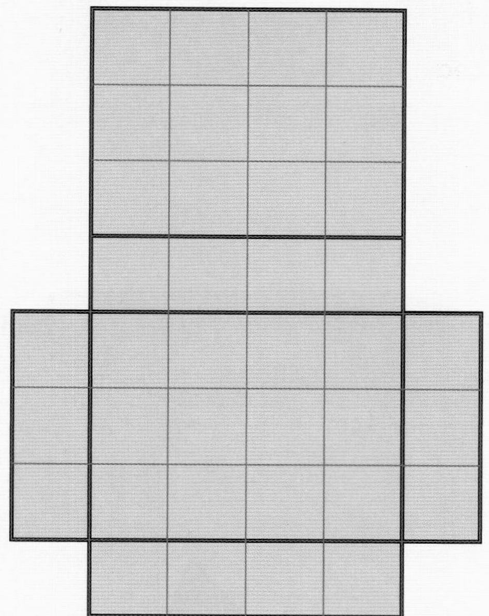

They are made of centimetre squares.
Compare the surface areas of the two cuboids.

Solution

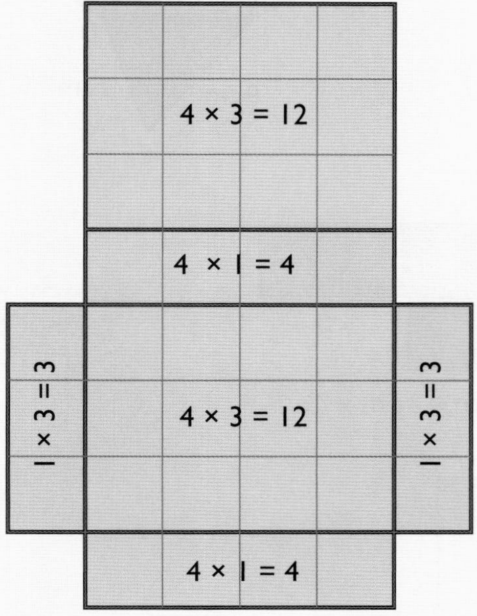

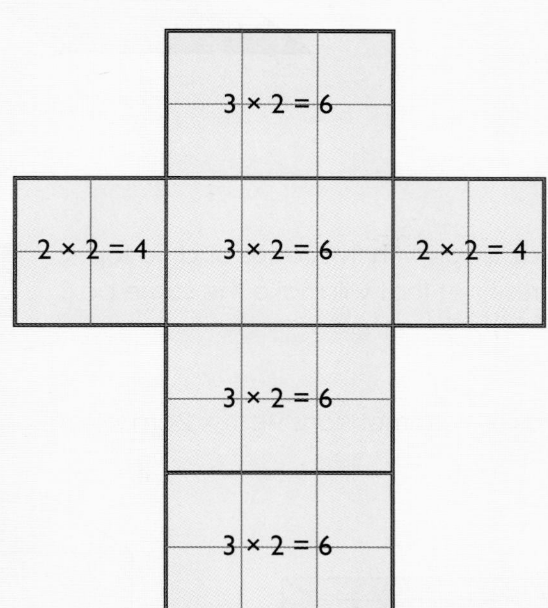

Surface area of pink cuboid = 12 + 4 + 12 + 4 + 3 + 3
= 38 cm^2

Surface area of blue cuboid = 6 + 6 + 6 + 6 + 4 + 4
= 32 cm^2

Remember:

- The shape and dimensions of the net sections must match the dimensions of the solid's faces.

Skills Practice A

1 **a** Design a net for this cuboid.

b Add flaps to your net so that the box can be assembled.

c How many flaps are needed?

d Could the flaps be placed on different edges of your net?

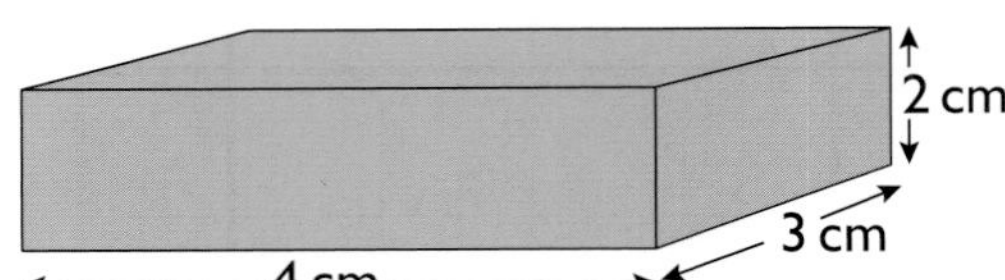

2 The diagram shows a square-based pyramid and its net.
Find the length of these sides on the net.

a BC

b CD

c FG

d GH

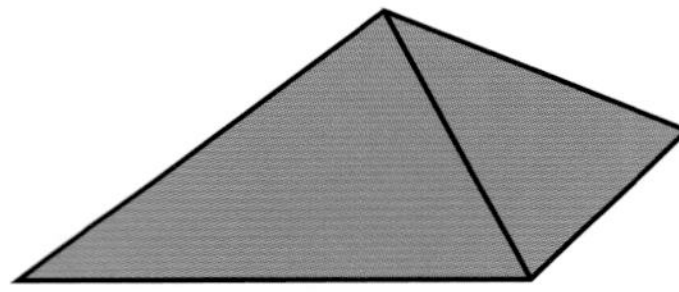

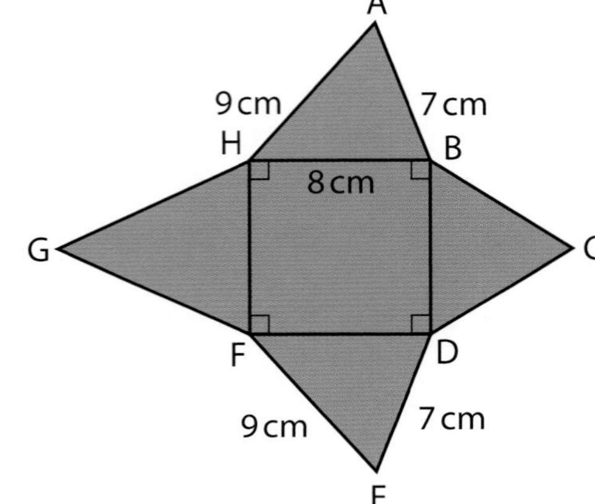

3 This net makes a box with five faces and no top.
Draw a different net that will make the same box.

4 These cuboids have dimensions 9 cm × 2 cm × 2 cm and 3 cm × 3 cm × 4 cm.

i

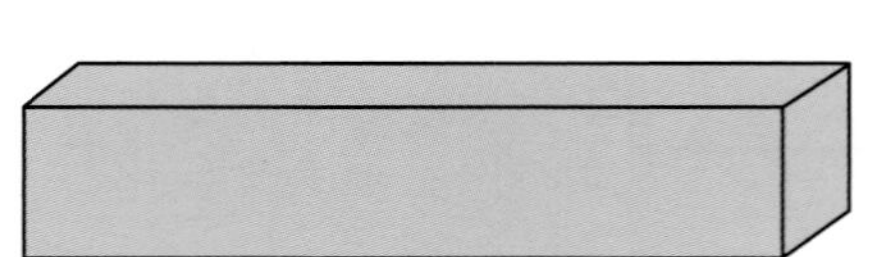

ii

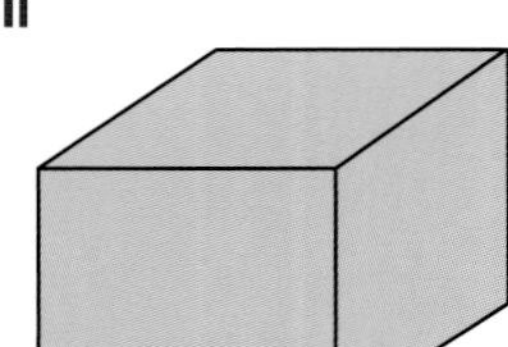

a Draw a net for each cuboid.

b Compare the area of material needed for each net.

5 Match these solids to their nets.

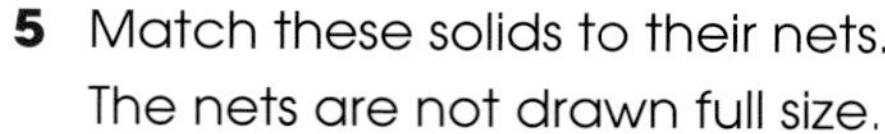

The nets are not drawn full size.

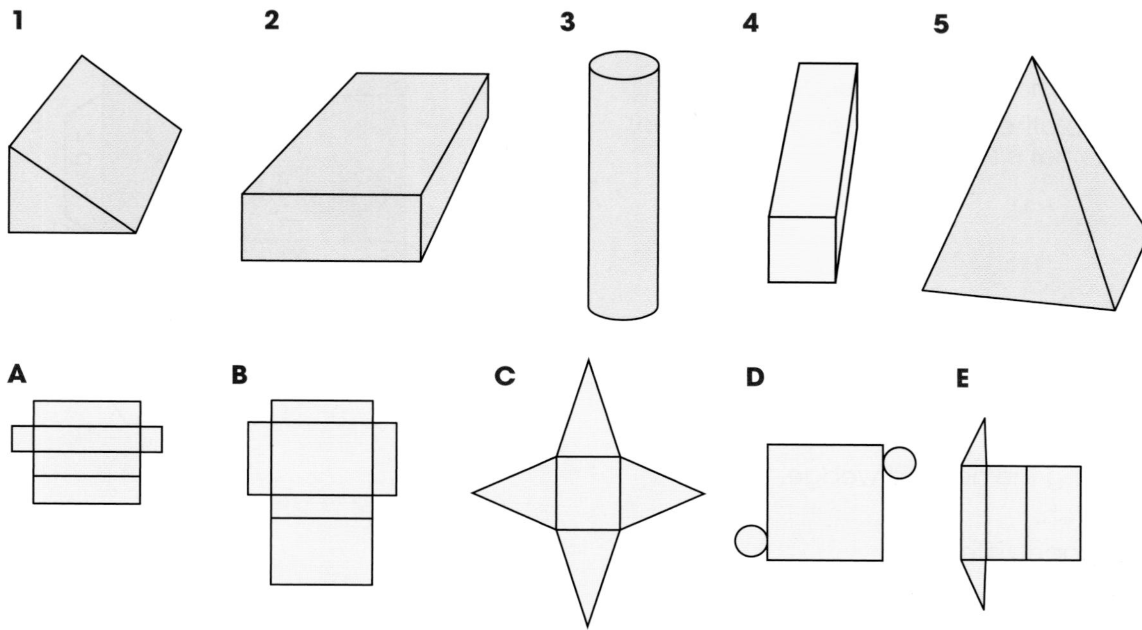

Reasoning

6 **a** Draw four equilateral triangles arranged to form the net of a regular tetrahedron.

b Add flaps to your net.
How many flaps do you need?
Does it matter where you place them?

Skills practice B

1 A cuboid is 9 cm long. It has square ends 2cm × 2cm.

a Draw a sketch of the cuboid.

b Draw a net for the cuboid.

c Find its surface area.

2 The diagram shows a cuboid and its net.

a When the net is made up, which points meet at E?

b Which points meet at G?

c Which points meet at the corner that the arrow points to in the picture?

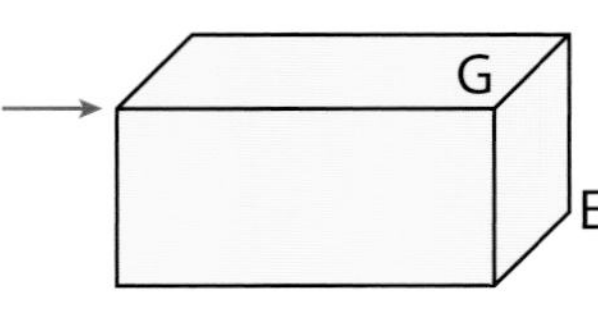

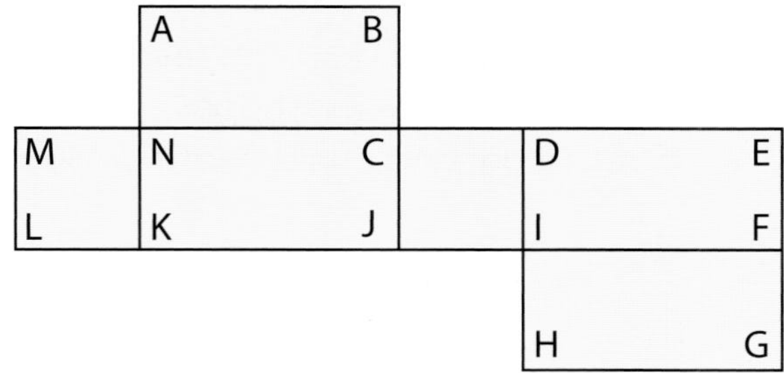

3 This is the net of a triangular prism.

a To which edge, A, B, C, D, E or F, do you glue flap 1?

b To which edge, A, B, C, D, E or F, do you glue flap 2?

c To which edge, A, B, C, D, E or F, do you glue flap 3?

d How many more flaps are required?

e Which two other corners does corner 1 meet when the net is folded?

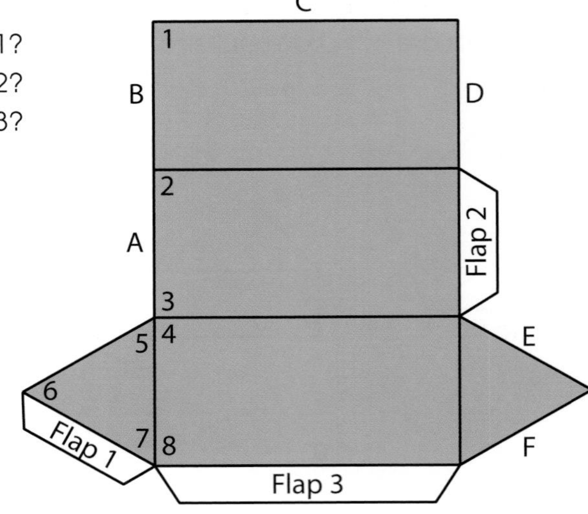

4 Sam is drawing the net of a wedge.

This is his sketch.

It is not very accurate.

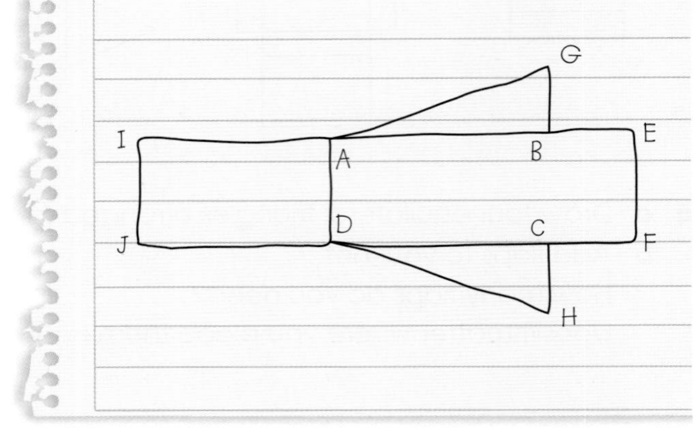

a Say whether these statements are true or false.

i Angle ABG should be a right angle.

ii AI should be the same length as AB.

iii BCFE should be a square.

iv When it is folded up J, H and F will all be the same point.

v AG and DH will fold on to the same edge.

b Draw a sketch of the solid shape.

5 A hexomino is made by connecting six congruent squares.

Draw as many different hexominoes as you can.

Which of your hexominoes are the nets of a cube?

6 The diagram shows a square-based pyramid VABCD.

The vertex V is directly above C.

AB = BC = CD = DA = 6 cm and VC = 8 cm.

a Draw a net for this shape.

b Cut it out and check that it works.

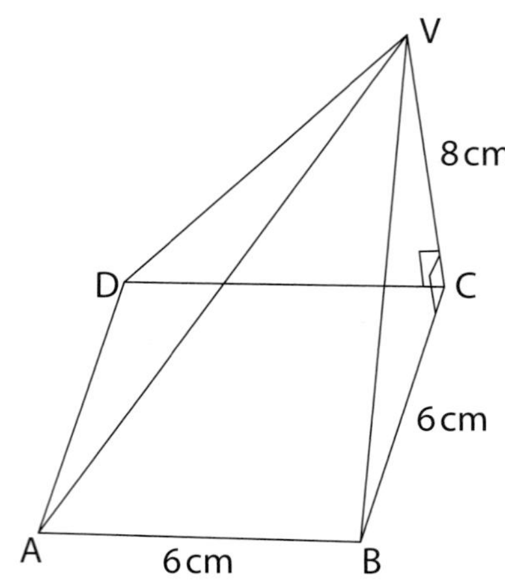

Reasoning

7 **a** How many different shapes can you make by sticking six congruent squares together? Draw as many as you can.

b Opposite faces of a die add up to 7.

For each of the arrangements in part **a** that will make a cube, number the squares so that the net will fold into a die.

Wider skills practice

Problem solving

1 Tariq has designed nets for these two small boxes.

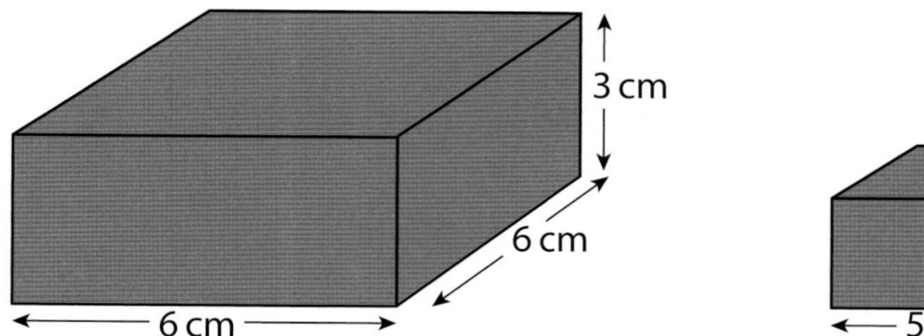

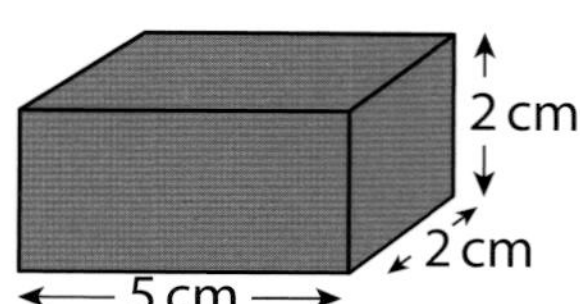

Stella has designed one box that is 8 cm × 4 cm × 4 cm.

She says it holds the same amount as both of Tariq's boxes.

a Is Stella correct?

b Who uses less cardboard?

How much cardboard is saved with this design?

2 Copy this net on to centimetre-squared paper, add flaps and cut it out.

Stick the cuboid together using the flaps.

Label the points shown in the diagram.

Label the mid-points of edges AE, BF, CG and DH with the letters Q, R, S and T.

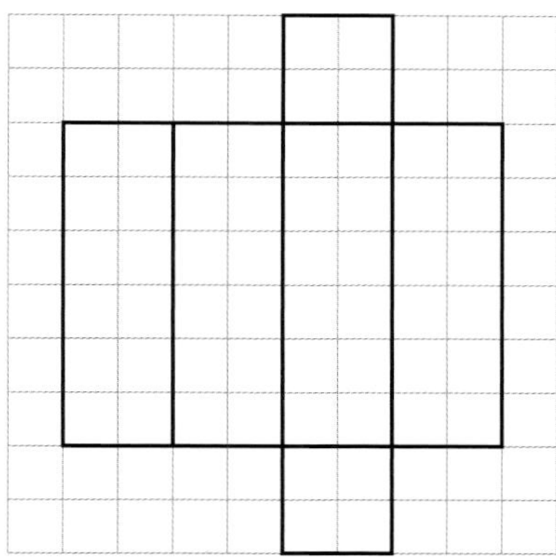

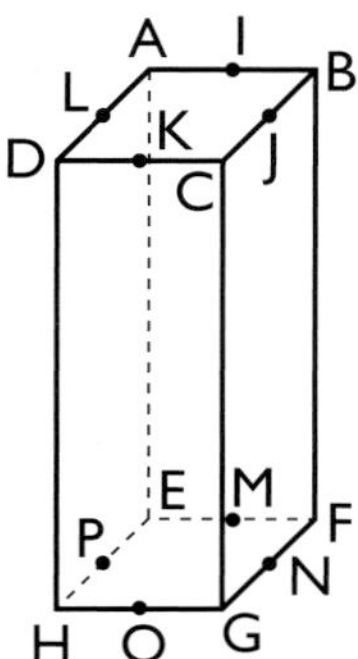

a Using letters, identify the horizontal plane of symmetry in this cuboid.

b This cuboid has four vertical planes of symmetry.

Using letters, identify them.

3 Ask your teacher for the nets of a tetrahedron and an octahedron.
Glue the nets to make the solids.

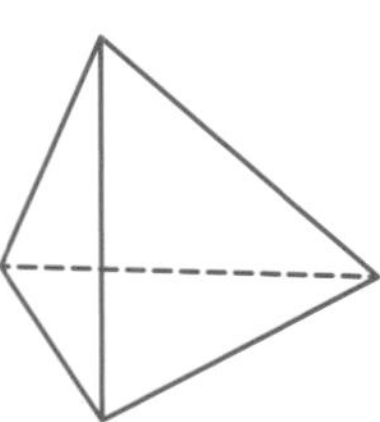

Tetrahedron

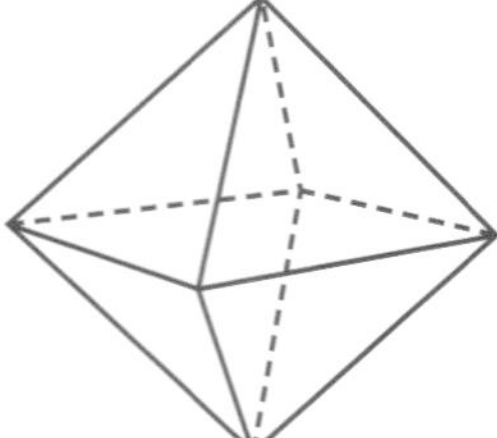

Octahedron

a These shapes are not prisms. Explain why.

b Count the number of faces, edges and vertices for each shape.
Do these shapes obey Euler's rule?

Applying skills

Problem solving

1 An octahedron is a solid with eight faces.
A regular solid has faces which are regular polygons.

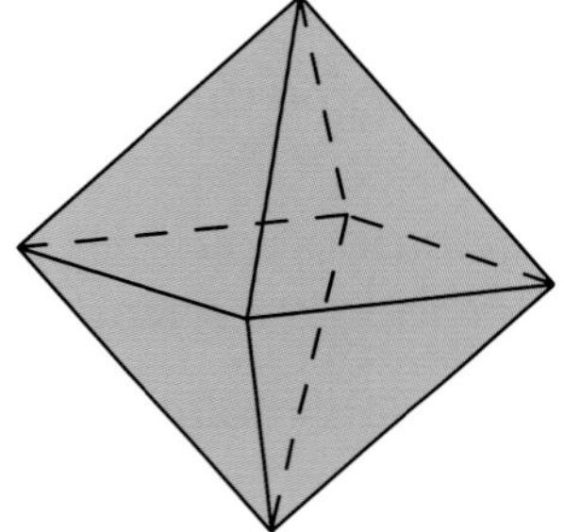

a Explain why this solid is a regular octahedron.

b Here is a net for the regular octahedron in part **a**.

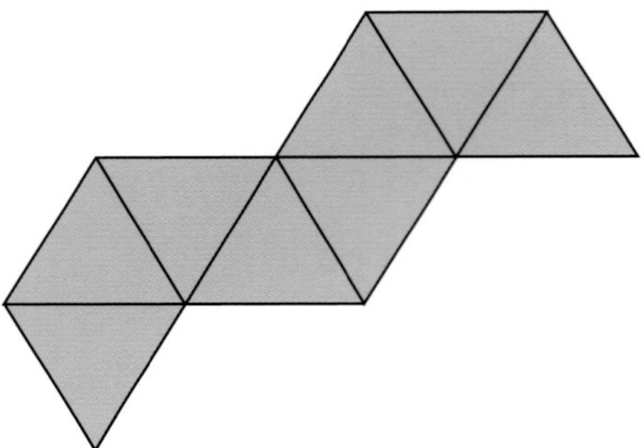

Draw the net on isometric paper.
Cut it out.
Fold along the lines to make the octahedron.

c There are ten different nets for a regular octahedron.
How many of them can you find?

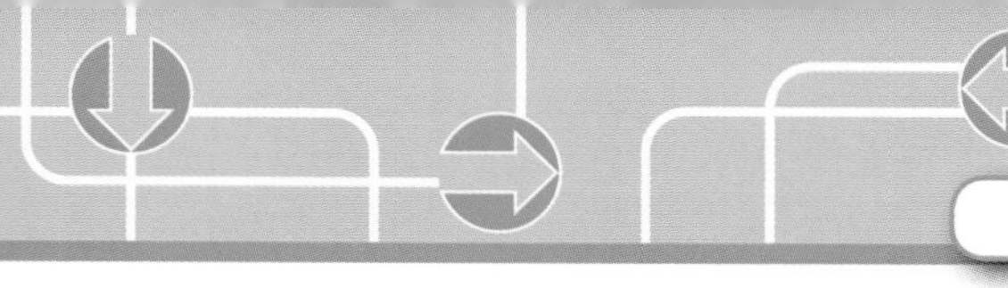

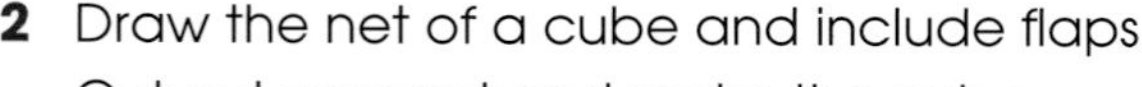

Problem solving

2 Draw the net of a cube and include flaps.
Cut out your net and make the cube.

a Choose one vertex.
Draw, in red, the diagonals of the three square faces that meet at that vertex.
These diagonals go to three new vertices.
Draw, also in red, the diagonals from these vertices.
Your red lines are the edges of a new three-dimensional shape.
What is it?

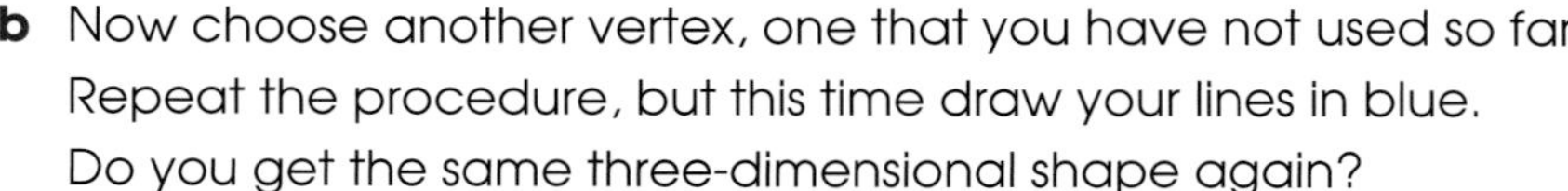

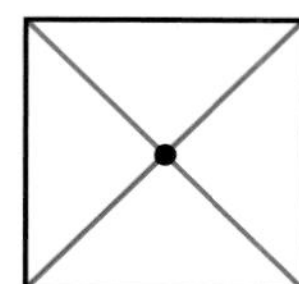

b Now choose another vertex, one that you have not used so far.
Repeat the procedure, but this time draw your lines in blue.
Do you get the same three-dimensional shape again?

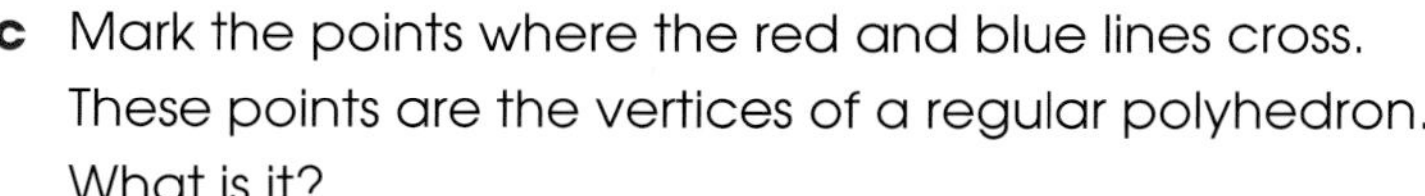

c Mark the points where the red and blue lines cross.
These points are the vertices of a regular polyhedron.
What is it?

Problem solving

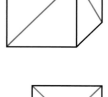

3 The London Chocolate Company is launching a new sweet.
It will be sold in pyramid-shaped boxes.
Two boxes are to be made out of a sheet of A3 card, 29.7 cm by 42.0 cm.
The base of each box is to be 20 cm × 20 cm.
The volume is to be as large as possible.
Show how they can do this.

Reviewing skills

1 **a** Draw the net of a cuboid which is 5 cm × 3 cm × 2 cm.
b Find the surface area of your net.
c Find the volume of the cuboid.

2 The ends of a prism are equilateral triangles with sides of 3 cm.
The prism is 7 cm long.
a Draw a net of the prism.
Add the minimum number of flaps.
b The net is to be drawn on a rectangular piece of cardboard.
Will it fit on a piece 13 cm × 9 cm?

Unit 3 • Volume and surface area of cuboids • Band f

Building skills

Example outside the Maths classroom

Packaging

Toolbox

The **volume** of a solid shape is the space inside it.

It is measured in cubic units such as cm^3 and m^3.

This is a centimetre cube.

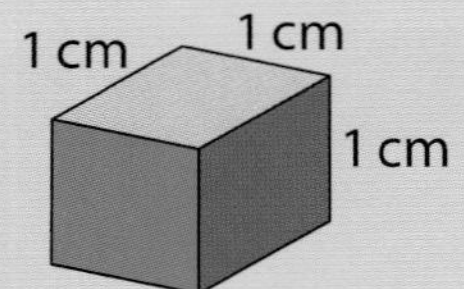

It has a volume of 1 cubic centimetre (1 cm^3).

You can sometimes find the volume of a cuboid by counting the number of cubes that fit into it.

This cuboid is made of 12 centimetre cubes.

So its volume is 12 cm^3.

Another way to find the volume of a cuboid is to use the formula

volume of a cuboid = length × width × height.

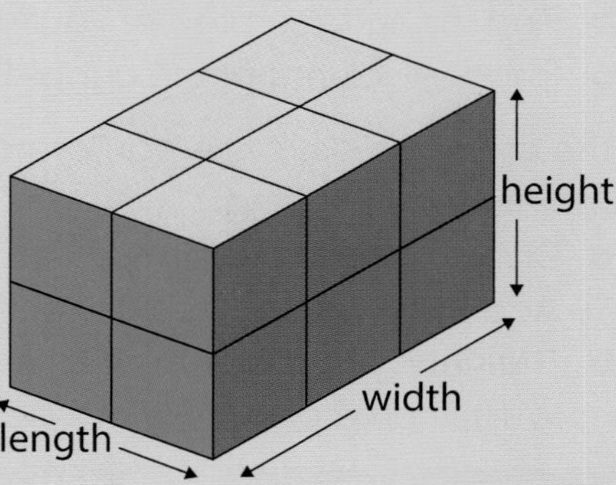

The surface area of a solid is the total area of all its faces.

In a cuboid, the front and back are the same; so are the top and bottom, and the two sides. So

surface area of a cuboid
= 2 × (area of base + area of front + area of one side)

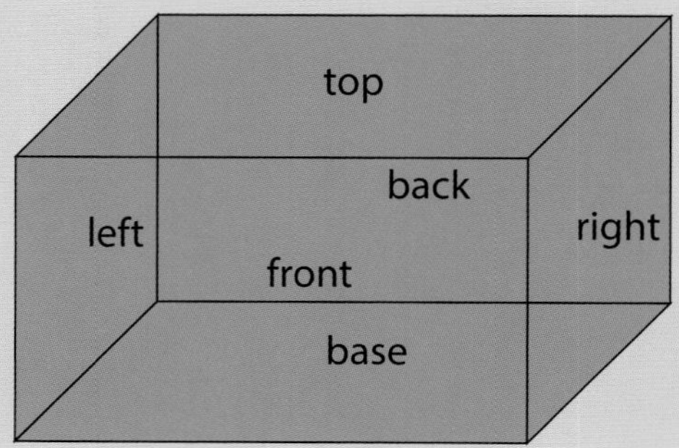

Example – Finding the volume and surface area of a cuboid

Find the volume and surface area of this cuboid.

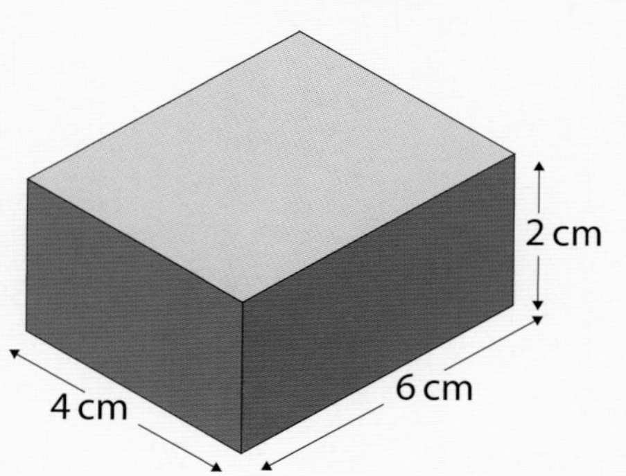

Solution

Volume of a cuboid = length × width × height

$= 4 \times 6 \times 2$

$= 48$

The volume of the cuboid is 48 cm^3.

The surface area of its faces is

Base:	$4 \times 6 = 24$	Top:	$4 \times 6 = 24$
Front:	$4 \times 2 = 8$	Back:	$4 \times 2 = 8$
Right:	$6 \times 2 = 12$	Left:	$6 \times 2 = 12$

So the total surface area $= 24 + 24 + 8 + 8 + 12 + 12$

$= 2 \times (24 + 8 + 12)$

$= 2 \times 44$

$= 88$

The surface area of the cuboid is 88 cm^2.

Example – Finding surface area from a net

Here is a net of a cuboid.

Find the surface area of the cuboid.

Each rectangle makes a face of the cuboid. The green rectangle is the bottom face.

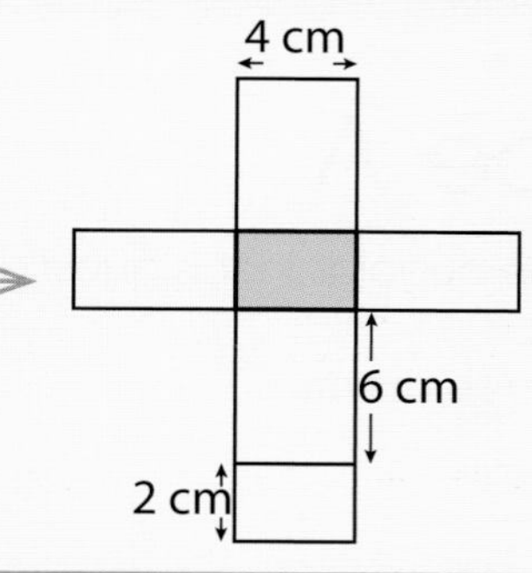

Solution

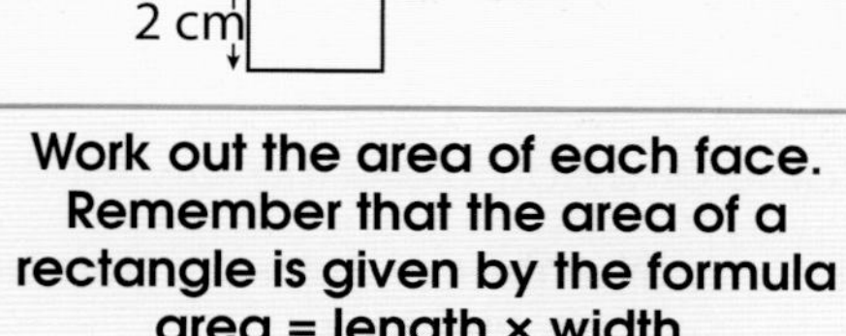

Work out the area of each face.
Remember that the area of a
rectangle is given by the formula
area = length × width.

The surface area of the faces is

Base:	$4 \times 2 = 8$	Top:	$4 \times 2 = 8$
Front:	$4 \times 6 = 24$	Back:	$4 \times 6 = 24$
Right:	$6 \times 2 = 12$	Left:	$6 \times 2 = 12$

Total surface area $= 2 \times (8 + 24 + 12)$

$= 88$

The surface area of the cuboid is 88 cm^2.

Example – Finding the surface area of an open box

Look at this open cardboard box.
It has no top.
Find its surface area.

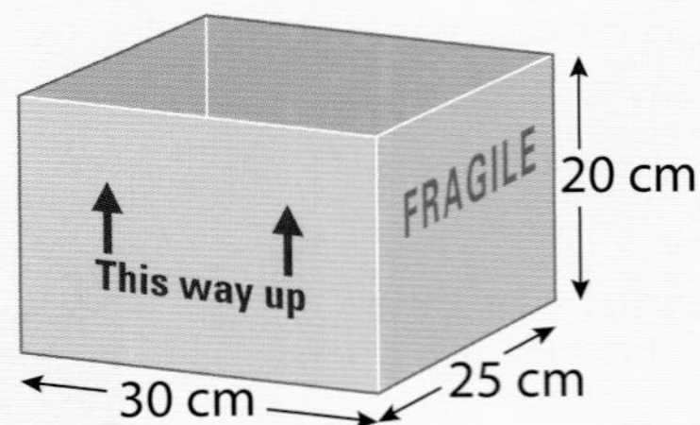

Solution

Area of bottom	$= 30 \times 25$	$= 750\,cm^2$
Area of front and back	$= 2 \times 30 \times 20$	$= 1200\,cm^2$ ← **The front and back have the same area.**
Area of both sides	$= 2 \times 25 \times 20$	$= 1000\,cm^2$ ← **Both sides have the same area.**
Total surface area	$= 750 + 1200 + 1000$	$= 2950\,cm^2$

Remember:

- You can calculate volume by counting cubes inside a shape or by using a formula.
- Volume is measured in cubic units such as cm^3.
- The surface area of a cuboid is the total of the areas of its six faces.
- To find the surface area of an open box, do not include the top.

Skills practice A

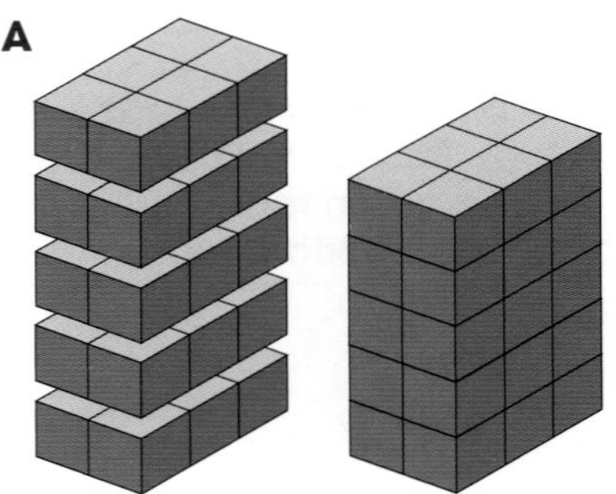

1 The cubes making up the cuboids in this question are centimetre cubes.

a Each layer of cuboid A is made of six cubes.
It has five layers.
What is its volume?
Copy this table and complete the row for cuboid A.

Cuboid	Number of cubes in one layer	Number of layers	Total number of cubes	Volume
A	6	5		
B	10			
C				
D				

b Each layer of cuboid B is made of ten cubes.
It has three layers.
Complete your table for cuboid B.

c Complete your table for cuboids C and D.

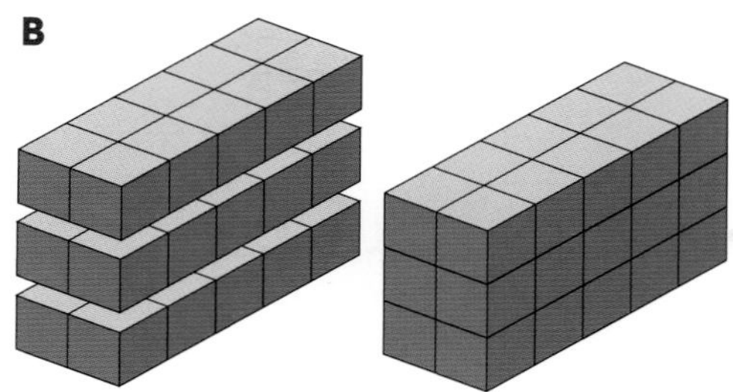

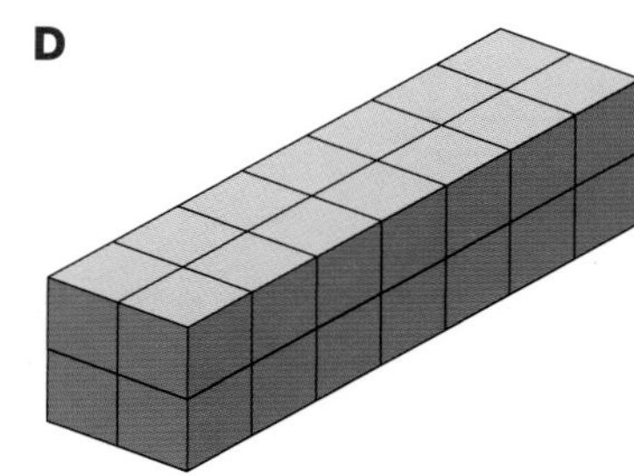

Reasoning

2 John has made these cuboids with building bricks.
Each brick is $1\,cm^3$.

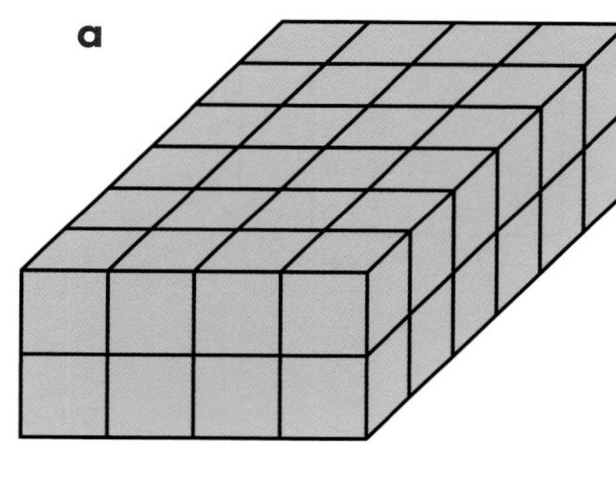

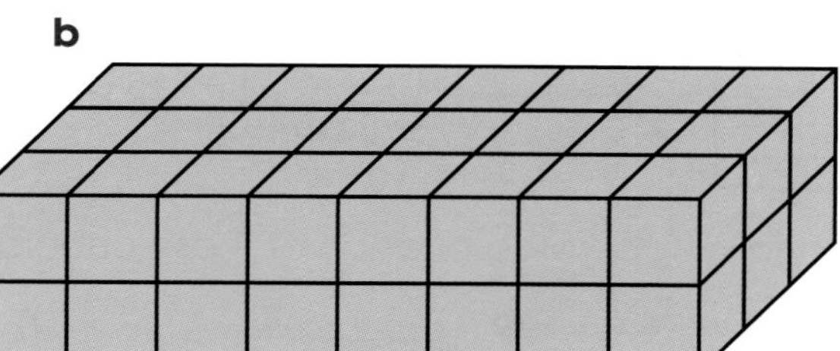

a Find the volume of each of these cuboids.
What do you notice?

b Describe two more cuboids with the same volume.

3 Find the volume of each of these cuboids.
They are not drawn to scale.

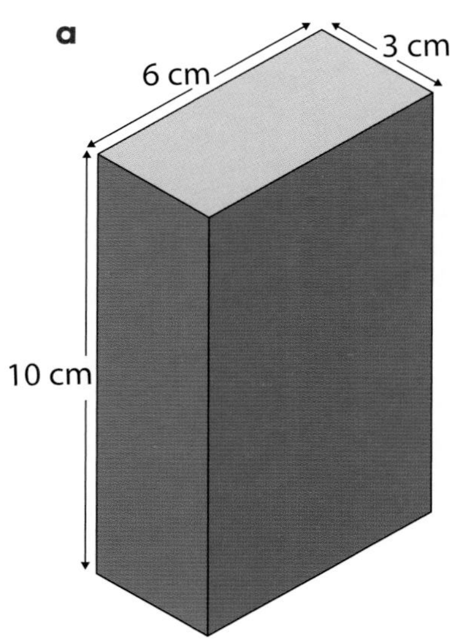

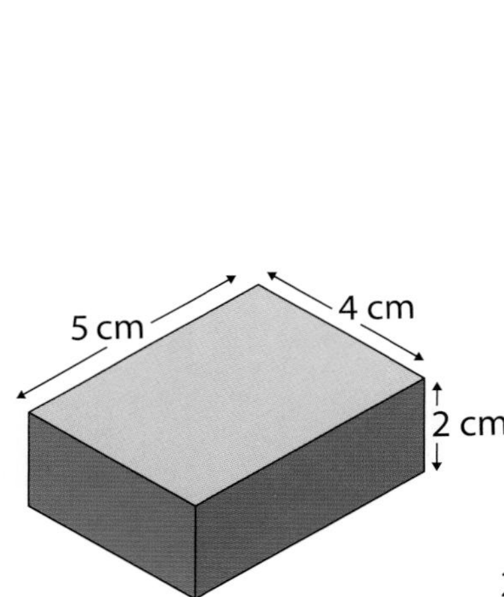

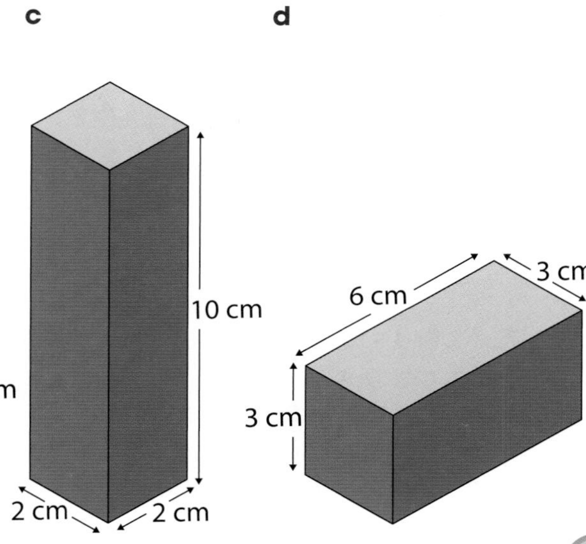

4 Calculate the surface area of each of these cuboids.

a 6 cm, 2 cm, 3 cm

b 10 cm, 3 cm, 4 cm

c 5 cm, 5 cm, 5 cm

d

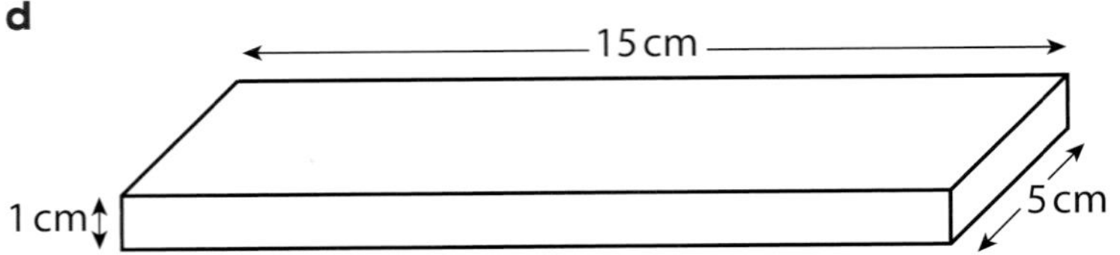

5 Find the surface area of each of these cuboids.

a 7 cm, 6 cm, 5 cm

b 8 cm, 15 cm, 10 cm

c

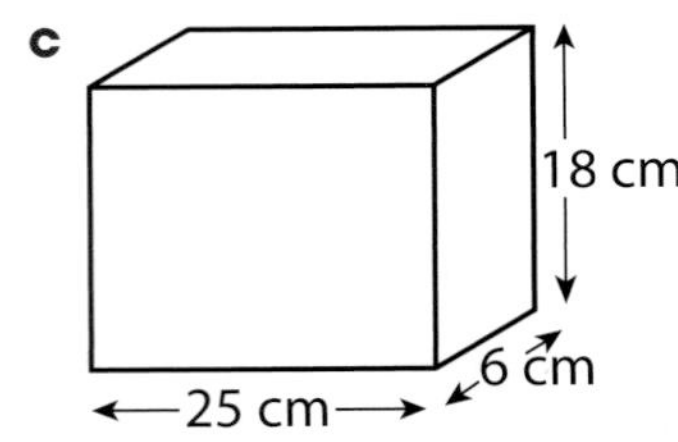

6 Find the volume and surface area of each of these cuboids.

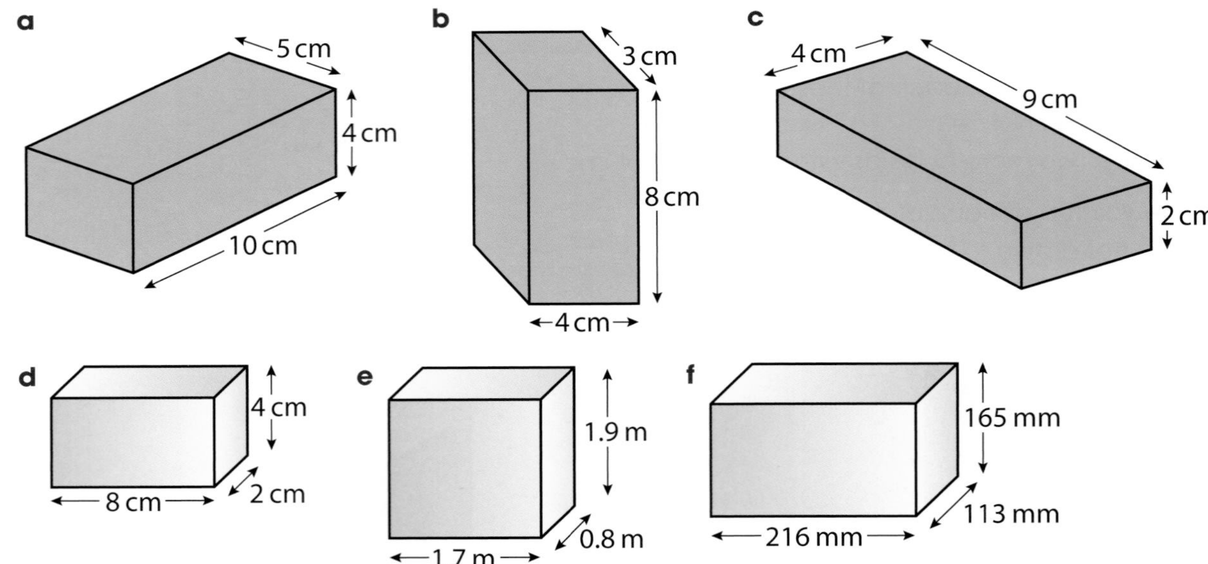

7 A child's toy brick is in the shape of a cube.
Each side is 4 cm long.

a Find the area of one face of the cube.

b Find the total surface area of the cube.

c Find the volume of the cube.

Skills practice B

1 How many of these matchboxes would fill a cube of side 10 cm?

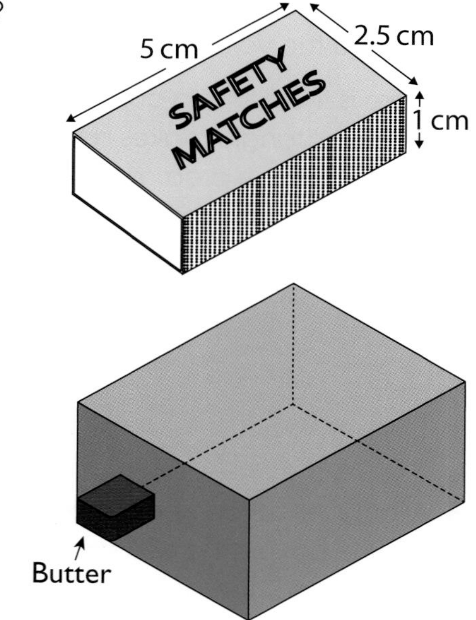

2 A pack of butter weighs 250 g and has measurements 6 cm × 4 cm × 10 cm.

Packs of butter are placed in a box with measurements 18 cm × 16 cm × 50 cm.

a How many packs of butter would fill the box?

b What is the weight of butter in the box?

Reasoning

3 **a** **i** How many sugar cubes fit in one layer of the box?

ii How many layers are there in the box?

iii How many cubes are there altogether?

iv What is the volume of the box?

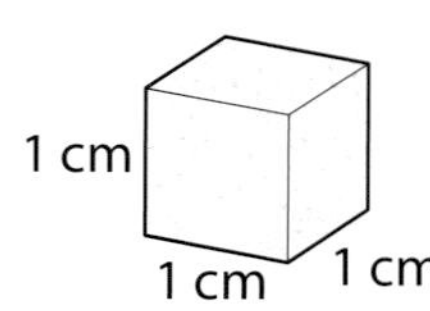

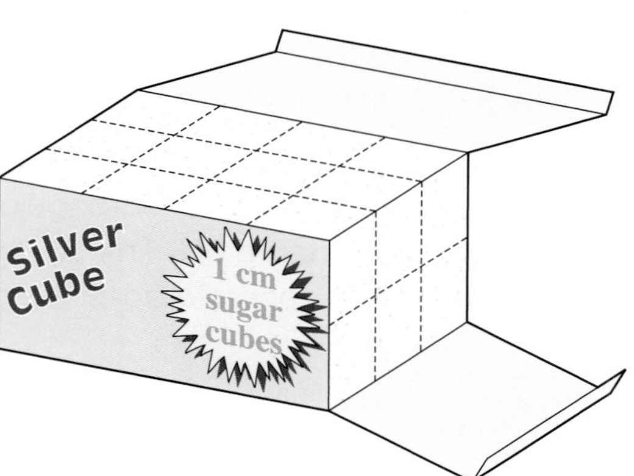

b **i** What is the volume of this box?

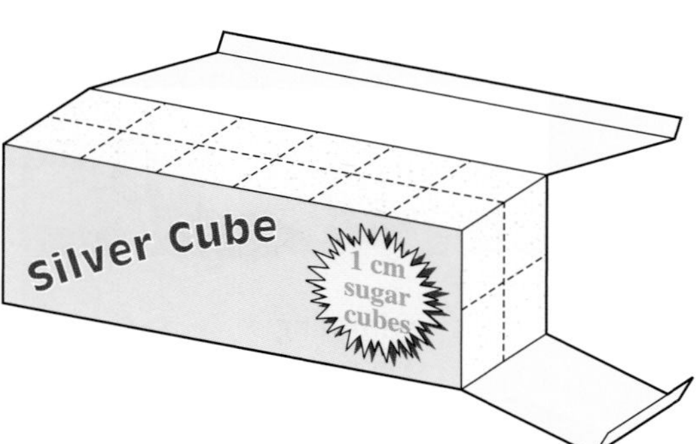

ii What do you notice?

c There are a lot of possible boxes that could hold 120 sugar cubes.

Choose one that you think will be suitable.

Explain your choice.

Draw your box.

4 The surface area of a cube is $150\,cm^2$.
 a Work out the area of one face.
 b Find the length of one edge.
 c Find the volume of the cube.

5 Here is the net of a cuboid.
Each rectangle makes a face of the cuboid.
 a What is the area of the bottom face?
 b What is the area of the top face?
 c What is the volume of the box?
 d What does the surface area tell you?

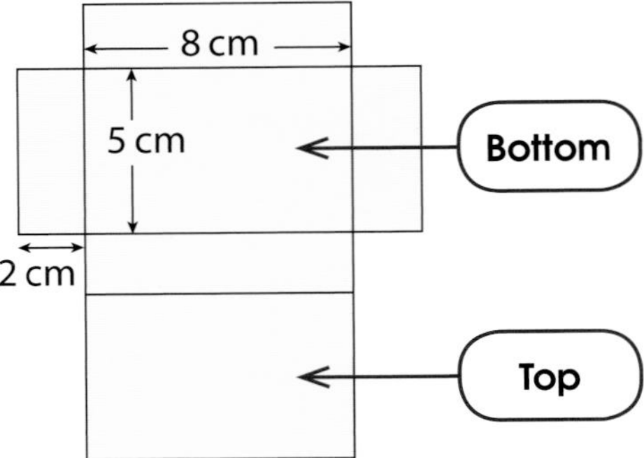

6 Martha is covering this box with coloured paper.
 a She covers the two largest faces with blue paper. Work out the area of blue paper that she uses.
 b She covers the two smallest faces with green paper. Work out the area of green paper that she uses.
 c She covers the other two faces with yellow paper. Work out the area of yellow paper that she uses.

8 cm
20 cm
30 cm

7 Brandon is making concrete.
He is going to concrete a rectangular hole of area $40\,m^2$ to a depth of 0.1 m.
What volume of concrete does he need?

8 A new outdoor swimming pool is planned.
This is one of the designs.
 a What is the volume of the paddling section?
 b What is the volume of the swimming section?
 c How many litres of water does the pool hold altogether?
 d Find the area to be covered with tiles.

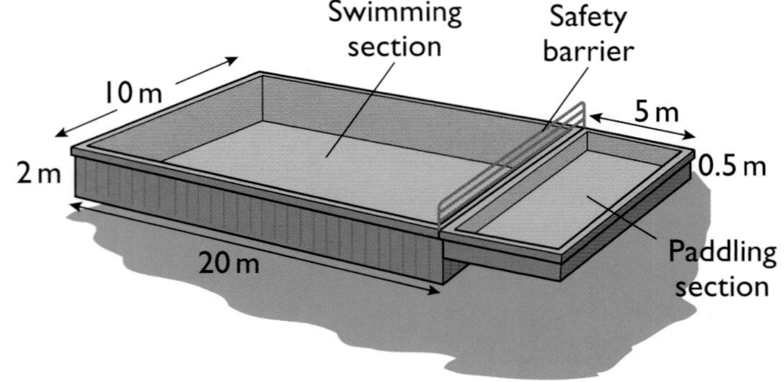

9 Lil is an artist.
She makes sculptures from marble.

a Work out the volume of each of these sculptures.

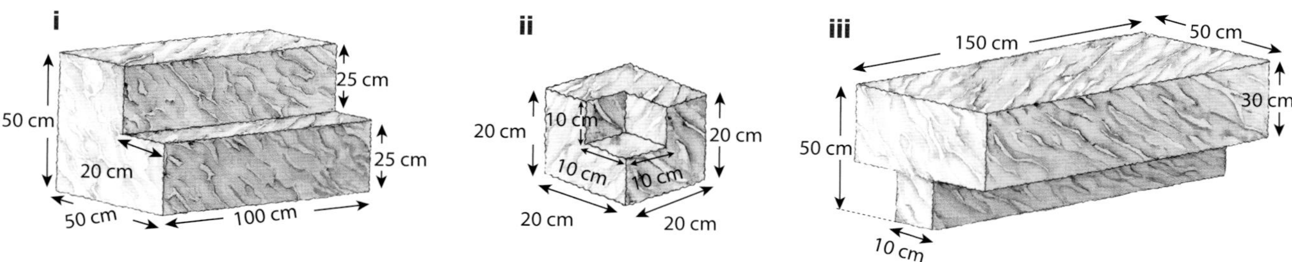

b How much marble is there in all the sculptures together?

c Lil has made another sculpture out of marble.
The hole goes right through the sculpture.
Work out the volume of this sculpture.

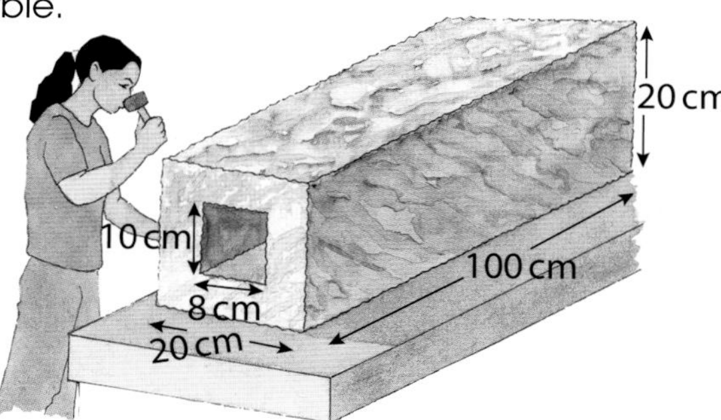

Reasoning

10 Would you expect these two boxes to weigh the same?
Explain your answer.

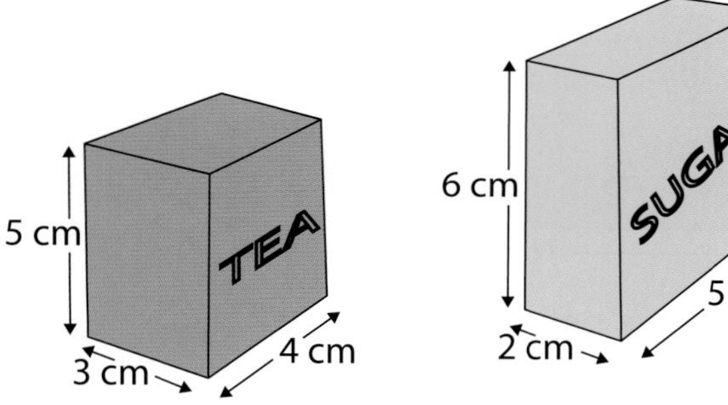

Reasoning

11 Pete is helping at his athletics club.

The rectangular long jump pit is 5 m long, 1 m wide and 50 cm deep. We'll need 250 m³ of sand to fill it.

Pete

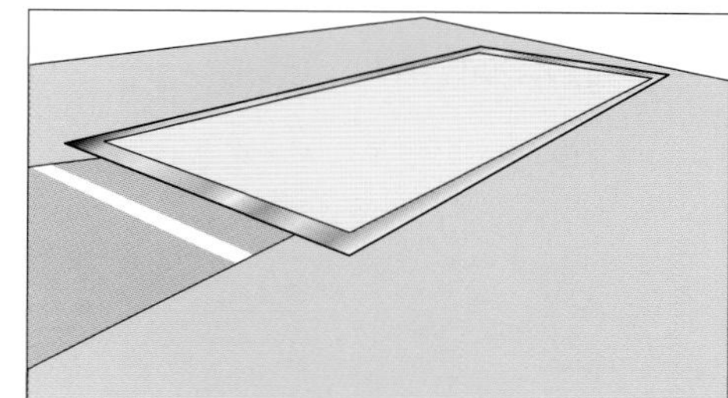

What is wrong with Pete's statement?

Wider skills practice

1 This is the swimming pool at Avonford Leisure Centre.

a What shape is the swimming pool?

b What is the volume of the swimming pool?

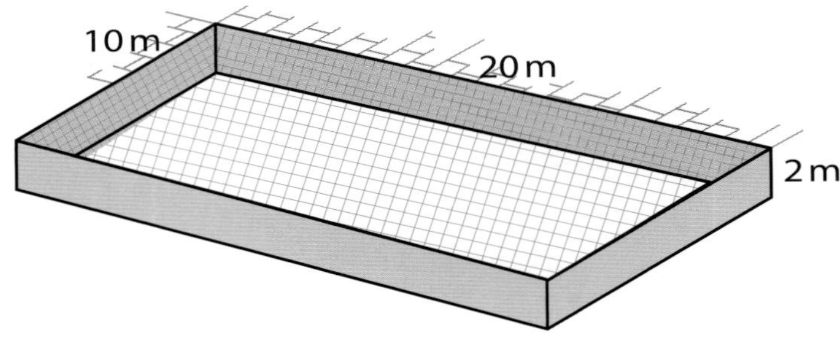

c How many litres of water does the swimming pool hold?

1 m³ = 1000 litres

d The bottom and sides of the pool are lined with tiles.
What area is tiled?

2 The diagram shows Natasha's water tank.
It is full of water.

a Work out the volume of water in the tank.

b Given that $1000\,\text{cm}^3 = 1$ litre, write down the volume in litres.

The tank is made of metal.
It has no top.

c Work out the area of metal used to make the tank.

3 A cuboid has length l cm, width w cm and height h cm.
Using l, w and h, write down the formula for

a its volume

b its surface area.

State the unit for each formula.

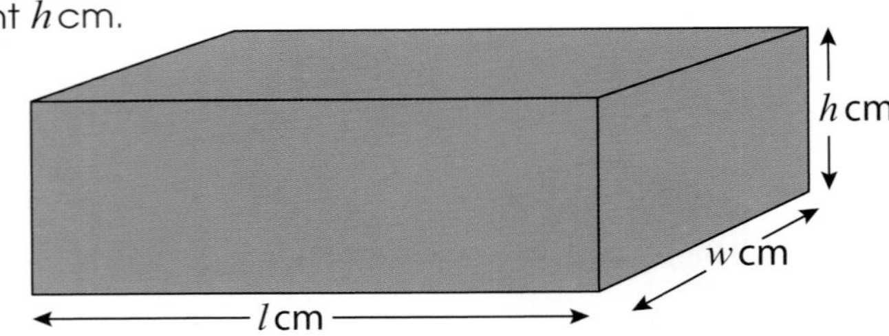

Reasoning

4 Match each of these objects with a suitable volume shown below.

a A shoe box

b A classroom

c A matchbox

d A large swimming pool

$2500\,\text{m}^3$	$1500\,\text{mm}^3$	$9000\,\text{cm}^3$	$50\,\text{m}^3$

Applying skills

Problem solving

1 This square of card is 10 cm by 10 cm.

The blue squares are 1 cm by 1 cm and are cut from each corner.

The sides are folded up to make an open box.

a What is the volume of the box?

b Experiment with different sized corner squares.

What is the maximum volume of box you can make?

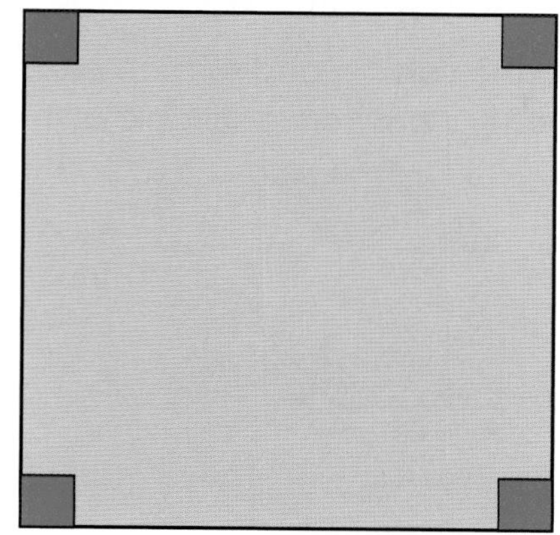

Problem solving

2 You are designing a fish tank.

It does not have a top.

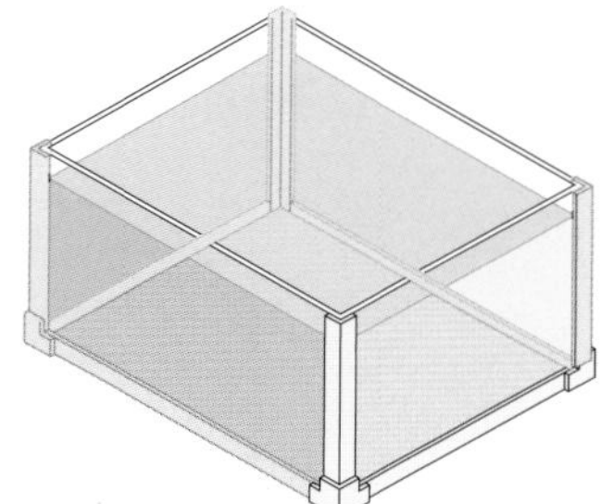

You have a sheet of clear plastic which is 2 m by 1 m.

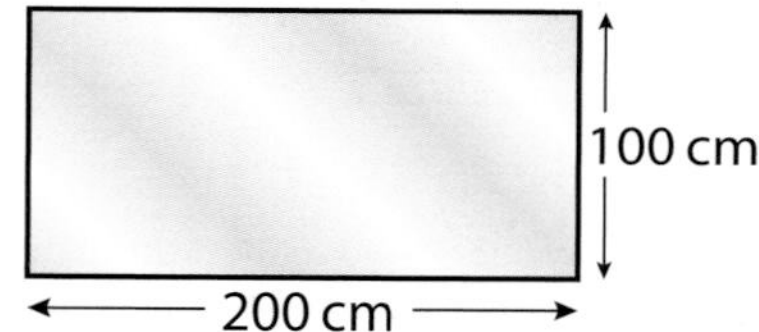

You also have edging strip and corner pieces.

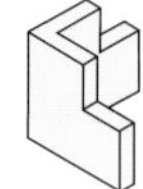

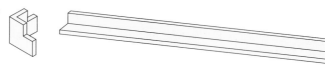

You have to decide on the length, width and height of your tank.

a Draw a sketch of your tank.

Mark on it the length, width and height that you have chosen.

b Show how you are going to cut up the plastic.

c What area is wasted?

d What length of edging strip do you need?

e How many corner pieces do you need?

f Work out the volume of your tank.

g what is the surface area of the fish tank

Reviewing skills

1 Find the volume and surface area of each of these cuboids.
Give your answers in suitable units.

a

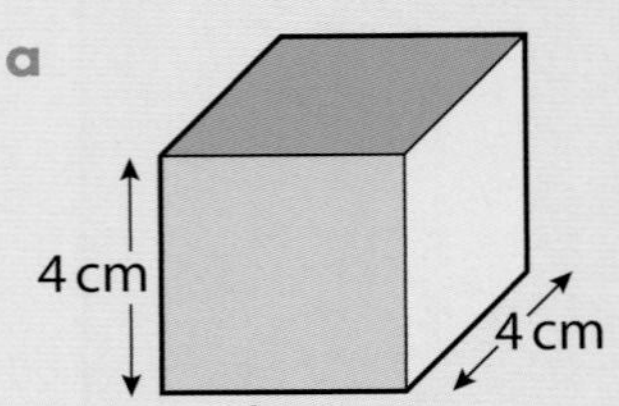

b

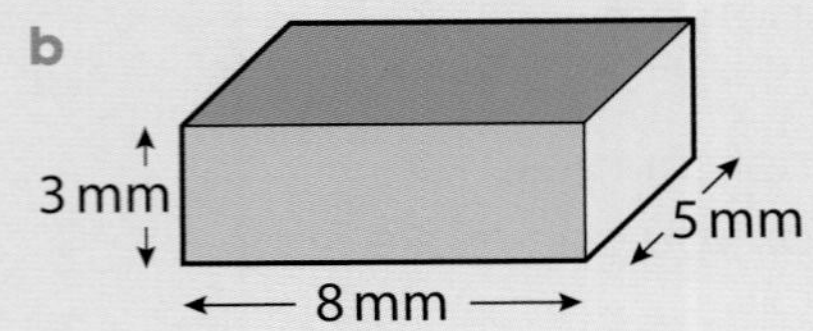

c

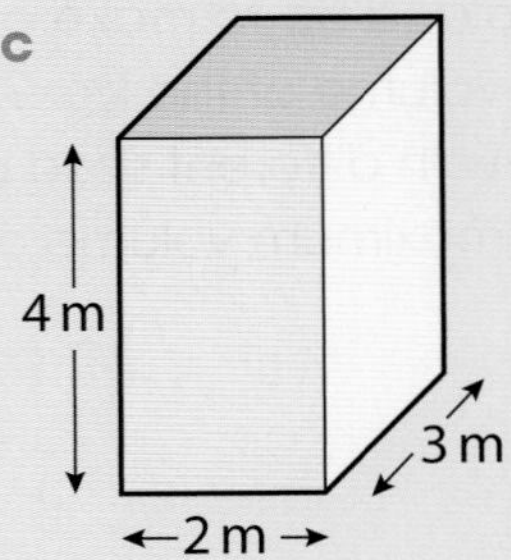

2 Find the volume and surface area of each of these solids.
Give your answers in suitable units.

a The lengths are in centimetres.

b The lengths are in millimetres.

i

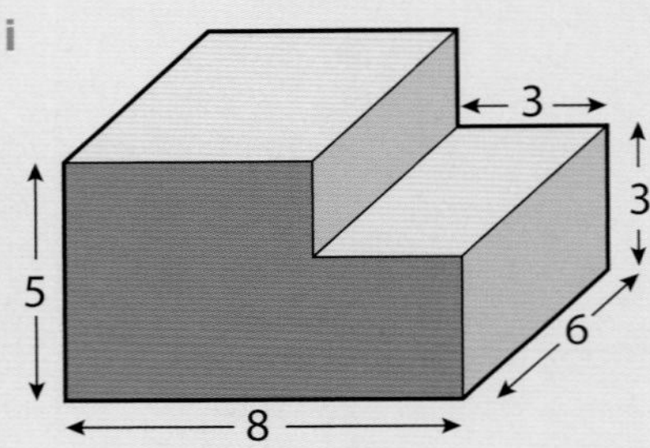

ii

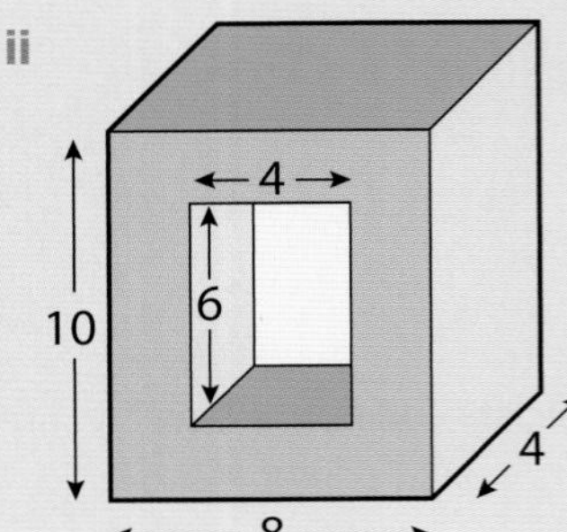

Unit 4 • 2-D representations of 3-D shapes • Band f

Building skills

Example outside the Maths classroom

Architects

Toolbox

You can represent a three-dimensional (3-D) object in two dimensions by drawing its plan and elevations separately.

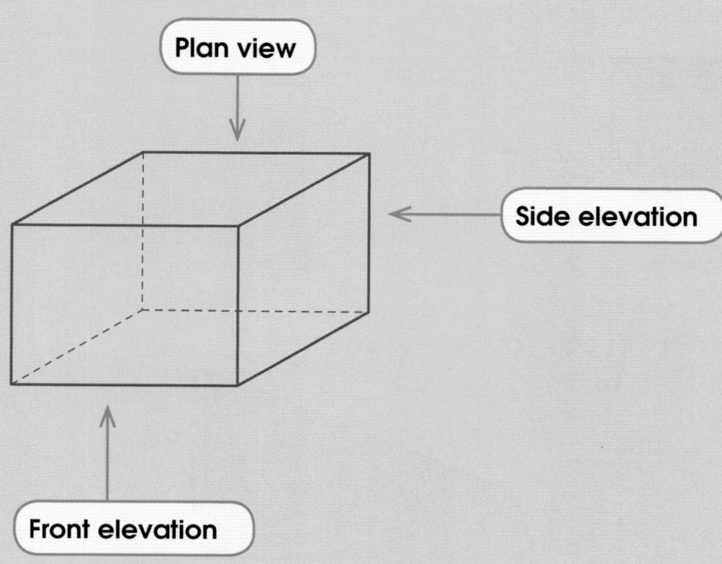

You can also use isometric paper to draw a representation of a 3-D object.

Example – Drawing plans and elevations

Draw the plan, front elevation and side elevation of this solid.

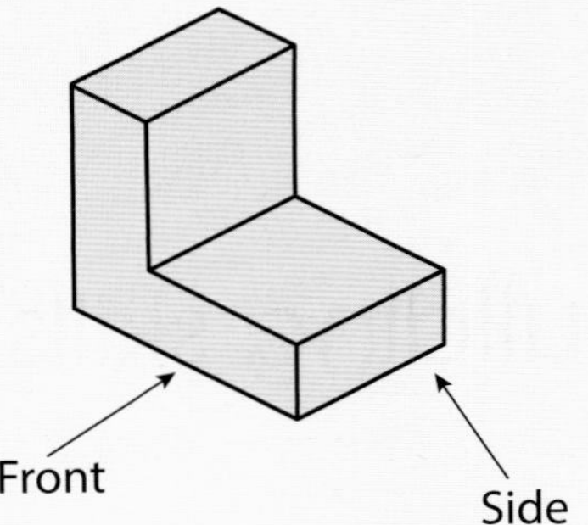

Solution

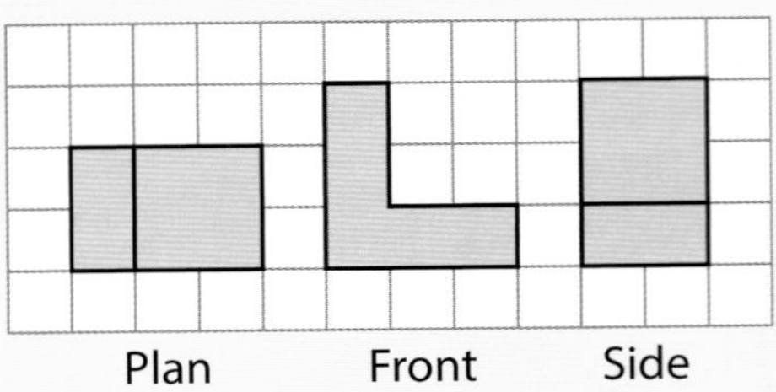

Remember:

- There are two side views and they might be different, so check which one you need.
- Solid lines on plans and elevations indicate where different surfaces meet.

Skills practice A

1 Draw the plan and front and side elevations of each of these solids.

a

b

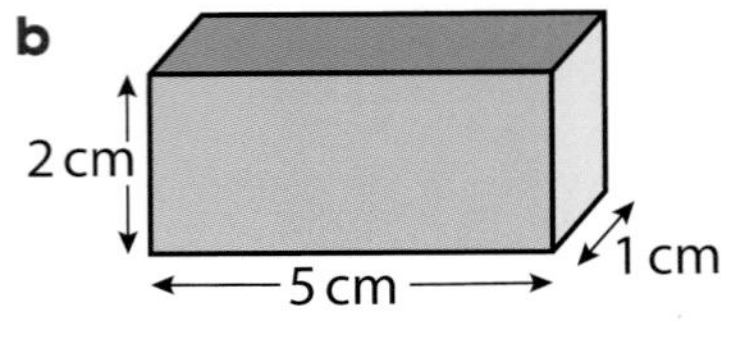

2 Draw the plan and front and both side elevations of each of these solids.

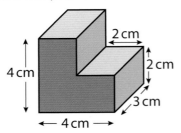

a

b

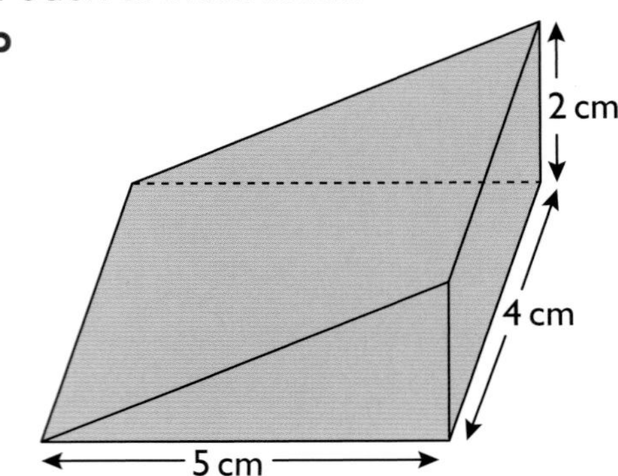

3 Here is a sketch of the plan and elevations of a solid.

Side elevation

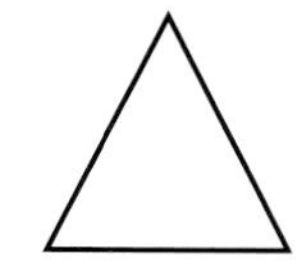

End elevation

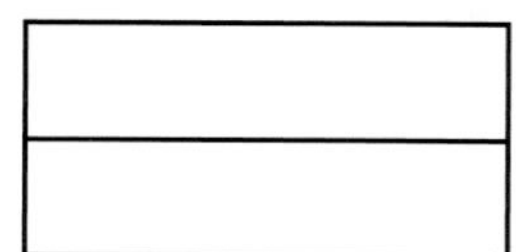

Plan

Name the solid and draw a sketch of it.

4 **i** What are these household items?

ii Where is the viewer standing to see each view?

a

b

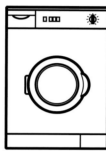

c

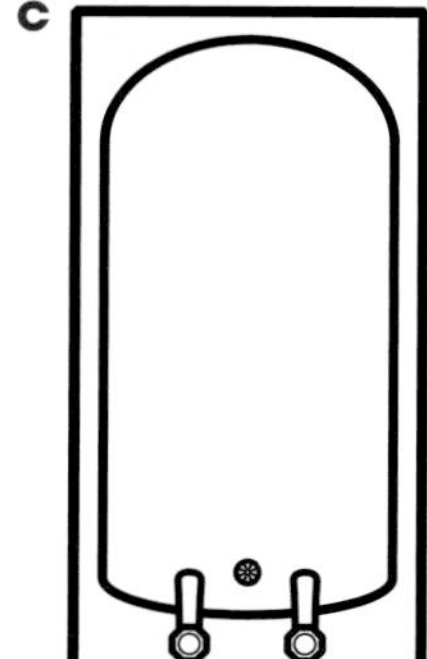

d

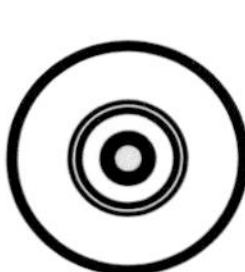

5 These drawings show different views of the racing car in the photograph.

a

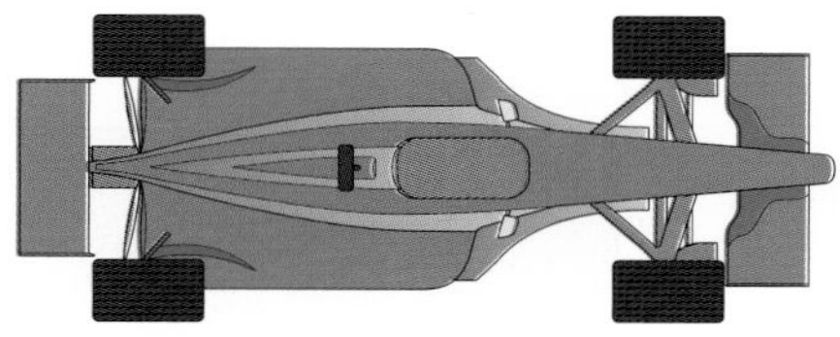

b

c

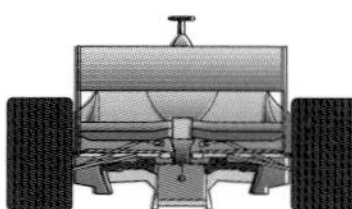

d

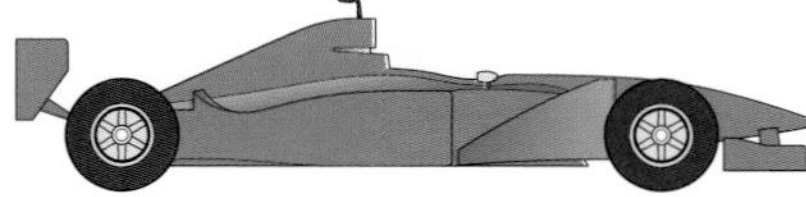

Match the drawings to these labels:

Front elevation	Side elevation	Plan	Back elevation

Skills practice B

Reasoning

1 Ms Roberts is showing her class the plan of a solid.

Who is correct?

2 Jo, Kim, Lucy and Meena go camping.
They view their tent from different directions.
It only has one window.

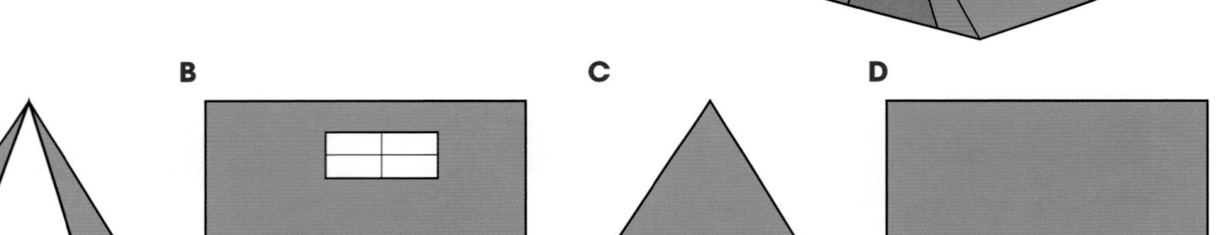

a Jo sees the front elevation.
Which view does she see?

b Kim sees the rear elevation.
Which view does she see?

c Lucy sees the side elevation without windows.
Which view does she see?

d Which view does Meena see?

e Draw a plan view of the tent.

3 Each of these shapes is made of four cubes.
Draw their plans and elevations.

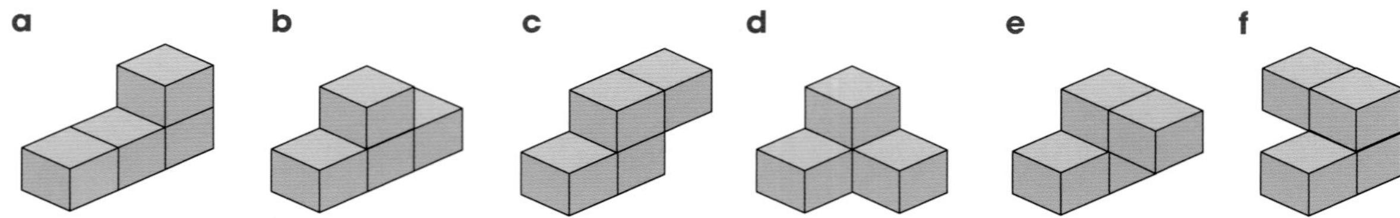

Reasoning

4 What solids could each of these plans be?

Sketch your answers.

a

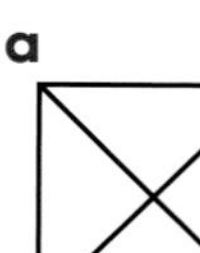
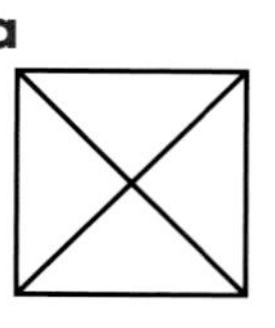

b

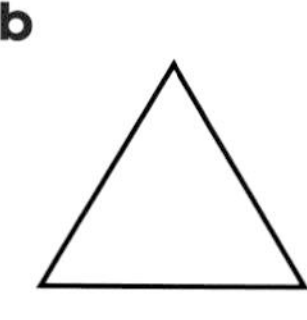

c

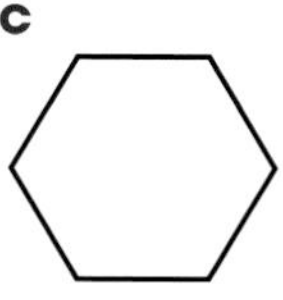

d

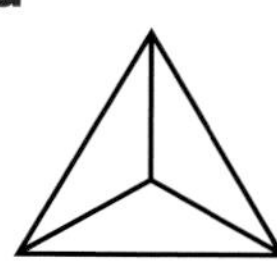

e

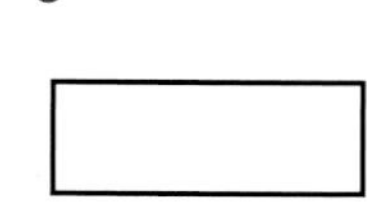

f

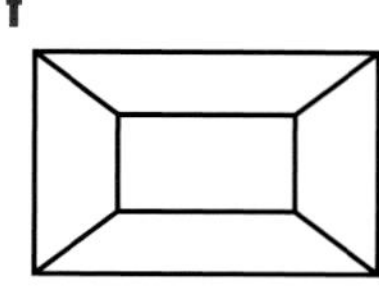

5 Draw the plan and front and side elevations of each of these solids.

a

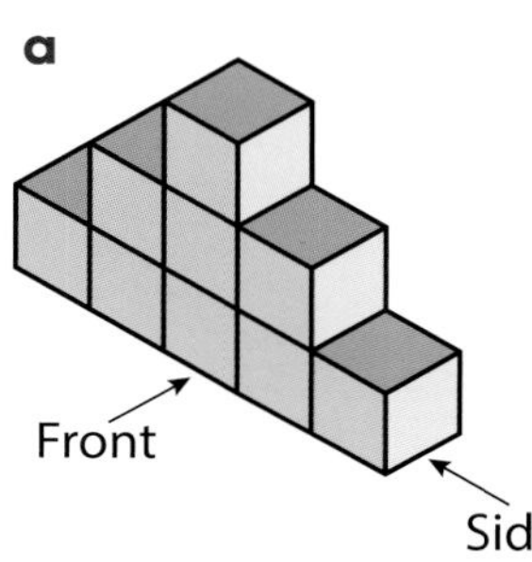

b

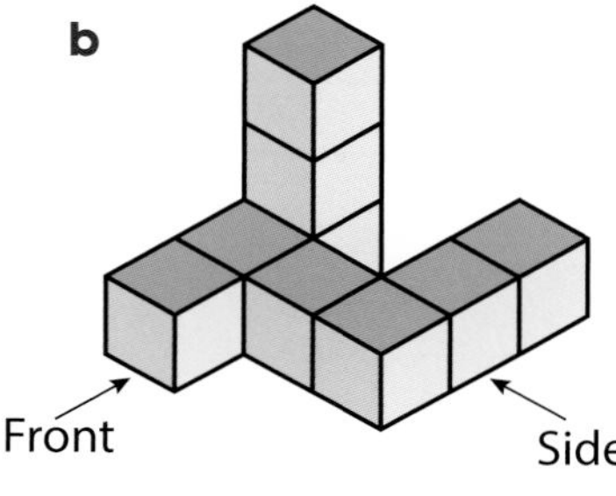

c

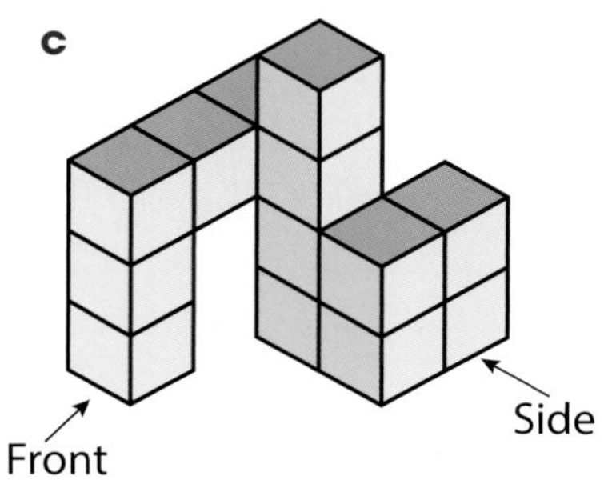

6 Draw a plan, front elevation and two side elevations of this object.

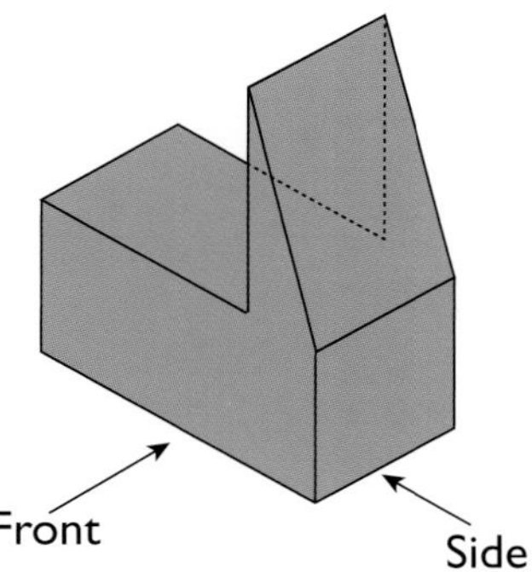

Wider skills practice

1 This is a pen-drive packet.

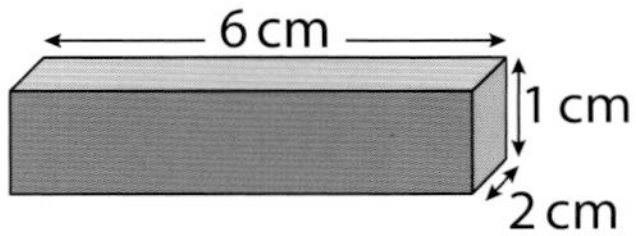

We need a box to hold six of these pen-drives to be sold on special offer.

Design a box that will hold six pen-drives.

Applying skills

1 This is a stand for athletes receiving medals.
It is 1.2 m high with equal steps, 3 m wide and 4 m wide.

a Draw its plan and elevations.
The horizontal surfaces that face upward are to be covered in red carpet.
The vertical faces will be painted green. One tin of paint covers 10 square metres.

b What area of red carpet will be needed?

c How many tins of green paint will be used?

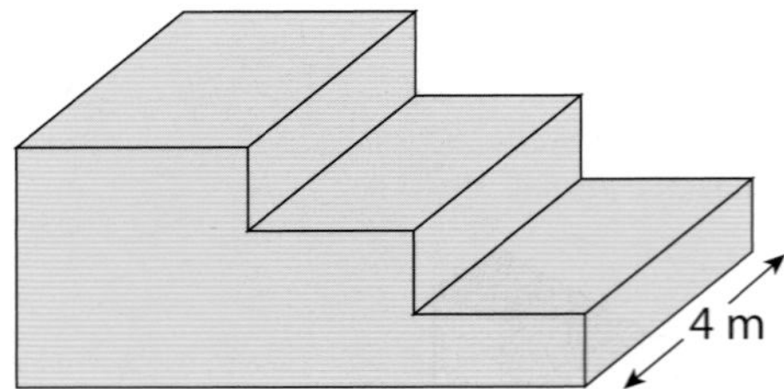

Reviewing skills

1 Sketch the solids that match these sets of plans and elevations.

a Plan; Front elevation; Side elevation

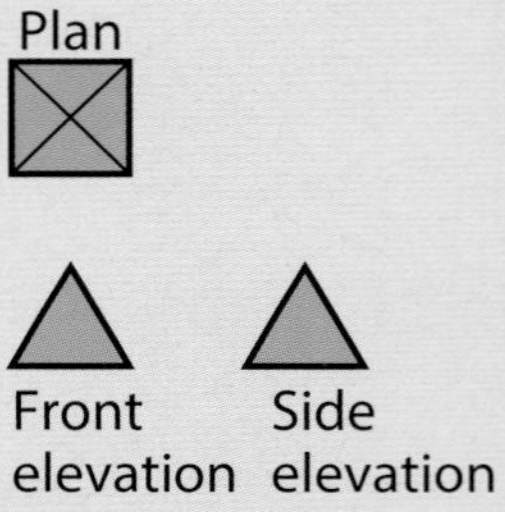

b Plan; Front elevation; Side elevation

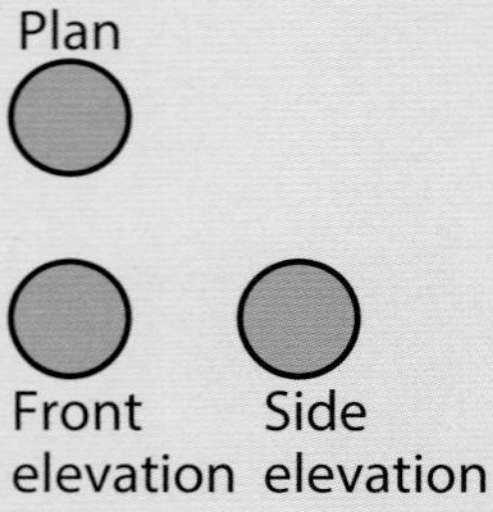

c Plan; Front elevation; Side elevation

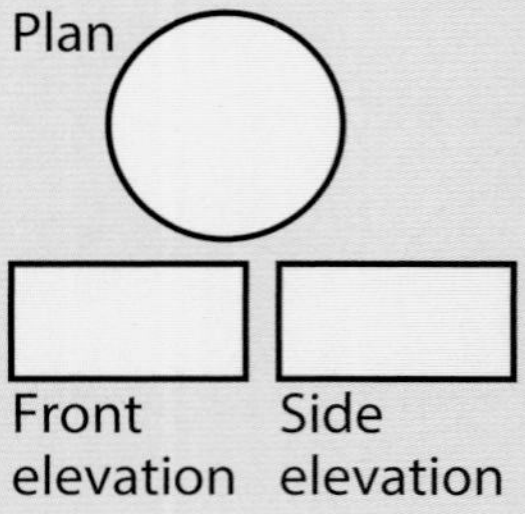

d Plan; Front elevation; Side elevation

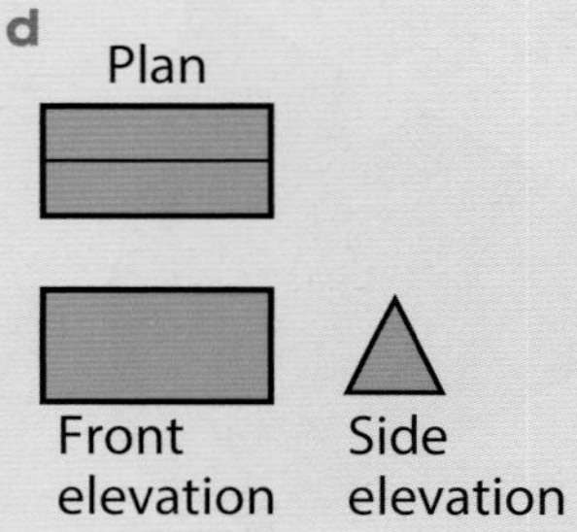

2 Draw the plan and front and side elevations of each of these solids.

a

b

c

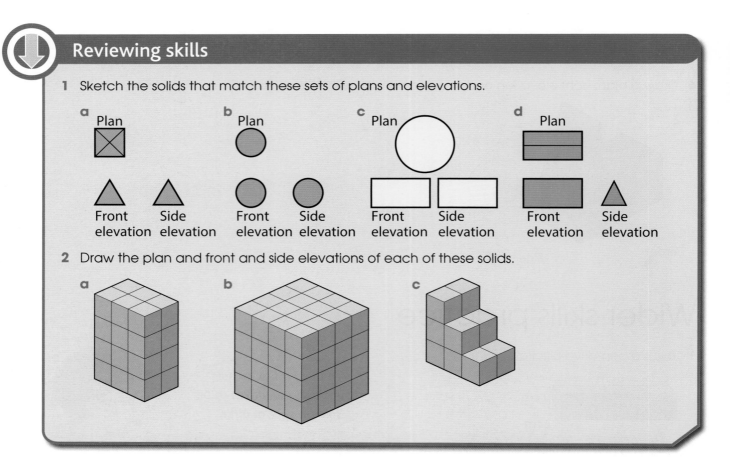

Unit 5 • Prisms • Band g

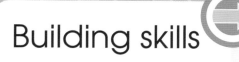

Building skills

Example outside the Maths classroom

Glass design

Toolbox

A **prism** is a three-dimensional shape with the same cross-section all along its length.

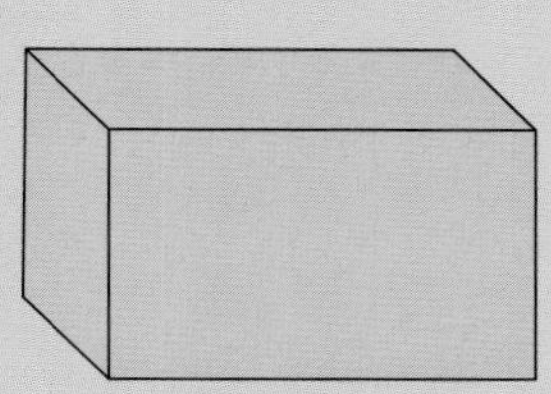

Cuboid

Triangular prism

Hexagonal prism

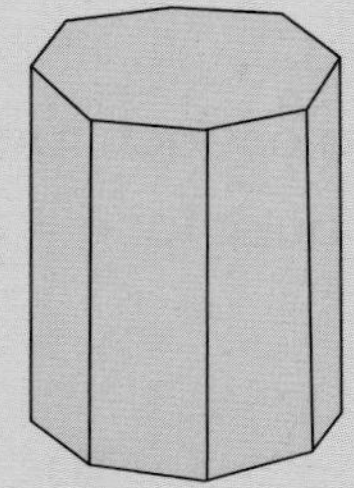

Octagonal prism

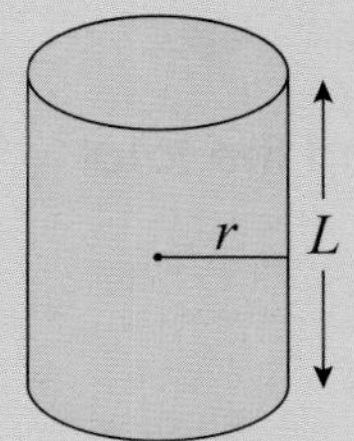

Cylinder

The cylinder is a prism with a circular base.
Its volume is $\pi r^2 L$.
Its surface area is $2\pi r L + \pi r^2 + \pi r^2 = 2\pi r L + 2\pi r^2$

To work out the **volume of a prism**, calculate the area of the cross-section and multiply by the height (or length).

The volume is measured in cubic units such as cubic centimetres (cm^3) or cubic metres (m^3).

The **surface area of a prism** is the total area of all the faces.

Example – Finding the volume of a prism

Work out the volume of this cylinder.

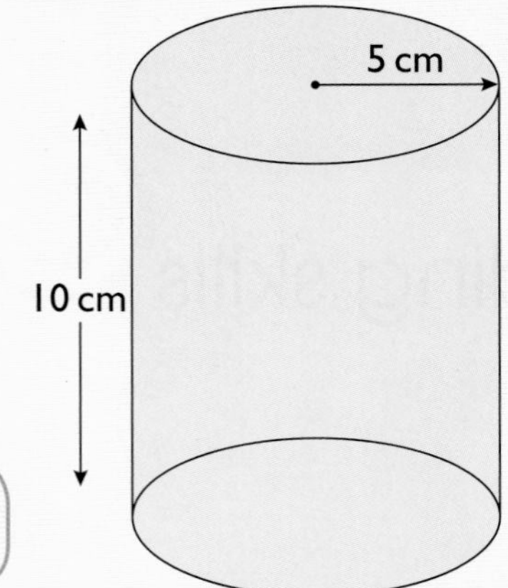

Solution

Area of circle = πr^2

$A = \pi \times 5^2$ ← **The cross-section is a circle with a radius of 5 cm.**

$= 78.539\ldots$

Volume of cylinder = area of cross-section × height

$= 78.539\ldots \times 10$ ← **The height of the cylinder is 10 cm.**

$= 785.398\ldots$

The volume of the cylinder is 785.4 cm^3 (to 1 d.p.).

Example – Finding the surface area of a prism

Calculate the surface area of this prism.

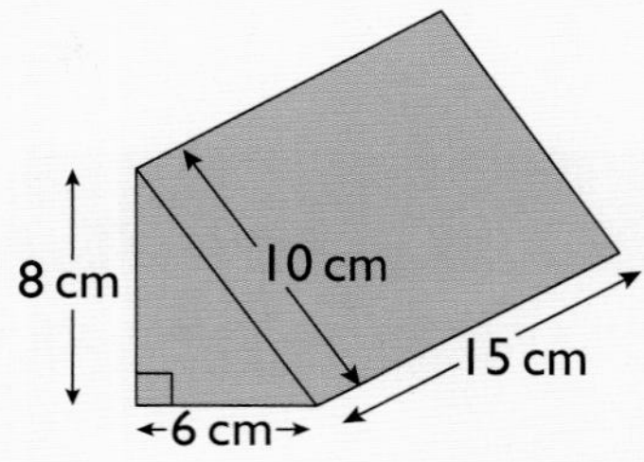

Solution

Work out the area of each face individually.

Area of one triangular face = $\frac{1}{2} \times$ base × height

$= \frac{1}{2} \times 6 \times 8$

$= 24\,cm^2$

Area of other triangular face = $24\,cm^2$ ←

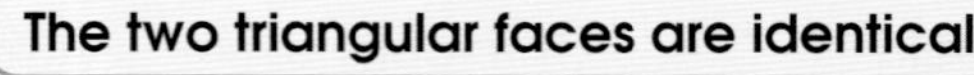

Area of rectangular base : $6 \times 15 = 90\,cm^2$ ← **The rectangular faces all have a length of 15 cm.**

Area of rectangular face : $8 \times 15 = 120\,cm^2$

Area of sloping rectangular face : $10 \times 15 = 150\,cm^2$

Total surface area = $24 + 24 + 90 + 120 + 150$ ← **Add the area of all the faces to find the total surface area.**

$= 408\,cm^2$

Remember:

- A prism has a cross-section which is constant throughout its length, and two congruent parallel faces.
- A cylinder is a prism with circular ends.
- A cuboid is a prism with rectangular ends.

Skills practice A

1 Work out the volume of each of these prisms. Each cube is 1 cm³.

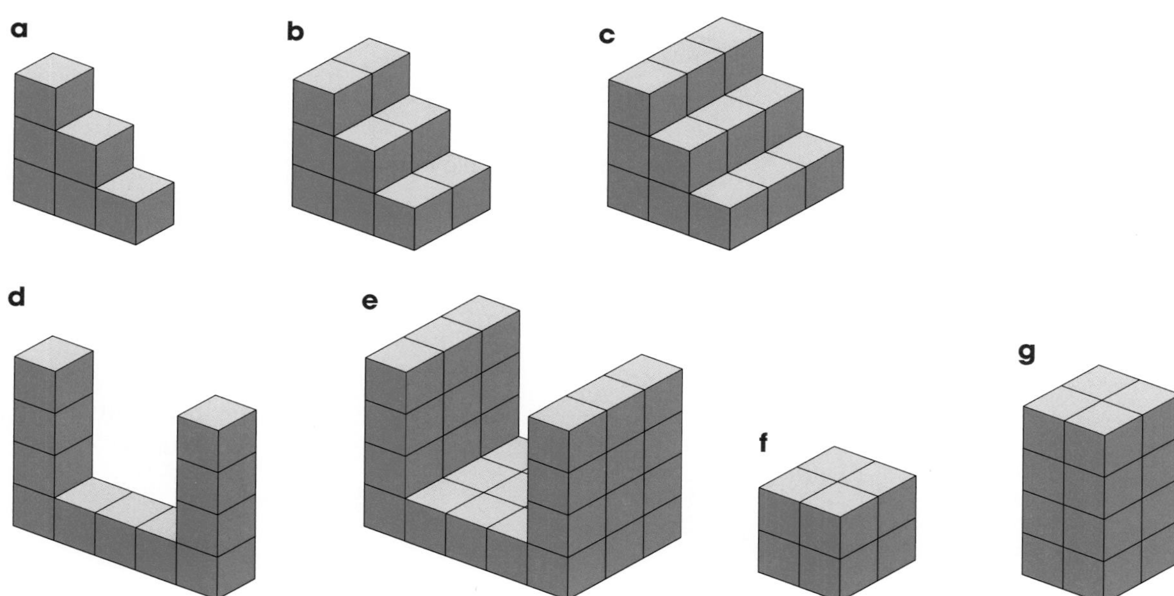

2 Sketch the net of this triangular prism and find its surface area.

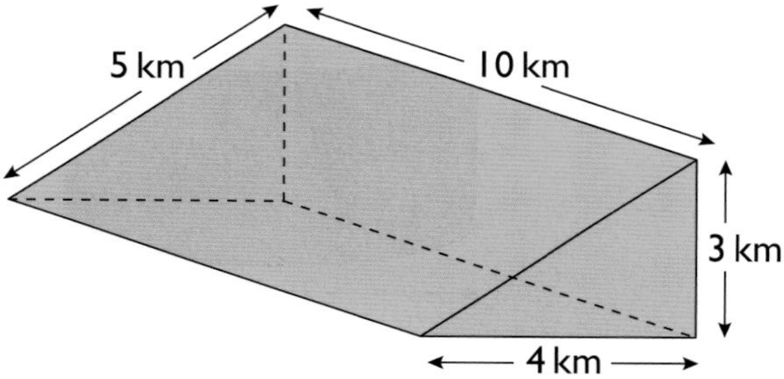

3 Calculate the volume of each of these prisms.

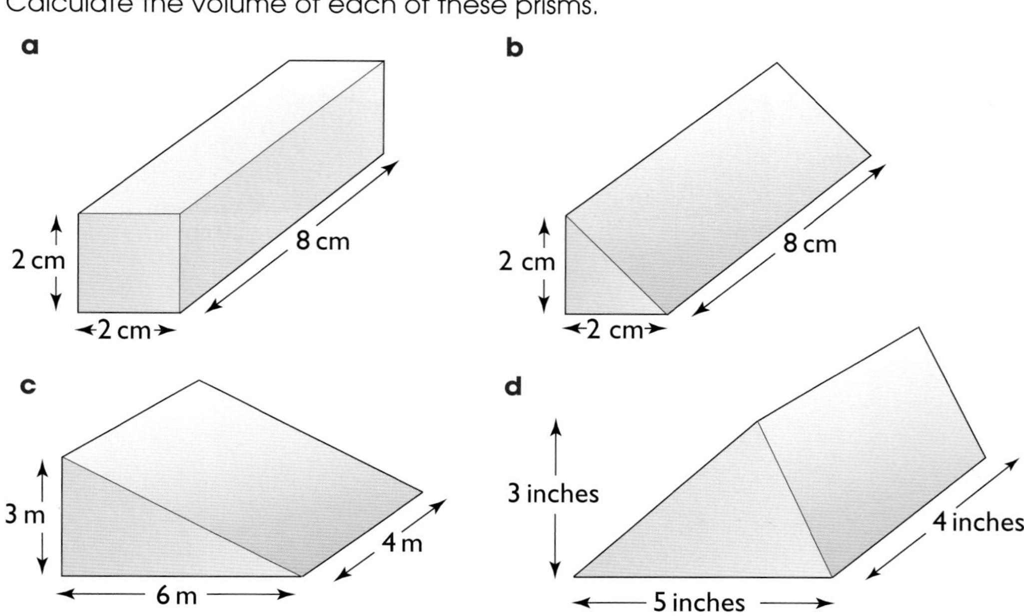

4 Work out the volume and the surface area of each of these triangular prisms.

a

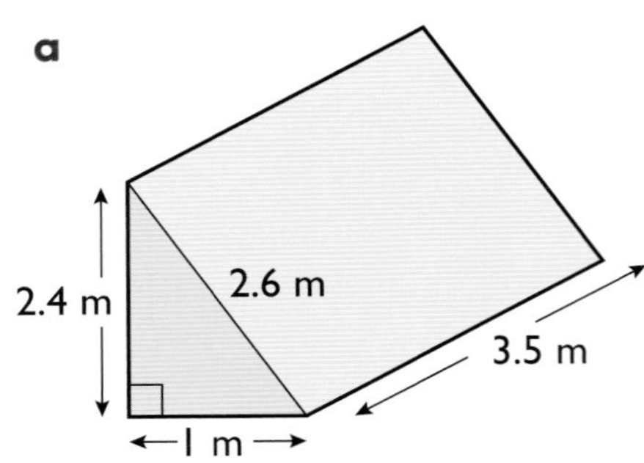

b

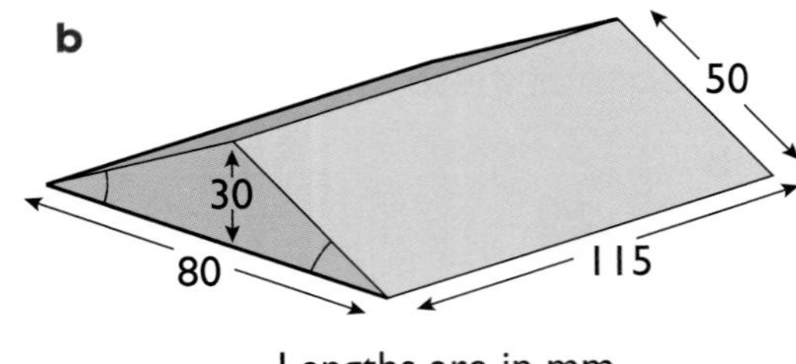

5 The area of one end of this tin is $50\,\text{cm}^2$.
What is the volume of the tin?

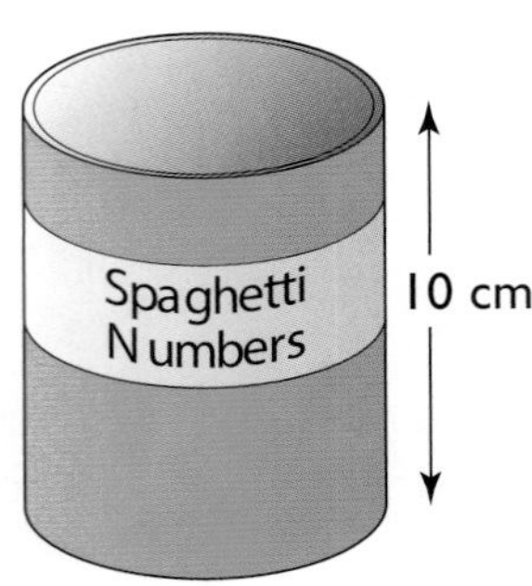

6 Find the curved surface area of this cylinder.

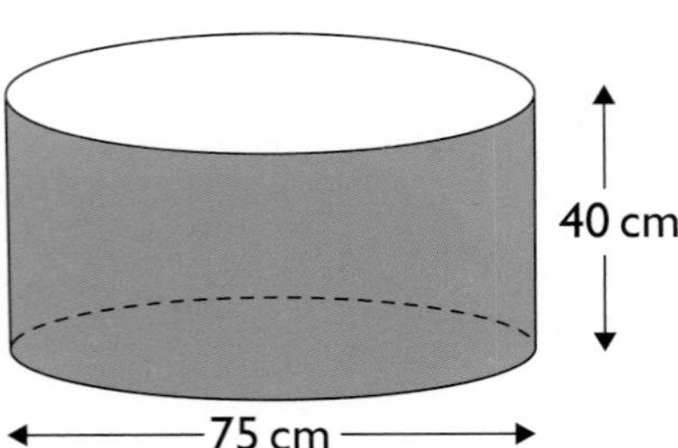

7 Find the volume of each of these cylinders.

a

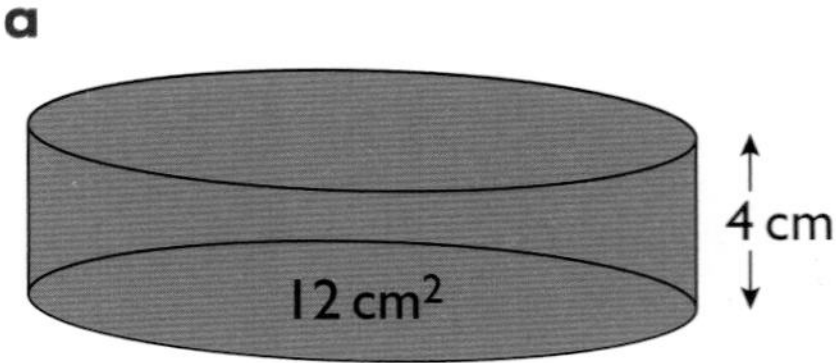

b

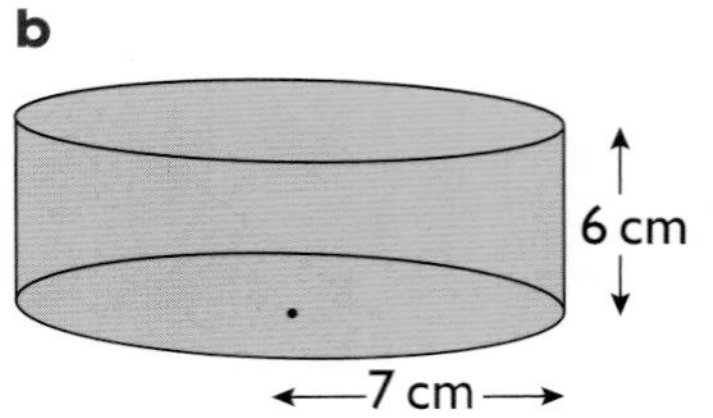

c

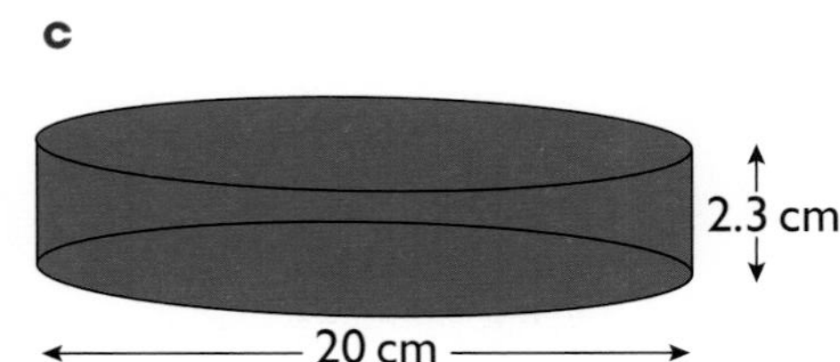

8 Find the curved surface area of each of these cylinders.

a

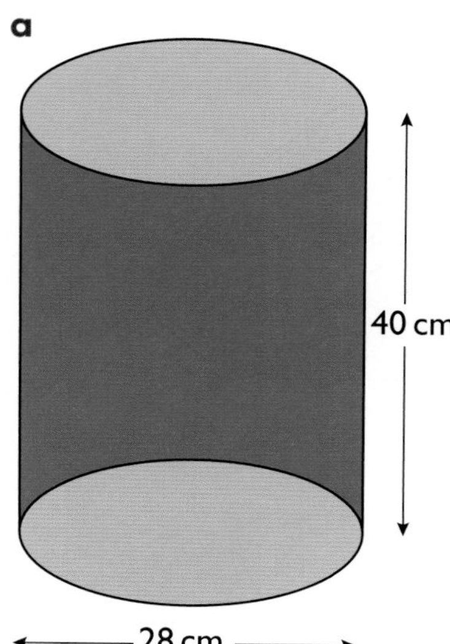

b

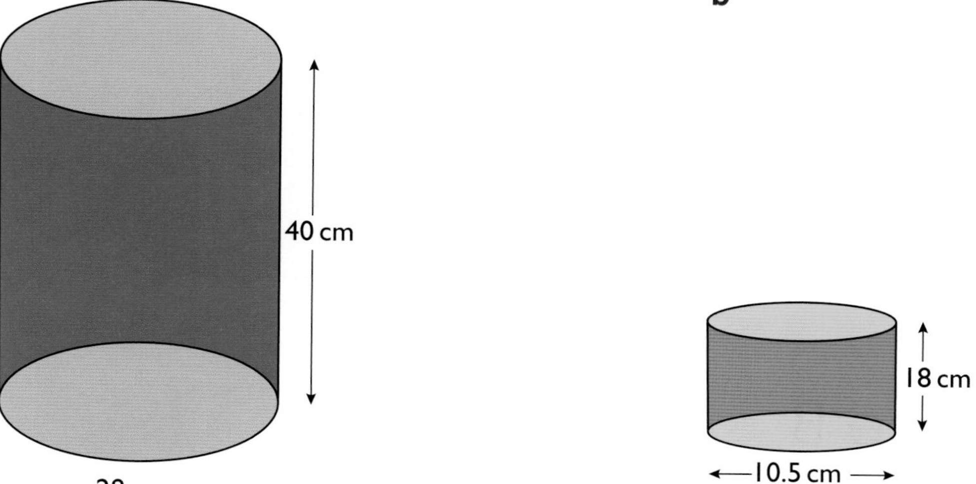

9 Compare the surface area and volume of each of these prisms.
Put them in order of size

a from largest to smallest surface area

b from largest to smallest volume.

i

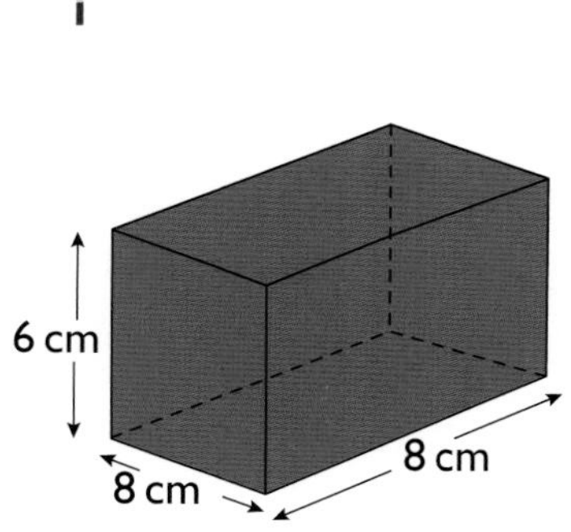

ii

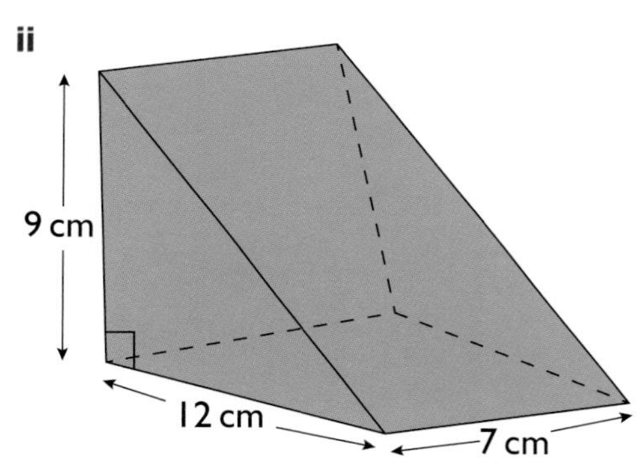

iii

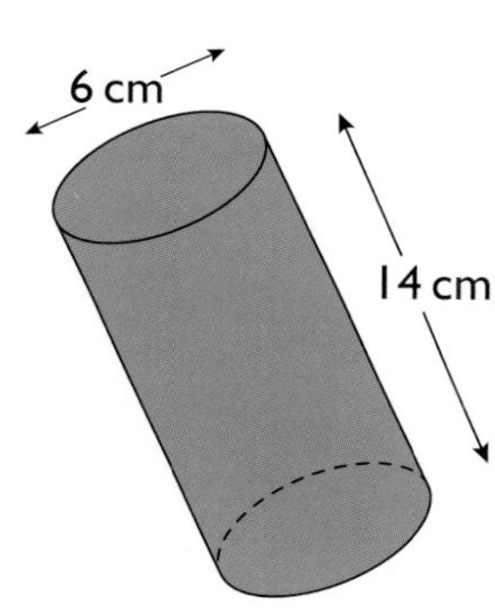

10 a At first sight, do you expect the volumes of these two cylinders to be equal?

i

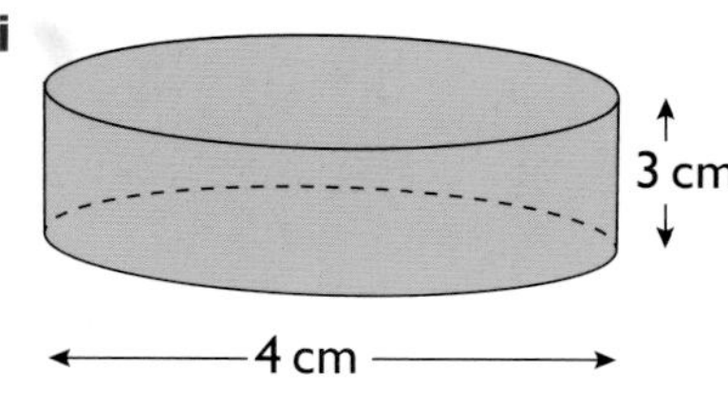

ii

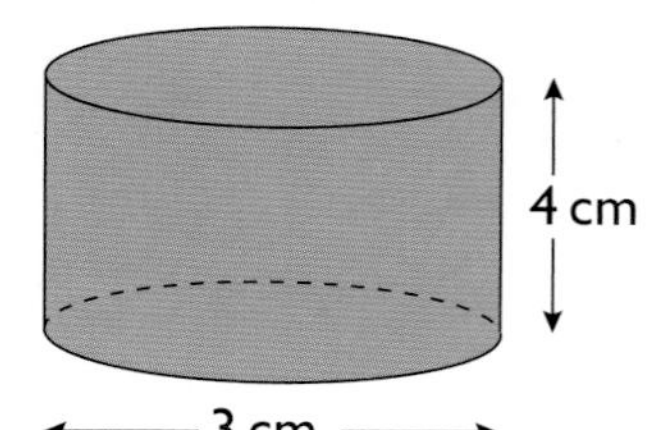

b **i** Calculate the volumes of the cylinders in part **a**.

ii Are the volumes equal?
Explain what you found.

11 Find **i** the volume and **ii** the surface area of each of these solids.

a

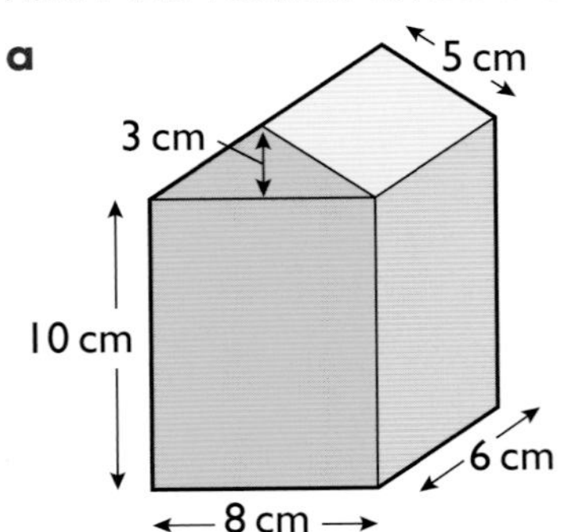

b

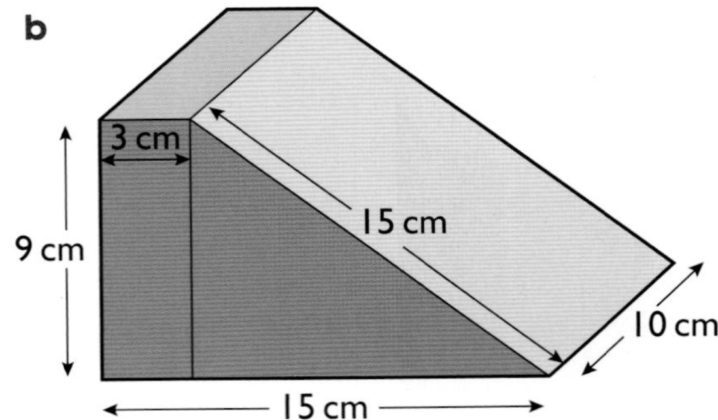

Skills practice B

1 For each of these prisms

i state what sort of prism it is

ii find its volume.

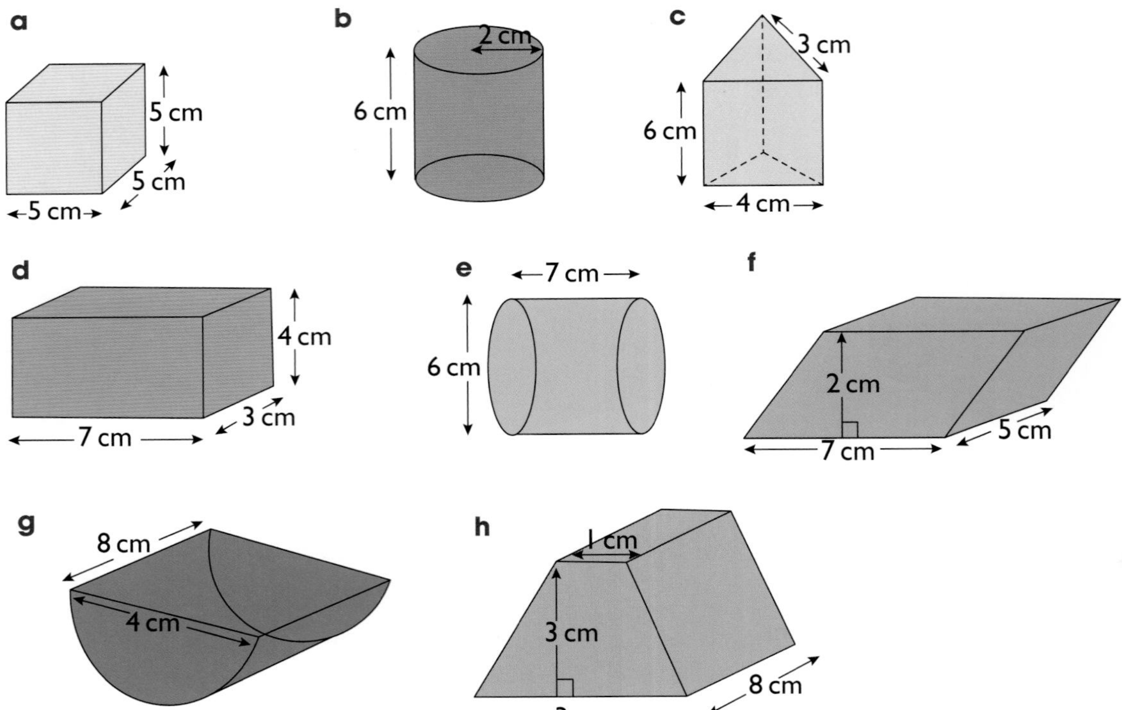

2 A cylinder has base area $16\,\text{cm}^2$ and volume $80\,\text{cm}^3$.
What is its height?

3 Look at this net.

a Describe and sketch the shape you get when you fold it up.

b Find the volume of the shape.

c Find the surface area of the shape.

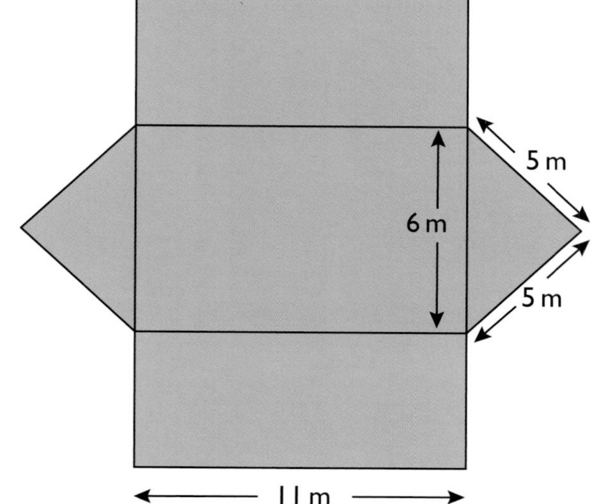

4 Find the volume of each of these prisms.

a

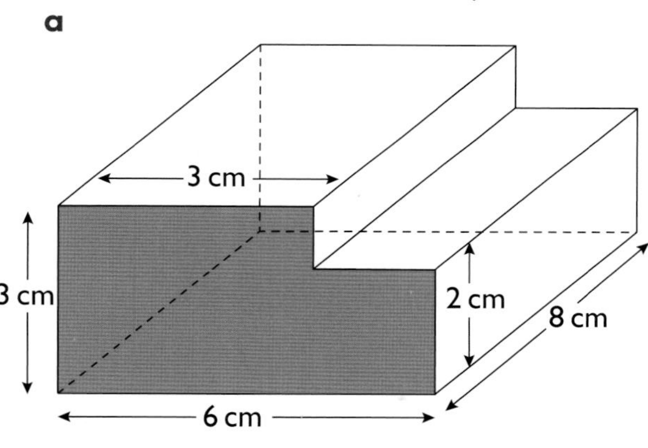

b

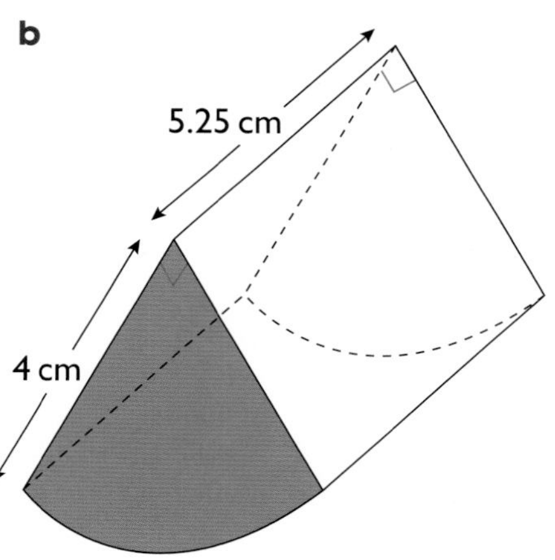

5 Find the volume and surface area of each of these prisms.

a

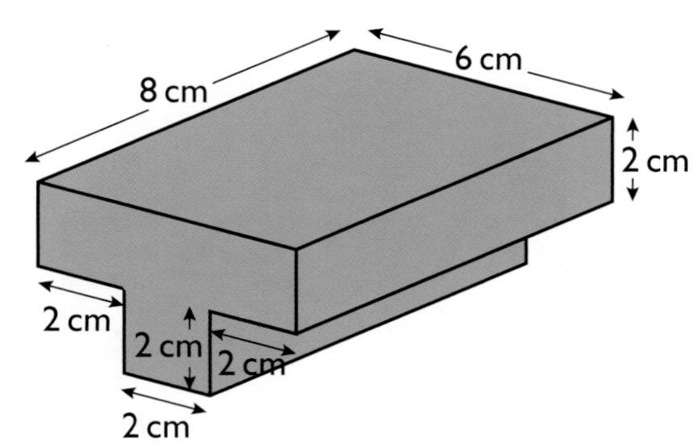

b

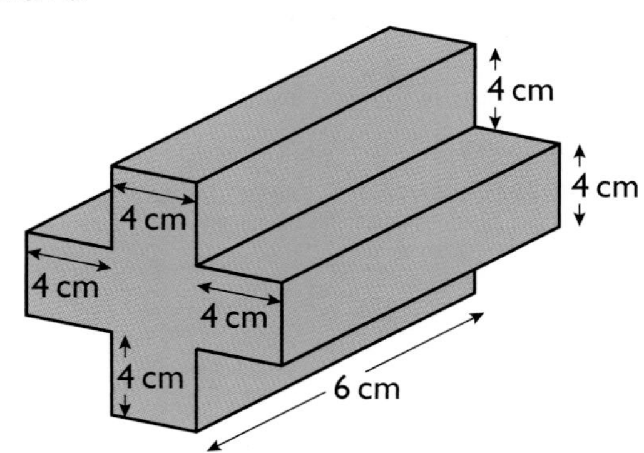

6 Calculate the surface area of each of these wooden letters.

a

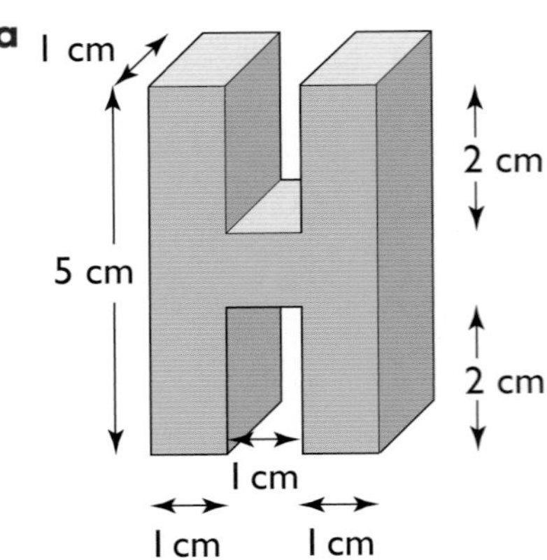

b

5 cm
3 cm
2 cm
4 cm
3 cm

7 What is the volume of this house?

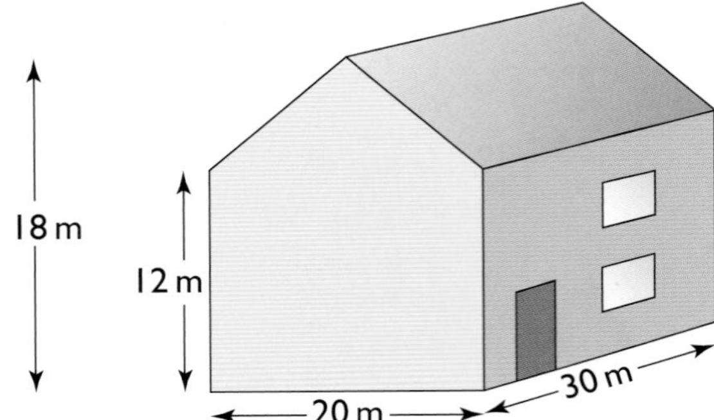

Reasoning

8 Jim is painting the outside of a ship's funnel.
The funnel is a cylinder of diameter 4 m and height 20 m.
A tin of paint covers 8 m^2.
How many tins of paint does Jim need?

9 A garden roller is 0.72 m in diameter and 1.1 m wide.
The roller does 40 complete revolutions.
What area of lawn has been rolled?

10 A fence consists of 30 cylindrical posts each with a diameter of 8 cm and a height of 72 cm.

a Find the volume of wood used to make this fence.

The curved surface and the top of each post are painted.

b Find the area of the fence that is painted.

11 Ann is a scientist.
She pours 50 ml of acid into a measuring cylinder of diameter 3 cm.
Find the depth of the acid in the cylinder.

12 A cylindrical candle has diameter 6 cm and length 15 cm.
Carlos packs it into a cuboid box in which it just fits.
Find the volume of air in the box.

13 A water pipe on the side of a house has a radius of 6 cm and a length of 4.5 m.
Harry has a tin of paint sufficient to cover 2 m^2.
Is this enough to paint the whole pipe?

Reasoning

14 A circular lawn has a path round it.

a Find the area of each of the following, giving your answers to two decimal places.

i The lawn

ii The lawn and path

iii The path

The path is covered in gravel 5 cm thick.

b Find the volume of the gravel.

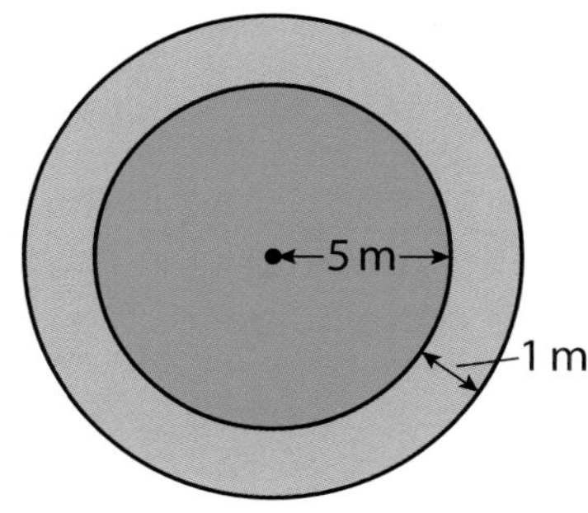

Wider skills practice

Reasoning

1 A cylindrical water tank has a diameter of 2.2 m.
The water level is 1.6 m high but the tank is not full.

a Calculate the volume of water in the tank.

b The tank's capacity is 7500 litres.
Calculate how full it is, giving your answer as a percentage.

2 A cylindrical tea urn has a diameter of 24 cm and a height of 60 cm.
A mug has a diameter of 6 cm and holds tea to the height 10 cm.
How many mugs of tea can be poured from the urn?

Reasoning

3 Look at Meena's new chicken shed.

a What is its volume?

b Each chicken needs 0.5 m^3 of space in the shed.
How many chickens can Meena have?

c Meena paints the outside of the shed (but not the roof) brown.
How many tins of paint does she buy?

Meena

4 This right-angled triangular prism has height h, width w and length l.

a Write an expression for the volume.

b Write an expression for the surface area.

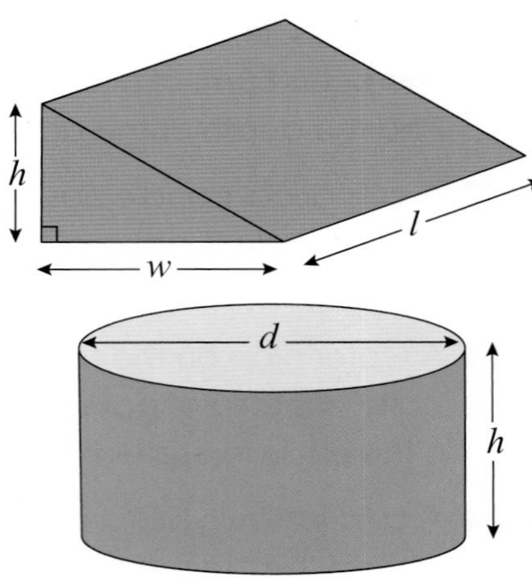

5 This cylinder has diameter d and height h.

a Show that the surface area of the cylinder can be expressed as $\frac{\pi d^2}{2} + \pi dh$.

b Write an expression for the volume.

Reasoning

6 A cylinder has volume $125\pi \text{ cm}^3$ and surface area $100\pi \text{ cm}^2$.
The radius and the height of the cylinder are equal.
Find the height of the cylinder.

Applying skills

Problem solving

1 Lisa is making a wedding cake with three cylindrical layers.
The radii are 6 cm, 10 cm and 20 cm.
The heights of the layers are 5 cm, 6 cm and 8 cm, respectively.

a Find the total volume of cake she needs to make.

She covers the top and sides of each layer with icing 3 mm thick.

b Find the volume of icing she needs to make.

Problem solving

2 Fatima measures rainfall for a geography project.
The rain is collected from a cylindrical tray which has radius 25 cm.
It then flows into a measuring cylinder underneath, which has diameter 10 cm.
What is the depth of water in the measuring cylinder after 0.4 cm of rain has fallen into the tray?

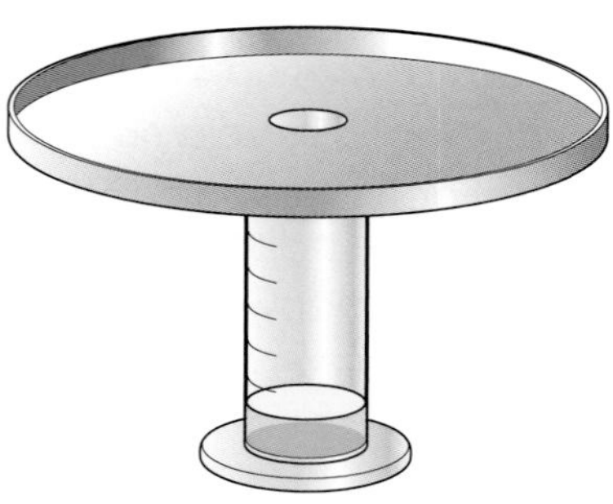

Problem solving

3 This glass display case is of length 60 cm and external diameter 24 cm.
The glass is 1 cm thick.
The case stands on a wooden base 1 cm thick.

a Find the volume of the wooden base.

b Find the volume of glass required to make the case.

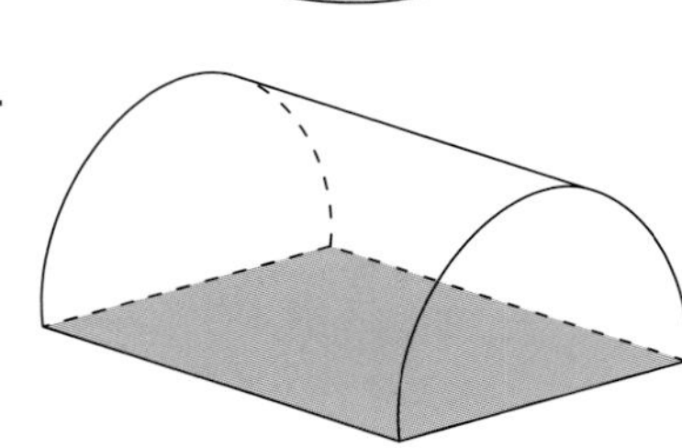

Problem solving

4 Adam draws a net of a cylinder on paper.
The sheet of paper measures 20 cm by 15 cm.
He cuts out the net and makes a cylinder.
Its radius is 2 cm and its height is 10 cm.
Show that the net will fit on the paper.

Problem solving

5 The net of a prism consists of two right-angled triangles and three rectangles.
The perimeter of each triangle is 24 cm.
The length of the prism is 20 cm.
The area of the largest face of the prism is 200 cm^2.
The shortest side of each of the triangles is 6 cm.

a Draw the net of the prism.

b Draw the prism.

Reviewing skills

1 Find the volume and surface area of each of these solids.
All lengths are in centimetres.
Remember to give the units for each of your answers.

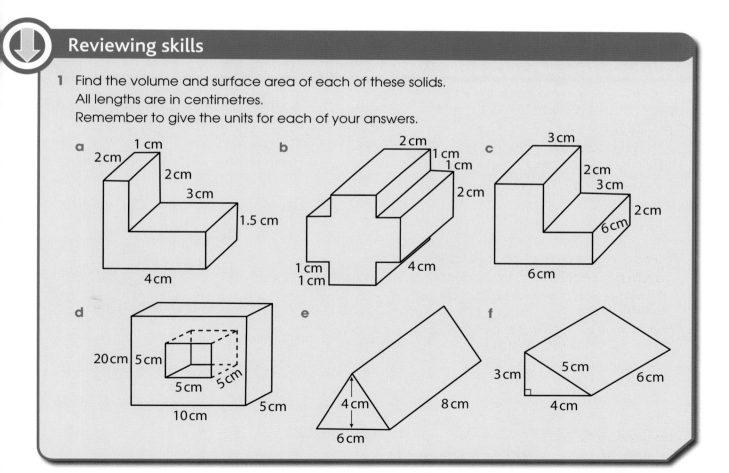

a 1 cm, 2 cm, 2 cm, 3 cm, 1.5 cm, 4 cm

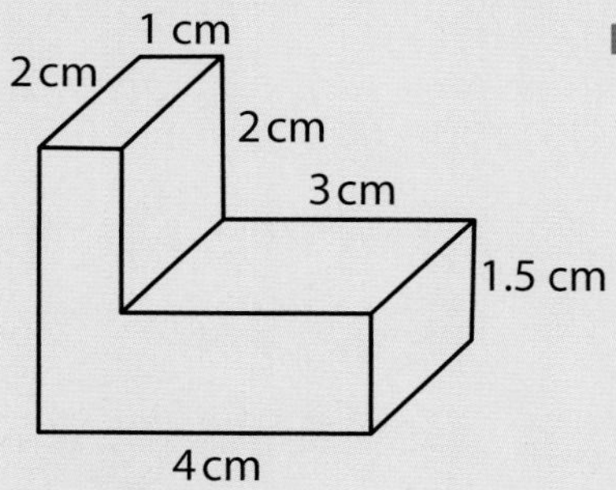

b 2 cm, 1 cm, 1 cm, 2 cm, 4 cm, 1 cm, 1 cm

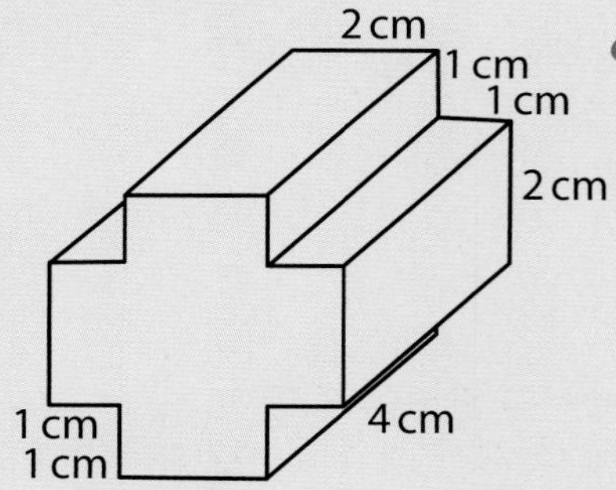

c 3 cm, 2 cm, 3 cm, 2 cm, 6 cm, 6 cm

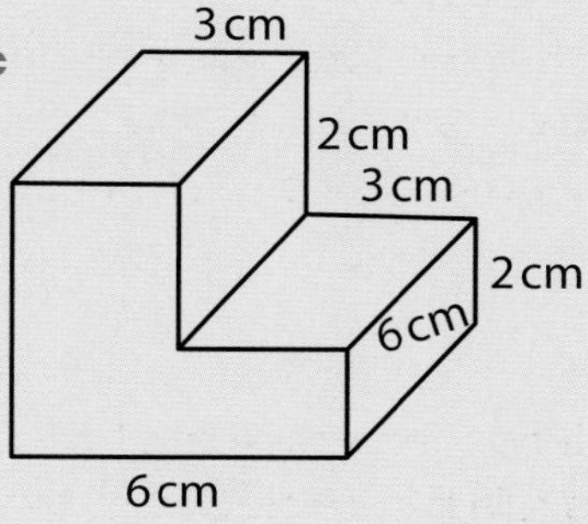

d 20 cm, 5 cm, 5 cm, 5 cm, 10 cm, 5 cm

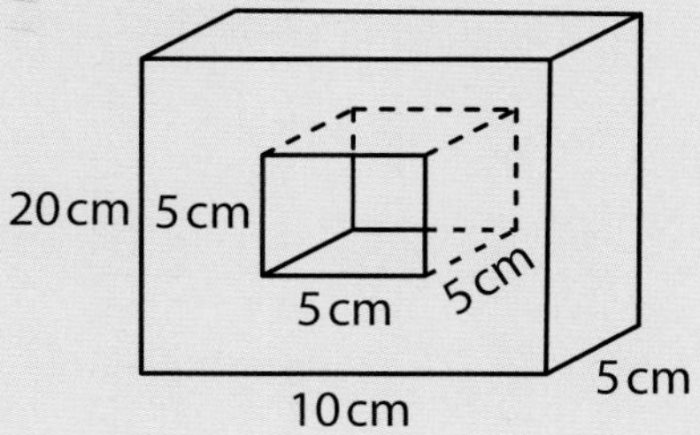

e 4 cm, 8 cm, 6 cm

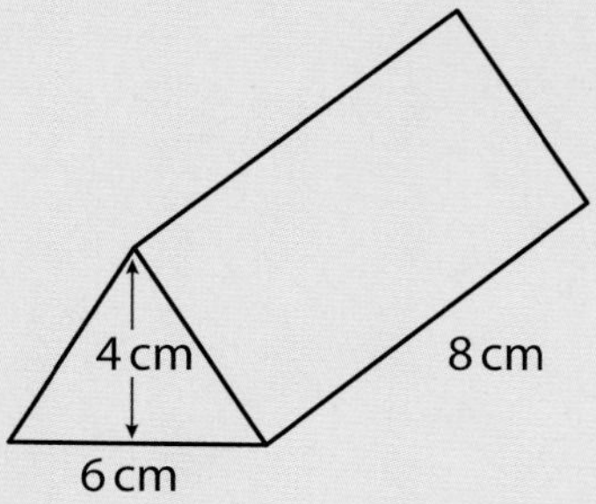

f 3 cm, 5 cm, 6 cm, 4 cm

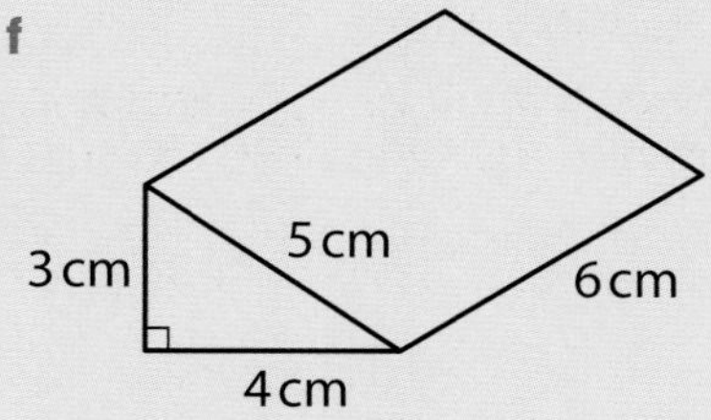

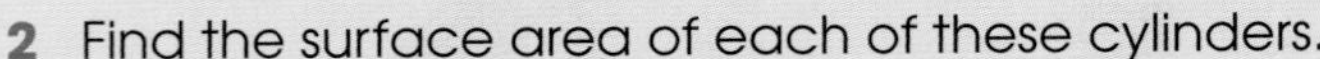

2 Find the surface area of each of these cylinders.

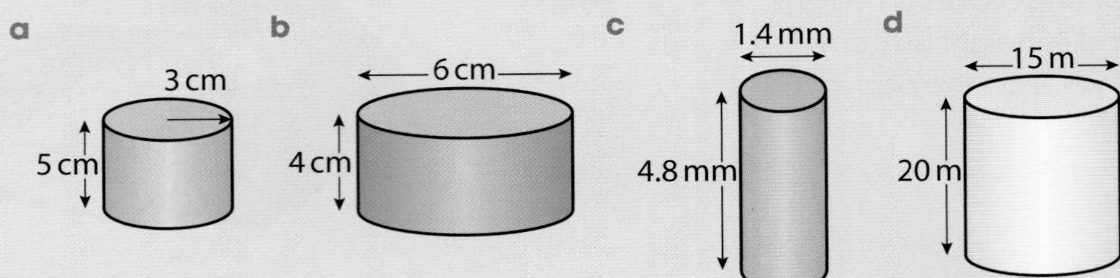

3 Samantha has two cylindrical mugs.
The first has diameter 6 cm and height 8 cm.
The second has diameter 8 cm and height 4.5 cm.
Both are filled with tea.
Which mug holds more tea and by how much?

4 A copper pipe has inner and outer radii of 2.6 cm and 3 cm.
It is 500 cm long.
Find the volume of copper in the pipe.